STRATHCLYDE UNIVERSITY
30125 00112

D1757986

NON-METALLIC INCLUSIONS IN STEEL

Non-metallic inclusions in steel

Part I *Inclusions belonging to the pseudo-ternary system $MnO–SiO_2–Al_2O_3$ and related systems*

Part II *Inclusions belonging to the systems $MgO–SiO_2–Al_2O_3$, $CaO–SiO_2–Al_2O_3$ and related oxide systems.*
Sulphide inclusions

Roland Kiessling and Nils Lange

Part III *The origin and behaviour of inclusions and their influence on the properties of steels*

Part IV *Supplement to parts I–III including literature survey 1968–1976*

Roland Kiessling

The Metals Society • London

Non-metallic inclusions in steel Second edition
Book No.194 published by
The Metals Society, 1 Carlton House Terrace, London SW1Y 5DB, UK

ISBN 0 904357-18-X

First edition published in three parts by The Iron and Steel Institute:
Part I 1964; Part II 1966; Part III 1968

Made and printed in Great Britain by Lund Humphries London and Bradford

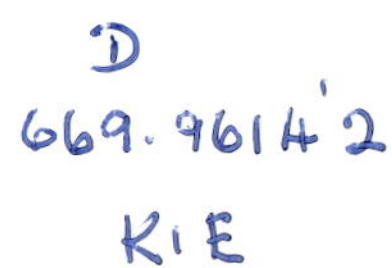

Contents

Index to Contents

Foreword to second edition

Part I Inclusions belonging to the pseudo-ternary system MnO–SiO_2–Al_2O_3 and related systems

Part II Inclusions belonging to the systems MgO–SiO_2–Al_2O_3, CaO–SiO_2–Al_2O_3 and related oxide systems. Sulphide inclusions

Part III The origin and behaviour of inclusions and their influence on the properties of steels

Part IV Supplement to parts I–III including literature survey 1968–1976

Errata to first edition parts I–III

Index to Contents

Note: The Roman numerals refer to the four parts of the volume, and the Arabic numerals to the page numbers. For example I:5 means part I page 5.

Foreword to second edition (xi)

Foreword to first edition, parts I, II, and III I:(i); II:(i); III:(i)

Introduction I:1

Experimental methods I:5, IV:1

A The MnO–SiO_2–Al_2O_3 system

A 1 General I:7
A 2 Corundum I:7, III:111, IV:5
A 3 Cristobalite I:17
A 4 Tridymite I:22
A 5 Quartz I:23, II:155
A 6 Manganosite I:26, III:113
A 7 Galaxite I:29, IV:5
A 8 Mullite I:37, II:156, III:113
A 9 Rhodonite I:44
A 10 Tephroite I:48
A 11 Spessartite I:52
A 12 Mn-anorthite I:55
A 13 Mn-cordierite I:57, IV:5
A 14 Transformations I:57
A 15 Summary and discussion I:61

B The $Fe_xMn_{1-x}O$–SiO_2–Al_2O_3 system

B 1 General I:67
B 2 Wüstite I:67, 89
B 3 Fayalite I:71
B 4 '$FeSiO_3$' (grunerite) I:73, III:113
B 5 Hercynite I:74, III:113, IV:5
B 6 Almandine I:76
B 7 Fe-cordierite I:78

C The $Fe_xMn_{1-x}O$–SiO_2–$Cr_yAl_{2-y}O_3$ system

C 1 General I:79
C 2 Cr_2O_3 (escolaite) I:80
C 3 Cr-galaxite I:83, IV:5
C 4 Chromite I:87

D The iron oxides

D 1 General I:89
D 2 Wüstite I:89
D 3 Magnetite I:93, IV:5
D 4 Haematite I:94

E Classification of inclusions according to steel type I:97

F The MgO–SiO_2–Al_2O_3 system

F 1 General II:1
F 2 Periclase II:6

F 3	Spinel	II:10, IV:5
F 4	Enstatite, protoenstatite, clinoenstatite	II:15
F 5	Forsterite	II:19
F 6	Cordierite	II:23
F 7	Pyrope	II:24
F 8	Sapphirine	II:25
F 9	Phase of osumilite type	II:26
F 10	Phase of petalite type	II:26

G The $CaO–SiO_2–Al_2O_3$ system

G 1	General	II:27
G 2	CaO	II:32
G 3	The calcium aluminates	II:36, III:113
G 4	The calcium silicates	II:48
G 5	Ternary oxide phases	II:72
G 6	Summary and discussion	II:81

H Oxide inclusions with the Group III, IV, and V transition metals

H 1	Group III (the lanthanides)	II:87, IV:29
H 2	Group IV (Ti, Zr, Hf)	II:87
H 3	Group V (V, Nb, Ta)	II:94

J Oxide inclusions: Tabular summary of phases II:96, IV:7

K Sulphide inclusions

K 1	General	II:97, IV:8
K 2	The iron–sulphur system	II:104
K 3	The manganese–sulphur system	II:114, IV:8
K 4	Solid solutions of the (Mn,Me) S type and double sulphides of the $MnS.B_2S_3$ type	II:128, IV:14
K 5	Sulphides of Group III (the lanthanides)	I:133: IV:15
K 6	Sulphides of Group IV (Ti, Zr, Hf)	II:137, IV:16
K 7	Sulphides of Group V (V, Nb, Ta)	II:139, IV:16
K 8	Sulphides of Group VI (Cr, Mo, W) and solid solutions of the $(Cr,Me)_{1-x}S$ type	II:140, IV:16
K 9	Sulphides of Group VIII (Co, Ni)	II:141
K 10	Sulphides of Group I (Cu, Ag, Au)	II:142
K 11	Sulphides of Group II (Mg, Zn, Cd, Hg, Ca, Ba, Sr)	II:142, IV:18
K 12	Sulphides of Group III, non-transition metals (Al)	II:143
K 13	Sulphides of Group IV, non-transition metals (Sn, Pb)	II:144
K 14	Sulphides with sulphur partly substituted by C, O, and N	II:144

L Inclusions with selenium and tellurium

L 1	Inclusions with selenium	II:146, IV:21
L 2	Inclusions with tellurium	II:149, III:114, IV:21

M Guidance for use II:150

N Additions to part I II:155

O The origin of non-metallic inclusions

O 1	Introduction	III:1
O 2	Inclusion size, quantity and distribution in the steel Quantity and size Distribution	III:4, IV:23

O 3 The influence of deoxidizing elements upon inclusion formation III:11, IV:29
The behaviour of oxygen in pure iron
The behaviour of oxygen in steel
Formation of inclusions during deoxidation by alloying elements
Carbon
Manganese
Silicon
Silicon–manganese
Aluminium
Chromium
Titanium, zirconium
Calcium
Rare-earth elements
Vacuum (carbon) deoxidation
O 4 The formation of inclusions during the steelmaking process III:31
O 5 Inclusions from ferroalloys III:36
O 6 Inclusions from furnace and ladle slag III:40
O 7 Inclusions from refractories III:43
Erosion products
Reaction products from refractories
Example I
Example II

P The behaviour of non-metallic inclusions in wrought steel
P 1 Introduction III:51
P 2 Behaviour during heat treatment of the steel III:52
Transformation to stable modifications
Crystallization from glassy phases
Precipitation from solid solutions
P 3 Deformability of steel inclusions III:54,IV:37
Plastic deformation of inclusions relative to the steel matrix ('index of deformability')
Influence of reduction and particle size
Deformability of FeO type inclusions
Deformability of corundum type inclusions
Deformability of spinel type inclusions
Deformability of calcium aluminates
Deformability of SiO_2 and silicates
Deformability of MnS
P 4 Sulphide shape control IV:141

Q The influence of non-metallic inclusions on the properties of steel
Q 1 Introduction III:74
Q 2 Influence on machinability III:74,IV:43
Machinability parameters
Effect of sulphide inclusions
Effect of oxide inclusions
Q 3 Inclusions and fatigue properties III:87,IV:43
Fatigue parameters
Effect of sulphide inclusions
Effect of oxide inclusions
Effect of size, shape and position of inclusions
The 'white spot'
Q 4 Hot-shortness (red-shortness) III:94
Q 5 Influence on various steel properties III:100
Surface finish
Weldability

Hot tearing
Q 6 Basic studies on the influence of inclusions on mechanical properties — III:101, IV:44
Tensile properties
Impact resistance
Crack formation
Fracture toughness
Precipitation phenomena and grain growth
Q 7 Influence on corrosion — IV:52

R Additions to parts I and II — III:111

References — I:100, II:157, III:116, IV:56

Appendixes
Names and ideal formulae of inclusion phases — I:102, II:160
Metal content of the inclusion phases in their ideal composition — I:103
Summary of tables — II:161, IV:60

Errata to parts I, II, and III — IV:59

Foreword

The original edition, parts I–III of this atlas of non-metallic inclusions, was published nearly ten years ago. Since then, rapid expansion in the use of special steelmaking processes such as AOD, ESR, and VAR, together with more effective deoxidation techniques, have reduced the number and size of oxide and sulphide inclusions present in special steels. The use of strong sulphide-forming elements, principally Ti, Zr, and Ca, and the rare-earth metals, has made possible modification of sulphides to give inclusion particle shape control. Plastic silicates and alumina clusters can now to a great extent be replaced by widely dispersed globular inclusions which do not deform during hot working. This has allowed greater control over the contribution to the anisotropy of steel properties resulting from the presence of deformed silicates, oxide stringers, and sulphides.

Much more is now known about the influence of non-metallic inclusions on the behaviour of steels under conditions of corrosion, fatigue, and fracture. New experimental techniques such as scanning electron microscopy and automatic inclusion counting methods are available for the study of inclusions. However, a steel containing as little as 1 ppm of oxygen and sulphur will still contain some 10^9–10^{10} non-metallic inclusions per tonne. Even given improvements in steel processing and techniques for inclusion study, the basic concept adopted in the first edition, namely that steel is a composite product comprising 'steel with inclusions', must be still kept in mind.

The fundamental information concerning oxide and sulphide inclusions, their structures, microscopical appearance, and properties, as given in parts I and II, is still valid. New structural and morphological information which has appeared since that time has been mainly of a complementary nature. This is particularly the case for certain sulphide systems and inclusions formed by the rare-earth metals.

The amount of new work on the behaviour of inclusions and their influence on different steel properties is, however, very large. Although this amount of work justifies a complete volume in its own right, the author has tried to cover the more important contributions to the field in an addition to this atlas, part IV. This has been written as a literature survey supplementing parts I–III. In this new edition parts I–IV are combined. Although the method of reprinting has prevented modifications to the original parts, the list of contents has been revised to incorporate new material. The opportunity has been taken to include at the end a list of errata *to parts I–III.*

The author is indebted to Mr Gunnar Grünbaum of Sandvik AB Research Centre for valuable help with the references for the period 1968–76 and also for constructive criticism. Thanks are due also to the following members of the staff of Sandvik AB: Mr Örjan Hammar for scanning electron micrographs of sulphide inclusions, Mr Stig Johansson for comments on inclusion counting methods, Mr Gösta Peterson for the drawings in part IV, and Mrs Ann Östberg for valuable help with editing and typing. Dr D. Dulieu (British Steel Corporation, Sheffield Laboratories), again assisted with the English text, and his help is gratefully acknowledged.

Roland Kiessling
Sandviken, Sweden, October 1977

Part I

Inclusions belonging to the pseudo-ternary system $MnO–SiO_2–Al_2O_3$ and related systems

Foreword

Non-metallic inclusions in steels have been investigated over the last few years at the Swedish Institute for Metal Research (Institutet för Metallforskning), Stockholm, using principally the technique of electron probe microanalysis. Part of this work has already appeared in English in the form of papers published in the Journal of The Iron and Steel Institute, *but a comprehensive selection of inclusion types has been collected and analysed since the papers appeared. The authors have therefore published a series of communications during 1964 in the Swedish journal* Jernkontorets Annaler *on the system* $MnO–SiO_2–Al_2O_3$. *Together these papers constitute an atlas of inclusions. The Editor of the* Journal of The Iron and Steel Institute *has found it of value to publish this atlas in English and the present work is a revised and enlarged edition of the Swedish papers. This constitutes the first volume which also contains an introduction describing the experimental procedure and the systematic basis of the atlas. Notes on some related oxide systems have been incorporated. A second part is planned which will deal with inclusions containing CaO and MgO compounds and the sulphide systems.*

Professor Roland Kiessling is director of the Swedish Institute for Metal Research, where Mr Nils Lange is research assistant. The authors are indebted to Miss Connie Helin and Mr Tore Malmberg for valuable help with X-ray diffraction methods and the preparation of synthetic slags. Mr Sten Bergh (of the Research Centre, Stora Kopparbergs Bergslags AB, Domnarvet, Sweden) has contributed much valuable criticism. Mr D. Dulieu assisted with the English text and gave valuable contributions to the systematic presentation of the material. The printing of the English edition has been greatly facilitated by the loan of the original blocks from Swedish publications. These have been provided by the editor of Jernkontorets Annaler, *whose help is gratefully acknowledged.*

Stockholm, September 1964

Introduction

The first major publication on non-metallic inclusions in steel, produced by Benedicks and Löfquist in 1930,[1] has for a long time been a very useful guide in this field for steel metallurgists. No equally comprehensive survey of the field has been published since, although a book in 1962 by Allmand,[2] and the well known paper by Rait and Pinder[3] give valuable contributions to the identification of inclusions. The experimental methods used in these studies were microscopy, petrography, and X-ray diffraction analysis.

The revolutionary method of electron probe analysis[4] has now become available to the metallurgist, making a quantitative analysis of inclusions possible *in situ* and important developments have also occurred in X-ray diffraction techniques (monochromatic X-ray techniques, micromethods). The authors have therefore developed a technique for studying steel inclusions using these additional methods in combination with optical microscopy. A great number of inclusions in various steels was investigated, making possible a systematic presentation with detailed information about different inclusion types. Much information is available concerning the steel in which the inclusions formed.

The present report is mainly intended to be used by ferrous metallurgists for the identification of inclusions. Therefore every effort has been made to present good and representative photomicrographs of the different inclusion types, and to collect as many data about them as possible.

The quantitative results on the composition of inclusions are of fundamental importance for the understanding of their formation and properties. The present work has shown that much detailed information concerning steelmaking practice is required before the complete history of an inclusion type can be given. As this information has rarely been available the authors have broadly limited the present review to a compilation of experimental results on the inclusion phases in steel. To augment this a short discussion on general aspects of inclusion formation and behaviour is given for the $MnO-SiO_2-Al_2O_3$ system. Salmon Cox and Charles[5] have made a detailed investigation of the distribution of inclusions in an ingot. Studies of this depth are the next stage in dealing with the general problem of inclusions. Experiments with controlled variations in steelmaking conditions are being conducted by the present authors and it is hoped to incorporate the results of this work in the later volume.

The present development of electron probe analysis has still some serious limitations for inclusion studies. The size of the focal spot for the electron beam is about 1 μm^2 but contribution to the excited X-ray radiation comes from a hemisphere with a diameter of 3–5 μm. Many inclusion phases are very fine precipitates and therefore great care must be taken in evaluating analytical results. It is also only possible to analyse for elements from sodium (atomic number 11) and upwards in the periodic table. A direct determination of oxygen, of fundamental importance as an inclusion component, is thus not possible at present with the standard electron probe analyser. Limits of detection for most other elements are between 0·1 and 0·5% of the total sample. This restricts the study of small amounts of different metals present in the phases, which is unfortunate since these are of value in tracing the origin of inclusions. The accuracy of the analyses for different metals varies but in general it is about $\pm$1 to 5% of the amount present, this therefore is also the approximate limit of error for values derived from probe analyses such as homogeneity ranges. These variations are relatively unimportant for the study of inclusions. A more serious error comes from the small number of points which were analysed within each inclusion and composition variations may sometimes have been overlooked.

Generally, however, the method of electron probe analysis, with its facilities for quantitative study of inclusions *in situ*, gives so much new information that the present limitations are of minor importance.

The phase analysis of inclusions has been established mainly from synthetic slags as described in the section on experimental methods. Direct phase identification of inclusions *in situ* was only possible for a few inclusions due to the rather large surface area irradiated in ordinary X-ray diffraction studies. A development of microdiffraction powder techniques should therefore be of great value and interesting results have recently been obtained with the camera described by Helin and Spiegelberg,[6] using a microbeam X-ray source, but the method needs further development before it can be used for inclusion studies.

A great number of inclusion types belong to the system $MnO-SiO_2-Al_2O_3$, independent of the steel class and analysis. This system has therefore been the basis for presentation of the inclusion types. Closely related to this system are those of the general formula $Fe_xMn_{1-x}O-SiO_2-Cr_yAl_{2-y}O_3$ and inclusions from these systems are also included in the present report, together with a short discussion of the different iron oxides. A later report will deal with oxide inclusions belonging to CaO and MgO systems as well as sulphides.

The inclusion types are largely independent of the steel analysis, and therefore no attempt has been made to give a systematic description of inclusions based on the steel composition. Such a classification would require a complete investigation of all the different inclusion types in the various steels and this has not been done with the present material. In order to give some information on the subject, however, reference to all the different examples arranged according to steel analysis has been made in the last section.

The steel type (basic or acid, electric, high-frequency, Kaldo, etc.), details of methods of steel melting and deoxidation and the teeming and casting operations all have a much greater influence on the formation of inclusions than the steel composition itself. The variations in melting practice could well be the basis of a systematic description of inclusion types, but not enough information about these methods was available for the samples investigated.

It should be borne in mind that the steels studied are representative products of the modern Swedish steel industry. Although a wide variety of steelmaking methods are used there is, of course, a basic pattern of ore sources which differs from those in other countries. An indication of Swedish plant type and capacity may be obtained from the Swedish Steel Manual[7] published by the Swedish Ironmasters Association (Jernkontoret).

The inclusion phases are usually well known to ceramists and mineralogists and much information has been found in the appropriate literature. General reference to ceramic phase diagrams is found in the summary of 1964 by Levin *et al.*[8] An excellent complement is the summary of physical properties and structure types for different inorganic compounds by Trojer,[9] which also has several references to investigations on ceramic systems.

The different inclusion phases observed have usually been named by mineralogists and these names have then been adopted by the metallurgist. These names and formulae are given in Appendix 1. A table, giving the percentage metal content of the different inclusion phases, arranged according to the different metals, is also included (Appendix 2). This may be of value for the identification of inclusion phases from electron probe analysis of their components.

The compilation of Wyckoff on crystal structures[10] was used as a general reference to structure types and the ASTM index[11] as a reference for the powder photographs.

Where trace elements were present in a specimen the phenomenon of catho-luminescence was frequently observed. This fluorescence is a useful qualitative indication of the phases present. Little is known of this effect and its exact dependence on the concentration of trace elements. Notes are given for some phases but this effect should be used with some reservation as a method of phase identification.

Experimental methods

A microscopic study of the inclusions *in situ* was the starting point for the present study and careful preparation of the samples was essential. Specimens were polished on wet silicon carbide papers followed by diamond paste on a velvet cloth with a very brief final polish using alumina. Silicon carbide particles have been reported in inclusion phases. It should be emphasized that contamination of the inclusions during polishing may easily occur; unless careful polishing techniques are used abrasive particles from papers may be accidentally retained. The inclusions were photographed by standard metallographic methods and the same samples were used for electron probe analysis and other investigations. Microscopy gives much general information about crystalline and amorphous phases present in the inclusions and their size, shape, and distribution.

The composition of the different phases observed was quantitatively determined by means of electron probe analysis according to the general methods indicated earlier.[12–15] Oxygen cannot be analysed with the present electron probe technique and this has been a difficulty for inclusion studies as the inclusions are mostly oxides. This drawback of the method is, however, not too serious provided that no other light elements are present at the same time. If the different elements in the inclusion phases are analysed with a high accuracy then general knowledge about inclusions usually permits a calculation of their composition, assuming that the metals and silicon, whose valencies are usually known, occur as oxides. This is general practice for inclusion analysis with other methods also. If the oxides add up to $100\% \pm 5\%$ the analyses are regarded as satisfactory. Nomograms have been made for the different oxide combinations using the intensity corrections for electron probe analysis as given by Philibert.[16] The wavelength of the characteristic X-ray spectrum is dependent on the bonding of the different atoms in the target. The wavelength shift due to differences in bonding type presents a considerable error in the correction factors for the light inclusion elements Mg, Al, and Si. To avoid this difficulty, standard samples of oxide mixtures close to the inclusion compositions have been prepared for these oxides according to the methods given below. These have then been used as standards for the electron probe analysis. It is essential for quantitative analysis of inclusions to use such oxide standards and not the metals themselves at least for the light elements. Without this method the present theoretical correction formulae are not developed to such an extent that accurate inclusion analysis is possible.

The mean composition of the inclusion was calculated for most of the examples shown, using planimetric evaluation.

The phase identification and structure type of the different phases was made by monochromatic X-ray diffraction methods. A Guinier-type camera with a ground and bent quartz crystal monochromator[17] was used, built at the Swedish Institute for Metal Research. The monochromatic X-ray methods are of great value for inclusion studies as they give a very low background and it is also possible to study low-angle reflections. These reflections are of special importance for the identification of the complex X-ray powder photographs of inclusion phases. It was not usually possible to carry out the phase analysis using the actual steel inclusions *in situ*, as they were frequently too small. Synthetic slags with different compositions were therefore prepared, using an electron beam melting furnace or, for the Cr_2O_3 slags, a plasma torch. The synthetic slag samples were then heat treated and studied by microscopy and electron beam analysis until they were identical with the actual steel inclusions *in situ*. The greater amount of material thus available was studied by monochromatic X-ray diffraction methods, making possible a correct phase analysis. In addition several synthetic oxide preparations were made to outline different binary or ternary oxide systems, where the information available has been incomplete regarding homogeneity ranges, solid solubility, etc. Finally also some *microhardness measurements* on inclusion phases were carried out.

By this combined technique it proved possible to undertake an almost complete determination of the different steel inclusions in the material available to the authors. This information, together with as many data about the steel as possible, has been included in the present report.

A. The $MnO-SiO_2-Al_2O_3$ system

A 1 General

This system is the most important for the study of inclusions in modern steel types and it will therefore be treated in more detail than the others. Many inclusion types in steels both low and high in aluminium or manganese as well as inclusions in chromium steels can be classified and discussed using this system as a starting point.

A summary of the different phases reported to exist in the $MnO–SiO_2–Al_2O_3$ system, together with their compositions according to the stoichiometric formula, is indicated schematically in Fig. 1 and Table I. The diagram is based on information from Levin[8] and Trojer.[9] The metallurgical aspects of the systems $MnO–SiO_2$, $SiO_2–Al_2O_3$, $Al_2O_3–MnO$, and $MnO–SiO_2–Al_2O_3$ are dealt with in 'Basic open hearth steel making'[18] and in the paper by Rait and Pinder.[3]

It is well known that many of the phases in this and following systems have an extended homogeneity range or a composition which differs from the reported stoichiometric formula. This has been confirmed in the present work, as may be seen by comparing the schematic diagram of Fig.1 with Fig.56 which depicts the analytical results for the system $MnO–SiO_2–Al_2O_3$.

In addition to the detailed discussion of the different phases given in sections A 2–A 13, some transformations observed in inclusions belonging to this system are discussed in section A 14.

Because of the importance of the $MnO–SiO_2–Al_2O_3$ system, the metallurgical significance of the results is considered in the closing section, A 15.

A 2 $MnO–SiO_2–Al_2O_3$: Corundum

Chemical formula Al_2O_3

Modifications

In addition to $\alpha–Al_2O_3$, corundum, the phases $\beta–Al_2O_3$ and $\gamma–Al_2O_3$ have been reported. Only $\alpha–Al_2O_3$ was observed in steel inclusions,

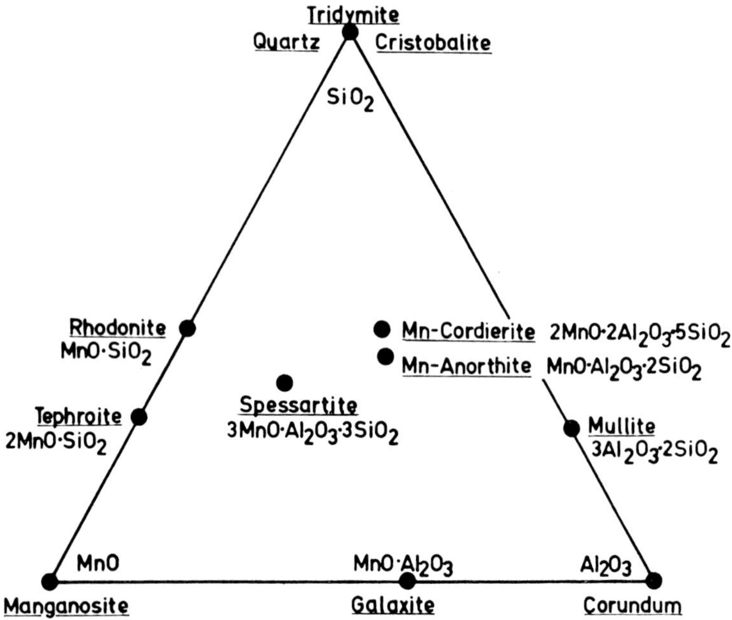

1 Schematic representation of the pseudoternary system $MnO–SiO_2–Al_2O_3$, showing the stoichiometric composition of the different phases which have been reported. The figure should be compared with Fig.56, summarizing the results of the present investigation

TABLE I Phases reported to exist in the $MnO–SiO_2–Al_2O_3$ system

Name	Chemical formula	Stoichiometric composition, wt-% MnO	SiO_2	Al_2O_3	Section
Corundum	Al_2O_3	...	...	100	A 2
Cristobalite	SiO_2	...	100	...	A 3
Tridymite	SiO_2	...	100	...	A 4
Quartz	SiO_2	...	100	...	A 5
Manganosite	MnO	100	...	...	A 6
Galaxite	$MnO.Al_2O_3$	41	...	59	A 7
Mullite	$3Al_2O_3.2SiO_2$	...	28	72	A 8
Rhodonite	$MnO.SiO_2$	54	46	...	A 9
Tephroite	$2MnO.SiO_2$	70	30	...	A 10
Spessartite	$3MnO.Al_2O_3.3SiO_2$	43	36	21	A 11
Mn-anorthite	$MnO.Al_2O_3.2SiO_2$	24	41	35	A 12
Mn-cordierite	$2MnO.2Al_2O_3.5SiO_2$	22	46	32	A 13

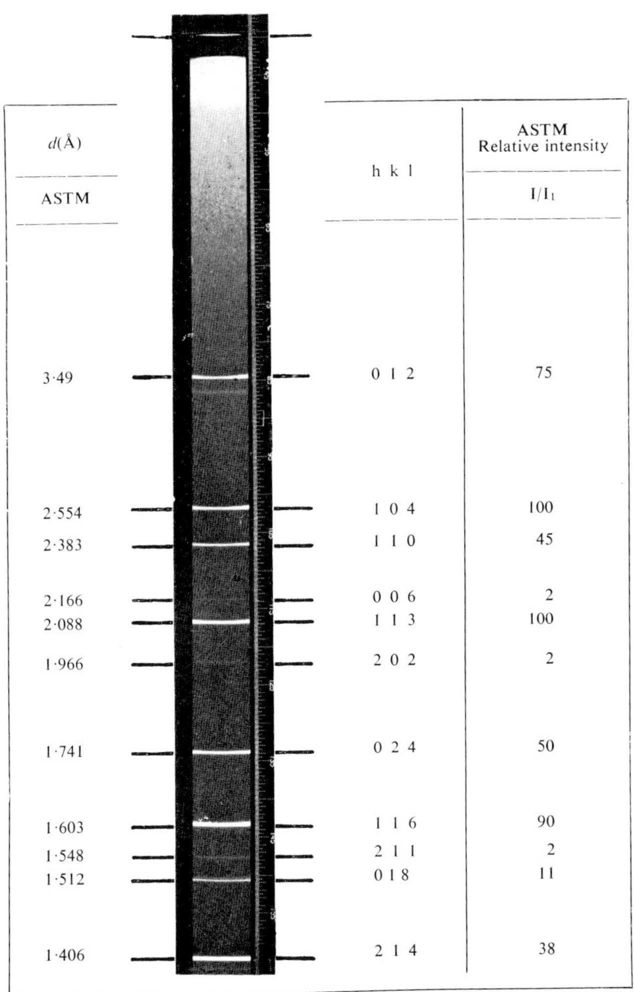

d(Å) ASTM	h k l	ASTM Relative intensity I/I_1
3·49	0 1 2	75
2·554	1 0 4	100
2·383	1 1 0	45
2·166	0 0 6	2
2·088	1 1 3	100
1·966	2 0 2	2
1·741	0 2 4	50
1·603	1 1 6	90
1·548	2 1 1	2
1·512	0 1 8	11
1·406	2 1 4	38

d-values for the different reflections (h k l) are quoted from the ASTM index The tabulated relative intensities are taken from the ASTM index and are not always comparable with those obtained with the focussing Guinier camera.

2 X-ray powder photograph of corundum; monochromatic Fe Kα-radiation; Guinier-type camera

but γ–Al_2O_3 which has a spinel type lattice forms solid solutions with the spinels of type $AO.B_2O_3$ (A 7). Inclusions with glassy Al_2O_3 were also observed.

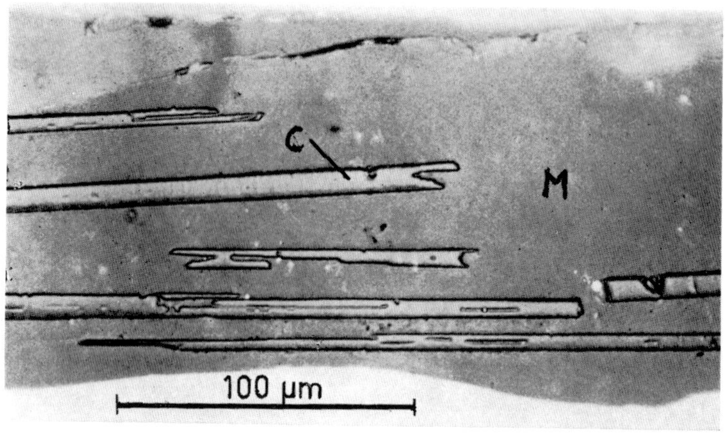

×400

STEEL TYPE Basic electric steel, Si-killed with a limited Al addition. Rolled billet
STEEL ANALYSIS, % 0·45 C, 0·22 Si, 0·70 Mn, 0·034 S, 0·002 Al, 39 ppm O
INCLUSION, % (C) Corundum with 100 Al_2O_3 (M) Matrix with 31 MnO, 45 SiO_2, 22 Al_2O_3
Mean analysis (calc.), % 27 MnO, 39 SiO_2, 34 Al_2O_3
COMMENT Spessartite and mullite were precipitated from the glassy matrix when the steel sample was heat treated for 2h at 1100°C (*see* A 14)

3 Oxide inclusion in steel: corundum in glassy matrix of manganese oxide–alumina–silica

Physical properties

Melting point: 2050°C
Density: 3·96
Microhardness: 3000–4500 kp/mm²
Fluorescence: Corundum often shows a red fluorescence in the probe analyser.

Structure type

Hexagonal; space group $R\bar{3}c$
a=4·759 Å, c=12·991 Å, c/a=2·73. ASTM data card 11–661
Cr_2O_3 type. Wyckoff I, chap. V, text p.4, table p.11
Monochromatic X-ray powder film Fig.2
Isostructural with Cr_2O_3 (C2) and Fe_2O_3 (D4)

Microscopy and electron probe analysis

A series of different inclusions in steels and ferroalloys is shown in Figs.3–9. These are typical examples of the appearance, *in situ*, of both the corundum and the glassy Al_2O_3 phases in inclusions. A summary of the analytical results, as well as some general information, is given in the captions.

The results have shown that the corundum phase in inclusions is

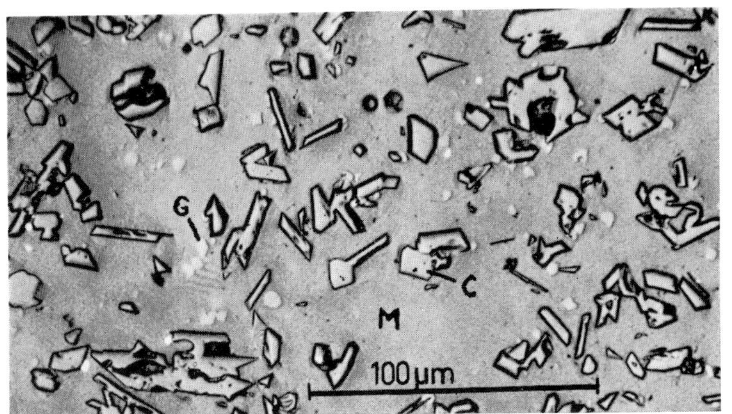

×400

STEEL TYPE Basic open hearth steel, ferro-silicon killed before tapping and aluminium killed in the ladle. From ingot
STEEL ANALYSIS, % 0·33 C, 0·34 Si, 0·60 Mn, 0·013 P, 0·018 S, 2·43 Cr, 0·12 Ni, 0·49 Mo, 0·05 Cu
INCLUSION (C) Corundum with 100 Al_2O_3 (G) Galaxite with 38 MnO, 42 Al_2O_3, 19 Cr_2O_3 (M) Matrix with 32 MnO, 46 SiO_2, 20 Al_2O_3. Mean analysis (calc.) 27 MnO, 39 SiO_2, 33 Al_2O_3
COMMENT The galaxite has considerable amounts of chromia in solid solution. The steel sample was later heat treated for 2 h at 1100°C which resulted in a precipitation of spessartite from the glassy matrix of the inclusion (*see* A 14)

4 Oxide inclusions in steel: corundum in glassy matrix of manganese oxide alumina–silica

usually pure, or contains only traces of most other metal oxides such as FeO, Fe_2O_3, MnO, and MgO. Only Cr_2O_3 was observed to dissolve as a solid solution in greater amounts in Al_2O_3 (Fig.8, *see also* section B).

The corundum phase is usually easy to identify in inclusions, being very hard and having characteristic crystal shapes. In polished but unetched microsections it is often observed as laths with split ends or idiomorphic plates (Figs.3–5). The plates are often rounded and more or less glassy (Figs.6 and 8). Corundum occasionally crystallizes as multiphase inclusions in a glassy matrix of manganese oxide–alumina–silica (Figs.3 and 4), or as isolated single phase inclusions (Figs.6, 8 and 9). Figure 5, showing inclusions in a rolled steel plate, illustrates how isolated single phase corundum crystals have been separated from the silicate matrix during the rolling operation. This is probably due to the great difference in plastic properties between the very hard corundum phase and the comparatively soft silicate matrix.

Spessartite (*see* A 11) is often observed together with corundum (Fig.7) in steel which has been heat treated below 1200°C, as, for instance, during rolling.

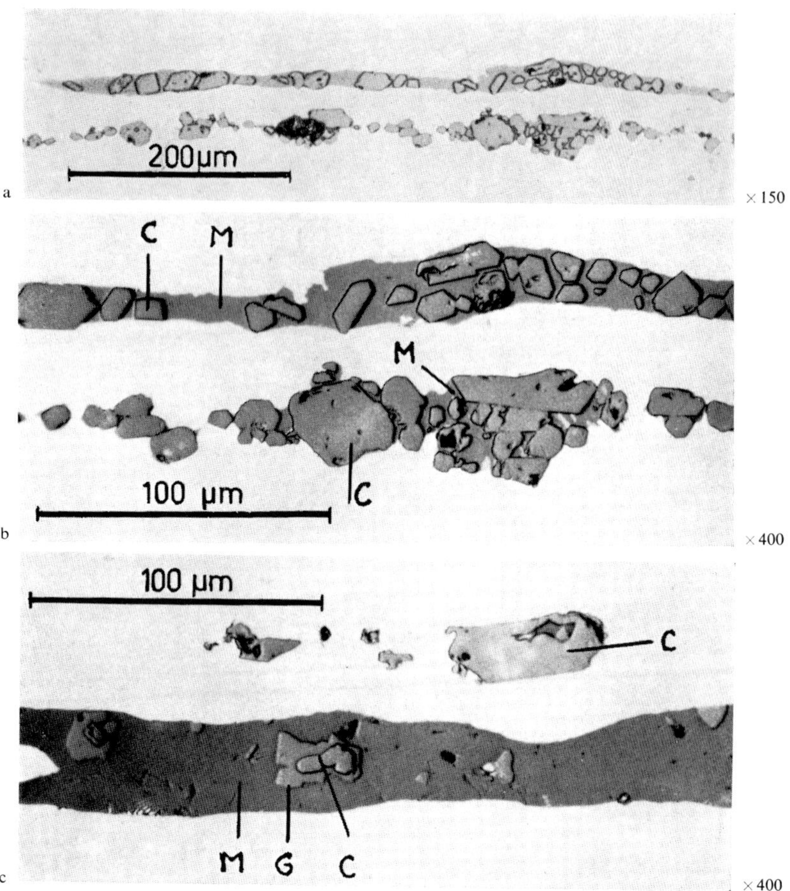

STEEL TYPE Basic open hearth steel, Al-killed. From rolled plate

STEEL ANALYSIS, % 0·11 C, 0·25 Si, 1·10 Mn, 0·004 P, 0·030 S, 0·006 N, 0·06 Cr, 0·05 Ni, 0·13 Cu, 0·022 Al (soluble), 0·04 Al (insoluble)

INCLUSIONS 5*a* and 5*b* (C) Corundum with 100 Al_2O_3, trace of MnO and FeO (M) Matrix with 43 MnO, 42 SiO_2, 16 Al_2O_3. Mean analysis (calc.) 20 MnO, 19 SiO_2, 61 Al_2O_3

5*c* (C) Corundum with 100 Al_2O_3 (G) Galaxite with 41 MnO, 57 Al_2O_3 (M) Matrix with 48 MnO, 39 SiO_2, 16 Al_2O_3. Mean analysis (calc.) 45 MnO, 33 SiO_2, 22 Al_2O_3

COMMENT The corundum phase has probably been mechanically separated from the plastic silicate during the working of the steel. The inclusion, shown in Fig.5*c*, is richer in manganese than those in Figs.5*a* and *b*. From the beginning it has probably had a higher mean alumina content and corundum has been the primary crystallization phase. Before its solidification the composition has changed due to increased MnO and SiO_2 additions from the steel phase, and its mean composition has changed to that part of the system, where galaxite is a stable phase. This has resulted in precipitation of galaxite on existing corundum nuclei (*see also* comments in sections A 2 and A 14)

5 **5a–c** Oxide inclusions in steel: corundum as single-phase inclusion particles, as precipitates in a manganese oxide–alumina–silica matrix and surrounded by galaxite

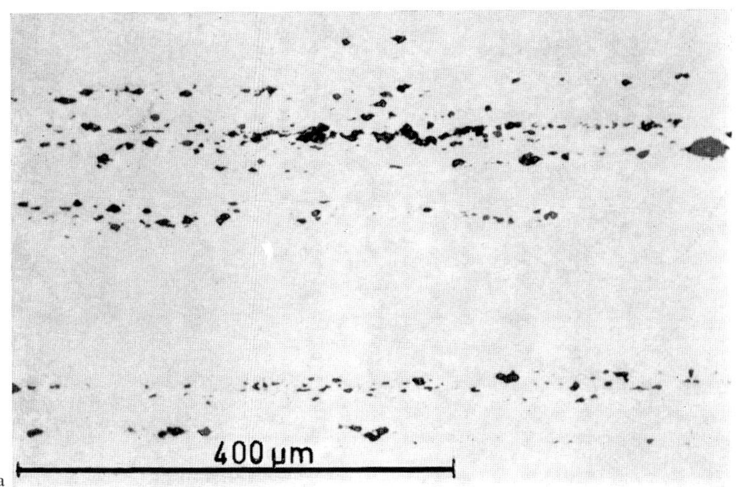

a ×150

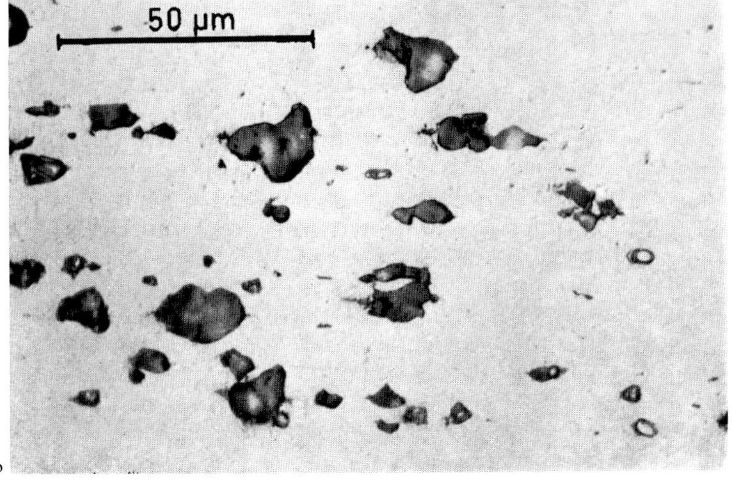

b ×750

STEEL TYPE Basic electric steel, Al- and ferro-silicon killed. From rolled billet
STEEL ANALYSIS, % 0·10 C, 0·32 Si, 0·46 Mn, 0·009 P, 0·006 S, 2·20 Cr, 0·9 Ni, 1·04 Mo
INCLUSIONS (C) Corundum with 100 Al_2O_3, trace of MgO (red fluorescence) (G) 82–86 Al_2O_3, 17–15 MgO (green fluorescence)

COMMENT The C-phase is corundum or glassy Al_2O_3. The G-phase is probably Mg-spinel. It has not been possible to distinguish between the two phases by the microscope, but their coloured fluorescent radiation in the electron probe analyser is different (red for C, green for G)

6 **6a–b** Oxide inclusions in steel: corundum (and glassy Al_2O_3) and Mg-spinel

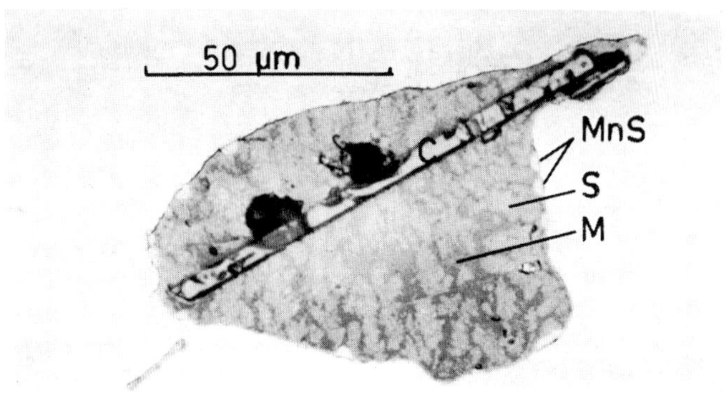

×600

STEEL TYPE Basic open hearth steel, ferrosilicon killed before tapping, after killed with aluminium. From a forged roll, normalized, air hardened and tempered
STEEL ANALYSIS, % 0·62 C, 0·26 Si, 0·79 Mn, 0·015 P, 0·018 S, 0·65 Cr, 0·07 Ni
INCLUSIONS (C) Corundum with 100 Al_2O_3 (S) Spessartite with 38 MnO, 45 SiO_2, 16 Al_2O_3 (M) Matrix with 32 MnO, 52 SiO_2, 18 Al_2O_3. Mean analysis (calc.) 32 MnO, 43 SiO_2, 25 Al_2O_3. MnS is visible as grey-white hemispheres at the boundary between the inclusion and the steel

7 Oxide inclusions in steel: corundum and spessartite in glassy manganese oxide–alumina–silica

Microscopically similar phases

Corundum bears some resemblance to the spinel type oxides (A 7) and they may be difficult to distinguish. An example is given in Fig.6, showing inclusions of both these phases in the same steel. Whereas they appear to be very similar in the microscope, the electron probe analysis shows the existence of two inclusion types; these being pure corundum inclusion particles and Al_2O_3 inclusions with an MgO content of 15–17%. They may be distinguished by the colour of the fluorescence stimulated by electron bombardment. That from the pure Al_2O_3 being red compared to the green colour of the MgO-containing particles. The two crystal types often grow together. The lower solid solubility limit reported for MgO in spinel (MgO. Al_2O_3) is 15–17%[3,19] which is fully supported by the present results. The inclusions in Fig.6 are markedly different from those shown in Fig.8 which depicts Al_2O_3 inclusions with a Cr_2O_3 content varying up to about 20%; these are all of the corundum type with Cr_2O_3 in solid solution.

A further example of the appearance of spinel and corundum together is shown in Fig.5*c*, where a nucleus of corundum is surrounded by galaxite phase. This inclusion also illustrates the possibility of transformations within the inclusions and will be discussed below and in section A 14.

A difference between corundum and the spinel type oxides is that

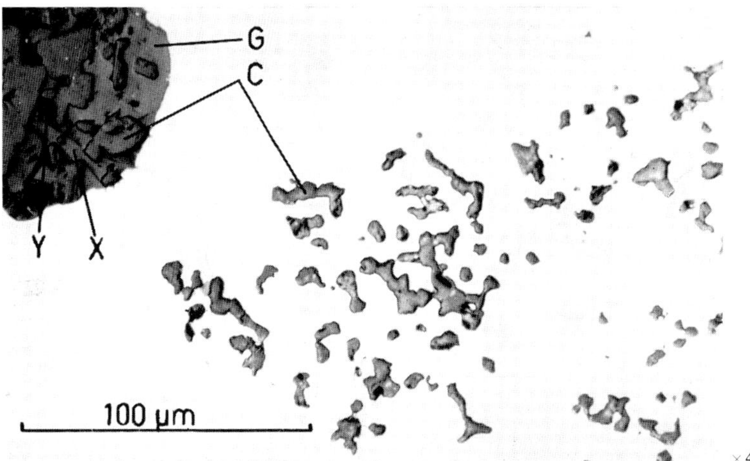

×400

MATERIAL TYPE Ferrochromium suraffiné
MATERIAL ANALYSIS, % 73 Cr, 0·03 C, 0·08 Si, balance Fe
INCLUSIONS The small single phase inclusions are of alumina with traces of chromia only. The larger single phase inclusions have 79 ±4 Al_2O_3 and 19 ±4 Cr_2O_3. As a complete solid solubility exists between alumina and chromia (*see also* B), all of these inclusions should have the corundum structure. Some of them are not completely crystalline but partly glassy.

The larger, multiphase inclusion also contains corundum (C) with chromia in solid solution. Three more phases (G, X and Y) are distinguished in this inclusion. The composition found for the different phases was (C) Corundum with 78 Al_2O_3, 19 Cr_2O_3, trace of MnO (G) Spinel (?) with 18 MnO, 26 Al_2O_3 54 Cr_2O_3, traces of SiO_2 and CaO (X) Unknown phase with 22 MnO, 30 Al_2O_3, 31 Cr_2O_3, 15 CaO (Y) Matrix with 42 SiO_2, 32 Al_2O_3, 24 CaO, traces of Cr_2O_3 and MnO

8 Oxide inclusions in ferroalloys: corundum with Cr_2O_3 in solid solution

the former is anisotropic whereas the latter are isotropic. The use of polarized light is therefore recommended.

Comments

Corundum is a common phase which may arise from both steelmaking additions and contamination of the liquid steel by refractory particles. A common source of indigenous corundum inclusions is the aluminium used for deoxidation. Single phase corundum particles often appear as clouds of small crystals as shown in Fig.6. Also, other deoxidizing alloys, such as ferrosilicon, often contain several percent of aluminium impurities.[20]

For aluminium and oxygen dissolved in liquid steel, the 'constant' $K=[\%A]^2 \cdot [\%O]^3$ is very low (10^{-14} at 1600°C).[21] Thus, inclusions often contain considerable amounts of Al_2O_3 although the steel is reported to be deoxidized by aluminium-free alloys, or has a very low amount of dissolved aluminium.

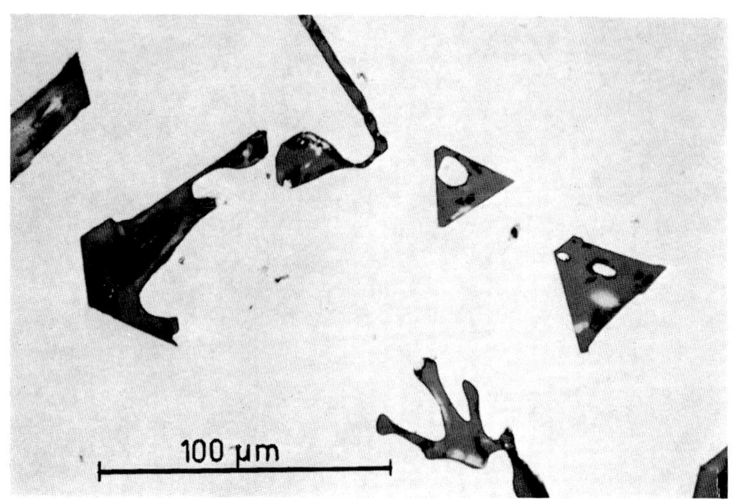

×400

MATERIAL TYPE Ferrovanadium
MATERIAL ANALYSIS, % 81·2 V, 0·15 C, 0·50 Si, 0·020 P, 0·020 S, 0·020 As, 0·36 Al, 2050 ppm O, balance Fe
INCLUSIONS Corundum with 100 Al_2O_3

9 Oxide inclusions in ferroalloys: corundum

Many ceramic materials, such as chamotte firebricks, contain a large amount of Al_2O_3 and are possible sources of exogenous inclusions. The familiar reaction between manganese and SiO_2 in steel[13,22] results in the formation of MnO which may further react with chamotte to form manganese–alumina–silicates, these then disperse into the steel bath. These silicates, together with ceramic particles such as those mechanically ground away from ladle and spout linings, are often nuclei for inclusion formation. They usually change composition during their subsequent life in the steel bath. Corundum or other alumina-containing phases, depending on the alumina concentration and the time of cooling, then appear in the inclusions. That shown in Fig.5*c* is thus probably an example of a transformation due to such a change in mean composition. An alumina-rich inclusion with primary corundum crystallization has changed its composition as more MnO and SiO_2 have been precipitated on the inclusion during the cooling of the liquid steel. The mean composition of the inclusion has then changed to the range of primary galaxite crystallization. As a result corundum has dissolved and galaxite has precipitated, often on the already existing nuclei of corundum.

A 3 $MnO–SiO_2–Al_2O_3$: Cristobalite

Chemical formula SiO_2

Modifications

Several silica modifications are known. The following phases are stable at ordinary pressure[23,24]

$$\text{quartz} \rightleftharpoons \text{high-quartz} \rightleftharpoons \text{tridymite} \rightleftharpoons \text{cristobalite} \rightleftharpoons \text{melt}$$

Quartz and high-quartz are also called α- and β-quartz but the nomenclature is sometimes reversed. For cristobalite and tridymite α usually refers to the low-temperature modification, but for these phases also this is not always the case. To avoid confusion the prefixes high- and low- are to be preferred.

The structure of all these silica modifications are three-dimensional molecules of SiO_2 tetrahedra in various spatial arrangements.[25] This arrangement is very similar in both quartz and high-quartz and these structures differ only slightly. Transformations between them are, therefore, fast and reversible, and such fast transformations also occur between different cristobalite and tridymite modifications in the supercooled state. Hence in inclusions containing cristobalite, tridymite or quartz, these phases always appear in their low-temperature modifications when studied at room temperature.

The transformations between high-quartz and tridymite, tridymite and cristobalite, as well as between all crystalline modifications and the melt are associated with more fundamental rearrangements of the SiO_2-tetrahedra. These transformations proceed more slowly with the result that cristobalite, tridymite and vitreous SiO_2 often appear as supercooled phases in steel inclusions.

Physical properties

Melting point (high-cristobalite): 1723°C
Stability range (high-cristobalite): 1470–1723°C
Transformation point (high-low cristobalite): 180–270°C
Density: 2·23 (high), 2·32–2·38 (low)
Microhardness: *circa* 1600 kp/mm^2 (low)
Fluorescence: cristobalite often shows a light-blue fluorescence in the electron probe analyser

Structure type

Tetragonal; space group $P4_12_12$
a=4·971 Å, c=6·918 Å. ASTM data card 11–695
Wyckoff I, chap. IV, text p.28
Monochromatic X-ray powder photograph, Fig.10

Microscopy and electron probe analysis

In non-metallic steel inclusions cristobalite often crystallizes as dendrites in a glassy or crystallized matrix of various metal silicates. The polished microsections of these dendrites often appear as rosettes

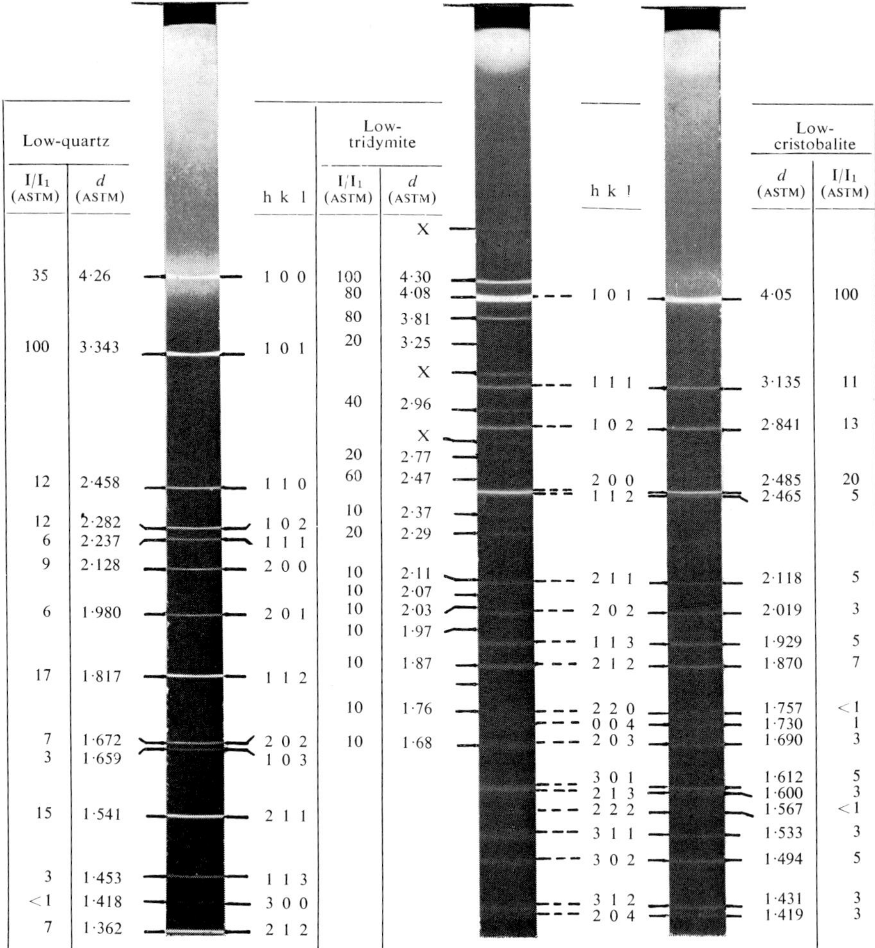

d-values for the different reflections are quoted from the ASTM index, and corresponding (h k l)-values have been given for low quartz and low-cristobalite, but not for low-tridymite for which the structure is unknown. The tridymite photograph also has some impurity diffraction lines from cristobalite and sodium tungstate (x)

10 X-ray powder photograph of low-quartz (left), low-tridymite (centre) and low-cristobalite (right); monochromatic Fe Kα-radiation; Guinier camera

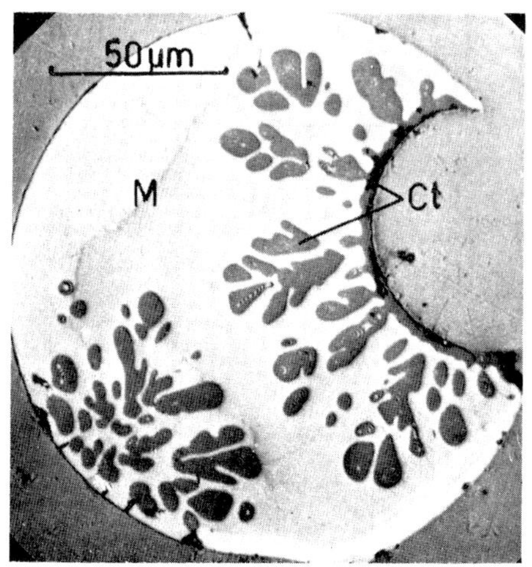

×400

STEEL TYPE Basic electric steel, Si-killed; inclusions isolated from the lower central part of a top cast 6 ton ingot
STEEL ANALYSIS, % 0·16 C, 0·35 Si, 0·67 Mn, 0·015 P, 0·017 S, 0·009 N, 100 ppm O, Al has not been found
INCLUSIONS (Ct) Cristobalite with 100 SiO_2, trace of MnO (M) Matrix with 51 MnO, 3 FeO, 40 SiO_2, traces of Ti, Al and S. Partly glassy, partly crystallized rhodonite (A 9). Mean analysis (calc. 34 MnO, 64 SiO_2. 2 FeO
COMMENT The well developed cristobalite dendrites and the very low Al_2O_3 content are an indication of an indigenous origin for this type of inclusion particle[13] (compare the inclusions shown in Fig.13)

11 Oxide inclusions in steel: cristobalite in manganese–silicate

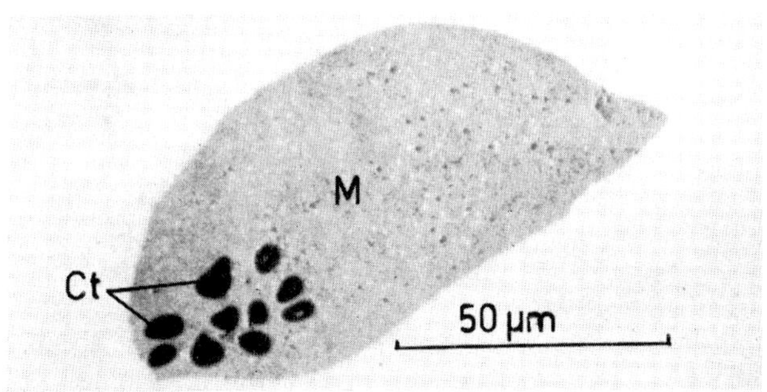

×600

MATERIAL TYPE Ferrochromium suraffiné
MATERIAL ANALYSIS, % 73 Cr, 0·03 C, 0·09 Si, balance Fe
INCLUSIONS (Ct) Cristobalite with 98 SiO_2, traces of Cr and Mn (M) Matrix with 14 MnO, 31 SiO_2, 57 Cr_2O_3. Mean analysis (calc.) 13 MnO, 33 SiO_2, 54 Cr_2O_3
COMMENT This inclusion *in situ* in ferrochromium is similar in appearance to the isolated inclusion from a carbon steel shown, in Fig.11. The matrix in this particle has, however, a large chromia content and is composed of two phases. It was not possible to carry out a separate analysis of each of the two phases because of the limited resolution of the electron probe analyser. *See also* Fig.70

12 Oxide inclusions in ferro-alloys: cristobalite in manganese oxide–chromia–silica

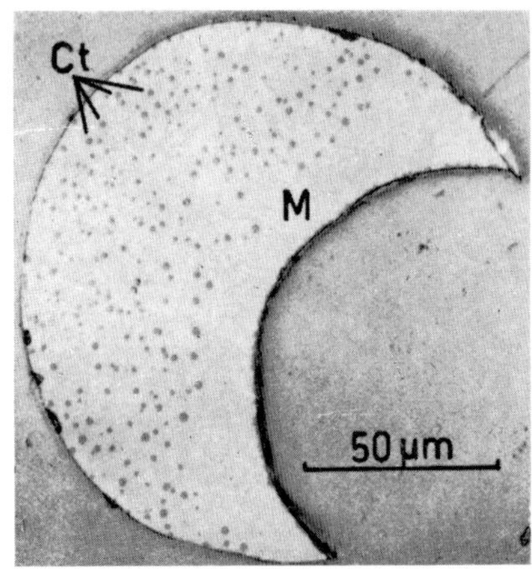

×400

STEEL TYPE From the same inclusion isolate as the particle shown in Fig.11
STEEL ANALYSIS *See* Fig.11
INCLUSIONS (Ct) Cristobalite with 93 SiO_2, 6 Al_2O_3, traces of Fe and Mn (M) Matrix with 46 MnO, 1 FeO, 44 SiO_2, 7 Al_2O_3, trace of Ti. Mean analysis (calc.) 30 MnO, 64 SiO_2, 6 Al_2O_3

COMMENT In this particle, the cristobalite phase is very finely distributed and the inclusion contains Al_2O_3. It is concluded, therefore, that the particle had an exogenous origin. The nuclei are probably a result of a reaction between the manganese in the steel and the Al_2O_3 ladle refractories.[22] Compare Fig.11 and discussion in section A 15

13 Oxide inclusion in steel: cristobalite in glassy manganese oxide–alumina–silica

(Figs.11, 12, 13, 17*a*). Cristobalite also occurs as a scale around the inclusions and lining the inside of the blowholes which are often visible in non-metallic silicate inclusions (Fig.11). This increase of silica concentration may result from reduction of the iron and manganese oxides at the surface of the non-metallic inclusions by silicon in the steel (Hultgren's mechanism[26]). This gives a silica-rich surface layer.

Other morphological forms of cristobalite have been observed in non-metallic inclusions. Thus, in comparison with Fig.11 showing precipitated cristobalite, Fig.14 shows an inclusion in ferrochromium with cristobalite as the matrix and chromia precipitated within and at the surface of the inclusion. In Fig. 18 cristobalite forms a porridge-like light grey structure similar to the cristobalite structure reported to be formed when quartz is heated to 1 500°C.[24] Examples of the phase are also shown in Fig.70.

The cristobalite phase can dissolve several percent of Al_2O_3, Cr_2O_3 and other metal oxides in solid solution. If the metallurgical history of the steel is known, these metal oxides are often valuable clues to the origin of the inclusions. An example is given in reference 13. Here an electron probe analysis for impurity elements in the cristo-

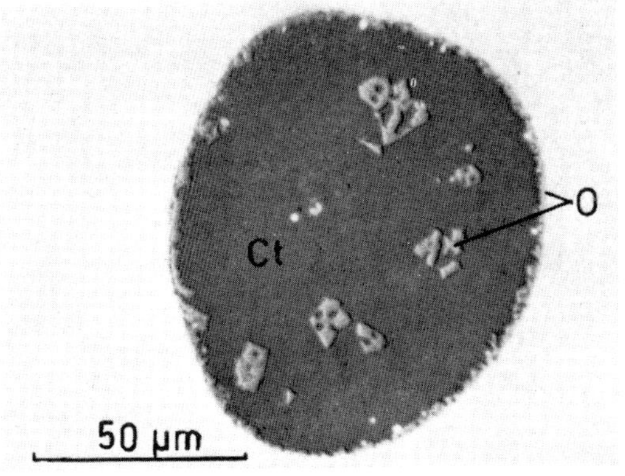

×500

MATERIAL TYPE Ferrochromium affiné
MATERIAL ANALYSIS, % 73 Cr, 0·03 C, 0·09 Si, balance Fe
INCLUSIONS (O) Cr_2O_3 with 98 Cr_2O_3, 5 MnO (Ct) Cristobalite with 96 SiO_2, 4 Cr_2O_3. Mean analysis (calc.) 60 SiO_2, 40 Cr_2O_3
COMMENT The inclusion is of chromia–silica type and chromia has been precipitated during cooling from the silica matrix

14 Oxide inclusions in ferroalloys: escolaite (Cr_2O_3) in cristobalite

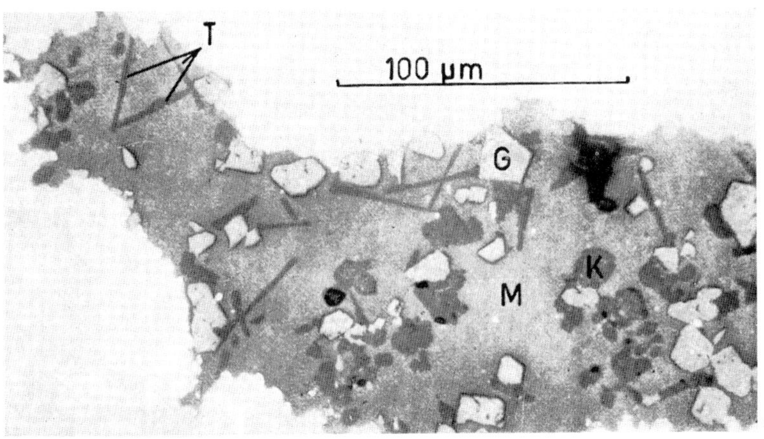

×400

STEEL TYPE Stainless heat resistant steel processed in a basic electric steel furnace and induction remelted. From a forged and annealed bar
STEEL ANALYSIS, % 0·18 C, 0·35 Si, 11·5 Cr, 0·60 Ni, 0·50 Mo, 0·25 V, 0·30 Nb
INCLUSIONS (T) Tridymite with 100 SiO_2 (K) Low quartz (?) with 100 SiO_2 (G) Chromium galaxite with 30 MnO, 59 Cr_2O_3, 3 Al_2O_3, 6 V_2O_5(C3) (M) Matrix with 37 MnO, 40 SiO_2, 8 Cr_2O_3, 5 Al_2O_3. Mean analysis (calc.) 36 MnO, 45 SiO_2, 14 Cr_2O_3, 4 Al_2O_3, 1 V_2O_5
COMMENT It is difficult to decide whether the silica modification K is low-quartz or low-cristobalite. Its appearance, however, indicates that the phase is low-quartz as may be seen by comparison with Fig.18. This inclusion probably had an exogenous origin (refractories) and reacted with alloying elements and deoxidizing products. The quartz has partly transformed to tridymite at the temperature of the steel, and low-tridymite is therefore present in the inclusion.

15 Oxide inclusions in steel: tridymite, quartz (?), and chromium-galaxite in glassy manganese oxide–chromia–alumina–silica

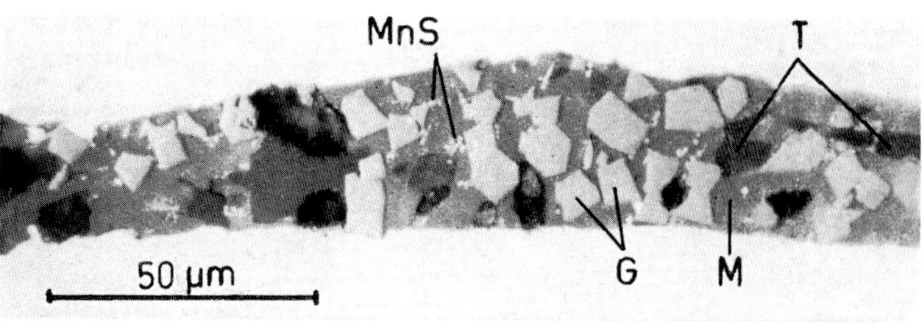

×600

STEEL TYPE From a rolled billet of austenitic stainless steel
STEEL ANALYSIS, % 0·05 C, 0·43 Si, 0·96 Mn, 0·023 P, 0·012 Si, 18·0 Cr, 9·8 Ni, 0·50 Mo
INCLUSIONS (T) Tridymite with 100 SiO_2, trace of Al (G) Chromium galaxite with 34 MnO, 66 Cr_2O_3, trace of Fe (A 7) (MnS) Manganese sulphide (M) Matrix with 45 MnO, 38 SiO_2, 3 Cr_2O_3, trace of Fe. Mean analysis (calc.) 42 MnO, 40 SiO_2, 18 Cr_2O_3
COMMENT This inclusion is very similar to that shown in Fig.15

16 Oxide inclusion in steel: tridymite, chromium–galaxite and manganese sulphide in glassy manganese oxide–chromia–silica

balite together with a morphological examination of its crystal habit were used successfully to trace the origin of the inclusions.

Microscopically similar phases
Glassy silicates, such as rhodonite (*see* A 9) are often mistaken for cristobalite, but their microhardness is lower. It is sometimes difficult to decide which of the SiO_2 modifications (cristobalite, tridymite, quartz, or vitreous silica) are actually present in steel inclusions. A knowledge of the earlier temperature cycle of the steel inclusion is often of value for a definite phase identification.

A 4 MnO–SiO_2–Al_2O_3: Tridymite

Chemical formula SiO_2
Modifications
See A 3, cristobalite. It is doubtful whether tridymite is a pure SiO_2-phase.[9,27] It may only exist if stabilized by traces of alkaline metal salts.

Physical properties
Melting point: 1670°C
Stability range: 870–1470°C
Transition temperature ranges between the four tridymite modifications: 100–150°C, 140–180°C, 180–220°C
Density: 2·26–2·30
Microhardness: *circa* 1600 kp/mm² (low tridymite)
Fluorescence: As for cristobalite (A 3)

Structure type
The crystal structures of the tridymite modifications are not completely known and the X-ray powder photographs are slightly different, depending on the chemical environments and the temperature conditions under which they were formed.[27] However, it has always been possible to identify tridymite in inclusions and ceramics by means of the *d*-values given on data card 3–0227 (*see also* 2–0242) in the ASTM index. Although a variation in the *d* values has often been observed, the powder photograph of tridymite differs distinctly from those of cristobalite and quartz. This is evident from Fig.10, where the indexed powder photographs of low-cristobalite, low-tridymite, and low-quartz are reproduced together.

Microscopy and electron probe analysis
The tridymite phase in steel inclusions often crystallizes as rather dark, thin plates which appear in polished sections as laths, often slightly split at the ends (Fig.15). Also, the dark SiO_2 plates of Figs.16 and 17*b* are probably tridymite. It was possible to produce similar tridymite plates in the synthetic slags but there is still a slight possibility that they are glassy SiO_2 or even other SiO_2 modifications.

According to the literature,[27] tridymite could only be stable if other metal ions are present in solid solution. The amounts necessary are, however, much lower than the present sensitivity limit for electron probe analysis. In the inclusions studied by the authors only pure SiO_2, or SiO_2 with traces of Al_2O_3, have been recorded for the tridymite phase.

A 5 MnO–SiO_2–Al_2O_3: Quartz

Chemical composition SiO_2
Modifications See A 3, cristobalite

Physical properties
Melting point: 1710°C
Stability ranges: < 573°C (low)
573–870°C (high)
Transformation temperature low-high: 573°C
Density: 2·651 (low)
2·518 (high at 600°C)
Microhardness: *circa* 1600 kp/mm^2 (low)

Structure type
Low-quartz; trigonal, space group $P3_121$
a=4·903 Å, c=5·393 Å. ASTM data card 5–0490
Wyckoff I, chap. IV, text p.26
Monochromatic X-ray powder photograph, Fig.10

a 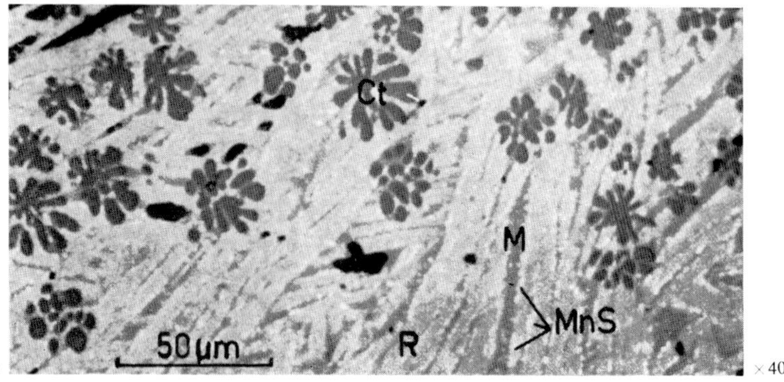

×400

b 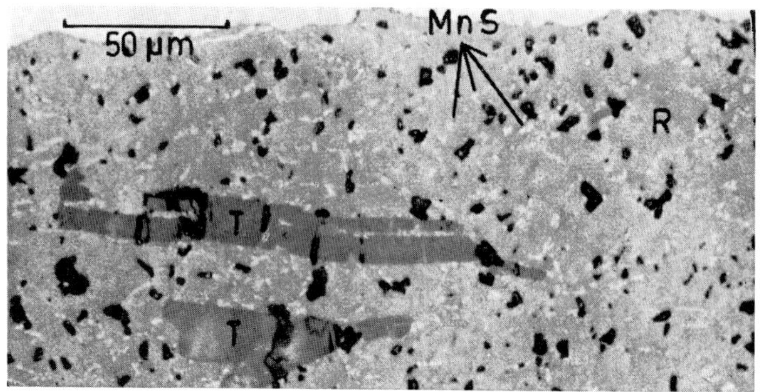

×400

17*a* STEEL TYPE Basic electric steel, Si-killed. Inclusion isolated from the lower central part of a top cast 6 ton ingot

STEEL ANALYSIS, % 0·16 C, 0·35 Si, 0·67 Mn, 0·015 P, 0·017 S, 0·009 N, 100 ppm O, no trace of Al

INCLUSIONS (Ct) Cristobalite with 96 SiO_2, 2 Al_2O_3, traces of Fe, Mn, Ti, and Ca (MnS) Sulphide of type MnS (R) Rhodonite with 48 MnO, 39 SiO_2, 5 Al_2O_3, 2 TiO_2, trace of Fe (A 9) (M) Glassy matrix, not analysed. Mean analysis (calc.) 45 MnO, 49 SiO_2, 5 Al_2O_3, 1 TiO_2

17*b* STEEL TYPE Steel ingot, bottom cast. From a hot rolled sheet

STEEL ANALYSIS, % 0·12 C, 0·20 Si, 0·70 Mn, 0·048 P, 0·046 S, 0·10 Cr, 0·19 Cu, 0·005 N

INCLUSIONS (T) Tridymite, cristobalite or glassy SiO_2 with 100 SiO_2 (MnS) Sulphide of MnS-type with Fe in solid solution (R) Rhodonite with 52 MnO, 43 SiO_2, trace of Fe (A 9). Mean analysis (calc.) 50 MnO, 49 SiO_2, 1 MnS. Black pores are also visible

COMMENTS In wrought steel from ingots with inclusions of the type shown in Fig.17*a*, the inclusions are deformed. If the working temperature is between 870° and 1470°C then transformation of cristobalite to tridymite will be more or less complete, the reaction rate being slow, as discussed in section A 3. It was not possible to establish beyond doubt that the phase T of Fig.17*b* is tridymite, but its general appearance is very similar to that of tridymite (*see* Figs.15 and 16). It also shows cracks which is in accordance with the brittleness reported for tridymite (and quartz) as compared with cristobalite[2] at the working temperatures of steel. The phase T may also be glassy silica

17 **17a–b** Oxide inclusions in steel: a comparison between silica in similar inclusions from a steel ingot (isolate) and a rolled steel sheet (*in situ*)

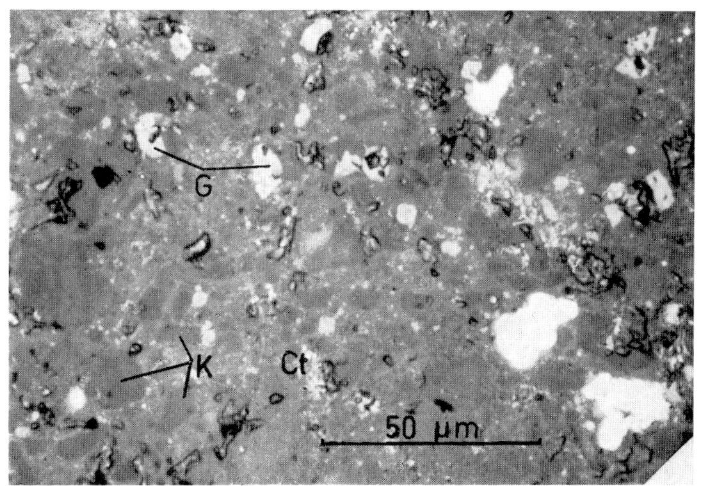

×600

STEEL TYPE Acid open hearth steel, killed (ballbearing steel). From a hot pressed, seamless tube
STEEL ANALYSIS, % 1·0 C, 0·30 Si, 0·30 Mn, 1·5 Cr
INCLUSIONS (K) Low-quartz with 100 SiO_2 (the darker phase) (Ct) Cristobalite with 95–100 SiO_2 (the lighter phase) (G) Chromium galaxite with 32 MnO, 14 Al_2O_3, 55 Cr_2O_3 (A 7). The white phase is steel
COMMENT The identification of low-quartz in this inclusion strongly points to an exogenous origin (refractories from the discharge side). An X-ray diffraction analysis of this inclusion revealed the presence of low-quartz, low-cristobalite, and chromium galaxite. A hypothesis for the formation of its structure is that a partial transformation of quartz (the original particle) to cristobalite has occurred at the temperature of the liquid steel. At the same time the inclusion has reacted with the steel and therefore the inclusion has a low content of chromia and small amounts of chromium galaxite have been formed

18 Oxide inclusions in steel: low-quartz, cristobalite, and chromium galaxite

Microscopy and electron probe analysis

Low-quartz in steel inclusions has been observed as dark grains which form more or less regular plates in polished sections (Fig.18). It is difficult to identify solely by microscopy and electron probe analysis. X-ray diffraction methods are often necessary in order to establish its presence as an inclusion phase. As mentioned in section A 3 the transformations of quartz to tridymite or cristobalite are often incomplete at temperatures of the ladle or the steel heat treatment. On the other hand, cristobalite and tridymite, once formed, do not readily transform to quartz during subsequent cooling of the steel. It is very important, therefore, to establish the presence or absence of low-quartz in steel inclusions. Its presence is a strong indication of the exogenous origin of the inclusion (refractories etc.).

Microscopically similar phases

See section A 3, cristobalite.

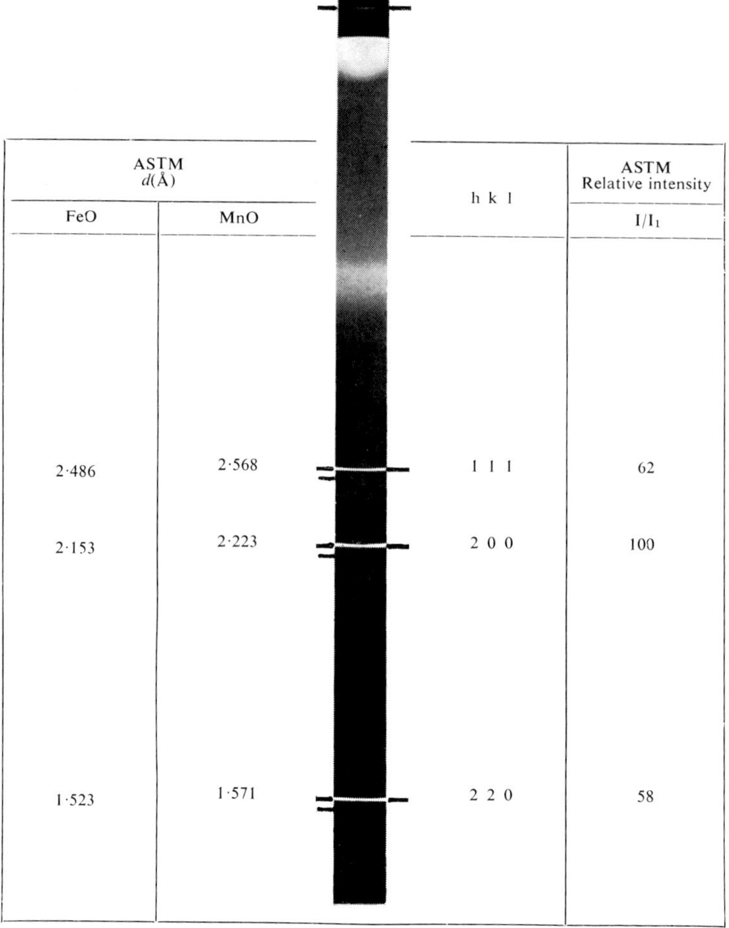

ASTM d(Å)		h k l	ASTM Relative intensity
FeO	MnO		I/I_1
2·486	2·568	1 1 1	62
2·153	2·223	2 0 0	100
1·523	1·571	2 2 0	58

d-values for the different reflections (h k l) are quoted from the ASTM index, and values for wüstite, FeO, which is isostructural with MnO, have also been quoted from the same index. Oxides of the composition $Fe_xMn_{1-x}O$ have d-values between these two oxides

19 X-ray powder photograph of manganosite, monochromatic Fe Kα-radiation; Guinier camera

A 6 $MnO–SiO_2–Al_2O_3$: Manganosite

Chemical formula MnO

Physical properties

Melting point: 1 850°C
Density: 5·365
Microhardness: 400 kp/mm^2

Structure type

Cubic; space group Fm3m
a=4·445 Å. ASTM data card 7–230
NaCl-type. Wyckoff I, chap. III, text p.1, table p.13
Monochromatic X-ray powder photograph, Fig.19
Isostructural with wüstite (*see* B2 and D2)

Solid solubility

A wide range of solid solubility exists between manganosite and several Me(II)-oxides. Of special importance for inclusions are the systems MnO–CaO, MnO–MgO and MnO–FeO. In the MnO–CaO and MnO–MgO systems a complete solid solubility exists. In the general reference[8] for the MnO–FeO system an immiscibility gap is reported to exist, but other studies[28] have indicated that this system also shows a complete solid solubility range.

Microscopy and electron probe analysis

No examples of the pure manganosite phase were found in any inclusions in the steel samples examined, although the phase has been reported to exist in inclusions.[1,2]

In synthetic MnO–SiO_2–Al_2O_3 slags, manganosite has been frequently observed, usually forming a characteristic dendritic structure (Fig.20). Although no examples of pure MnO were found, several examples of the phase $Fe_xMn_{1-x}O$ with different x-values down to about 0·3 were observed (Figs.21–23, 76). The manganosite phase

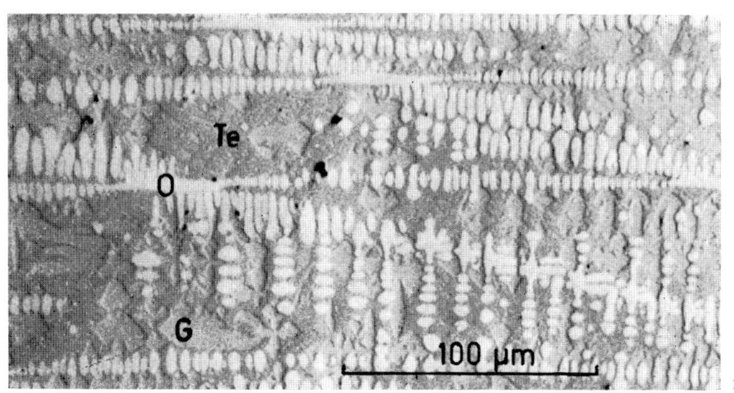

×350

SLAG ANALYSIS, % (O) Manganosite with 100 MnO (G) Galaxite (Te) Tephroite. Mean analysis (by weight) 72 MnO, 9 SiO_2 and 19 Al_2O_3

20 Synthetic slag, prepared from manganese oxide–silica–alumina mixture by electron beam melting at about 2500°C

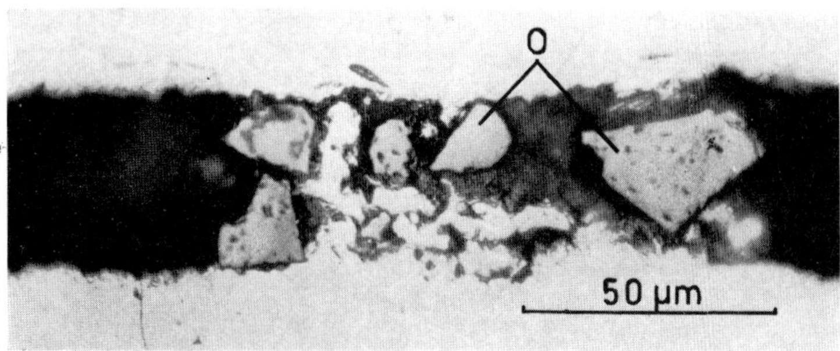

×600

STEEL TYPE Rimming deep-drawing steel. From cold rolled sheet
STEEL ANALYSIS, % 0·05 C, 0·29 Mn, 0·007 P, 0·030 S, 0·15 Cu, 0·07 Cr, 0·019 Sn

INCLUSIONS (O) Manganosite with 68 MnO and 32 FeO
COMMENT Surface crack

21 Oxide inclusions in steel: manganosite containing iron oxide

reported earlier[1,2] is probably of the type $Fe_xMn_{1-x}O$, but it has proved impossible to establish the presence of FeO by earlier methods. Pure manganosite probably only exists in steels high in manganese or in molten manganese metal, and such a material has not been investigated.

Comments

Most of the $Fe_xMn_{1-x}O$ inclusions, which were observed during this investigation, were situated in the vicinity of cracks (Figs.21, 22, 76). They were therefore probably formed as a result of oxidation of the

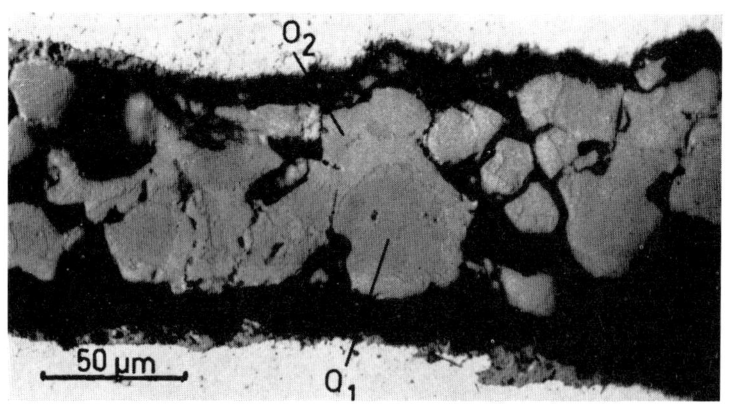

×400

STEEL TYPE Carbon steel
STEEL ANALYSIS, % 0·46 C, 0·22 Si, 0·67 Mn, 0·028 S

INCLUSIONS (O_1) 96 FeO, 3–7 MnO (dark) (O_2) 97 Fe_3O_4, 4 MnO (light)
COMMENT Surface crack

22 Oxide inclusions in steel: wüstite and magnetite containing manganese oxide

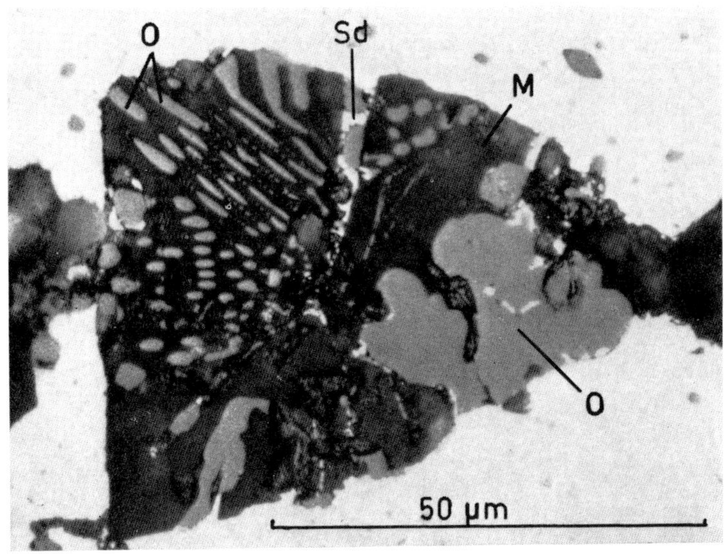

×1200

STEEL TYPE Unknown
STEEL ANALYSIS, % 0·15 C, 0·05 Si, 0·84 Mn, 0·009 P, 0·033 S, 0·03 Cr, 0·05 Ni, 0·12 Cu, 0·004 N

INCLUSIONS (O) wüstite with 89 FeO and 11 MnO (M) Matrix with 23 MnO, 50 FeO and 23 SiO_2 (Sd) FeS

23 Oxide inclusion in steel: manganosite in manganese–iron–silicate

steel surface during the heat treatment of the steel. The oxide was left in surface cracks or pushed into surface defects of the billets during rolling or forging. The multiphase inclusion in Fig.23 with $Fe_xMn_{1-x}O$ dendrites, is not of this type; it had an exogenous origin.

A 7 $MnO–SiO_2–Al_2O_3$: Galaxite

Chemical formula $MnO{\cdot}Al_2O_3$

Physical properties

Melting point: 1 560°C (disintegration)
Density: 4·23
Microhardness: 1 500–1 700 kp/mm²
Fluorescence: Galaxite shows no fluorescence in the electron probe analyser.

Structure type

Cubic; space group Fd3m
a=8·271 Å. ASTM data card 10–310
Spinel type. Wyckoff II, chap. VIII, text p.16, table p.41

Monochromatic X-ray powder photograph, Fig.24
Isostructural with hercynite (B5), chromium galaxite (C3), chromite (C4), and magnetite (D3).

Solid solubility

Spinels are double oxides[23,25,29] usually composed of oxides of two

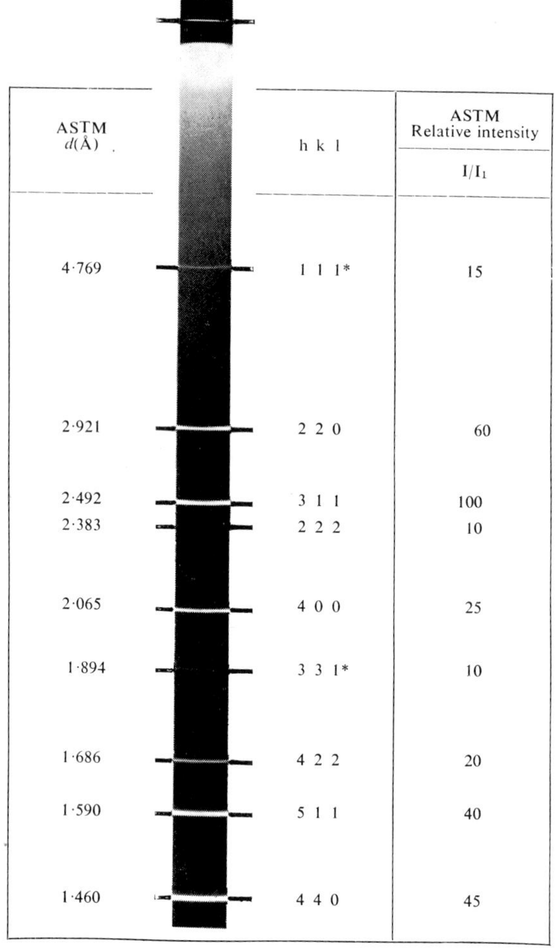

ASTM d(Å)	h k l	ASTM Relative intensity I/I_1
4·769	1 1 1*	15
2·921	2 2 0	60
2·492	3 1 1	100
2·383	2 2 2	10
2·065	4 0 0	25
1·894	3 3 1*	10
1·686	4 2 2	20
1·590	5 1 1	40
1·460	4 4 0	45

d-values for the different reflections (h k l) are quoted from the ASTM index. The reflections (1 1 1) and (3 3 1), marked by an asterisk, are not reported in this index but were observed during the present investigation

24 X-ray powder photograph of galaxite; monochromatic Fe Kα-radiation: Guinier camera

TABLE II Double oxides of spinel type which are related to the galaxite phase and may occur in steel inclusions

Ideal formula	Name	Stoichiometric composition, wt-% AO	B_2O_3	Section
$MnO.Al_2O_3$	Galaxite	41	59	A 7
$MnO.Cr_2O_3$	Cr-galaxite	32	68	C 3
$MnO.Fe_2O_3$	Jacobite	31	69	(A 7)
$MnO.Mn_2O_3$	Hausmannite	31	69	(A 7)
$FeO.Al_2O_3$	Hercynite	41	59	B 5
$FeO.Cr_2O_3$	Chromite	32	68	C 4
$FeO.Fe_2O_3$	Magnetite	31	69	D 3
$MgO.Al_2O_3$	Spinel	28	72	(A 7)
$MgO.Cr_2O_3$	Picrochromite	21	79	(A 7)
$MgO.Fe_2O_3$	Mg-ferrite	20	80	(A 7)

different metals A (valency II) and B (valency III) and of the general formula $AO.B_2O_3$. They take their name from the mineral spinel $MgO{\cdot}Al_2O_3$. Spinel-type lattices are formed by a great number of two- or three valency metals, and slightly different lattices exist.[29] Metals which form solid solutions with galaxite and are of interest in steel inclusions are: Fe (II) and Mg which may substitute Mn, also Cr and Fe (III), which may substitute Al. A summary of those double oxides which are related to galaxite is given in Table II. Wide ranges of solid solubility exist between these different double oxides, but their extent is not completely known.

The various double oxides of spinel type all have extended homogeneity ranges, especially to the side of the ternary metal oxide. They therefore often contain a higher amount of this oxide than is indicated by the ideal stoichiometric formula given in Table II. Galaxite has also been found to take higher amounts of Mn in solid solution than that indicated by its stoichiometric formula. It is not known if these additional Mn atoms form bi- or trivalent ions, but in Tables III and IV these additional ions were assumed to be binary. Thus galaxite on this basis shows a homogeneity range on both sides of the ideal $MnO{\cdot}Al_2O_3$ composition.

Microscopy and electron probe analysis

As already mentioned a great number of double oxides exist which are of the same structure type as galaxite and are closely related to this phase. The present knowledge of homogeneity ranges and solid solubilities is small. The results from the present investigation are summarized in Tables III and IV. The values given were obtained from steel inclusions as well as from synthetic slags. They are not

TABLE III Solid solubility ranges in steel inclusions for double oxides of the spinel type which are related to the galaxite phase (*cf.* Table IV)

Oxide system	Solid solubility range, wt-%	Fig.
Galaxite, $MnO.Al_2O_3$ ($41MnO.59Al_2O_3$)	$35MnO.65Al_2O_3$–$66MnO.34Al_2O_3$	25–27
$MnO.Al_2O_3$–$MnO.Cr_2O_3$	Complete solid solubility. When chromia is substituted for alumina, the solubility limit on the MnO-side is shifted to lower MnO-values (15%MnO). Some indications of an immiscibility range have been observed (*see* A 7 and C 2)	28–31, 71
$MnO.Al_2O_3$–$FeO.Al_2O_3$ and $MnO.Al_2O_3$–$MnO.Fe_2O_3$	The Fe-content of the galaxite phase in inclusions is usually low. The Fe-ions are most probably of valency II (substituting Mn) but could also have the valency III (substituting Al)	25, 26, 29
$MnO.Al_2O_3$–$MgO.Al_2O_3$	Only small amounts of Mg have been found in the galaxite of inclusions of the manganese oxide–alumina–silica type (max 5%MgO). In inclusions with CaO however the spinel phase ($MgO.Al_2O_3$) has been observed with small amounts of MnO in solid solution	6, 27, 33
$MnO.Al_2O_3$–SiO_2	No solubility of silica in galaxite has been observed	

results from complete studies of all the different oxide systems but are to be regarded as representative for those temperatures and equilibrium conditions which obtain in steel melting practice. In this connexion it may be mentioned that galaxite was never found to contain any SiO_2, a common inclusion component.

Galaxite was observed frequently in inclusions from different steel types as well as from ferroalloys. In Figs.25–27, characteristic precipitates of galaxite in silica inclusions from carbon steels are shown. In microsections the grey galaxite phase often has a cloverleaf appearance. Galaxite is also a common inclusion phase in chromium steels, but the alumina is usually substituted to a varying extent by chromia (*see* C3). Such chromium galaxite in silica inclusions with a gradually increasing chromia content is shown in Figs.28–30 and 72. The colour of chromium galaxite is usually whiter than that of pure galaxite and the crystals are often idiomorphic with regular sections.

TABLE IV Composition of the double oxide phase in different inclusion types from steel and ferroalloys; Fe has been assumed to have the valency II and is written as FeO

Inclusion type	Steel type	Double oxide, wt-%						Inclusion, mean analysis, wt-%					Fig.
		MnO	FeO	MgO	Al_2O_3	Cr_2O_3	TiO_2	MeO	SiO_2	Al_2O_3	Cr_2O_3	Others	
Mn-silicate	Carbon steel	34	1	5	54	...	...	36	41	23	...	...	25
Mn-silicate	Carbon steel	32	8	...	55	...	...	36	36	28	...	...	26
Mn-silicate	Carbon steel	34	2	3	61	...	...	36	26	33	...	5	27
Mn-silicate	Alloyed	38	...	...	42	19	...	...	...	...	...	...	...
Mn-silicate	Alloyed	36	1	...	25	39	...	31	19	22	24	4	29
Mn-silicate	Alloyed	35	...	...	10	52	5	35	22	11	27	5	28
Mn-silicate	Alloyed	35	1	...	7	57	...	...	...	...	...	...	...
Mn-silicate	Alloyed	34	...	...	...	66	...	42	40	...	18	...	30
Mn-silicate	Alloyed	25	12	...	...	64	...	...	...	...	...	...	...
Chromite	Alloyed	3	14	...	...	83	...	17	...	...	83	...	73
Ca-silicate	Alloyed	25	1	5	46	12	6	...	...	...	...	...	...
Ca-silicate	Carbon steel	6	...	27	67	...	...	...	...	...	...	...	...
Ca-silicate	Carbon steel	+	+	31	68	...	...	...	...	...	...	...	...
Ca-aluminate	Low-alloyed	...	...	31	68	...	...	...	...	69	...	31	33
Galaxite+sulphide	Alloyed	30	2	...	...	70	...	...	...	...	...	...	32
Al_2O_3+Spinel	Low-alloyed	...	...	15	85	...	...	...	...	...	...	...	6
Cr-galaxite	Ferrochromium	15	...	...	...	85	...	15	...	...	85	...	31 *b*
Cr_2O_3+Cr-galaxite	Ferrochromium	16	...	...	...	84	...	7	...	...	93	...	31 *a*

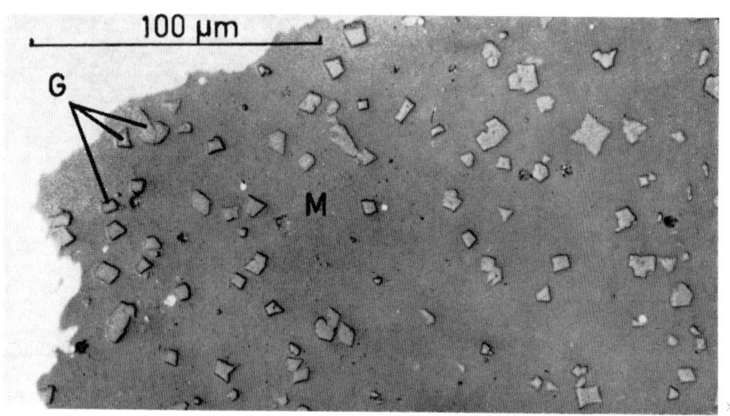

×400

STEEL TYPE Carbon steel ingot, cast from the bottom. From a secondary pipe in a rolled bar

STEEL ANALYSIS, % 0·24 C, 0·38 Si, 1·39 Mn, 0·030 P, 0·038 S

INCLUSIONS (G) Galaxite with 34 MnO, 1 FeO, 54 Al_2O_3, 5 MgO (M) Matrix with 36 MnO, 1 FeO, 44 SiO_2, 21 Al_2O_3. Mean analysis (calc.) 35 MnO, 41 SiO_2, 23 Al_2O_3, balance 1

COMMENT The steel sample was later heat treated for 2 h. The galaxite phase was dissolved and precipitation of corundum and spessartite occurred (Fig.52)

25 Oxide inclusion in steel: galaxite in glassy manganese oxide–alumina–silica

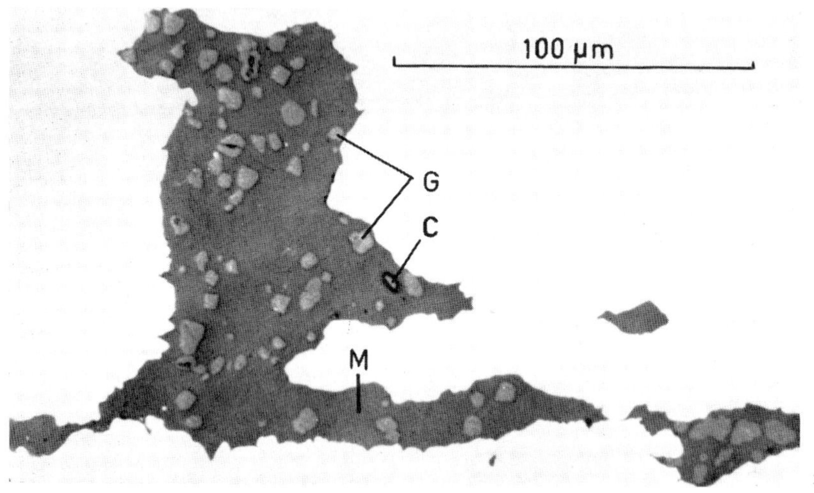

STEEL TYPE Carbon steel. From a forged chain link

STEEL ANALYSIS, % 0·33 C, 0·25 Si, 0·60 Mn, 0·011 P, 0·023 S, 0·009 N

INCLUSIONS (G) Galaxite with 32 MnO, 8 FeO, 55 Al_2O_3 (C) Corundum with 100 Al_2O_3 (M) Matrix with 33 MnO, 2 FeO, 43 SiO_2, 21 Al_2O_3. Mean analysis (calc.) 33 MnO, 3 FeO, 36 SiO_2, 28 Al_2O_3

COMMENT This inclusion is localized below the steel surface but connected with a surface crack with iron oxides

26 Oxide inclusion in steel: galaxite and corundum in glassy manganese oxide–alumina–silica

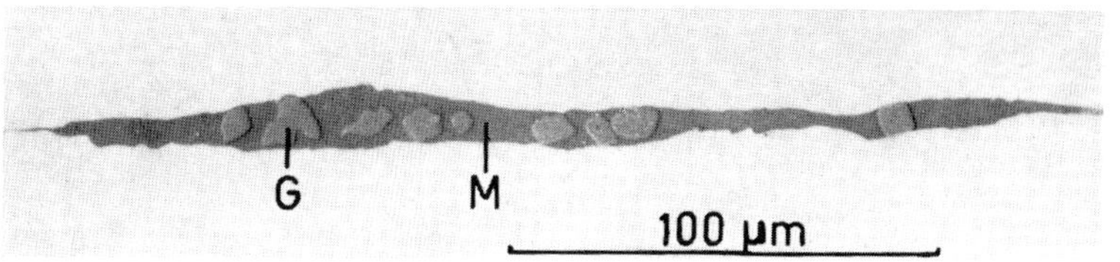

×400

STEEL TYPE Carbon steel. From a hot-rolled billet
STEEL ANALYSIS, % 0·60 C, 0·14 Si, 1·11 Mn, 0·029 P, 0·022 S, 0·006 N
INCLUSIONS (G) Galaxite with 34 MnO, 2 FeO, 61 Al_2O_3, 3 MgO (M) Matrix with 37 MnO, 2 FeO, 35 SiO_2, 22 Al_2O_3, 2 CaO. Mean analysis (calc.) 36 MnO, 26 SiO_2, 33 Al_2O_3, other oxides 5
COMMENT A comparison between the inclusions shown in Figs.25 and 26 and the inclusion shown in this figure clearly indicates how the plastic silicate matrix is deformed when the steel is worked whereas the hard galaxite phase retains to shape

27 Oxide inclusion in steel: galaxite in glassy manganese oxide–alumina–silica

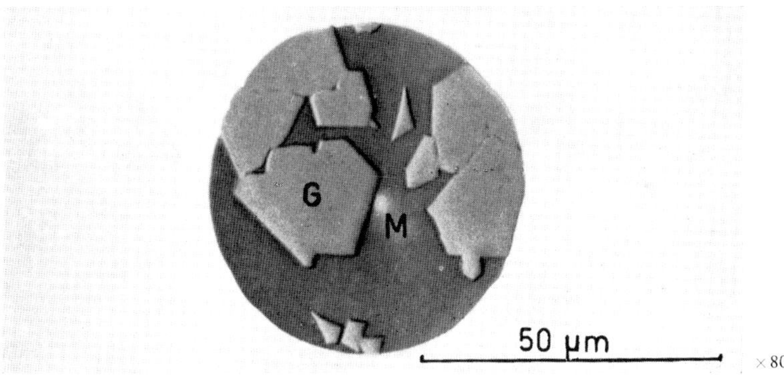

×800

STEEL TYPE Austenitic stainless steel, vacuum cast. From a scoop sample, taken from the mould
STEEL ANALYSIS, % 0·04 C, 0·40 Si, 1·34 Mn, 18·5 Cr, 10·1 Ni, 0·50 Mo, 0·020 P, 0·005 S, 0·029 N
INCLUSIONS (G) Galaxite with 35 MnO, 10 Al_2O_3, 52 Cr_2O_3, 5 TiO_2 (M) Matrix with 34 MnO, 42 SiO_2, 12 Al_2O_3, 4 Cr_2O_3, 2 TiO_2, 2 CaO. Mean analysis (calc.) 35 MnO, 22 SiO_2, 11 Al_2O_3, 27 Cr_2O_3, other oxides 5
COMMENT This inclusion in a steel sample from an ingot should be compared with the inclusion in Fig.29 in a rolled steel sample

28 Oxide inclusion in steel: chromium galaxite in glassy manganese oxide–chromia–alumina–silica

In ferrochromium different inclusion types with chromium galaxite have been observed. Figure 31*b* shows a single-phase chromium galaxite inclusion. Two types of two-phase inclusions have also been found. Those in Fig.31*a* consist of Cr_2O_3 (escolaite) and chromium galaxite, whereas those in Fig.31*d* show two chromium galaxite phases in the same inclusion but with different alumina content.

Three-phase inclusions with Cr_2O_3 and two different chromium galaxites are shown in Fig.31*c*.

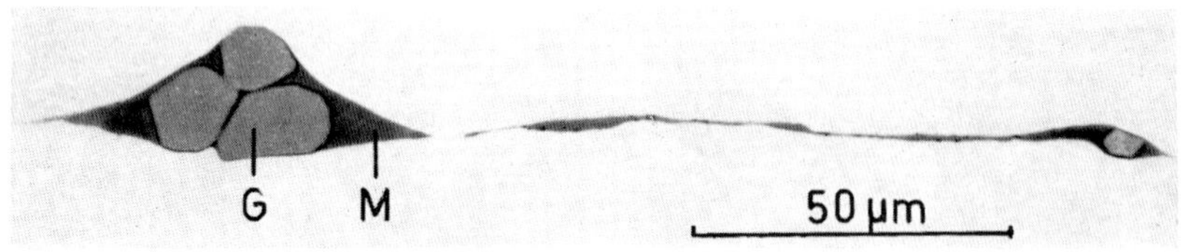

×

STEEL TYPE Austenitic stainless steel. From a hot rolled strip
STEEL ANALYSIS, % 0·05 C, 0·54 Si, 1·27 Mn, 18·3 Cr, 9·1 Ni, 0·25 Mo, 0·035 P, 0·009 S, 0·050 N
INCLUSIONS (G) Galaxite with 36 MnO, 1 FeO, 25 Al_2O_3, 39 Cr_2O_3 (M) Matrix with 23 MnO, 1 FeO, 43 SiO_2, 18 Al_2O_3, 4 Cr_2O_3, 8 CaO. Mean analysis (calc.) 31 MnO, 19 SiO_2, 22 Al_2O_3, 24 Cr_2O_3, other oxides 4
COMMENT A comparison between the inclusion shown in Fig.28 and the inclusion, shown in this figure clearly indicates how the plastic silicate matrix is deformed when the steel is worked whereas the hard galaxite phase retains its shape

29 Oxide inclusion in steel: chromium galaxite in glassy manganese oxide–chromia–alumina–silica

×1200

STEEL TYPE Austenitic stainless steel. From a rolled billet
STEEL ANALYSIS, % 0·05 C, 0·43 Si, 0·96 Mn, 0·023 P, 0·012 S, 18·0 Cr, 9·8 Ni, 0·50 Mo
INCLUSIONS (G) Chromium galaxite with 34 MnO, 66 Cr_2O_3, trace of Fe (T) Tridymite with 100 SiO_2, trace of Al (Sd) MnS (M) Matrix with 45 MnO, 38 SiO_2, 3 Cr_2O_3, trace of Fe. Mean analysis (calc.) 42 MnO, 40 SiO_2, 18 Cr_2O_3

30 Oxide inclusion in steel: chromium galaxite, tridymite and manganese sulphide in glassy manganese oxide–chromia–silica

It is of interest to note that inclusions in ferrochromium with two different chromium galaxite phases having different alumina contents were observed. This is an indication of an immiscibility gap in the chromia–alumina system. On the other hand no such gap is reported in the literature and has not been detected in chromium steel inclusions, where apparently a continuous variation of the ratio alumina–chromia in the system galaxite–chromium galaxite exists (Table IV, Figs.28–30). Some further comments are given in section (C2) and Fig.71 shows an example of such a gap.

Chromium galaxite has also been found in calcium–silica inclusions (Table IV) and together with a sulphide (Fig.32).

Pure spinel ($MgO.Al_2O_3$), sometimes with MnO in solid solution, is a common phase in calcia–silica and calcia–alumina inclusions (Fig.33) and has also been observed in two-phase corundum-spinel inclusions (Fig.6).

Microscopically similar phases
Galaxite can easily be mistaken for corundum. Cr_2O_3 (escolaite) and chromium galaxite are also microscopically similar and the single-phase chromium galaxite inclusions of the type shown in Fig.31*b* have often been reported as chromia. The greater hardness of corundum and chromia as compared with the galaxites is of value when the different phases have to be identified. Both corundum and chromia are anisotropic, whereas the cubic spinels are isotropic. An investigation of these inclusion types using polarized light, which reveals the difference, is therefore always of value (*see also* section A 2).

Comments
It is of interest to note that the composition of the galaxite phase in the different steel inclusions has always been on the alumina side of the homogeneity range for this phase (35–40%$Fe_xMn_{1-x}O$, Table IV), whereas the *synthetic* slags have a continuously variable MnO content between 35 and 66%MnO. Some further comments are made in section A 15.

If steel with galaxite inclusions is deformed, for instance by rolling or forging, the hard galaxite phase stays intact or is crushed, whereas deformation of the plastic silica phase of the inclusions occurs. This is evident from a comparison between the inclusions in Figs.26 and 27 as well as between Figs.28 and 29 (*see also* Fig.55).

A 8 $MnO–SiO_2–Al_2O_3$: Mullite

Chemical formula $3Al_2O_3.2SiO_2$
The Al_2O_3–SiO_2 system
In this pseudobinary system only one stable phase, mullite, exists at normal pressure. The equilibrium diagram has recently been revised by Aramachi and Roy.[30] They have shown that mullite melts congruently and has an extended homogeneity range between about 25 and 28 wt-%SiO_2. Depending on temperature and prehistory of the mullite phase, however, the extension of the homogeneity range shows a considerable variation. The phases cyanite, andalusite, and sillimanite,[31] reported to have the composition $Al_2O_3.SiO_2$, have been found in clays but they are only stable in the earth's crust and decompose to mullite and silica when heated. Earlier investigations

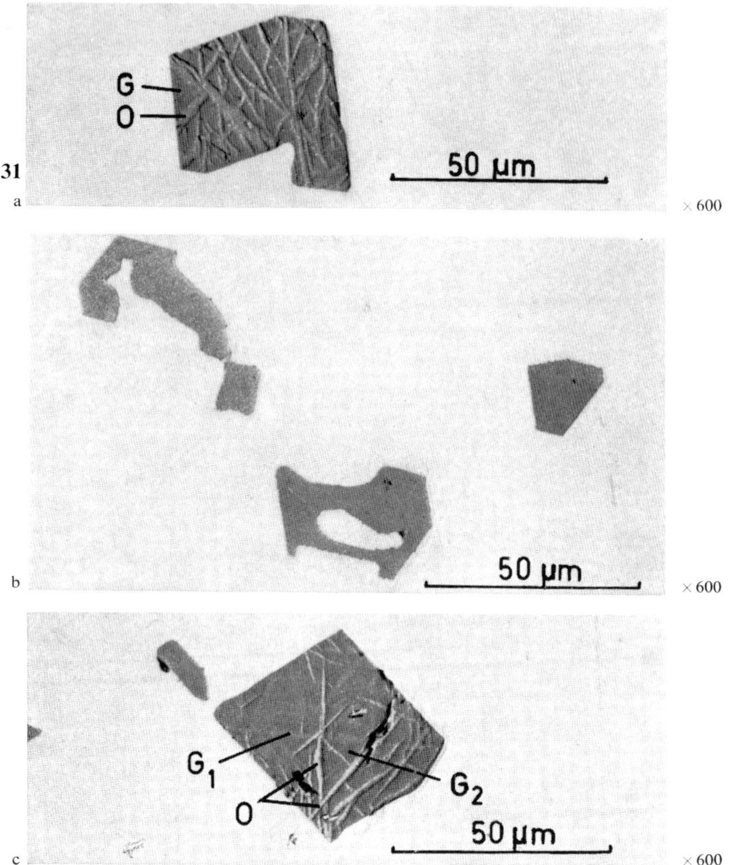

31 a ×600

b ×600

c ×600

have indicated an extension of the homogeneity range of mullite to sillimanite (about 37 % SiO_2), but the recent study[30] mentioned above disagrees with this large extension. The electron probe analyses also indicate a narrow homogeneity range for mullite around 26–30 % of SiO_2.

Physical properties

Melting point: 1 850°C

Density: 3·156

Microhardness: 1 500 kp/mm²

Fluorescence: Mullite shows a blue-white fluorescence in the electron probe analyser

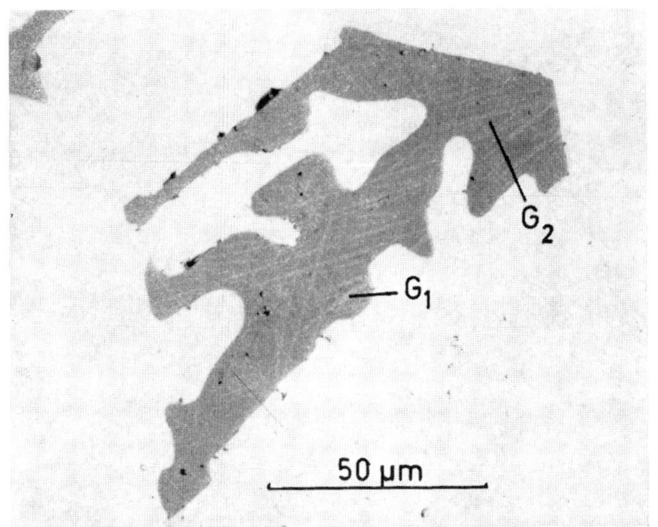

d ×600

MATERIAL TYPE Ferrochromium suraffiné
MATERIAL ANALYSIS, % 73 Cr, 0·03 C, 0·08 Si, balance Fe
31*a* Escolaite (Cr_2O_3) in chromium galaxite
INCLUSIONS (G) Chromium galaxite with 16 MnO, 84 Cr_2O_3 (O) Cr_2O_3. Mean analysis (calc.) 7 MnO, 93 Cr_2O_3
31*b* Single-phase chromium galaxite
INCLUSIONS Chromium galaxite with 15 MnO, 85 Cr_2O_3
COMMENT This inclusion type is easily confused with pure chromia inclusions
31*c* Escolaite (Cr_2O_3) and two different galaxites
INCLUSIONS (G_1) Al-poor chromium galaxite (light) (G_2) Al-rich chromium galaxite (dark) (O) Cr_2O_3. Mean analysis (calc.) 10 MnO, 10 Al_2O_3, 80 Cr_2O_3
31*d* Two different galaxites
INCLUSIONS (G_1) Al-poor chromium galaxite (light) (G_2) Al-rich chromium galaxite (dark). Mean analysis (calc.) 20 MnO, 15 Al_2O_3 65 Cr_2O_3

31 **31a–d** Oxide inclusions in ferro-chromium: galaxite in different inclusion types with decreasing chromia and increasing alumina content

Structure type

Orthorhombic; space group Pbam

a=7·537 Å, b=7·671 Å, c=2·878 Å (typical). ASTM data card 10–394

Wyckoff II, chap. XII, text p.16

Monochromatic X-ray powder photograph, Fig.34.

The unit cell of mullite is slightly variable. The d values are easily mistaken for those of sillimanite, but the a axis of the latter compound is shorter[32] than that of mullite (about 7·49 as compared with 7·54).

Solid solubility

Mullite is reported to dissolve small amounts of the oxides Cr_2O_3, Fe_2O_3, TiO_2, and V_2O_5 in the lattice.[9] This is of importance for steel inclusions. The authors have not found any indication of solid solubility of MnO in mullite even in inclusions rich in manganese. It should be noted that the common X-ray method of estimating solid solubility by means of variation in *d* values can not be used, due to the variation in *d* values for different mullites, mentioned above. Electron probe analysis must be employed.

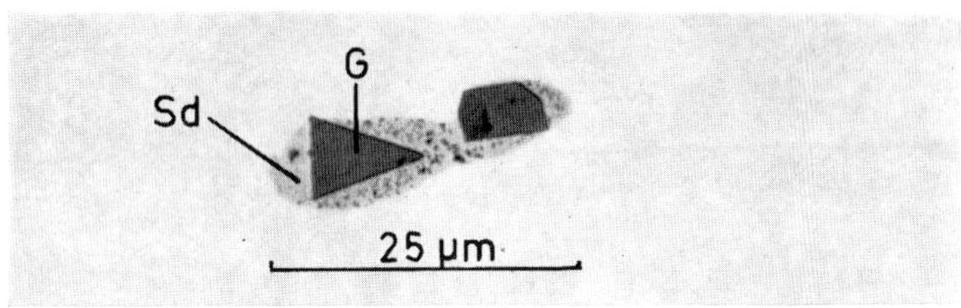

×1250

STEEL TYPE Basic electric steel. From billet
STEEL ANALYSIS, % 0·18 Cr, 0·13 Si, 0·39 Mn. 0·023 P, 0·14 S, 13·6 Cr, 0·66 Ni
INCLUSIONS (G) Chromium galaxite with 30 MnO, 2 FeO, 70 Cr_2O_3 (Sd) Manganese-chromium sulphide with 38 Mn, 25 Cr, 37 S
COMMENT A finely dispersed dark phase of unknown composition is precipitated in the sulphide phase

32 Sulphide inclusions in steel: chromium galaxite in manganese-chromium sulphide

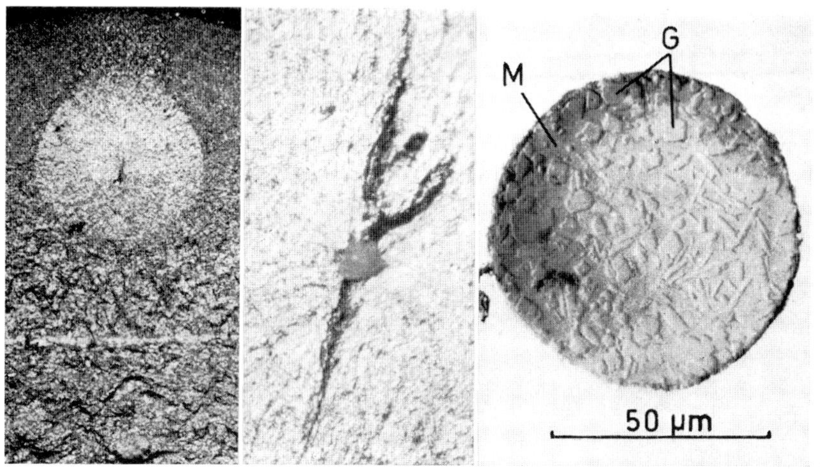

×

STEEL TYPE Low-alloyed chromium steel, heat treated 925°C/12 h→830°C/0·5 h→ oil 50°C. From a broken fatigue test sample
STEEL ANALYSIS, % 0·15 C, 0·29 Si, 0·75 Mn, 0·86 Cr, 1·45 Ni, 0·08 Mo
INCLUSIONS (G) Spinel with 68 Al_2O_3, 31 MgO, 2 CaO, traces of FeO and MnO (M) Matrix of 71 Al_2O_3, 29 CaO and trace of FeO Mean analysis (calc.) 69 Al_2O_3, 13 MgO, 18 CaO
COMMENT The figures show at an increasing magnification how the fatigue failure has started at an inclusion crack. The high CaO and MgO content of the inclusion is a strong indication of its origin from ladle or furnace slag

33 Oxide inclusions in steel: Mg-spinel in calcium aluminate

Microscopy and electron probe analysis

In microsections of steel inclusions with mullite the phase is usually found as characteristic rhombic or rhombohedral crystals, often with a glassy residue in their centres. Careful microscopic preparation is necessary however, as the optical properties of mullite are rather similar to those of glassy silica and it is easy to fail to dis-

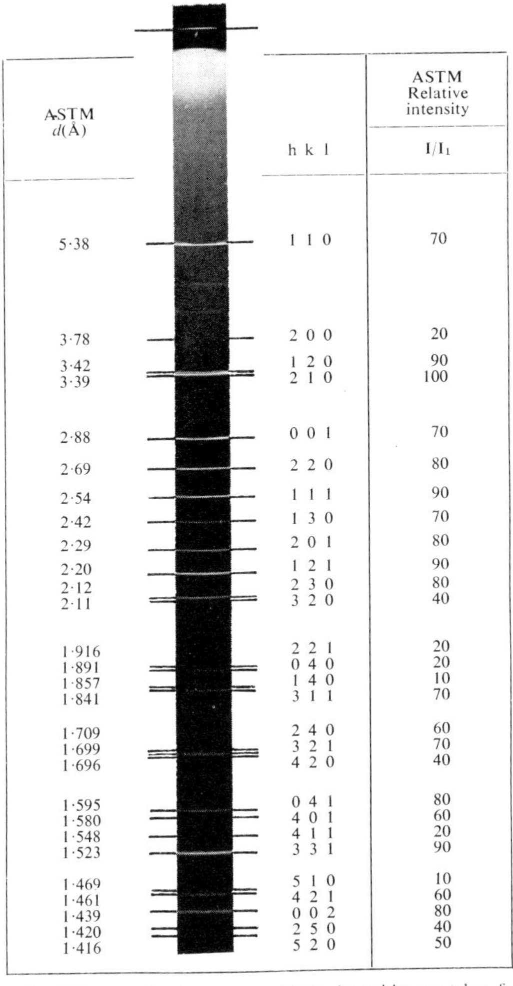

ASTM d(Å)	h k l	ASTM Relative intensity I/I₁
5·38	1 1 0	70
3·78	2 0 0	20
3·42	1 2 0	90
3·39	2 1 0	100
2·88	0 0 1	70
2·69	2 2 0	80
2·54	1 1 1	90
2·42	1 3 0	70
2·29	2 0 1	80
2·20	1 2 1	90
2·12	2 3 0	80
2·11	3 2 0	40
1·916	2 2 1	20
1·891	0 4 0	20
1·857	1 4 0	10
1·841	3 1 1	70
1·709	2 4 0	60
1·699	3 2 1	70
1·696	4 2 0	40
1·595	0 4 1	80
1·580	4 0 1	60
1·548	4 1 1	20
1·523	3 3 1	90
1·469	5 1 0	10
1·461	4 2 1	60
1·439	0 0 2	80
1·420	2 5 0	40
1·416	5 2 0	50

d-values for the different reflections are quoted from the ASTM index. The tabulated relative intensities are taken from the same index

34 X-ray powder photograph of mullite; monochromatic Fe Kα-radiation; Guinier camera

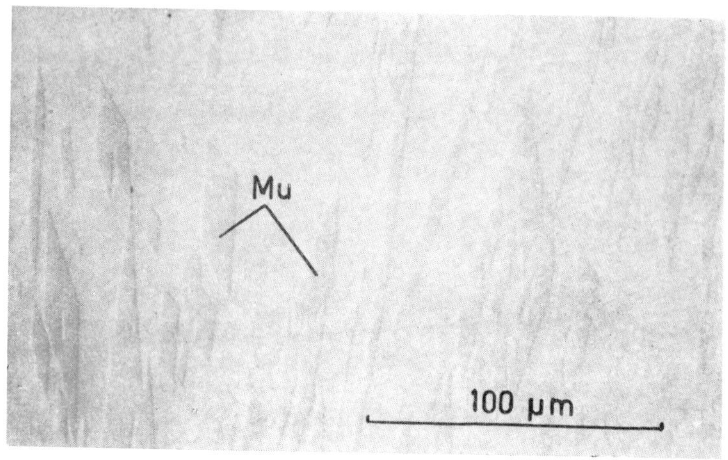

×400

SLAG ANALYSIS, % (Mu) Mullite with 32 SiO_2, 67 Al_2O_3, trace of MnO. Matrix has a composition of 33 MnO, 52 SiO_2, 15 Al_2O_3. Mean analysis (by weight) 23 MnO, 59 SiO_2, 18 Al_2O_3

35 Synthetic slag, prepared from manganese oxide–alumina–silica mixture by electron beam melting at about 2500°C, followed by a heat treatment for 2 h at 1200°C

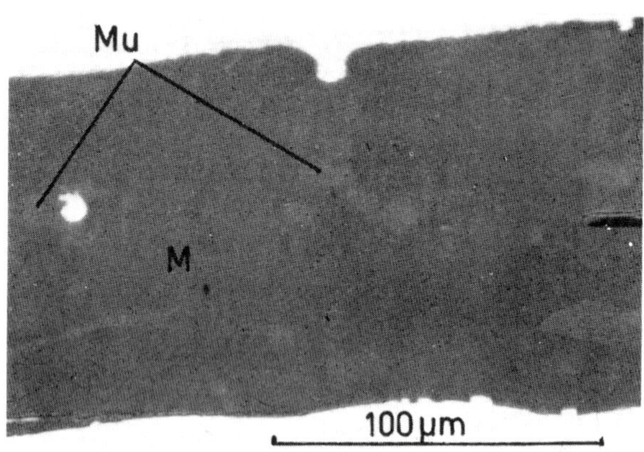

×400

STEEL TYPE Basic electric steel, Si-killed with a limited Al-addition. From a rolled billet
STEEL ANALYSIS, % 0·45 C, 0·22 Si, 0·70 Mn, 0·034 S, 0·002 Al, 39 ppm O
INCLUSIONS (Mu) Mullite with 34 SiO_2 and 68 Al_2O_3 (C) Corundum with 100 Al_2O_3. White sphere is steel
COMMENT Part of the same inclusion as shown in Fig.3, but the steel sample has been heat treated for 2 h at 1100°C, resulting in precipitation of mullite from the glassy matrix

36 Oxide inclusion in steel: mullite and corundum in glassy manganese oxide–alumina–silica

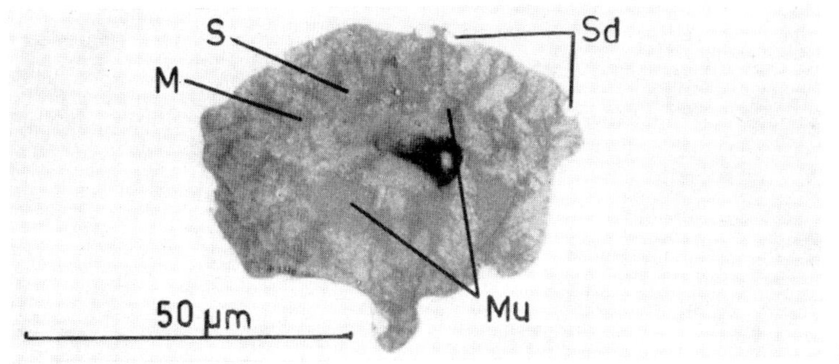

×800

STEEL TYPE Basic open-hearth steel, ferro-silicon killed before tapping, after killed with aluminium. From a forged roll, normalized, air-hardened and tempered
STEEL ANALYSIS, % 0·62 C, 0·26 Si, 0·79 Mn, 0·015 P, 0·018 S, 0·65 Cr, 0·07 Ni
INCLUSIONS (Mu) Mullite with 26 SiO_2 and 66 Al_2O_3 (S) Spessartite with 38 MnO, 45 SiO_2, 16 Al_2O_3 (Sd) Sulphide of MnS-type (M) Matrix, not analysed

37 Oxide inclusion in steel: mullite and spessartite in glassy matrix of manganese oxide–alumina–silica

tinguish the mullite phase in a microscopic investigation. Typical mullite crystals in a synthetic slag are shown in Fig.35. The phase is not very common in indigenous inclusions but it sometimes precipitates from the matrix in alumina–silica inclusions, especially if the steel has been heat treated or worked (Figs.36 and 37). Mullite commonly crystallizes together with spessartite (see A 11) as in Fig. 37. Most mullite inclusions have been found to have a silica content of about 30%, that is, on the silica side of the homogeneity range for mullite.

Figure 38 gives an example of a type of mullite inclusion which differs considerably both in appearance and composition from those shown in Figs.35–37. The mullite is found as a very fine-grained precipitate together with corundum and its existence in considerable amounts has been verified by X-ray diffraction analysis of the actual inclusion. Electron probe analysis, which in this case gives a mean value between the precipitates and the matrix, has shown a silica content of 31–45%SiO_2. It is of interest to note that the matrix in this inclusion is completely free from MnO, whereas MnO was present in considerable amounts in the matrix of the inclusions shown in Figs.36 and 37. It is concluded, therefore, that the inclusion in Fig.38 has an *exogenous* origin. The mean analysis of the inclusion is in close agreement with analytical results from different ladle refractories.

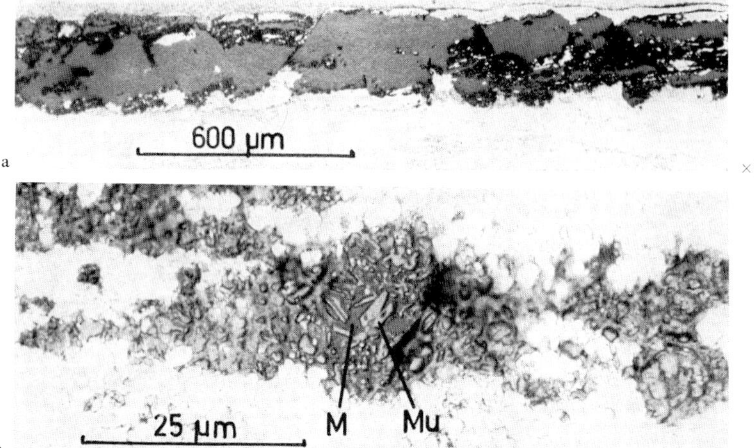

The inclusion is shown at both low magnification (*a* 50×) and high magnification (*b*, 1200×)
STEEL TYPE Basic high-speed steel. From billet
STEEL ANALYSIS, % 0·8 C, 0·34 Si, 0·35 Mn, 4·30 Cr, 4·90 Mo, 6·40 W, 2·03 V
INCLUSIONS (Mu) Mullite and corundum, mean analysis 31–45 SiO_2 (M) Matrix with 72 SiO_2 and 25 Al_2O_3
COMMENT The mean analysis of this inclusion was very similar to the analysis of the bottom smearing from the ladle. This inclusion is free from manganese which is a further proof of its exogenous origin

38 **38a–b** Oxide inclusion in steel: large mullite–sillimanite inclusion of exogenous origin

A 9 MnO–SiO_2–Al_2O_3: Rhodonite

Chemical formula

$MnO.SiO_2$. The mineral $CaO.4MnO.5SiO_2$ is also called rhodonite but metallurgists and ceramists now seem to have adopted the name rhodonite for the calcium-free manganese silicate.

Modifications

Rhodonite exists in two modifications, α and β.[33] The low-temperature modification β is the phase usually called rhodonite, α being an unstable high-temperature modification.

Physical properties

Melting point: 1291°C (forming tridymite and melt)
Density: 3·72
Microhardness: 750 kp/mm^2
Fluorescence: Rhodonite and other manganese silicates show a dark-red fluorescence in the electron probe analyser if the inclusions are free of calcium. If they contain calcium the fluorescent colour is more yellow.

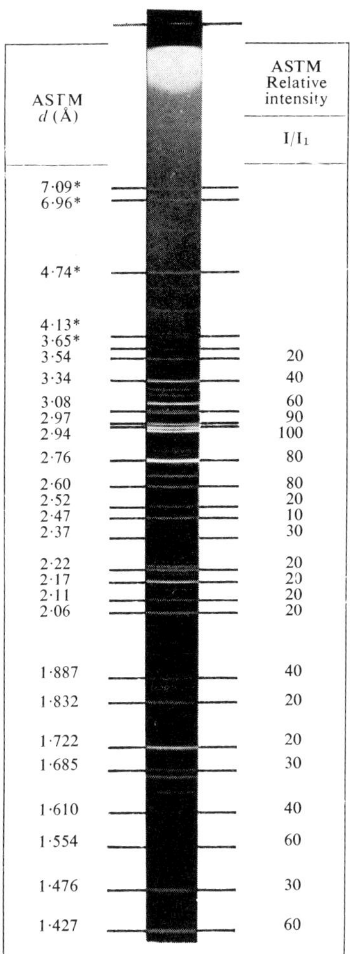

d-values and tabulated relative intensities are quoted from the ASTM index. A feature of the focussing Guinier camera is the accurate resolution of low-angle reflections. A number of high *d*-value reflections have been identified. These are marked with an asterisk and have not been reported in the ASTM index

39 X-ray powder photograph of rhodonite; monochromatic Fe Kα-radiation; Guinier camera

Structure type

Triclinic

$a=7{\cdot}77$ Å, $b=12{\cdot}20$ Å, $c=6{\cdot}70$ A, $\alpha=85°15'$, $\beta=94°00'$, $\gamma=111°$ $29'$.

ASTM data card 5–0614

The structure is described in reference 33.
Monochromatic X-ray powder photograph Fig.39.

Solid solubility
The MnO of rhodonite can be substituted to a great extent by the oxides MgO and FeO (*see also* B4) and to some extent by CaO.[9] Rhodonite is a common phase in steel inclusions. Single-phase rhodonite inclusions, sometimes partly glassy, are often found as round droplets (Fig.40) which are easily mistaken for pure silica (*see* below). At high SiO_2 contents of the inclusions silica is precipitated as cristobalite in a more or less crystallized rhodonite matrix (Figs.11 and 41, left). In sections of multiphase inclusions the rhodonite phase is

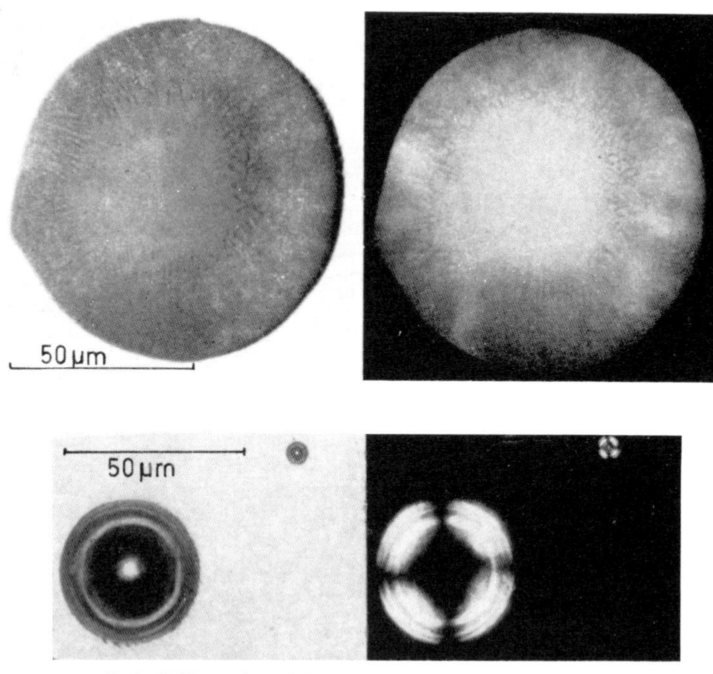

STEEL TYPE Basic Kaldo steel, semi-killed. From ingot
STEEL ANALYSIS, % 0·14 C, 0·07 Si, 1·00 Mn
INCLUSIONS 60 MnO, 2 FeO, 37 SiO_2, 2 Al_2O_3, trace of CaO
COMMENT This inclusion type, which is often partly glassy, is sometimes reported as pure silica. It is rather difficult to distinguish between the two phases with the microscope only, but the microhardness of rhodonite is lower than that of silica. Two different inclusions are shown both in ordinary and in polarized light. The larger is crystallized, the smaller is glassy, but both have the rhodonite composition

40 Oxide inclusion in steel: single-phase rhodonite

frequently found as broad, rather bright laths (Fig.41, right). If the mean inclusion composition is close to that of rhodonite these laths form a dense structure, often with a fine precipitate of MnS in the grain boundaries (Figs.42 and 17).

×400

STEEL TYPE Basic electric steel, Si-killed. The inclusions have been isolated from the lower central part of a top cast 6 ton ingot
STEEL ANALYSIS, % 0·16 C, 0·35 Si, 0·67 Mn, 0·015 P, 0·017 S, 0·009 N, 100 ppm O, no trace of Al
INCLUSIONS (*left*) (R) Rhodonite with 51 MnO, 3 FeO, 40 SiO_2, (Ct) Cristobalite with 100 SiO_2 (M) Matrix with 51 MnO, 3 FeO, 40 SiO_2, traces of Ti and Al. Mean analysis (calc.) 34 MnO, 64 SiO_2, 2 FeO
(*right*) (R) Rhodonite with 49 MnO, 45 SiO_2, 3 Al_2O_3, 5 TiO_2 (M) Matrix with 31 MnO, 51 SiO_2, 13 Al_2O_3 4 TiO_2. The particle is surrounded by a scale of MnS which is also finely precipitated in the matrix. Mean analysis (calc.) 44 MnO, 46 SiO_2, 6 Al_2O_3, 4 TiO_2 (sulphide not included)

41 Oxide inclusions in steel: cristobalite in matrix (partly glassy, partly crystalline) of rhodonite composition (*left*) and rhodonite in glassy matrix of manganese oxide–alumina–silica, surrounded by sulphide scale (*right*)

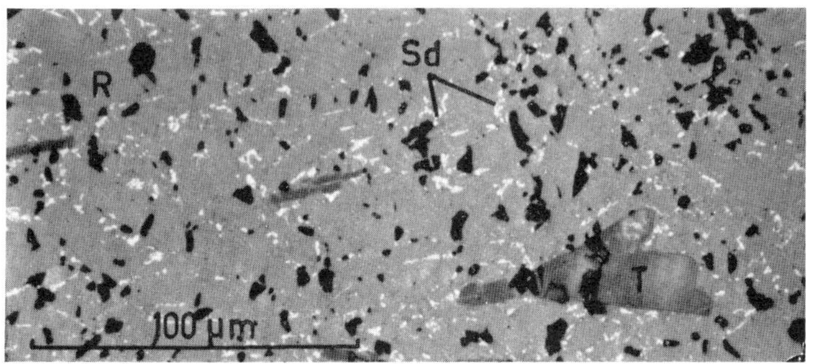

×400

STEEL TYPE Steel ingot, bottom cast. From a hot rolled sheet
STEEL ANALYSIS, % 0·12 C, 0·20 Si, 0·70 Mn, 0·048 P, 0·046 S, 0·005 N
INCLUSIONS (R) Rhodonite with 52 MnO, 43 SiO_2, trace of Fe (T) Tridymite (?) with 100 SiO_2 (A 4) (Sd) 100 MnS. Mean analysis (calc.) 50 MnO, 49 SiO_2, 1 MnS. Black pores are also visible

42 Oxide inclusion in steel: rhodonite, tridymite (?) and manganese sulphide

Microscopically similar phases
Glassy silicate inclusions with the rhodonite composition have been reported as consisting of pure silica and it is difficult to distinguish between these two phases by the microscope only.

The microhardness of glassy silicate inclusions is, however, much lower than for pure silica (*see* A 3).

A 10 MnO–SiO_2–Al_2O_3: Tephroite

Chemical formula 2 $MnO.SiO_2$
Physical properties
Melting point: 1 345°C. According to the Levin monograph[8] the tephroite phase melts incongruently forming MnO, but later investigations[34] have shown it to have a congruent melting point.
Density: 4·04
Microhardness: 950 kp/mm^2

Structure type
Orthorhombic; space group Pbnm*
a=4·862 Å, b=10·62 Å, c=6·221 Å. ASTM data card 9–485
Chrysoberyl type. Wyckoff II, chap. XII, text p.6
Monochromatic X-ray powder photograph, Fig.43
Isostructural with fayalite (B3)

Solid solubility
MnO in tephroite can be completely substituted by FeO (fayalite B3) but an immiscibility gap is reported to exist for small FeO-contents.[36] Mixtures of tephroite and fayalite in this range are often called knebelite in older literature (ref.1, p.99). MnO is also completely substituted by MgO and to about 50 wt-% by CaO.[9] According to the present observations tephroite has a solid solubility limit for Al_2O_3 of about 15 wt-% and the homogeneity range is between about 65 and 73 wt-% of MnO (stoichiometric value 70%).

Microscopy and electron probe analysis
In microsections of synthetic slags, the tephroite phase often forms a characteristic banded structure (Figs.44 and 20). No inclusions with pure tephroite have been found, but in Fig.45 a silicate inclusion with 54%MnO and 17%FeO is shown, having the banded

* According to the new international tables[35] the space group notation should be Pnma and the parameter notations as well as the (hk1)-values in Fig. 43 should be changed accordingly. The ASTM index data card has, however, the older notations which therefore have also been used here.

structure typical of tephroite in synthetic slags. As an immiscibility gap is reported for low FeO contents (*see* above), it is not possible to decide if the phase is tephroite with FeO or fayalite (B3) with MnO. The composition is close to that of ‘knebelite’.

Comment

No inclusions with pure tephroite or manganosite were found. These two phases are those with the highest manganese content in the

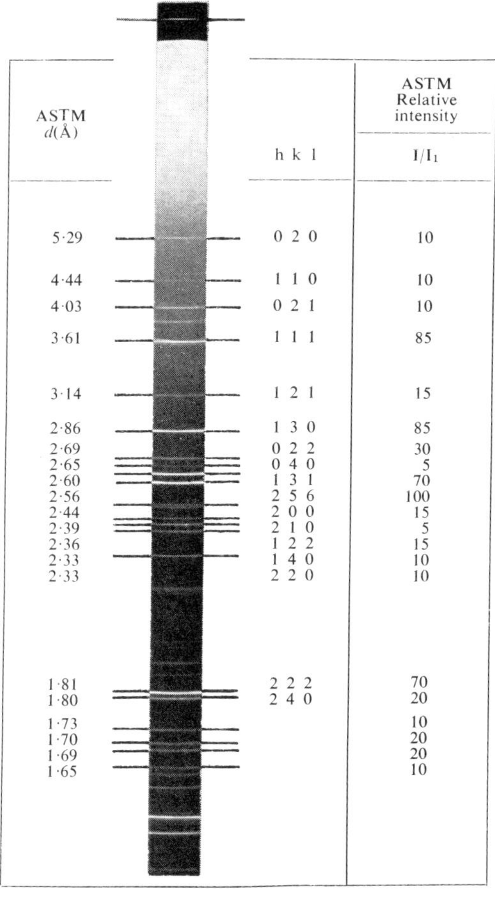

d-values and relative intensities for the different reflections are quoted from the ASTM index

43 X-ray powder photograph of tephroite; monochromatic Fe Kα-radiation; Guinier camera

system $MnO–SiO_2–Al_2O_3$. The reason may be that most of the steel samples were taken from deoxidized steels, where the deoxidant has a much higher oxygen affinity than manganese. The steels were also all rather low in manganese with not more than about 1% of that metal.

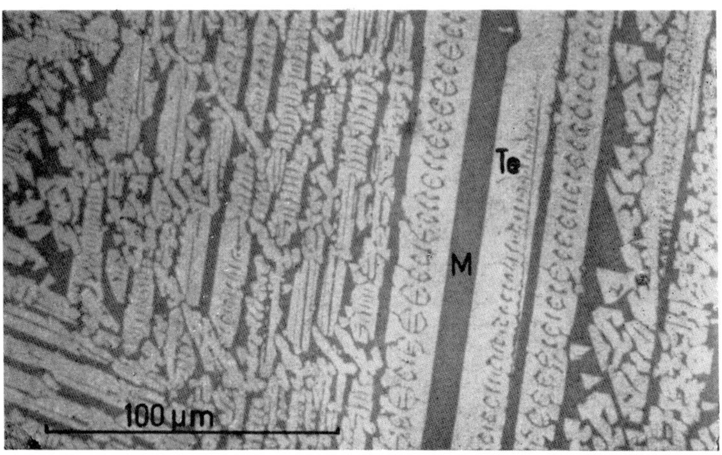

×400

SLAG ANALYSIS, % (Te) Tephroite with 73 MnO, 27 SiO_2 (M) Matrix with 46 MnO, 36 SiO_2, 15 Al_2O_3. Mean analysis (by weight) 60 MnO, 35 SiO_2, 5 Al_2O_3

44 Synthetic slag, prepared from manganese oxide–alumina–silica mixture by electron beam melting at about 2500°C

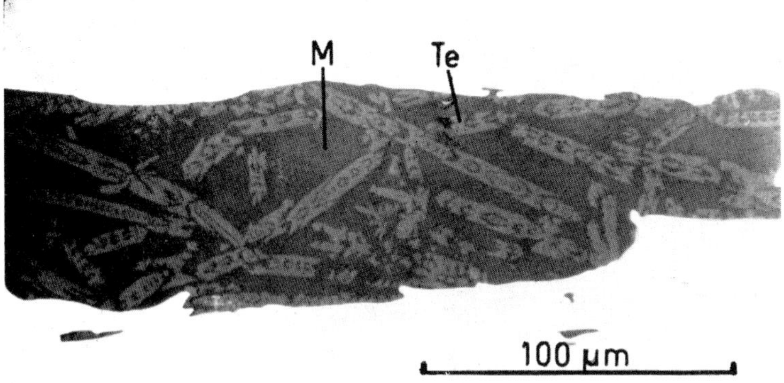

×400

STEEL TYPE Basic Kaldo steel, semi-killed. From ingot
STEEL ANALYSIS, % 0·16 C, 0·06 Si, 1·00 Mn, 0·03 P, 0·015 S, 0·004 N
INCLUSIONS (Te) Tephroite with 54 MnO, 17 FeO and 31 SiO_2 (M) Matrix with 27 MnO, 9 FeO, 48 SiO_2, 19 Al_2O_3 and 4 CaO. Mean analysis (calc.) 58 $Fe_xMn_{1-x}O$, 35 SiO_2, 6 Al_2O_3, 1 others

45 Oxide inclusion in steel: tephroite in glassy matrix of manganese oxide–alumina–silica

Microscopically similar phases
Tephroite and spessartite (*see* A 11) often have a similar crystallization pattern. As neither their microhardness nor composition differ considerably the identification, if possible, should be done by X-ray methods or with an accurate electron probe analysis. Tephroite and fayalite (B3) both crystallize with the same characteristic banded structure and also have to be identified by electron probe analysis.

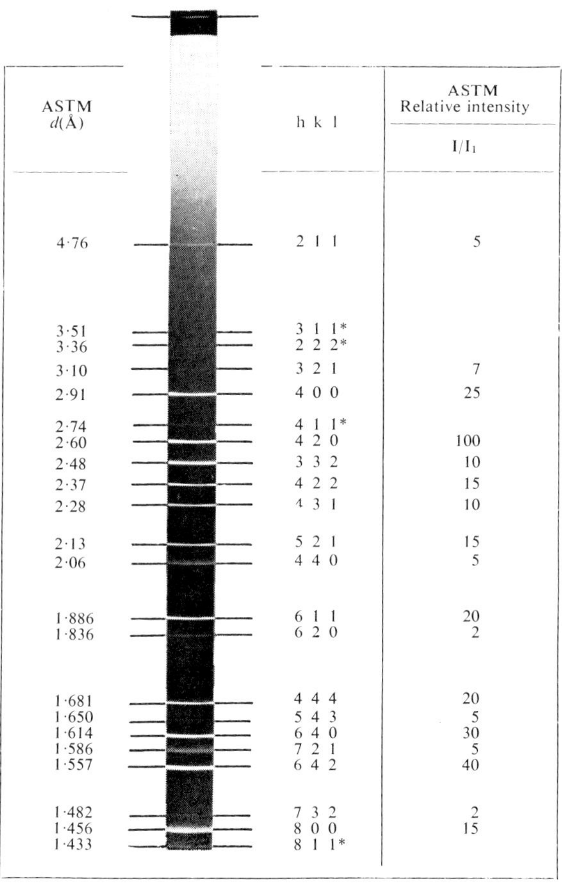

ASTM d(Å)	h k l	ASTM Relative intensity I/I_1
4·76	2 1 1	5
3·51	3 1 1*	
3·36	2 2 2*	
3·10	3 2 1	7
2·91	4 0 0	25
2·74	4 1 1*	
2·60	4 2 0	100
2·48	3 3 2	10
2·37	4 2 2	15
2·28	4 3 1	10
2·13	5 2 1	15
2·06	4 4 0	5
1·886	6 1 1	20
1·836	6 2 0	2
1·681	4 4 4	20
1·650	5 4 3	5
1·614	6 4 0	30
1·586	7 2 1	5
1·557	6 4 2	40
1·482	7 3 2	2
1·456	8 0 0	15
1·433	8 1 1*	

d-values and relative intensities for the different reflections are quoted from the ASTM index. A feature of the focussing Guinier camera is a high sensitivity and a number of weak reflexions have been identified. These are marked with an asterisk and have not been reported in the ASTM index

46 X-ray powder photograph of spessartite; monochromatic Fe Kα-radiation; Guinier camera

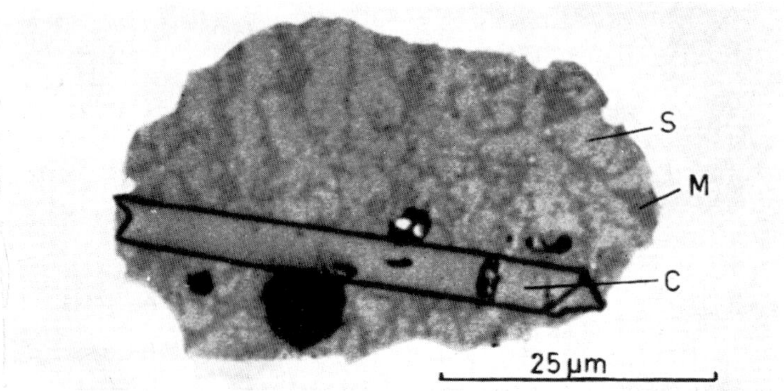

× 1200

STEEL TYPE Basic open hearth steel, ferro silicon killed before tapping, afterkilled with aluminium. From a forged roll, normalized, air hardened and tempered
STEEL ANALYSIS, % 0·62 C, 0·26 Si, 0·79 Mn, 0·015 P, 0·018 S, 0·65 Cr, 0·07 Ni

INCLUSIONS (S) Spessartite with 38 MnO, 45 SiO_2, 16 Al_2O_3 (C) Corundum with 100 Al_2O_3 (M) Matrix with 32 MnO, 52 SiO_2, 18 Al_2O_3. Mean analysis (calc.) 29 MnO, 41 SiO_2, 30 Al_2O_3

47 Oxide inclusion in steel: spessartite and corundum in glassy manganese oxide–alumina–silica

A 11 MnO–SiO_2–Al_2O_3: Spessartite

Chemical formula 3 MnO·Al_2O_3·3SiO_2

Physical properties

Melting point: 1 195°C
Density: 4·18
Microhardness: 1000–1 100 kp/mm^2

Structure type

Cubic. Space group Ia3d
a=11·63 Å. ASTM data card 10–354
Granate type. Wyckoff II, chap. XII, text p.3
Monochromatic X-ray powder photograph, Fig.46
Isostructural with almandine (B6)

Solid solubility

Several isostructural minerals are known, where MnO is substituted by FeO (almandine), MgO (pyrope), and CaO (grossularite), and a wide solid solubility between these different compounds exists.

Microscopy and electron probe analysis

Spessartite forms a featherlike, branched structure in microsections of inclusions (Figs.7, 37, 47–49). It crystallizes slowly and has the lowest melting point (1 195°C) of the different phases in the MnO–

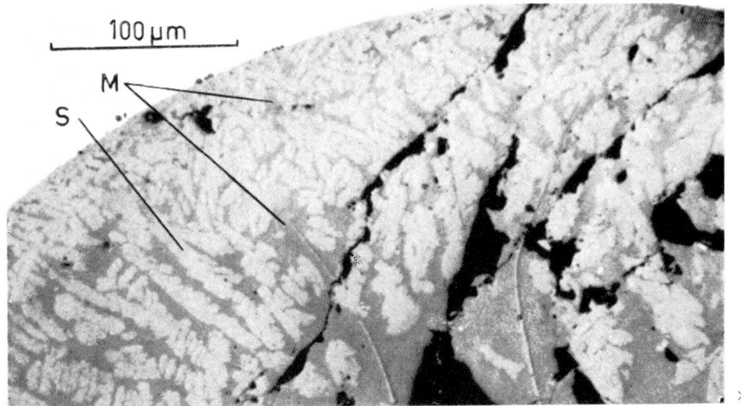

×250

STEEL TYPE Basic open-hearth steel. From a cast crank prepared from a 6 ton ingot which was normalized, turned and tempered
STEEL ANALYSIS, % 0·15 C, 0·44 Si, 0·84 Mn, 0·023 P, 0·018 S, 0·21 Cr, 0·08 Cu
INCLUSIONS (S) Spessartite with 31 MnO, 47 SiO_2, 19 Al_2O_3 (M) Matrix with 38 MnO, 39 SiO_2, 19 Al_2O_3. Mean analysis (calc.) 35 MnO, 45 SiO_2, 20 Al_2O_3. Several cracks and pores (black) are visible

48 Oxide inclusion in steel: spessartite in glassy manganese oxide–alumina–silica

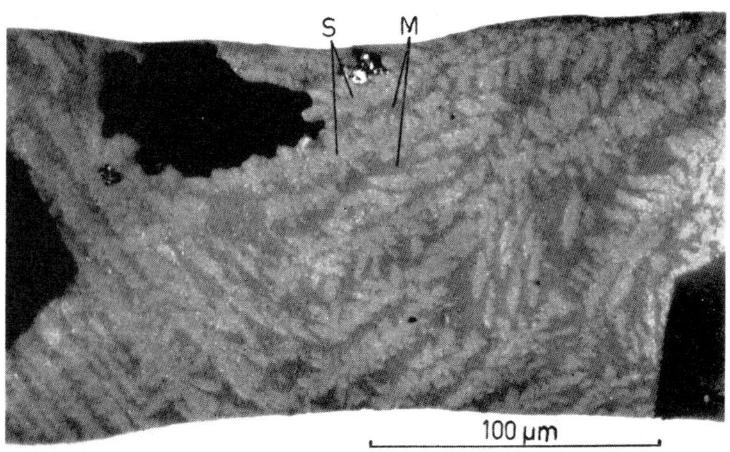

×400

STEEL TYPE Basic electric steel, Si-killed with a limited Al-addition. Rolled billet
STEEL ANALYSIS, % 0·45 C, 0·22 Si, 0·70 Mn, 0·034 S, 0·002 Al, 39 ppm O
INCLUSIONS (S) Spessartite with 42 MnO, 44 SiO_2, 13 Al_2O_3 (M) Matrix with 16 MnO, 55 SiO_2, 31 Al_2O_3
COMMENT Part of the same inclusion which has been shown in Fig.3, but the steel sample was heat treated for 2 h at 1 100°C, resulting in a precipitation of spessartite from the glassy matrix of the inclusion

49 Oxide inclusion in steel: spessartite in glassy manganese oxide–alumina–silica

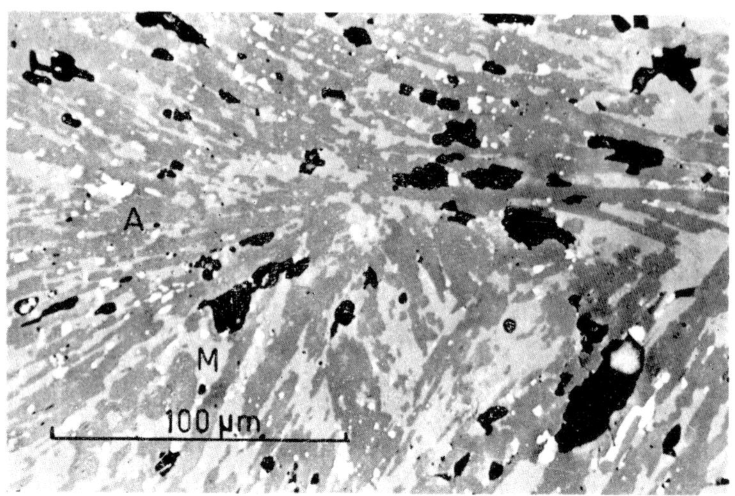

×400

The sample was taken from the slag lining of the ladle, remelted in the electron beam furnace at about 2 500°C and heat treated for 2 h at 1 150°C

SLAG ANALYSIS, % (A) Anorthite with 15 MnO, 50 SiO_2, 8 Al_2O_3, 14 CaO (G) Galaxite not analysed (M) Matrix with 47 SiO_2, 38 Al_2O_3, 16 CaO, trace of MnO

50 Slag sample: anorthite and galaxite in glassy matrix of calcia–alumina–silica

SiO_2–Al_2O_3 system. This phase was only observed in those steel inclusions which had been hot worked (Figs.7, 37, and 47) or cooled slowly (Fig.48). If steel samples with inclusions having a glassy matrix near the spessartite composition range are heat treated below 1 195°C, spessartite often crystallizes in the matrix. Examples are shown in Fig.49 and also in Figs.52–54, section A 14. Spessartite often appears together with corundum (Figs.7 and 47) and sometimes with mullite (Fig.37).

Spessartite has a homogeneity range around its stoichiometric composition as shown by the summary of the analyses in Table V. From these results there seems to be a tendency for spessartite in inclusions always to have a low MnO and a high SiO_2 content within the homogeneity range.

The spessartite phase in inclusions apparently has only a small solid solubility for FeO, MgO or CaO, as only traces of these oxides were found in the phase.

Microscopically similar phases
See tephroite, section A 10.

TABLE V Composition of the spessartite phase in inclusions and in synthetic slags, wt-%

Inclusion type	MnO	SiO_2	Al_2O_3	Sum	Fig.
Ideal formula	43	36	21	100	
Synthetic 1	46	30	18	94	
2	47	40	13	100	
3	49	37	16	102	
4	40	42	15	97	
Corundum, spessartite, glass	38	45	16	99	7, 47
Mullite, spessartite, glass	38	45	16	99	37
Spessartite, glass	31	47	19	97	48
Corundum, spessartite, glass	37	42	19	98	52
Spessartite, glass	42	44	13	99	49
Corundum, spessartite, galaxite, glass	37	40	24	101	54

A 12 MnO–SiO_2–Al_2O_3: Mn-anorthite

Chemical formula $MnO \cdot Al_2O_3 \cdot 2SiO_2$

Physical properties

Not reported

Structure type

No information about Mn-anorthite was found in literature other than a reference to its existence and probable structural similarity to anorthite.[3] The latter mineral with the chemical formula $CaO \cdot Al_2O_3 \cdot 2SiO_2$ belongs to the felspars and is triclinic. Its structure type is reported in Wyckoff II, chap. XII, text p.95, and anorthite is also included in the ASTM index, data card 9–464. The three strongest diffraction lines are reported to have the d values of 3·20 (100), 3·26 (80) and 3·17 (80) and several diffraction lines with high d values are also reported, namely d=6·56 (40), 4·70 (60) and 4·04 (70). The Mn-anorthite in some inclusions and slags had a similar powder pattern. The Ca-compound anorthite (δ) is known to exist in three other modifications α, β, and γ with different structure.[9]

Microscopy and electron probe analysis

The authors were not able to prepare Mn-anorthite synthetically from pure oxides. A Ca–Mn-anorthite was identified as dark crystals in a ladle slag which was heat treated at 1 100°C. The heat treatment resulted in an anorthite crystallization from the glassy matrix (Fig. 50). Mn-anorthite (but with traces of CaO) was also precipitated from the glassy matrix of some different steel inclusions with a composition close to the anorthite range, if the steel samples were heat treated. Such an example is shown in Fig.51. Rait and Pinder[3] as well as Bruch[37] have observed the phase in macroinclusions. Bruch has reported that he had only found this phase in large inclusions, where the matrix of the inclusions has a composition close to the chemical formula of Mn-anorthite and where the inclusion matrix has been heavily deformed, such as by severe deformation of the steel.

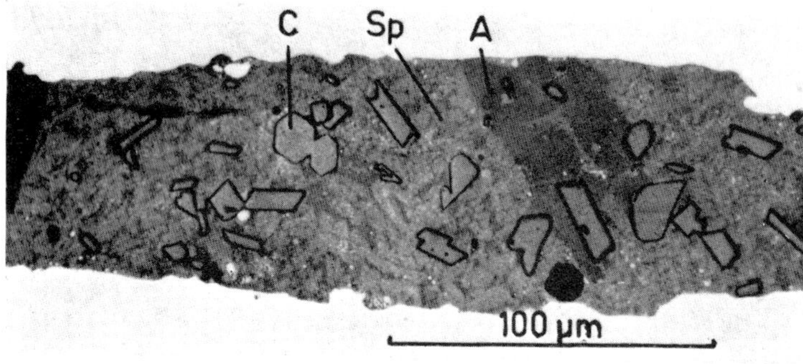

a ×400

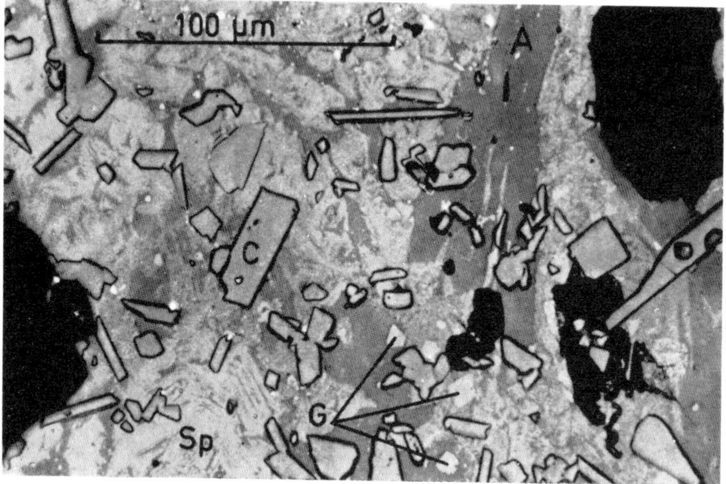

b ×400

STEEL TYPE Basic open-hearth steel, ferrosilicon killed before tapping and aluminium killed in the ladle. From ingot
STEEL ANALYSIS, % 0·33 C, 0·34 Si, 0·60 Mn, 0·013 P, 0·018 S, 2·43 Cr, 0·12 Ni, 0·49 Mo, 0·05 Cu
INCLUSIONS (A) Anorthite with 14 MnO, 48 SiO_2, 39 Al_2O_3, trace of CaO (C) Corundum with 100 Al_2O_3 (G) Chromium galaxite with 38 MnO, 42 Al_2O_3, 19 Cr_2O_3 (M) Matrix with 37 MnO, 40 SiO_2, 24 Al_2O_3
COMMENT Part of the same inclusion as shown in Fig.4, but the steel sample has been heat treated for 1 h at 1100°C, resulting in crystallization of anorthite

51 **51a–b** Oxide inclusion in steel: anorthite, corundum and chromium-galaxite in glassy matrix of manganese oxide–alumina–silica

Comments
The authors are of the opinion that Mn-anorthite, as well as Mn-cordierite (*see* A 13), are phases which only exist if stabilized by calcia. They therefore only exist in those inclusions which contain a certain amount of calcia. This is usually the case for inclusions with nuclei from the furnace or ladle slags which then may have grown indigenously. In fact, the few examples of Mn-anorthite found in inclusions by the authors contain calcium (for instance in Fig.51) and it was not possible to prepare synthetically Mn-anorthite or Mn-cordierite from the pure oxides MnO, SiO_2, and Al_2O_3.

A 13 MnO–SiO_2–Al_2O_3: Mn-cordierite

Chemical formula 2 MnO·2Al_2O_3·5SiO_2
Physical properties
Not reported

Structure type
Not much information about Mn-cordierite is given in the literature. Trojer[9] gives a comment about its structure being analogous to α-cordierite (2 MgO·2Al_2O_3·5SiO_2) and to Fe-cordierite (2 FeO·2Al_2O_3·5SiO_2) and Rigby and Green[38] have also mentioned its existence. The structure type of α-cordierite has been reported[39] and the ASTM index includes powder data for α-cordierite (data card 9–326), Fe-cordierite (data card 9–473) as well as a Mg–Fe-cordierite (data card 9–472). In addition the authors have identified a Ca-cordierite (unpublished). The powder patterns of all these phases have their strongest diffraction line with the high *d*-value of 8·54–8·58 (100). Therefore Mn-cordierite, if it exists, should be identifiable by this strong reflection. Other strong diffraction lines have *d*-values 4·06–4·09 (80), 3·37–3·43 (80), 3·13–3·18 (80) and 3·03–3·07 (80).

Microscopy and electron probe analysis
No example of Mn-cordierite was observed among the steel inclusions available. Bruch[37] has reported that he has observed the phase in macroinclusions and also Rait and Pinder[3] have confirmed the existence of the phase. Analytical data about its composition are not given; it is therefore not known if calcia was present in their inclusions. Apparently slow cooling and the presence of calcia accelerate the crystallization of the phase.

A 14 MnO–SiO_2–Al_2O_3: Transformations

Structure changes were produced in the inclusions on several occasions by heat treatment of the steel samples and transformations

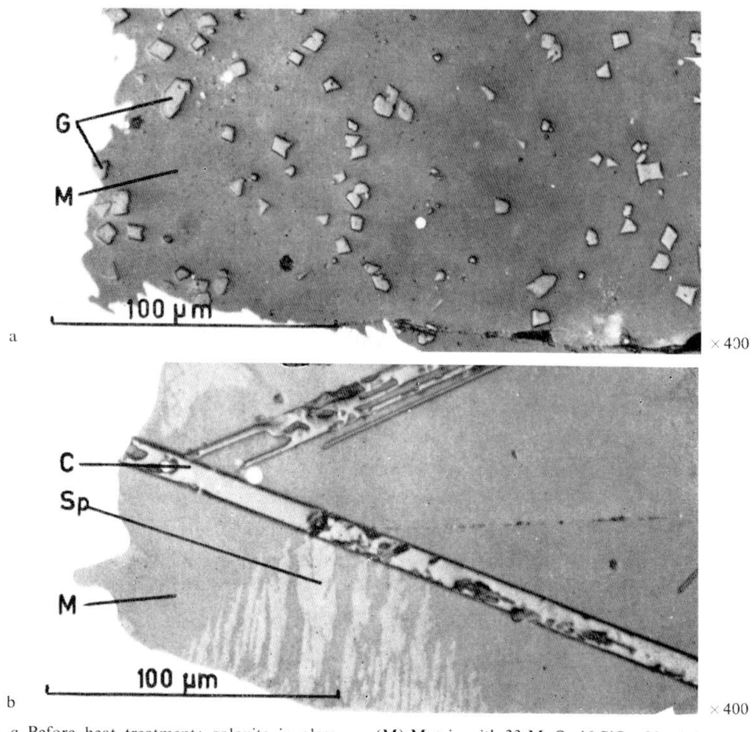

a Before heat treatment: galaxite in glassy matrix of manganese oxide–alumina–silica (G) Galaxite with 34 MnO, 1 FeO, 54 Al_2O_3, 5 MgO (M) Matrix with 36 MnO, 1 FeO, 44 SiO_2, 21 Al_2O_3
b After heat treatment corundum and spessarite in glassy matrix of manganese oxide–alumina–silica
(C) Corundum with 100 Al_2O_3 (Sp) Spessartite with 37 MnO, 42 SiO_2, 19 Al_2O_3, 2 MgO (M) Matrix with 33 MnO, 46 SiO_2, 20 Al_2O_3
COMMENT The mean analysis of the inclusion was found to be 35 MnO, 41 SiO_2, 23 Al_2O_3 and 1 other oxides. Its composition is thus in the stability range for the phases reported in *b*. The heat-treatment therefore results in solution of the metastable galaxite and a precipitation of the stable phases corundum and spessartite. Information about the steel is given in the caption to Fig.25

52 **52a–b** Transformation of a steel inclusion *in situ* as a result of heat treatment of the steel for 2 h at 1 100°C

were accelerated by the effect of working at high temperatures. Different types of inclusion transformation are possible.

1. The inclusions may contain phases which are unstable at the working temperatures of the steel and which transform to stable phases during working or heat treatment. An example of this is given in Fig.17, where cristobalite transformed to tridymite during hot working of the steel.

2. Both in the liquid steel and during its cooling there is a continuous change in composition of many inclusions. Therefore other

a 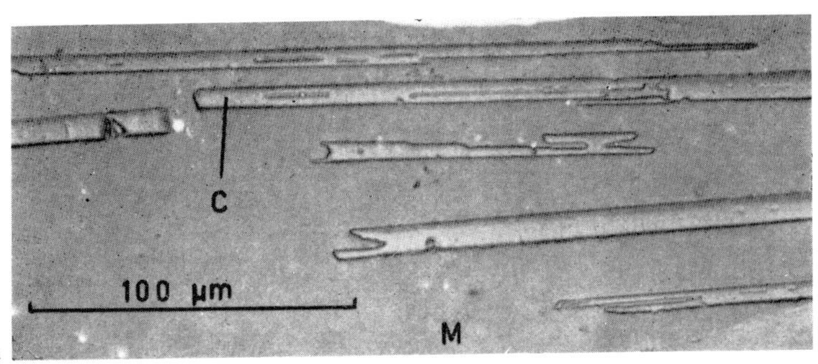

×400

b 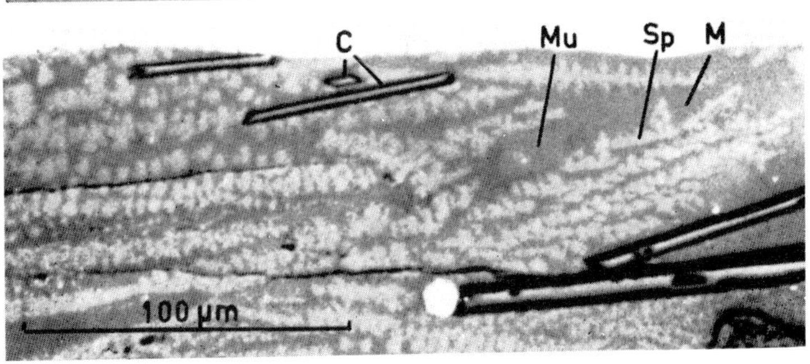

×400

a Before heat treatment: corundum in glassy matrix of manganese oxide–alumina–silica (C) Corundum with 100 Al_2O_3 (M) Matrix with 31 MnO, 45 SiO_2, 22 Al_2O_3
b After heat treatment: corundum, spessartite and mullite in glassy matrix of manganese oxide–alumina–silica (C) Corundum with 100 Al_2O_3 (Sp) Spessartite with 42 MnO, 44 SiO_2, 13 Al_2O_3 (Mu) Mullite with 34 SiO_2, 68 Al_2O_3 (M) Matrix with 16 MnO, 55 SiO_2, 31 Al_2O_3
COMMENT Information about the steel is given in the caption to Fig.3

53 **53a–b** Transformation of a steel inclusion *in situ* as a result of heat treatment of the steel for 2 h at 1 100°C

phases than those which were first formed will be the equilibrium phases within the inclusions. As the nucleation of these new phases is often incomplete, however, the original phases are left in the inclusions of the solidified steel, but now in a metastable condition. A further combination of heat treatment and working of the steel may produce new nuclei and results in a transformation to the stable structure within the inclusions. Thus the metastable phases are dissolved. An example is shown in Fig.5*c*, where corundum is dissolved, forming galaxite, which is the equilibrium phase at the mean composition of the inclusion. A similar example is shown in Fig.52 with a transformation of galaxite in glass to corundum and spessartite in glass after heat treatment of the steel sample. The later phases are those in equilibrium at the mean composition of the inclusion.

a

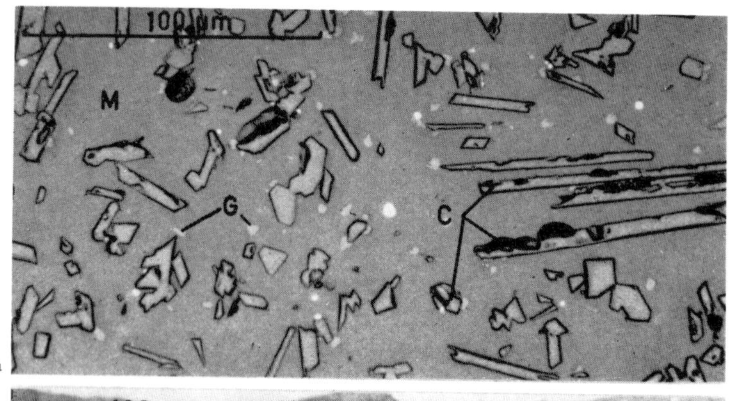

×400

b

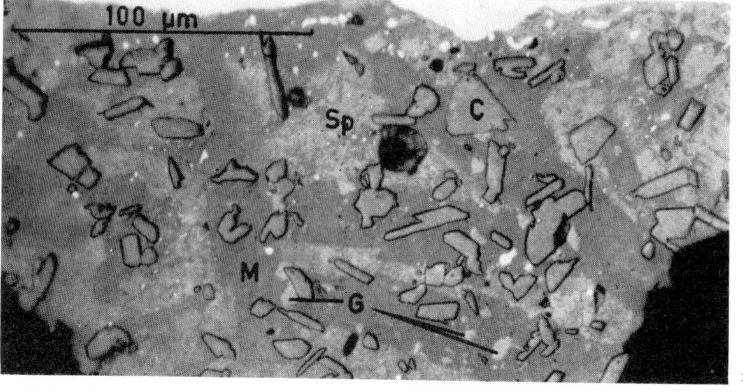

×400

a Before heat treatment: corundum and galaxite in glassy matrix of manganese oxide–alumina–silica
(C) Corundum with 100 Al_2O_3 (G) Galaxite, not analysed (M) Matrix with 31 MnO, 46 SiO_2, 20 Al_2O_3
b After heat treatment: spessartite and anorthite have crystallized in the matrix
(C) Corundum with 100 Al_2O_3 (Sp) Spessartite with 37 MnO, 40 SiO_2, 24 Al_2O_3 (G) Galaxite, not analysed (M) Dark, large crystals are anorthite with 14 MnO, 48 SiO_2, 39 Al_2O_3, trace of CaO
COMMENT Information about the steel is given in the caption to Fig.4

54 **54a–b** Transformation of a steel inclusion *in situ* as a result of heat treatment of the steel for 2 h at 1 100°C

3. Several silicates have a slow crystallization rate and silicate inclusions are often glassy. If the steel is heat treated and deformed, the nucleation rate is often increased, and different phases may be precipitated. An example of this is shown in Figs.40 and 41 with rhodonite crystallizing from a glassy matrix. Further examples are given in Fig.53 with spessartite and mullite crystallizing from a glassy silicate matrix after heat treatment of the steel and also in Fig.54.

4. Important changes in structure and composition of non-

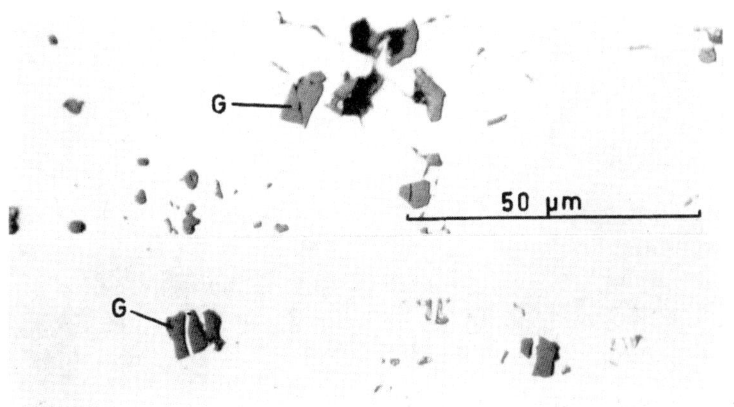

×800

STEEL TYPE Austenitic stainless steel. From cold rolled sheet
STEEL ANALYSIS, % 0·05 C, 0·66 Si, 1·03 Mn, 17·8 Cr, 9·0 Ni
INCLUSIONS (G) Chromium galaxite

55 Chromium galaxite inclusions which have been crushed during deformation of the steel

metallic inclusions in steel may also result from reactions between the inclusions and the steel phase. Such examples are given by Fischer and Fleischer.[40] For instance the FeO:MnO ratio in wüstite inclusions changes during slow cooling, or a reaction between wüstite inclusions and phosphorus from the steel may occur during annealing at 800°C.

5. Inclusion transformations in a wider sense also include physical changes through deformation of the steel during its working. This is illustrated in several figures. For example, from Figs.5*a* and *b* it is evident that the plastic silicate matrix has been more or less completely separated from the hard corundum phase during the steel rolling operation. A comparison between Figs.26 and 27, as well as 28 and 29, shows deformation of the plastic silicate phase in the inclusions during the rolling of the steel, whereas the harder galaxite phase has remained undeformed. A more severe working of the steel (cold rolling of 18:8 austenitic steel) may result in crushing of hard and fragile inclusions as shown by Fig.55.

A 15 $MnO–SiO_2–Al_2O_3$: Summary and discussion

Nearly all the different phases in the $MnO–SiO_2–Al_2O_3$ system have extended homogeneity ranges, as already mentioned in the introduction. During the present investigation additional information about the extension of these homogeneity ranges was obtained, as

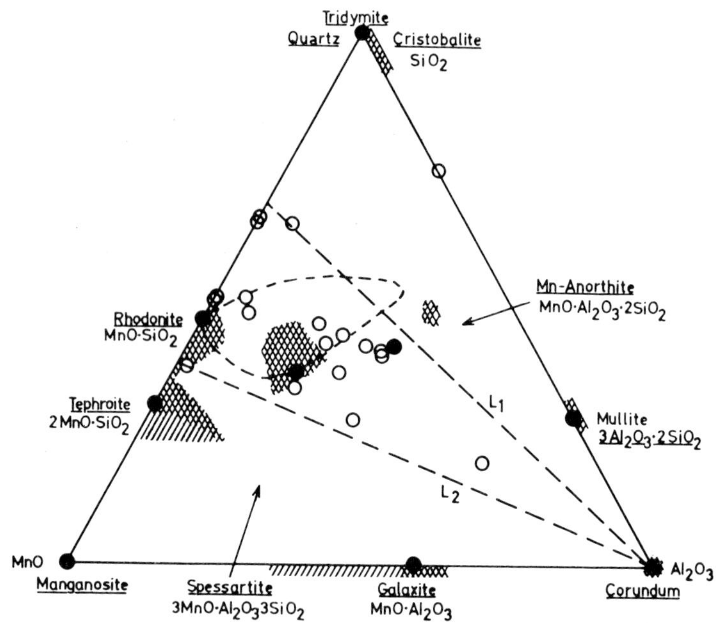

56 Summary of the analytical results regarding occurrence and homogeneity ranges for the different phases in the MnO–SiO_2–Al_2O_3 system, based on analyses of inclusions as well as synthetic slags (compare Fig.1). The stoichiometric composition of the different phases has been represented by dark circles and the homogeneity ranges observed are marked by shadowed areas. Lighter shadows represent homogeneity areas observed for synthetic slags only, whereas a darker shadow indicates that part of the homogeneity range within which phase compositions for both inclusions and synthetic slags have been observed. The different open circles represent the mean composition analyses obtained for the different inclusions analysed during the present work. This composition area is limited by the lines L_1 and L_2, through inclusions from steels with a Mn:Si ratio of 1·9 and 17 respectively. A third line, representing a maximum alumina content of 40 wt-% is a further limitation, but is not shown. The extension of the composition range for the glassy matrix has been indicated by the dotted curve in the central part of the diagram

reported for each phase in the different sections. A schematic summary of these results is given in Fig.56, where the stoichiometric composition (*cf.* Fig.1) for the different phases is also indicated. The homogeneity areas for each phase were obtained by summarizing all the different composition analyses obtained for each phase from synthetic slags as well as from different inclusion types. The results include compositions from quenched, as well as from heat treated steels and synthetic slags, and the different steels and slags have passed through a variety of temperature cycles. The diagram therefore does not represent any isothermal section through the $MnO–SiO_2–Al_2O_3$ system. It merely gives general information about the composition ranges which may be found for the different inclusion phases in steel.

The composition ranges for synthetic slags which give general information about homogeneity ranges are not consistent with those for inclusion phases. The two areas have therefore been indicated differently in the diagram and comments are given below.

Most of the inclusions have a glassy manganese oxide–silica–alumina matrix and the composition range found for this matrix is also outlined in Fig.56.

Finally also the mean composition of each of the different inclusions, as calculated by planimetric analysis, is marked in the diagram.

The present material, though mainly collected to describe different inclusion types in steel, also permits some general conclusions to be drawn about inclusions.

1. All the mean compositions of the inclusions from the $MnO–SiO_2–Al_2O_3$ system are situated in that part of the ternary diagram (Fig.56) which is limited by the lines L_1 and L_2. A third line representing a maximum alumina content of 40 wt-% is a further limitation, but is not shown in Fig.56. The only exceptions are the inclusion shown in Fig.38 (a ceramic particle from an exogenous source which was not molten and in reaction with the steel) the two $Fe_xMn_{1-x}O$ inclusions in Figs.21 and 22, formed by surface oxidation of the steel and the inclusion in Fig.5a–b. This concentration of the mean inclusion composition to a limited part of the composition diagram is a consequence of the relationship which seems to exist between the Mn:Si ratio in the steel phase and the $MnO:SiO_2$ ratio in the inclusions at least for carbon steels (table VI). The limits are the lines L_2 through the inclusion in Fig.45 with a Mn:Si ratio in the steel phase of 17 and L_1 through the inclusion in Fig.13 with Mn:Si being 1·9. The corresponding $MnO:SiO_2$ ratios in the inclusions were found to be 1·7 and 0·47 respectively. The influence of the Mn:Si ratio in steel on the $MnO:SiO_2$ ratio in macroscopic inclusions has been mentioned by Bruch.[37] A discussion of the deoxidation mechanism if both Mn and Si are present in the steel is given in reference 18 and the present results are in support of the mechanism proposed.

2. The reaction between Mn in solution in the liquid steel and silica from the surrounding ceramics seems to be an important source for inclusions, as already mentioned by several authors.[13,22] Many inclusions studied in the present investigation appeared to be

TABLE VI **Comparison between the Mn:Si ratio in the steel phase and the $MnO:SiO_2$ ratio in the inclusions for carbon steels studied**

Mn:Si	$MnO:SiO_2$	Fig.
1·9	0·47	13
1·9	0·53	41 *a*
1·9	0·54	11
1·9	0·78	48
1·9	0·92	17 *a*
1·9	0·95	41 *b*
2·4	0·92	26
3·2	0·69	3
3·5	1·0	17 *b*
3·5	1·0	42
3·7	0·85	25
4·4	1·1	5 *a, b*
4·4	1·4	5 *c*
7·9	1·4	27
14	1·7	40
17	1·7	45

products of such a reaction, for example the inclusion shown in Fig.5*c*.

The reaction proceeds in the two steps

$$2[Mn]+SiO_2 \rightarrow 2\ MnO+[Si] \text{ and}$$

$$2\ MnO+\text{chamotte} \rightarrow \text{liquid (Mn, Al) silicates.}$$

A suggested mechanism for inclusion formation would then be that a liquid manganese oxide–alumina–silica is formed which is emulsified in the liquid steel. When deoxidation proceeds and the temperature of the steel bath gradually decreases, more MnO and SiO_2 are added to the inclusion droplet with a ratio depending on the Mn:Si ratio of the steel. The resulting inclusion thus forms a manganese oxide–silica–alumina particle.

3. Indigenous manganese oxide–silica inclusions free from Al_2O_3 are also formed with a $MnO:SiO_2$ ratio related to the Mn:Si ratio of the steel.

4. Several inclusions were analysed which have a high Al_2O_3 content although no Al-deoxidation was used for the steel and although no Al was detected by an analysis of the steel. This is probably explained by the observations by d'Entremont *et al.*,[21] that the 'constant' $K=[\%Al]^2.[\%O]^3$ for aluminium and oxygen in solution in the steel phase is considerably lower than given by earlier investigators and now has been given a value of about 10^{-14} at 1600°C. The alumina content of inclusions may therefore have a high value even for very low aluminium concentrations in the liquid steel. The

source for such alumina in inclusions, if no aluminium deoxidation has been used, is probably the surrounding ceramics (*see* point 2). Also deoxidation alloys have to be considered as sources of aluminium impurities.

5. The homogeneity ranges of the different phases in the MnO–SiO_2–Al_2O_3 system are usually much wider than the compositions found for inclusion phases. Thus the galaxite phase in inclusions was always found to have a composition on the alumina side of the homogeneity range. No examples of inclusions with manganosite were found, and only one steel gave a sample with tephroite. These two phases are those richest in manganese in the system studied. It therefore seems as if the combination of effective deoxidants and comparatively low manganese content of the steels studied in this investigation decreased the oxygen content of the steel to such a low value that no manganese-rich phases were formed.

6. Some observations regarding the different phases of the MnO–SiO_2–Al_2O_3 system may be mentioned.

The corundum phase in inclusions is always free from all other oxides of interest, the only exception being Cr_2O_3.

The homogeneity range of spessartite forms part of the glassy manganese oxide–silica–alumina range and heat treatment or working of steels below about 1 200°C often results in spessartite precipitation from the matrix.

No traces of the phases Mn-anorthite and Mn-cordierite were found in synthetic MnO–SiO_2–Al_2O_3 slags. Mn-anorthite with traces of calcia was identified in inclusions and so was cordierite (the CaO-compound). Ca–Mn-anorthite has been found in slags from the ladle. Rait and Pinder[3] as well as Bruch[37] have reported the presence of the phases in macro-inclusions. (The former authors also indicate their presence in synthetic melts which have crystallized very slowly.) As analyses *in situ* were not possible in the earlier investigations, it is not known if the inclusion phases reported also contained any calcium. The formation of these phases seems to be greatly facilitated by Ca, and they are therefore most likely to occur in those inclusions which originate from furnace or ladle slags and which have cooled slowly. It is doubtful if the pure Mn-anorthite and Mn-cordierite exist.

The identification of different silica modifications in inclusions is of special value in deciding whether or not inclusions are originating from furnace lining, ladle or nozzle refractories, etc.

7. The detailed metallurgical steel practice has a much greater influence on the formation of inclusions than the steel analysis. In particular the conditions of deoxidation and the teeming operation

as well as the method of casting ingots (from the bottom or from the top) seem to have a great influence on the number, size and types of inclusions found. Ladle slags from earlier charges often adhere to the ladle wall, contaminating subsequent charges. Inclusions originating from furnace or ladle slags usually contain varying amounts of calcium, and are therefore treated in a later section.

Structure changes in the steel inclusions are a common phenomena if the steel ingots are heat treated or worked in different ways. The influence of the inclusions on such properties of the steel as its machinability, its sensitivity to crack formation or the surface quality of rolled stainless steel is at present a subject for much research.[41,42,42a] It is therefore of importance to pay special attention to the frequent occurrence of structure changes and therefore also to changes in physical properties of the inclusions.

B. The $Fe_xMn_{1-x}O$-SiO_2-Al_2O_3 system

B 1 General

Fe (II) is closely related to Mn (II), and FeO and MnO show a complete range of solid solubility.[28] Therefore a substitution of all or part of the MnO in steel inclusions belonging to the MnO–SiO_2–Al_2O_3 system by FeO is common. In fact, in most of the inclusion examples given in section A, part of the MnO has been substituted against FeO. The pure FeO–SiO_2–Al_2O_3 system (x = 1) is similar to the MnO–SiO_2–Al_2O_3 system, the main difference being that a phase corresponding to rhodonite (MnO.SiO_2) has never been found. A summary of the different phases reported to exist in the system and their composition according to the reported stoichiometric formula is indicated schematically in Fig.57 and Table VII.

In the present section some comments are given about those phases in the pure FeO–SiO_2–Al_2O_3 system which were not discussed in section A. The main conclusions are based on synthetic slags, and a special slag series was prepared to study the extension of the rhodonite phase in the FeO·SiO_2–MnO·SiO_2 system. Most modern Swedish steels contain a certain amount of manganese, and as this element has a stronger affinity for oxygen than iron, it was difficult to find many examples of steel inclusions belonging to the pure FeO–SiO_2–Al_2O_3 system.

B 2 FeO–SiO_2–Al_2O_3: Wüstite

Chemical formula $Fe_{1-x}O$

Physical properties

Melting point: 1 370°C
Stability range: 560°C–1 370°C
Density: 5·745

Structure type

Cubic; space group Fm3m

a=4·310 Å. ASTM data card 6–0615
NaCl-type. Wyckoff I, chap. III, text p.1, table p.13
Monochromatic X-ray powder photograph, Figs.19 and 75
Isostructural with manganosite (A 6)

Solid solubility and stability range
See A 6, manganosite. Pure FeO does not exist even in the presence of metallic iron. Part of the iron has the valency III, and therefore FeO is always reported with $FeO{\cdot}Fe_2O_3$ in solid solution.[43,44] As the electron probe method does not permit any direct determination of oxygen, the wüstite phase in the inclusions discussed has, however, always been given as FeO although its formula should have been given as $Fe_{1-x}O$. Wüstite is only stable above 560°C and decomposes to α-iron and magnetite (Fe_3O_4) below this temperature (Fig.74).

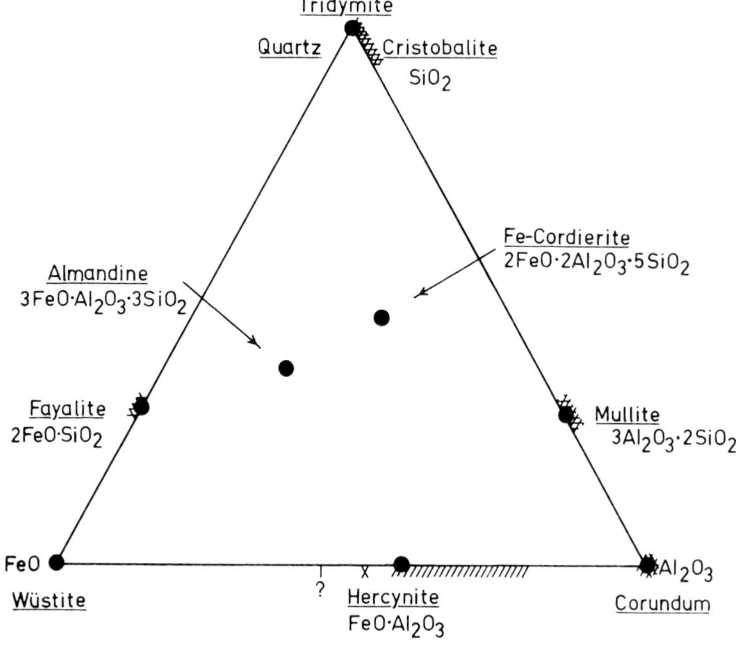

57 Schematic representation of the pseudoternary system $FeO–SiO_2–Al_2O_3$, showing the stoichiometric composition of the different phases which have been reported. The homogeneity ranges for the different phases as formed from synthetic slags and inclusions have been indicated. It is not known if the Fe in hercynite has the valency II or III for those double oxides which have a higher Fe-content than corresponding to the ideal formula $FeO{\cdot}Al_2O_3$. The extension of its composition range is to 55 FeO, if the valency is II, and this has been indicated in the diagram

TABLE VII Phases reported to exist in the $FeO–SiO_2–Al_2O_3$ system

Name	Chemical formula	Stoichiometric composition, wt-% FeO	SiO_2	Al_2O_3	Section
Corundum	Al_2O_3	...	...	100	A 2
Cristobalite	SiO_2	...	100	...	A 3
Tridymite	SiO_2	...	100	...	A 4
Quartz	SiO_2	...	100	...	A 5
Wüstite	FeO	100	...	...	B 2, D 2
Mullite	$3Al_2O_3.2SiO_2$	...	28	72	A 8
(Grunerite)	$(FeO.SiO_2)$	55	45	...	B 4
(Knebelite)	$2Fe_xMn_{1-x}O.SiO_2$	...	...	...	(B 3)
Fayalite	$2FeO.SiO_2$	71	29	...	B 3
Hercynite	$FeO.Al_2O_3$	41	...	59	B 5
Almandine	$3FeO.Al_2O_3.3SiO_2$	43	37	20	B 6
Fe-cordierite	$2FeO.2Al_2O_3.5SiO_2$	22	47	31	B 7

During rapid cooling this transformation takes place only partially or not at all and therefore wüstite can also exist at room temperature.

Microscopy and electron probe analysis

Wüstite with magnetite in indigenous silicate steel inclusions often forms a dendritic structure (Fig.58), similar to that of manganosite (Fig.20). Wüstite inclusions (with magnetite), a product from iron oxidation, are often found below the surface of solidified steel cast

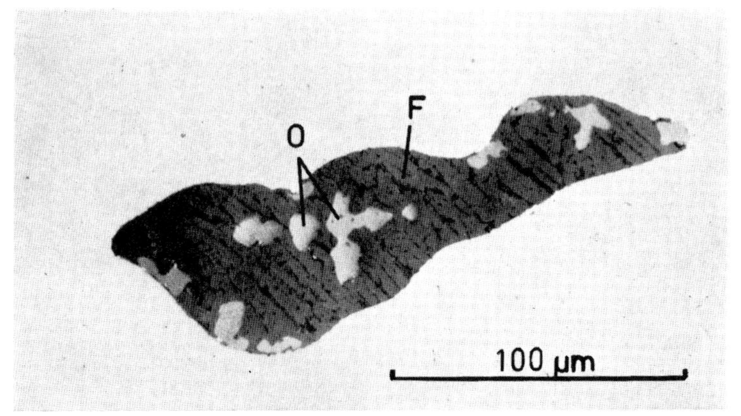

×400

STEEL TYPE Museum specimen. From an old iron anchor
STEEL ANALYSIS Not known

INCLUSIONS, % (O) wüstite with 100 FeO (F) Fayalite with 70 FeO, 28 SiO_2

58 Oxide inclusion in steel: wüstite and fayalite in glassy matrix of iron silicate

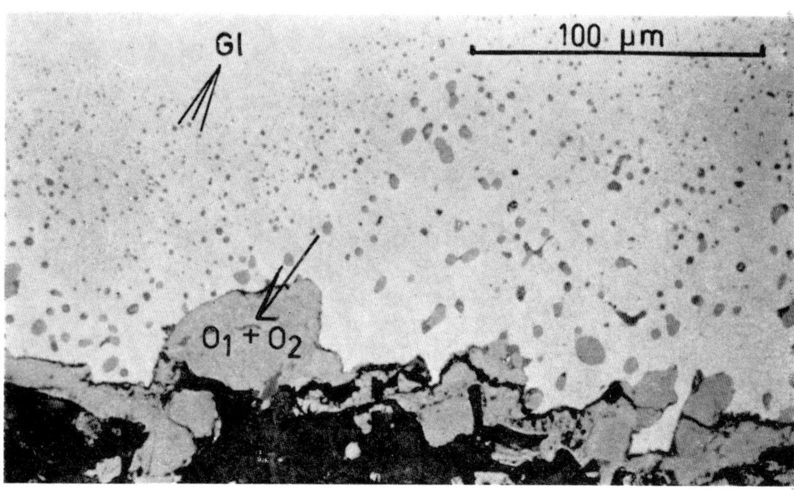

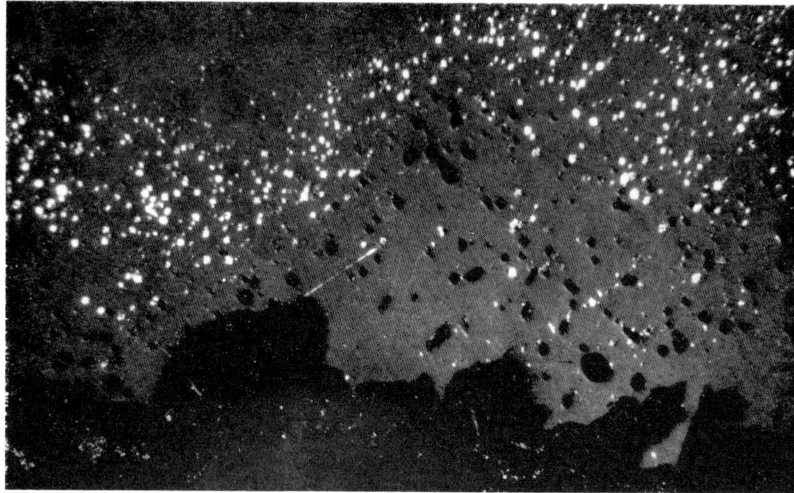

STEEL TYPE Basic electric steel. From a surface of crack in the ingot
STEEL ANALYSIS, % 0·33 C, 0·25 Si, 0·60 Mn, 0·011 P, 0·023 S, 0·009 N
INCLUSIONS Surface layer: (O_1) Wüstite (dark) (O_2) Magnetite (light)
INCLUSIONS (O) near the surface are wüstite and magnetite with 94 FeO, 2 MnO. The MnO content of these inclusions rises with increasing distance from the surface, and an increase in silica content was also observed. The small inclusions (GI) about 100 μm below the surface are rich in manganese oxide and silica. They are also glassy which is evident from the lower photograph, showing the same area in polarized light

59 Surface oxidation and oxide inclusions in steel: wüstite, magnetite, and manganese silicate

from the top (Fig.59). Multiphase oxide inclusions with wüstite together with the other iron oxides are frequently observed in or around surface cracks or places where the steel surface has been oxidized. Such wüstite inclusions also often have varying amounts of MnO in solid solution (Figs.21, 22, and 76).

Comments
Inclusions with composition $Fe_xMn_{1-x}O$ have been reported to consist of two phases in spite of the complete solid solubility but the two-phase character of these inclusions has been explained as being due to their specific formation process and the cooling cycle of the steel.[45]

Oxide inclusions of the $Fe_xMn_{1-x}O$ type are often not in equilibrium with the steel. As shown by Fischer and Fleischer[40] $Fe_xMn_{1-x}O$ inclusions are frequently rich in Fe when precipitated from the melt, but increase their Mn content at lower temperature if the cooling time is very slow. A further possibility is a reaction between Mn-deficient oxide (II) inclusions and phosphorus in rimmed steel. Inclusions of the type $Fe_xMn_{1-x}O$ may therefore change their composition due to reactions between the inclusions and the steel during heat treatment and their composition will depend on the cooling time of the steel. (*See also* section D 2)

B 3 FeO–SiO_2–Al_2O_3: Fayalite

Chemical formula $2FeO \cdot SiO_2$

Physical properties
Melting point: 1205°C (incongruently)
Density: 4·32

Structure type
Orthorhombic; space group Pbnm
a=4·80 Å, b=10·59 Å, c=6·16 Å. ASTM data card 9–307
Crysoberyl type. Wyckoff II, chap. XII, text p.6
Monochromatic X-ray powder photograph, Fig.60
Isostructural with tephroite, A 10

Solid solubility
FeO in fayalite can be substituted by MnO to about 90%.[36] (The name knebelite has been used in older literature for such a Mn-substituted fayalite.[1]) It is completely substituted by MgO and to 50 at % by CaO, forming $FeO \cdot CaO \cdot SiO_2$.

A narrow homogeneity range has been reported; this is also supported by results from synthetic slag melts in the present investigation, which all gave the same analysis for the phase.

Microscopy and electron probe analysis

Fayalite is usually only found in steel free from MnO and its appearance in inclusions from these steels is very similar to that of tephroite (*see* A 10). It forms a characteristically banded structure in microsections (Figs.58 and 61). Also furnace slags from, for example, Lancashire iron frequently contain fayalite (Fig.62) with a similar characteristic crystallization pattern.

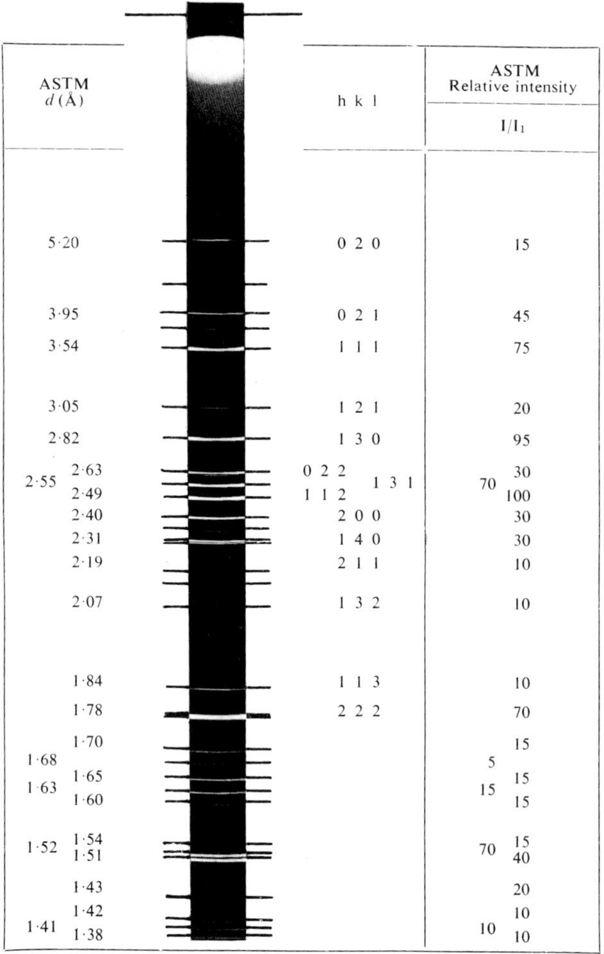

ASTM d (Å)		h k l		ASTM Relative intensity I/I₁	
	5·20	0 2 0			15
	3·95	0 2 1			45
	3·54	1 1 1			75
	3·05	1 2 1			20
	2·82	1 3 0			95
2·55	2·63	0 2 2	1 3 1	70	30
	2·49	1 1 2			100
	2·40	2 0 0			30
	2·31	1 4 0			30
	2·19	2 1 1			10
	2·07	1 3 2			10
	1·84	1 1 3			10
	1·78	2 2 2			70
	1·70				15
1·68				5	
	1·65				15
1·63				15	
	1·60				15
1·52	1·54			70	15
	1·51				40
	1·43				20
	1·42				10
1·41	1·38			10	10

d-values and relative intensities are quoted from the ASTM index

60 X-ray powder photograph of fayalite; monochromatic Fe Kα-radiation; Guinier camera

Microscopically similar phases
See A 10, tephroite.

B 4 $FeO–SiO_2–Al_2O_3$: Grunerite, '$FeSiO_3$'

Due to the close relationship between FeO and MnO the existence of a monosilicate with FeO, isostructural to rhodonite (*see* A 9),

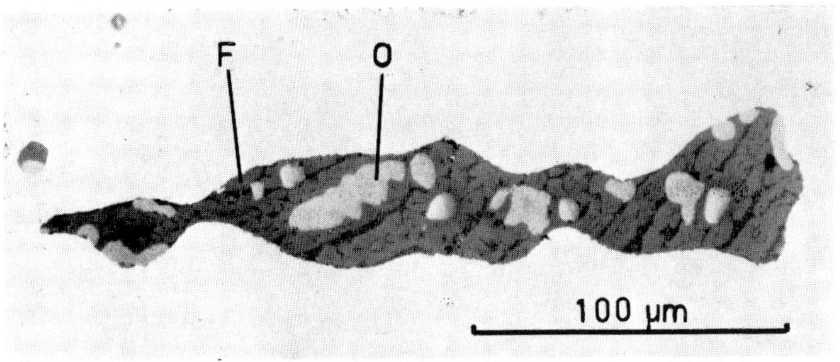

×400

STEEL TYPE Museum specimen. From an old iron anchor
STEEL ANALYSIS Not known
INCLUSIONS, % (F) Fayalite with 70 FeO, 29 SiO_2 (O) Iron oxide

61 Oxide inclusion in steel: fayalite and iron oxide in glassy matrix of iron silicate

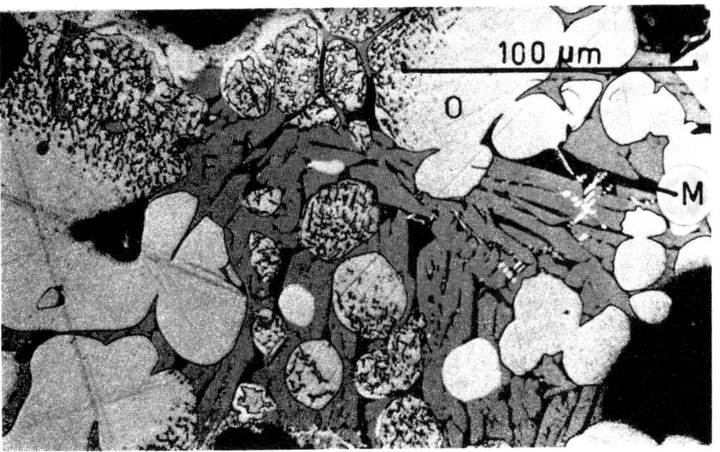

×400

STEEL TYPE Lancashire iron. Slag from tapping bar
STEEL ANALYSIS, % 0·55 C, 0·07 Si, 0·04 Mn, 0·018 P, 0·013 S
INCLUSIONS (F) Fayalite with 64 FeO, 3 MnO, 30 SiO_2 (O) Iron oxides (*see* Fig.77) (M) Matrix with FeO, Al_2O_3, SiO_2 and small amounts of CaO and MnO. Etched in nital
COMMENT For an identification of the different iron oxides in this type of inclusion, *see* Fig.78

62 Oxide inclusion in steel: fayalite and iron oxides in iron oxide–alumina–silica

would be expected. However, it has been reported that $FeSiO_3$ is unstable in the pure condition.[36] A mineral, grunerite, with the composition $FeSiO_3$ has been reported, but a reliable analysis for other elements is not given.[1] CaO and MgO monosilicates with high contents of FeO have also been reported[9] and it could be said, therefore, that $FeSiO_3$ exists if stabilized by certain amounts of CaO and MgO. According to Trojer[9], MnO in rhodonite should be nearly completely substituted by FeO, but no references are given. As the solid solubility limit is of some importance for inclusion studies, the authors have prepared a series of synthetic slags in the range $Fe_xMn_{1-x}O\cdot SiO_2$ with varying values of x. These slags were prepared in an argon arc melting furnace at about 2000°C and studied by X-ray diffraction analysis as prepared and after heat treatment for 2 h at 1 100°C. The results showed that silicates of composition $(Fe_{0\cdot5}Mn_{0\cdot5})O\cdot SiO_2$ were isostructural to rhodonite and consisted of one phase. Those with composition $(Fe_{0\cdot75}Mn_{0\cdot25})O\cdot SiO_2$ showed reflections of both fayalite and rhodonite. The results therefore indicate that the substitution of FeO for MnO in rhodonite is possible only to a maximum percentage of about 75. This is in agreement with the indications of Benedicks[1] (p.100).

B 5 $FeO–SiO_2–Al_2O_3$: Hercynite

Chemical formula $FeO\cdot Al_2O_3$

Physical properties

Melting point: 1 780°C

Density: 4·05

Structure type

Cubic; space group Fd3m

a=8·15 Å. ASTM data card 3–0894

Spinel type. Wyckoff II, chap. VIII, text p.16, table p.42

Monochromatic X-ray powder photograph, Fig.63

Isostructural to galaxite (A 7), chromium galaxite (C 3), chromite (C 4) and magnetite (D 3)

Solid solubility

Hercynite is a double oxide of the spinel type and some comments about this structure were given in section A 7. Those spinels which are of interest in relation to inclusion phases are indicated in Table II.

FeO in hercynite is reported to be substituted to some extent by several divalent oxides, of which MnO and MgO are of interest for inclusions, but no detailed information appears in the literature.

Al_2O_3 can be substituted by Cr_2O_3 (C 3) and Fe_2O_3 (D 4), both of importance in inclusions. For the system $FeO{\cdot}Al_2O_3$–$FeO{\cdot}Fe_2O_3$ it is reported that complete solid solubility exists above 850°C but that an immiscibility gap is formed with increasing width with lower temperature.[46]

The homogeneity range for hercynite was found to be from 20%

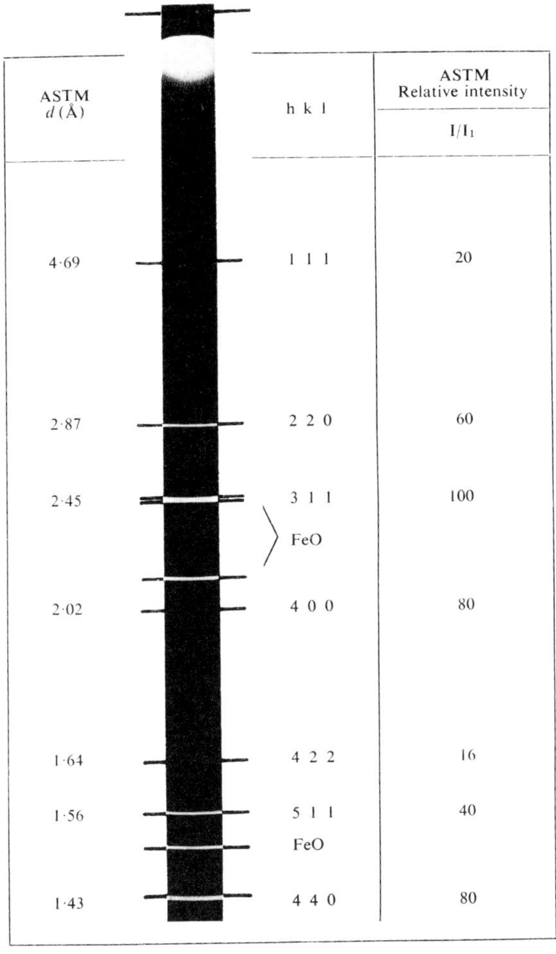

d-values and relative intensities are quoted from the ASTM index

63 X-ray powder photograph of hercynite; monochromatic Fe Kα-radiation; Guinier camera

FeO to 55%FeO for synthetic $FeO–Al_2O_3$ slags and thus lies on both sides of the stoichiometric composition 41 $FeO·59Al_2O_3$. It is not known if part of the iron for compositions higher than 41%FeO has the valency III. All iron is reported as FeO in this investigation.

The homogeneity range is wider than that previously reported in the Levin monograph,[8] according to which hercynite has a constant composition.

Microscopy and electron probe analysis

In *synthetic* slags, hercynite usually has a crystallization pattern similar to galaxite (A 7) and a characteristic microsection is shown in Fig.64. This structure is an example of hercynites of different composition appearing in the same sample.

Only one example of steel inclusions with hercynite was found. These inclusions were situated in a weld between two plates of carbon steel (Fig.65) but the probable origin of the inclusions was the welding electrode. The presence of hercynite in an iron oxide–alumina–silica matrix was confirmed by electron probe analysis on inclusions, but unfortunately the polishing for photography was unsuccessful. The small single phase inclusions shown in Fig.65, however, are also hercynite but are not very characteristic in appearance.

B 6 $FeO–SiO_2–Al_2O_3$: Almandine

Chemical formula $3FeO·Al_2O_3·3SiO_2$

Physical properties

Density: 4·32

Structure type

Cubic; space group Ia3d

a=11·526

Granate type. Wyckoff II, chap. XII, text p.3

A monochromatic X-ray powder photograph of the isostructural spessartite is shown in Fig.46.

Solid solubility

In the mineral almandine, FeO can be completely substituted by MnO (spessartite, A 11), MgO (pyrope) and CaO (grossularite).

Comment

The phase has not been observed in steel inclusions, and it is reported[47] that it cannot be precipitated from the liquid phase in the $FeO–SiO_2–Al_2O_3$ system.

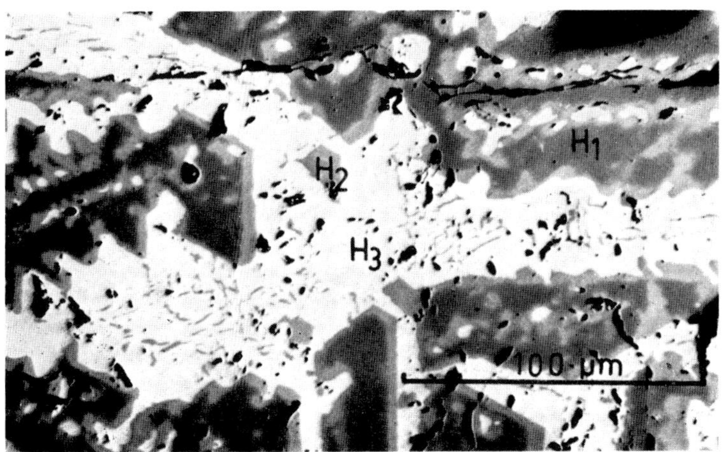

×400

SLAG ANALYSIS, % (H_1) Hercynite 48 FeO, 52 Al_2O_3 (dark grey) (H_2) Hercynite 60 FeO, 40 Al_2O_3 (grey) (H_3) Hercynite 86–97 FeO, balance Al_2O_3 (white). Mean analysis (by weight) 80 FeO, 20 Al_2O_3

COMMENTS Characteristic crystallization pattern of synthetic hercynite. Immiscibility ranges between hercynites of different composition appear during cooling. It is not possible to determine if all the iron has the valency II, or if part of it has the valency III. Here the hercynite should be written FeO·$(Fe_xAl_{2-x})O_3$, but all the iron in the analyses has been reported as FeO

64 Synthetic slag: hercynite. Prepared by melting in electron beam at about 2500°C

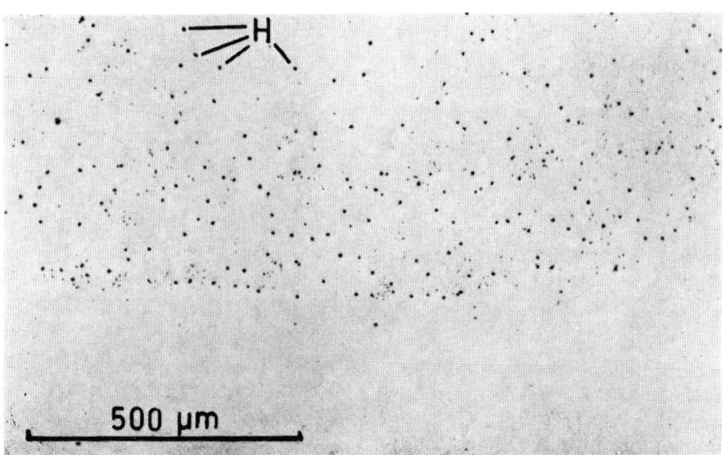

×75

STEEL TYPE Not known

STEEL ANALYSIS Not known

INCLUSIONS, % (H) Hercynite with 45 FeO, 52 Al_2O_3

COMMENT The inclusions probably originate from the welding electrode as no inclusions of the same type were found in the steel or at any other place than in the weld. The figure shows the border between welding zone (above) and steel (below)

65 Oxide inclusions in steel: small hercynite inclusions in a weld

B 7 FeO–SiO_2–Al_2O_3: Fe-cordierite

Chemical formula $2FeO \cdot 2Al_2O_3 \cdot 5SiO_2$

Physical properties

Not reported

Structure type

Isostructural with α-cordierite (Trojer[9] p.361)

Powder data are given in the ASTM index, data card 9–473.

Comments

No examples have been found among the inclusions investigated. According to Rigby and Green[38] this phase is sometimes found in the reaction zone between furnace slags and refractories. (*See also* Mn-cordierite, A13)

C. The $Fe_xMn_{1-x}O$-SiO_2-$Cr_yAl_{2-y}O_3$ system

C 1 General

Al_2O_3 (corundum) and Cr_2O_3 (escolaite) are isostructural and the pseudobinary system shows a complete solid solubility at higher temperatures,[8] but at lower temperatures an immiscibility gap exists. There are many similarities between the pure MnO–SiO_2–Cr_2O_3 system and the MnO–SiO_2–Al_2O_3 system, but also some differences. The Cr_2O_3–SiO_2 system has no intermediate phase similar to mullite in the Al_2O_3–SiO_2 system. No ternary phases have been reported, but it is not possible to decide whether such phases are absent or are undetected since the system has not been completely investigated. The authors have made an outline study of the system by melting synthetic oxide mixtures in a plasma furnace. The main emphasis has been to prepare synthetic slags under similar temperature and composition conditions as those of inclusion formation in steel. Although this was not a complete study of the system, no ternary phases were found. The schematic ternary composition diagram for the MnO–SiO_2–Cr_2O_3 system indicated by this work is given in Fig.66.

In the present section some comments are given about those phases in the MnO–SiO_2–Cr_2O_3 system which were not discussed in section A and which are of importance for steel inclusions as well as their relation to the corresponding alumina compounds.

The system FeO–SiO_2–Cr_2O_3 may be briefly considered since the substitution of part of the MnO by FeO is common. As already mentioned in section B the FeO–SiO_2 system has only one intermediate phase as compared with two for the MnO–SiO_2 system. A schematic diagram of the composition of the phases in the FeO–SiO_2–Cr_2O_3 system is shown in Fig.67, but knowledge of the ternary range is small. The most important compound for this system is chromite, FeO·Cr_2O_3. This phase has also been observed in steel inclusions and will be discussed below.

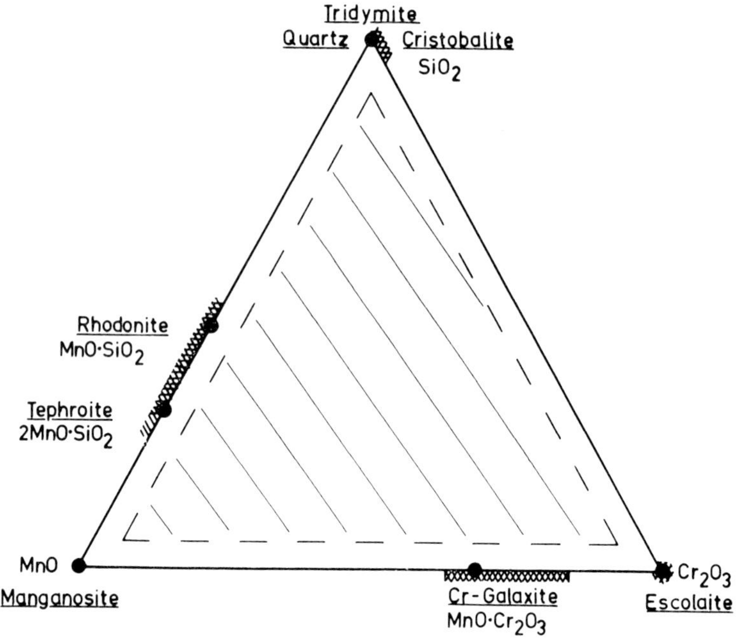

66 Schematic representation of the pseudoternary system MnO–SiO_2–Cr_2O_3, showing the stoichiometric composition of the different phases which have been reported. The homogeneity ranges of the phases which have been studied in synthetic slags and inclusions are also indicated. The system was not completely investigated regarding ternary phases

C 2 MnO–SiO_2–Cr_2O_3: Cr_2O_3 (escolaite)

Chemical formula Cr_2O_3

Physical properties

Melting point: 2265°C
Density: 5·215

Structure type

Hexagonal; space group $R\bar{3}c$
a=4·9591 Å, c=13·599 Å, c/a=2·74. ASTM data card 6–0504
Cr_2O_3-type. Wyckoff I, chap. V, text p.4, table p.11
Monochromatic X-ray powder photograph, Fig.68
Isostructural with corundum (A 2) and haematite (D 4)

Solid solubility

At higher temperatures Cr_2O_3 (escolaite) shows a complete solid

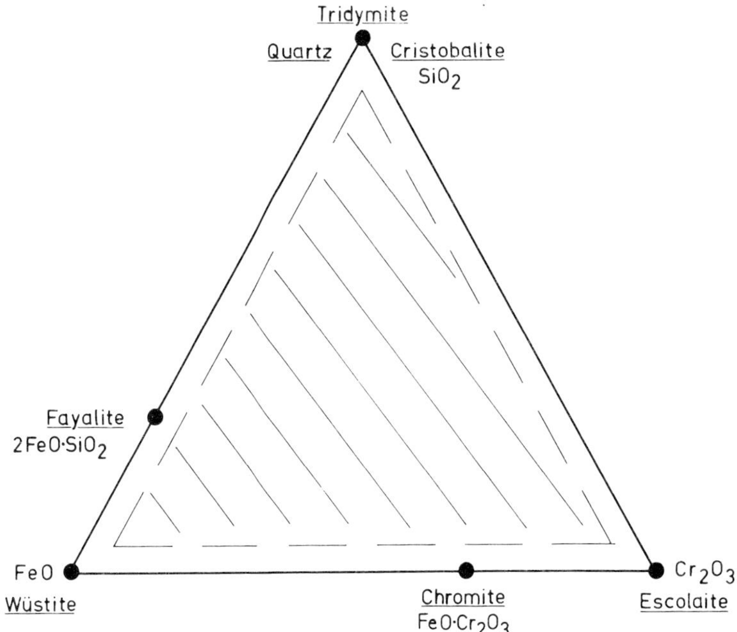

67 Schematic representation of the pseudoternary system $FeO-SiO_2-Cr_2O_3$, showing the stoichiometric composition of the different phases which have been reported. The system was not completely investigated regarding ternary phases

solubility with corundum (α-Al_2O_3) and haematite (α-Fe_2O_3) but an immiscibility gap exists at lower temperatures with corundum.

It is also reported[48] that Cr_2O_3 dissolves Cr at higher temperatures. This solid solution is not stable at lower temperatures and therefore different changes occur during cooling of the supersaturated Cr_2O_3 phase in Cr-rich steels.

Microscopy and electron probe analysis

The Cr_2O_3 phase is common in ferrochromium inclusions. Single-phase inclusions are idiomorphic with plane surfaces, hexagonal or regular in section (Fig.69). In Fig.70*a* as well as in Fig.14, the phase forms branched dendrites, often with a cross-shaped section. The inclusions in Figs.70*a* and *b*, 12, and 14 are all examples of inclusions from the chromia–silica system. Primary Cr_2O_3 or cristobalite, depending on the composition of the inclusion, crystallize in a eutectic matrix. The matrix can also be glassy.

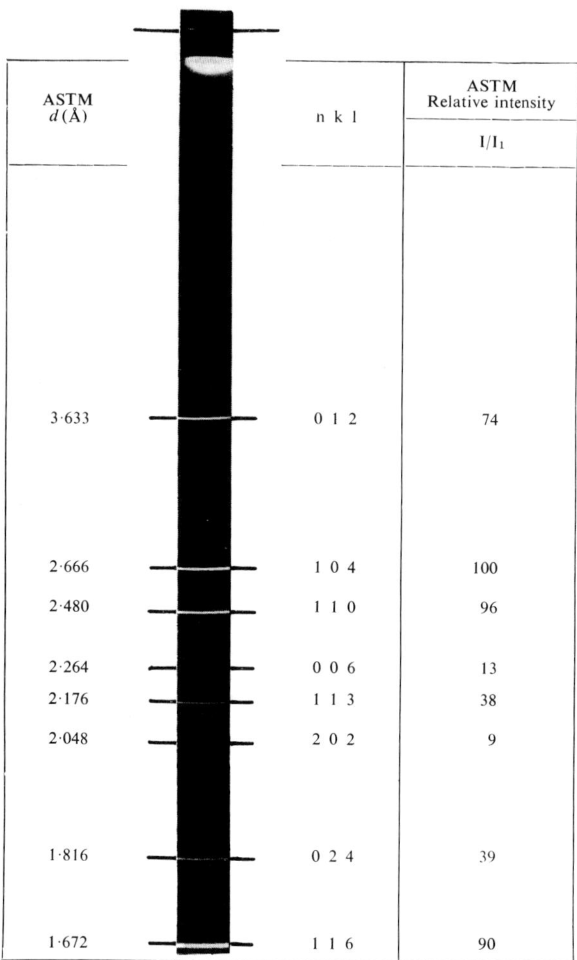

ASTM d(Å)	n k l	ASTM Relative intensity I/I_1
3·633	0 1 2	74
2·666	1 0 4	100
2·480	1 1 0	96
2·264	0 0 6	13
2·176	1 1 3	38
2·048	2 0 2	9
1·816	0 2 4	39
1·672	1 1 6	90

d-values and relative intensities for the different reflections are quoted from the ASTM index

68 X-ray powder photograph of escolaite (Cr_2O_3); monochromatic Cr Kα-radiation; Guinier type camera

Inclusions with Cr_2O_3 and galaxite have been shown in Figs.31*a–d* with the Cr_2O_3 phase crystallizing as plates or laths. In Fig.71, the formation of two oxide phases with different Al_2O_3:Cr_2O_3 ratios from a single phase solid solution in the Al_2O_3–Cr_2O_3 system are shown. A synthetic slag was quenched and heat treated, and the primary Al_2O_3–Cr_2O_3 crystals have reacted with the alumina rich matrix forming an outer zone of different composition.

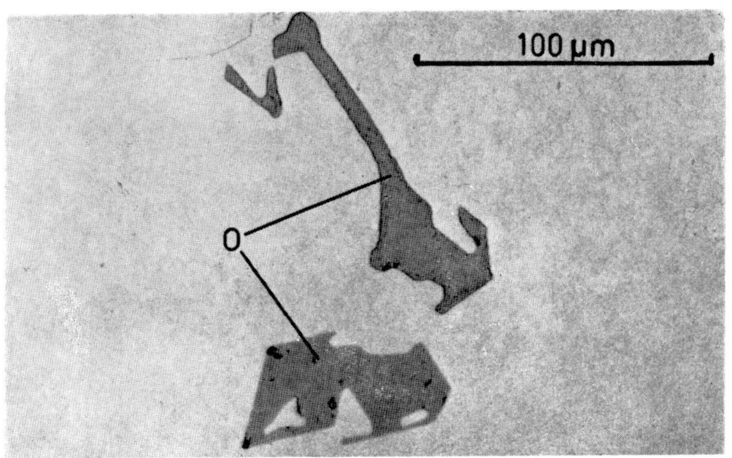

×400

MATERIAL TYPE Ferrochromium suraffiné
MATERIAL ANALYSIS, % 0·028 C, 0·20 Si, 66·0 Cr, 0·016 P, 0·004 S, 2850 ppm O
INCLUSIONS (O) 100 Cr_2O_3, 2 MnO

COMMENT This type of inclusion is very similar to chromium galaxite inclusions, but harder (*c.f.* Fig.31*b*)

69 Oxide inclusions in ferrochromium: Cr_2O_3 (escolaite)

Microscopically similar phases
Cr_2O_3 and chromium galaxite are microscopically very similar, but the former is harder and anisotropic (*see* A 7).

C 3 MnO–SiO_2–Cr_2O_3: Chromium galaxite*

Chemical formula MnO·Cr_2O_3
Physical properties
Melting point: unknown
Density: 4·87

Structure type
Cubic; space group Fd3m
a=8·51 A
Spinel type. Wyckoff II, chap. VIII, text p.16, table p.42
Monochromatic X-ray powder photograph similar to galaxite, Fig.24
Isostructural with galaxite (A 7), hercynite (B 5), chromite (C 4) and magnetite (D 3).

* The name chromium galaxite has been given to this compound by the authors, in order to emphasize its close relationship with galaxite, MnO·Al_2O_3. The double oxide, which is little known, was observed by Yearian *et al.*[49] in oxide scales on chromium steels and has also been reported but not named by Trojer.[9]

a 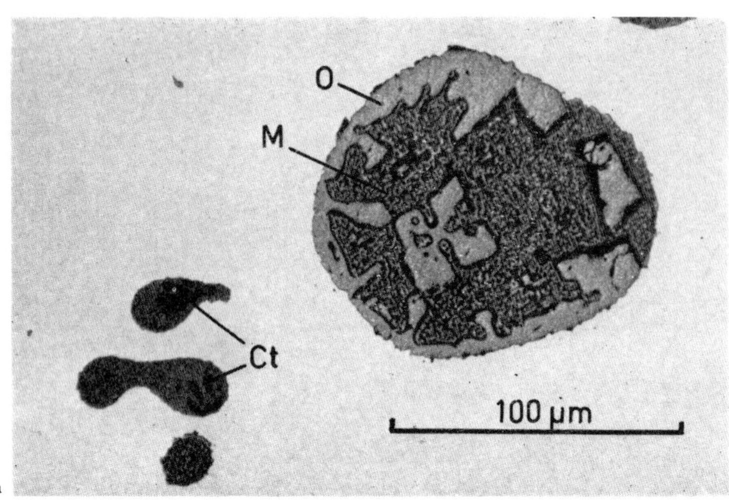

×400

b 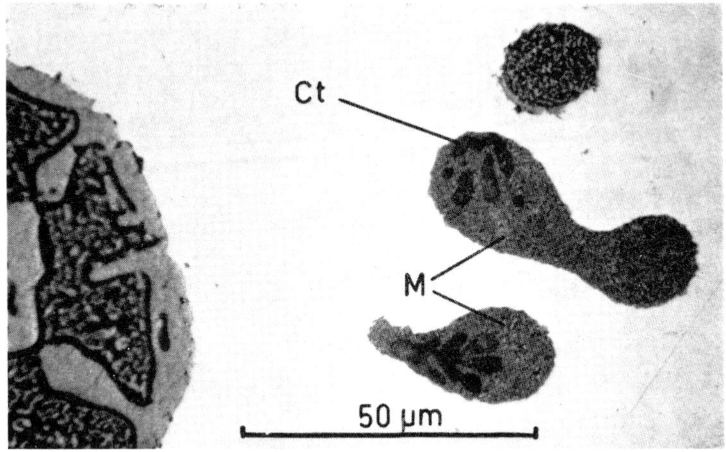

×800

MATERIAL TYPE Ferrochromium suraffiné
MATERIAL ANALYSIS, % 0·028 C, 0·57 Si, 68·6 Cr, 0·014 P, 0·03 S, 1620–1750 ppm O
INCLUSIONS *a* (O) Escolaite with 98 Cr_2O_3, 5 MnO (M) Matrix with a finely dispersed structure (eutectic) of Cr_2O_3 and cristobalite *b* (Ct) Cristobalite with 100 SiO_2 (M) Matrix with a finely dispersed structure (eutetic) of Cr_2O_3 and cristobalite
COMMENT The binary system Cr_2O_3–SiO_2 has no intermediate phases but a eutectic at high SiO_2 contents.[8] The mean composition of these two particle types is slightly different and on different sides of the eutectic composition, resulting in a primary crystallization of Cr_2O_3 (a) and cristobalite (b) respectively. The inclusion in Fig.70*a* should be compared with the similar inclusion in Fig.14, showing Cr_2O_3 in supercooled cristobalite with chromia in solid solution. The inclusion in Fig.70*b* should be compared with Fig.12, showing cristobalite in a very finely dispersed matrix

70 **70a–b** Oxide inclusions in ferrochromium: Cr_2O_3 (escolaite) and cristobalite in different types of inclusions

a

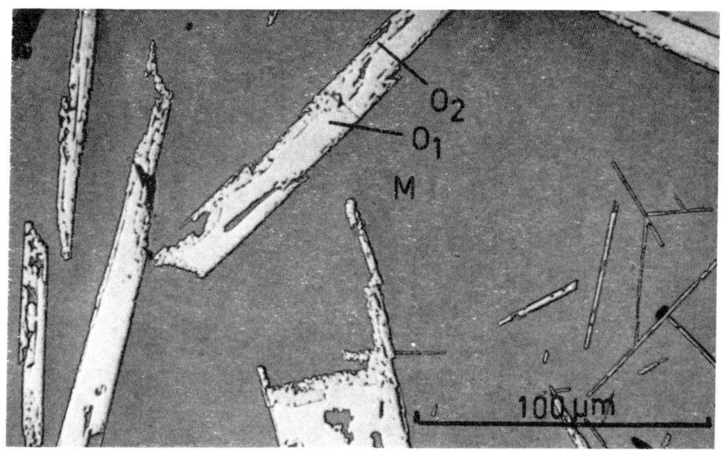

×400

b

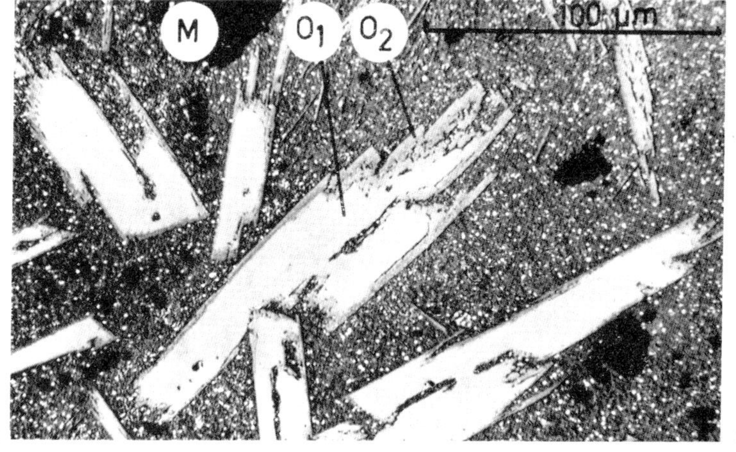

×400

MATERIAL TYPE Synthetic slag, prepared by melting alumina–chromia–silica mixture in a plasma furnace at about 2000°C
a Slag after rapid cooling
b Slag after heat treatment for 1 h at 1100°C
MATERIAL ANALYSIS, % 27 Al_2O_3, 41 Cr_2O_3, 32 SiO_2 (by weight)
SLAG ANALYSIS *a* (O) Oxide with Cr_2O_3 structure, 25–30 Al_2O_3, 75–60 Cr_2O_3 (M) Glassy matrix with 35–45 Al_2O_3, 0–13 Cr_2O_3, balance SiO_2. *b* (O_1) Oxide with Cr_2O_3 structure, 20 Al_2O_3, 80 Cr_2O_3 (light) (O_2) Oxide with Cr_2O_3 structure, 55–40 Al_2O_3, 45–60 Cr_2O_3 (dark) (M) Matrix with fine grained precipitate
COMMENT Immiscibility gaps exist at lower temperatures in several Me_2O_3 systems of importance for steel inclusions, such as Al_2O_3–Cr_2O_3, Al_2O_3–Fe_2O_3, Cr_2O_3–Fe_2O_3 and Al_2O_3–Cr_2O_3–Fe_2O_3

71 **71a–b** Generation of immiscibility gap in alumina–chromia crystals

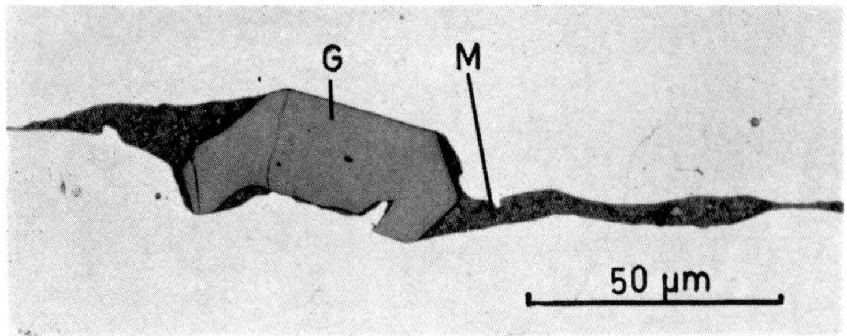

× 600

STEEL TYPE Austenitic stainless steel. From hot rolled sheet
STEEL ANALYSIS, % 0·043 C, 0·40 Si, 1·34 Mn, 0·020 P, 0·005 S, 18·5 Cr, 10·1 Ni, 0·50 Mo

INCLUSIONS (G) Chromium galaxite with 35 MnO, 64 Cr_2O_3 (M) Matrix with 44 MnO, 45 SiO_2, 9 Cr_2O_3, 3 CaO, trace of sulphur

72 Oxide inclusion in steel: chromium galaxite in manganese oxide–chromia–silica

Solid solubility

In chromium galaxite Cr_2O_3 can be completely substituted by Al_2O_3, and a complete solid solubility range with galaxite exists at higher temperatures. Observations on inclusions and synthetic MnO–Al_2O_3–Cr_2O_3 slags indicate, however, that an immiscibility gap is formed at lower temperatures (*see* below).

The binary oxide MnO can be completely substituted by FeO and MgO.

The homogeneity range extends, according to the present investigation, from about 15%MnO·85%Cr_2O_3 to 32%MnO·68Cr_2O_3 and is thus on the side of the ternary oxide (*see also* A 7).

Microscopy and electron probe analysis

Chromium galaxite is a common phase in chromium steel inclusions. It is usually easily identified in microsections as white, regular idiomorphic crystals (Fig.72). Several examples of pure chromium galaxite and galaxite with varying chromia contents have already been shown in section A 7 (Figs.28–30) and analytical results have been given in Table IV.

A series of inclusions in ferrochromium in section A 7 (Figs.31*a–d*) are an indication of an immiscibility gap in the MnO·Al_2O_3–MnO·Cr_2O_3 system. Here two galaxite phases have been observed in the same inclusion, having different Al_2O_3:Cr_2O_3 ratios. This is further supported by studies on synthetic slags. As shown in Fig.71 an immiscibility gap exists in the alumina–chromia system which should also give a gap for the galaxites.

Microscopically similar phases
Chromium galaxite and Cr_2O_3 (escolaite) are microscopically similar, but the former has a lower microhardness and is isotropic (*see* A 7).

C 4 FeO–SiO_2–Cr_2O_3: Chromite

Chemical formula $FeO{\cdot}Cr_2O_3$

Physical properties

Melting point: 1850°C

Density: 5·1

Structure type

Cubic; space group Fd3m

a=8·36 A. ASTM data card 4–0759

Spinel type. Wyckoff II, chap. VIII, text p.16, table p.42

Monochromatic X-ray powder photograph similar to galaxite, Fig.24

It has been reported[48] that a tetragonal deformation of the lattice occurs for higher Cr_2O_3 contents than the stoichiometric composition. This is visible as a splitting of the reflections 511 and 440 as well as 311 and 400.

Isostructural with galaxite (A 7), hercynite (B 5), chromium galaxite (C 3), and magnetite (D 3).

Solid solubility
At high temperatures, FeO as well as Cr_2O_3 may be replaced by several oxides. For slag inclusions, the binary oxides MnO (A 7) and MgO, and the ternary oxides Al_2O_3 (B 5) and Fe_2O_3 (D 3) are of special importance. A wide range of solid solubility exists, but at lower temperatures immiscibility gaps sometimes occur.

The homogeneity range is wide, extending on both sides of the stoichiometric composition.[9] The Cr_2O_3-rich chromite inclusions observed in Cr-rich steels, are reported to deform tetragonally when the steel is quenched.[48] This solid solution is unstable however, and transforms during further annealing of the steel. The final structure is cubic chromite but deformed spinels are formed as intermediate phases during annealing.

Microscopy and electron probe analysis
The chromite phase in steel inclusions is similar in appearance to the other spinels discussed forming crystals with cubic or hexagonal sections, limited by plane surfaces. For example Fig.25–33 in section A 7. Also clover-leaf sections are common. Single phase inclusions of Cr_2O_3-rich chromite have been observed in chromium-rich steels

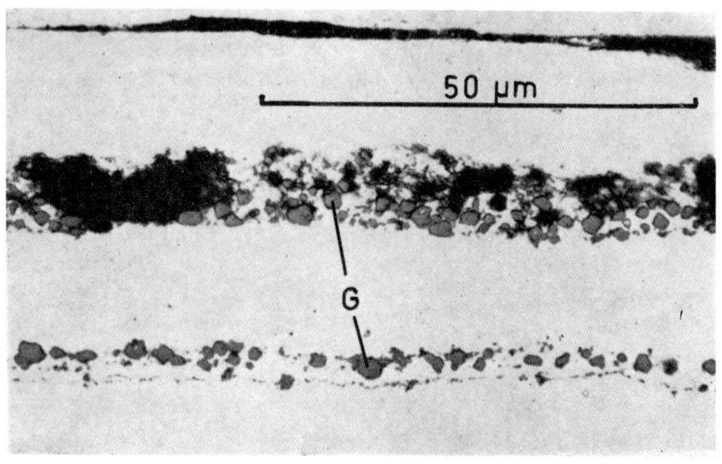

×1200

STEEL TYPE Basic electric austenitic stainless steel, deoxidized in the ingot mould. From hot rolled sheet
STEEL ANALYSIS, % 0·053 C, 0·54 Si, 1·30 Mn 0·019 P, 0·006 S, 18·6 Cr, 9·2 Ni

INCLUSIONS (G) Chromite with 14 FeO, 3 MnO, 83 Cr_2O_3
COMMENT The inclusions are situated below a surface defect of the sheet

73 Oxide inclusions in steel: chromite

and such inclusions from a rolled steel sheet are shown in Fig.73. Excellent pictures of isolated inclusions of chromite with varying chromium content have been given in the paper by Koch *et al.*,[48] which also shows diffraction patterns of the deformed chromites.

Microscopically similar phases
See A 7 and C 3.

D. The iron oxides $Fe_{1-x}O$, Fe_3O_4, and Fe_2O_3

D 1 General

Three iron oxides, wüstite ($Fe_{1-x}O$), magnetite (Fe_3O_4), and haematite (Fe_2O_3) exist as shown by the iron–oxygen system constitution diagram (Fig.74). Scale formation during the heating of steel is an important source for these oxides and a metallographic study of their microstructure has been published by Modin.[50] This reference contains several plates showing the microscopic appearance of the different iron oxides as well as useful comments about their properties and identification in steel. The iron oxides are sometimes found as steel inclusions, for instance in ingots cast from the top, in Lancashire iron, or in rolled, forged, or pressed steel products where the surface layer may be incorporated in the bulk material. They are closely related structurally to the phases discussed in the sections A–C; therefore some comments on their properties, appearance and analytical composition in steel inclusions will be given in this section. The electron probe method is not very sensitive to differences in oxygen content. The combination of iron analyses for the different phases and general metallographic knowledge has, however, made a positive identification of the phases possible. The probe has also given some information about other metals in solid solution in the oxide phases.

D 2 The iron oxides: Wüstite

See section B 2. Monochromatic X-ray photograph Fig.75.

Comment

It is sometimes difficult to distinguish between wüstite and magnetite which are both cubic and often grow in close contact, but etching with 4%HNO_3 in alcohol gives a darker colour for the wüstite phase. An example of this (oxides in a surface crack) is shown in Fig.76. A general study of the relative position of the phases in relation to the oxidized iron is often useful. An additional help is

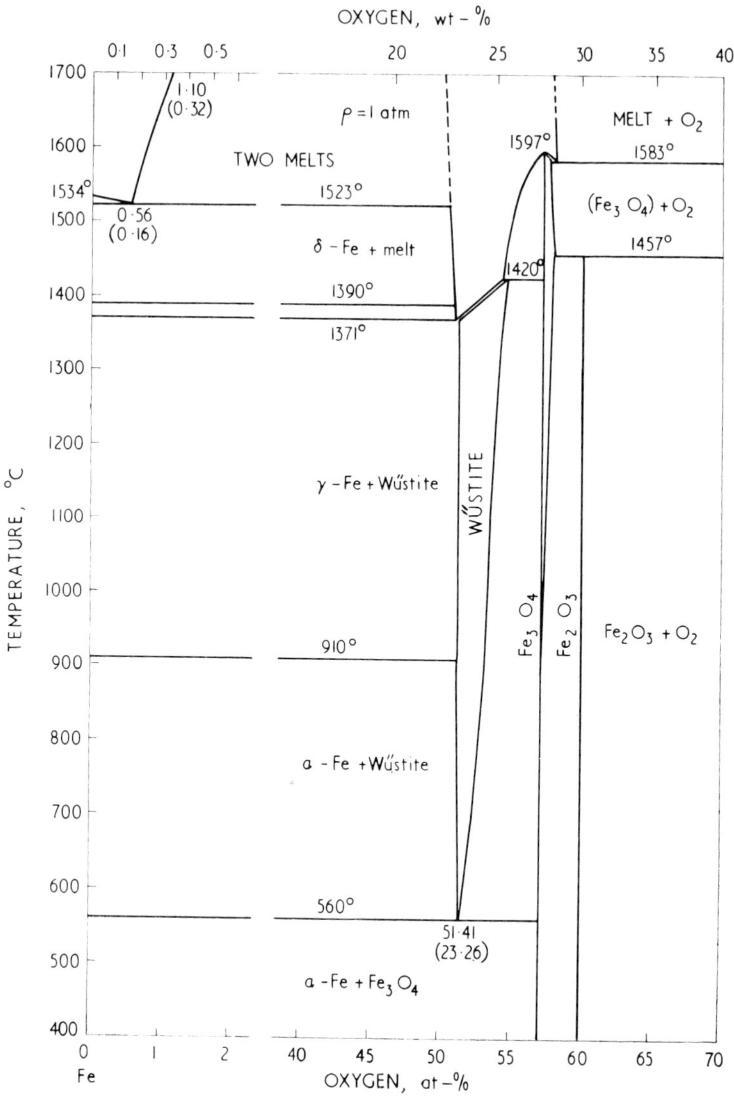

74 The iron–oxygen system, constitution diagram (*from M. Hansen "Constitution of binary alloys", 1958, McGraw-Hill*)

the observation[50] that the wüstite phase shows a weak anisotropic effect in scales in iron and steel. This probably depends on a tetragonal deformation of the cubic wüstite lattice, occurring when the phase becomes supersaturated with oxygen on cooling. (*See also* Figs.77 and 78).

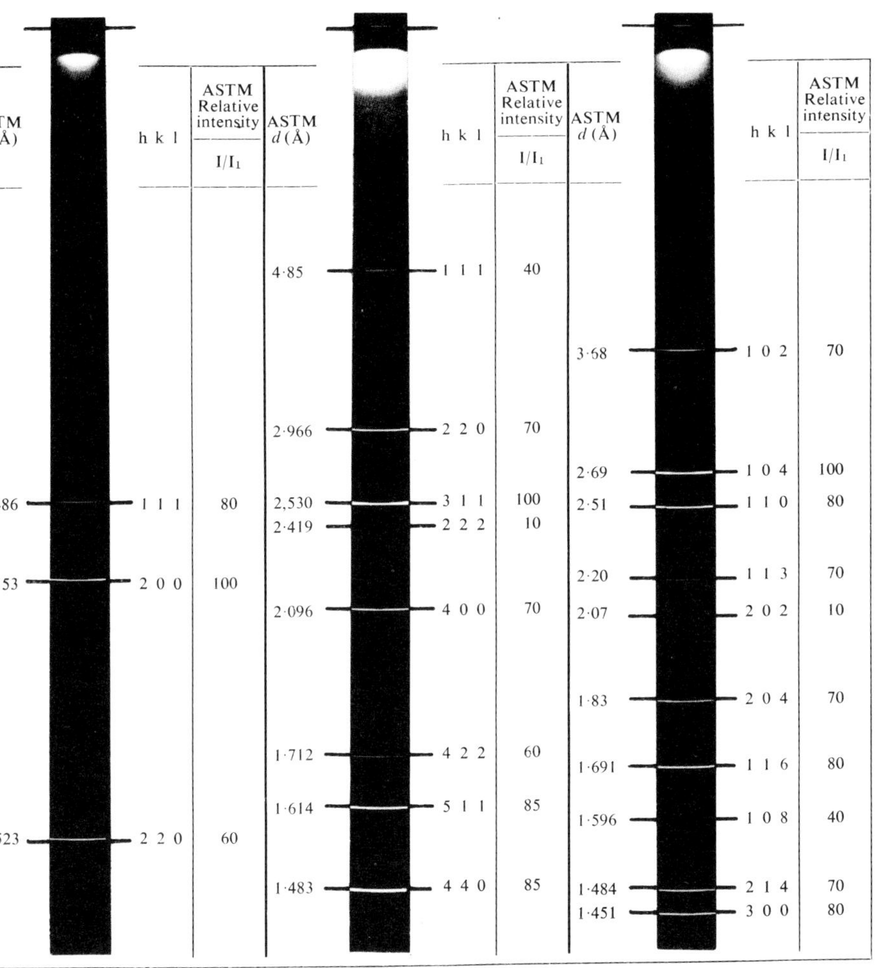

d-values and relative intensities for the different reflections have been quoted from the ASTM index

75 X-ray powder photographs of wüstite (*left*), magnetite (*middle*) and haematite (*right*); monochromatic Fe Kα-radiation; Guinier camera

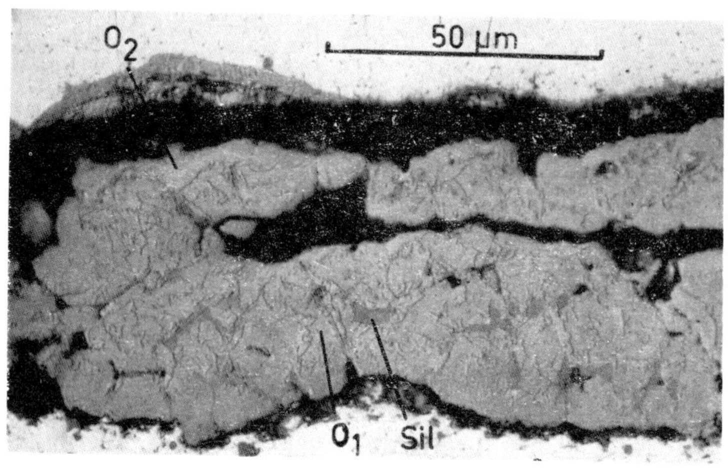

×750

STEEL TYPE Carbon steel
STEEL ANALYSIS, % 0·46 C, 0·22 Si, 0·67 Mn, 0·028 S
INCLUSIONS (O_1) Wüstite with 96 FeO, 3–7 MnO (dark grey) (O_2) Magnetite with 94 Fe_3O_4, 4 MnO (light) (Sil) Dark phase is a silicate. Etched in nital
COMMENTS Surface crack

76 Oxide inclusion in steel: wüstite, magnetite and iron silicate

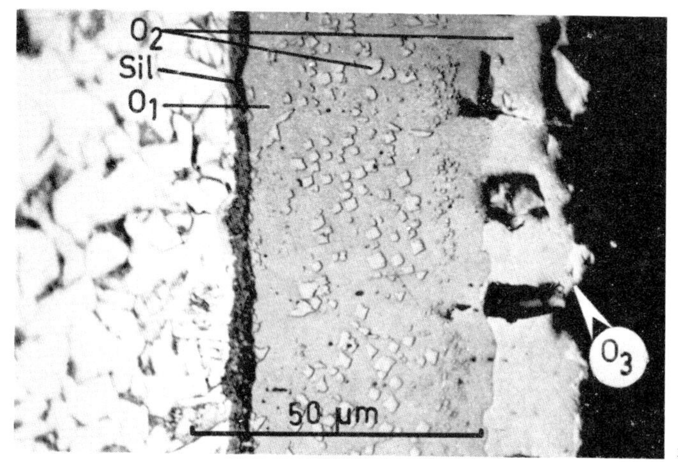

×800

STEEL TYPE Carbon steel, killed. Oxidized 20 min at 900°C in air, cooled in nitrogen. Section through steel surface
STEEL ANALYSIS, % 0·22 C, 0·24 Si, 0·50 Mn, 0·032 P, 0·029 S
SCALE (O_3) Haematite, Fe_2O_3 (O_2) Magnetite, Fe_3O_4 (O_2+O_1) Magnetite+wüstite, FeO (Sil) Dark layer rich in silica. Etched in nital

77 Scale on steel: the different iron oxides

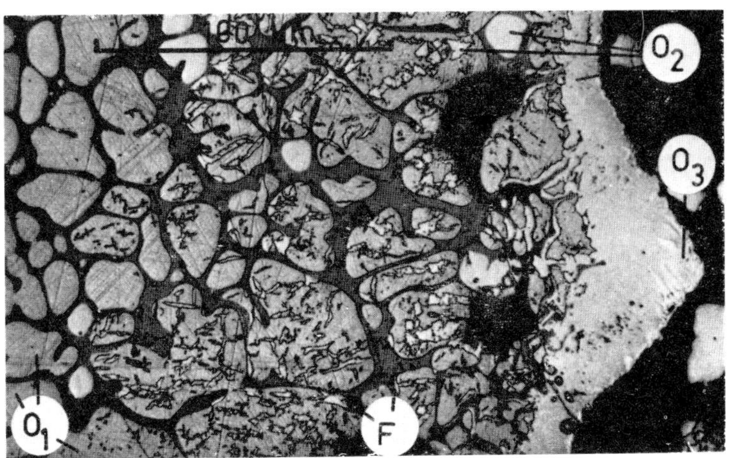

×400

STEEL TYPE Lancashire iron. Inclusion from the fining process
STEEL ANALYSIS, % 0·55 C, 0·07 Si, 0·04 Mn, 0·018 P, 0·013 S

INCLUSIONS (O_3) Haematite, Fe_2O_3 (O_2) Magnetite, Fe_3O_4 (O_1) Wüstite, FeO (O_2+O_1) Magnetite + wüstite, FeO (F) Fayalite, $2FeO{\cdot}SiO_2$. Etched in nital

78 Oxide inclusion in steel: slag of the different iron oxides from oxide scale

D 3 The iron oxides: Magnetite

Chemical formula $FeO{\cdot}Fe_2O_3$

Physical properties

Melting point: 1 597°C

Density: 5·125

Structure type

Cubic; space group Fd3m

a=8·374 Å, ASTM data card 11–614

Spinel type. Wyckoff II, chap. VIII, text p.16, table p.42

Monochromatic X-ray powder photograph Fig.75

Isostructural with galaxite (A 7), hercynite (B 5), chromium galaxite (C 3), and chromite (C 4).

Solid solubility

As for the other double oxides (A 7, B 5, C 3, and C 4) a wide range of solid solubility exists for $FeO.Fe_2O_3$ and both FeO and Fe_2O_3 may be substituted by several other metal oxides over a wide composition range. Of importance for steel inclusions are the binary oxides MnO and MgO and the ternary oxides Al_2O_3 and Cr_2O_3. Immiscibility gaps sometimes appear at lower temperatures.

The magnetite phase usually has a higher content of the ternary oxide than corresponds to the ideal formula. The composition depends on the oxygen partial pressure. If finely dispersed magnetite is oxidized at temperatures below 350°C, then oxidation of all Fe (II) to Fe (III) occurs and metastable γ–Fe_2O_3 is formed (D 4).

Microscopy and electron probe analysis
Magnetite is an important constituent of the oxide scale on iron and steel. It is therefore often present on surface layers, in surface cracks or in inclusions from slags or from scales. A scale on a carbon steel, heated at 1200°C, is shown in Fig.77. Single phase magnetite forms the grey layer below the thin surface layer of haematite (Fe_2O_3). Magnetite is also present together with wüstite (FeO) in the two-phase layer below the pure magnetite. This layer was only one phase, FeO, at the oxidation temperature, but on cooling the solubility of oxygen in the FeO phase decreases. The isotropic magnetite phase, which is richer in oxygen, was precipitated. (The wüstite in this layer was weakly anisotropic, depending on a tetragonal deformation of the cubic wüstite phase due to oxygen supersaturation.) Inclusions in Lancashire iron which originate from surface scales frequently contain magnetite and other iron oxide (Fig.78).

Magnetite with MnO has been observed in surface cracks (Fig.22). Magnetite is, together with α-iron, a transformation product of wüstite below 560°C, but this transformation is often only partial or incomplete.

D 4 The iron oxides: Haematite

Chemical formula Fe_2O_3
Modifications and stability range
In addition to α-Fe_2O_3, haematite, the metastable γ-Fe_2O_3 exists. The latter is formed by oxidation of magnetite below 350°C and it has a defect spinel type lattice. At higher temperatures the Fe_2O_3-phase loses oxygen and Fe_3O_4 is formed. The oxygen pressure over Fe_2O_3 is 0·5 mmHg at 1150°C, increasing to 1 atm at 1457°C. Above this temperature, therefore, no formation of Fe_2O_3 occurs if steel is oxidized at atmospheric pressure.

Structure type
Hexagonal; space group $R\bar{3}c$
a=5·034 Å, c=13·749 Å. ASTM data card 6–0502
Cr_2O_3-type. Wyckoff I, chap. V, text p.4, table p.11
Monochromatic X-ray powder photograph Fig.75
Isostructural with corundum (A 2) and Cr_2O_3 (C 2)

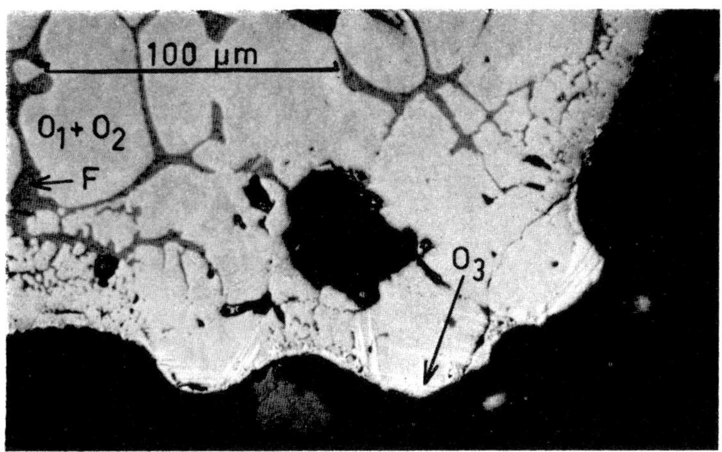

×400

STEEL TYPE Lancashire iron. Inclusion from the fining process
STEEL ANALYSIS, % 0·55 C, 0·07 Si, 0·04 Mn, 0·018 P, 0·013 S
INCLUSIONS (O_3) Haematite, Fe_2O_3 ($O_2 + O_1$) Magnetite, Fe_3O_4 + wüstite, FeO (F) Fayalite, 2 FeO·SiO_2. Unetched
COMMENT The haematite is strongly anisotropic and therefore easy to recognize

79 Oxide inclusion in steel: haematite on surface of inclusion from oxide scale

Physical properties

Melting point: 1 730°C (at 53 atm O_2)
Density: 5·24

Solid solubility

Haematite and Cr_2O_3 (escolaite) are reported to have a complete solid solubility range at all temperatures above 20°C. Haematite also dissolves considerable amounts of corundum. The solid solubility limit depends on temperature and oxygen pressure.[9]

Microscopy and electron probe analysis

Haematite is the only iron oxide which shows a marked anisotropy and is therefore easily recognizable in the microscope. In Fig.79 showing inclusions in Lancashire iron, the haematite phase is visible at the inclusion surface as small, white needles on and just below the surface. Haematite is that iron oxide which is richest in oxygen, and therefore forms the outermost scale when iron and steel are oxidized. The Fe_2O_3 layer is usually thin (Figs.77, 78) and disappears completely above 1 457°C at atmospheric pressure (*see* above).

E. Classification of inclusions according to steel type

In the present report systematic description and discussion of steel inclusions has been based on the inclusion systems regardless of the steel types and analyses. This has been found to be a more adequate principle than to use the steel analysis which is of less importance, or the steel type, where the information available has been incomplete. As a guidance for those who are more interested in inclusion types appearing in a particular steel, all the different steels with the inclusions reproduced in the present report are also arranged according to steel analysis in Table VIII. The steels and ferroalloys are grouped according to the main alloying elements and in each group arranged in order of increasing carbon content. It must be emphasized, however, that no attempts were made to make a complete study of all the different inclusion types present in each steel. The inclusions chosen for the illustrations have been selected only for the purpose of studying the different inclusion phases. Unfortunately the information available is insufficiently complete to arrange a similar table with the steels grouped according to steel type (e.g. basic or acid steel, Kaldo, or LD steel, deoxidation practice, etc.), as the detailed metallurgical treatment seems to be the most important factor in determining the inclusion types.

(Table VIII *overleaf*)

TABLE VIII **The $MnO–SiO_2–Al_2O_3$ system: Inclusions classified according to steel analyses. The steels are grouped according to main alloying elements and within each group arranged with increasing carbon content; the general type of inclusion is given, but the reader is referred to the captions in the different Figures given for each steel.**

Steel type	Steel analysis, wt-% C	Si	Mn	Cr	Ni	Mo	Inclusion type	Fig.	Table	Section
Carbon steel										
Rimmed deep-drawing steel, from sheet	0·05	...	0·29	0·07	...	...	Manganosite	21	...	A 6
Basic open hearth, Al-killed, from plate	0·11	0·25	1·10	0·06	0·05	...	Corundum	5 *a*, *b*	...	A 2
Basic open hearth, Al-killed, from plate	0·11	0·25	1·10	0·06	0·05	...	Corundum and galaxite	5 *c*	...	A 2
Bottom cast ingot, from sheet	0·12	0·20	0·70	0·10	...	...	Tridymite, rhodonite, and sulphide	17 *b*, 42	...	A 4
Basic Kaldo semi-killed, from ingot	0·14	0·07	1·00	...	...	...	Rhodonite	40	...	A 9
Basic Kaldo semi-killed, from ingot	0·16	0·06	1·00	...	...	...	Tephroite	45	...	A 10
Basic electric, Si-killed. Isolate	0·16	0·35	0·67	...	...	...	Cristobalite and rhodonite	11, 41	...	A 3
Basic electric, Si-killed. Isolate	0·16	0·35	0·67	...	...	...	Cristobalite, rhodonite, and sulphide	17 *a*	...	A 3
Basic electric, Si-killed. Isolate	0·16	0·35	0·67	...	...	...	Cristobalite	13	...	A 3
Carbon steel, killed	0·22	0·24	0·50	...	...	...	Scale of iron oxides	77	...	D
Bottom cast ingot, from rolled bar	0·24	0·38	1·39	...	...	...	Galaxite	25	IV	A 7
Carbon steel, from forged link	0·33	0·25	0·60	...	...	...	Galaxite and corundum	26	IV	A 7
Basic electric, surface crack in ingot	0·33	0·25	0·60	...	...	...	Wüstite and magnetite	59	...	D 2, 3
Basic electric, Si-killed, from billet	0·45	0·22	0·70	...	...	...	Mullite and corundum	36	...	A 8
Basic electric, Si-killed, from billet	0·45	0·22	0·70	...	...	...	Spessartite	49	V	A 11
Carbon steel	0·46	0·22	0·67	...	...	...	Wüstite and magnetite	22	...	D 2, 3
Carbon steel, surface crack in ingot	0·46	0·22	0·67	...	...	...	Wüstite, magnetite, and fayalite (?)	76	...	D 2, 3
Lancashire iron	0·55	0·07	0·04	...	...	...	Fayalite and iron oxides	62	...	B 3
Lancashire iron	0·55	0·07	0·04	...	...	...	Iron oxides	78	...	D
Lancashire iron	0·55	0·07	0·04	...	...	...	Haematite	79	...	D 4
Carbon steel, from billet	0·60	0·14	1·11	...	...	...	Galaxite	27	IV	A 7
Low-alloy steel										
Basic electric, killed, from billet	0·10	0·32	0·46	2·20	0·9	1·04	Corundum and spinel	6	IV	A 2
From broken fatigue-test sample	0·15	0·29	0·75	0·86	1·45	0·08	Spinel	33	IV	(A 7)

Basic open hearth, killed, from ingot	0·33	0·34	0·60	2·43	0·12	0·49	Corundum	4	...	A 2
Basic open hearth, killed, from ingot	0·33	0·34	0·60	2·43	0·12	0·49	Anorthite, corundum, and Cr-galaxite	51	...	A 12
Basic open hearth, killed, from forging	0·62	0·26	0·79	0·65	0·07	...	Corundum, spessartite, and sulphide	7, 47	V	A 2
Basic open hearth, killed, from forging	0·62	0·26	0·79	0·65	0·07	...	Mullite, spessartite, and sulphide	37	V	A 8
Acid open hearth, killed, from tube	1·0	0·30	0·30	1·5	...	...	Quartz, cristobalite, and Cr-galaxite	18	...	A 5
Alloy steel										
From vacuum cast ingot	0·04	0·40	1·34	18·5	10·1	0·50	Cr-galaxite	28	IV	C 3
From hot-rolled sheet	0·04	0·40	1·34	18·5	10·1	0·50	Cr-galaxite	72	IV	C 3
From rolled billet	0·05	0·43	0·96	18·0	9·8	0·50	Tridymite, Cr-galaxite, and sulphide	16, 30	IV	A 4
From hot-rolled strip	0·05	0·54	1·27	18·3	9·1	0·25	Cr-galaxite	29	IV	C 3
Basic electric, killed, from sheet	0·05	0·54	1·30	18·6	9·2	...	Chromite	73	...	C 4
From cold-rolled sheet	0·05	0·66	1·03	17·8	9·0	...	Cr-galaxite, crushed	55	...	C 3
Basic electric, from billet	0·18	0·13	0·39	13·6	0·66	...	Cr-galaxite and sulphide	32	IV	C 3
Basic electric, from forging	0·18	0·35	...	11·5	0·60	0·50	Tridymite, quartz, and Cr-galaxite	15	...	A 4
High-speed steel										
From billet	0·81	0·34	0·35	4·30	...	4·90	Mullite	38	...	A 8
Cast steel										
Basic open hearth, from forging	0·15	0·44	0·84	0·21	...	...	Spessartite	48	V	A 11
Ferroalloys										
Ferrochromium	0·03	0·08	...	73	...	...	Corundum with Cr_2O_3	8	...	A 2
Ferrochromium	0·03	0·09	...	73	...	...	Cristobalite	12	...	A 3
Ferrochromium	0·03	0·09	...	73	...	...	Cr_2O_3 and cristobalite	14	...	C 2
Ferrochromium	0·03	0·08	...	73	...	...	Cr_2O_3 and galaxite	31 *a*	IV	C 2
Ferrochromium	0·03	0·08	...	73	...	...	Cr-galaxite	31 *b*	IV	C 3
Ferrochromium	0·03	0·08	...	73	...	...	Cr_2O_3 and two galaxites	31 *c*	...	C 2
Ferrochromium	0·03	0·08	...	73	...	...	Two galaxites	31 *d*	...	C 3
Ferrochromium	0·03	0·20	...	66·0	...	...	Cr_2O_3	69	...	C 2
Ferrochromium	0·03	0·57	...	68·6	...	...	Cr_2O_3 and cristobalite	70 *a*, *b*	...	C 2
Ferrovanadium (81·2%V)	0·15	0·50	...	...	...	...	Corundum	9	...	A 2

REFERENCES

1. C. Benedicks and H. Löfquist: 'Non-metallic inclusions in iron and steel', 1930, London, Chapman and Hall.
2. T. R. Allmand: 'Microscopic identification of inclusions in steel', 1962, London, BISRA.
3. J. R. Rait and H. W. Pinder: *JISI*, 1946, **154,** 371–398.
4. L. S. Birks: 'Electron probe microanalysis', Chemical Analysis Vol. XVII, 1963, New York/London, Interscience.
5. P. H. Salmon Cox and J. A. Charles: *JISI*, 1963, **201,** 863–872.
6. C. Helin and P. Spiegelberg: *Jernkont. Ann.*, 1962, **146,** 297–306.
7. 'Swedish Steel Manual', 1962, Stockholm, Jernkontoret.
8. E. M. Levin *et al.*: 'Phase diagrams for ceramists', 1964, Columbus, Ohio, Amer. Ceram. Soc.
9. F. Trojer: Die oxydischen Kristallphasen der anorganischen Industrieprodukte, 1963, Stuttgart, Schweizerbart'sche Verlagsbuchhandlung.
10. R. W. G. Wyckoff: 'Crystal structures', Vol. I–II, 1951–57, New York/London, Interscience.
11. 'X-ray powder data file', ASTM Spec. Techn. Publ. No.48-J, 1960, Philadelphia.
12. R. Kiessling, S. Bergh, and N. Lange: *JISI*, 1962, **200,** 914–921.
13. *ibid.*, 1963, **201,** 509–515.
14. *ibid.*, 965–967.
15. R. Kiessling and N. Lange: *ibid.*, 1016–1024.
16. R. Theisen: 'Metallurgy and ceramics', Nov. 1961, Brussels, Comm. Europ. l'Energie Atomique EUR-I-1.
17. A. Taylor: 'X-ray metallography', 1961, New York/London, Wiley.
18. 'Basic open-hearth steelmaking'. 1951, ed. W. O. Philbrook *et al.*, New York, AIME.
19. G. Hägg and G. Söderholm: *Z. Phys. Chem.*, (B), 1935, **29,** 88–94.
20. H. Willners and K. A. Ottander: *Medd. Jernkont. Tekn. Råd.*, (190), 1953, 787–817.
21. J. C. d'Entremont *et al.*: *Trans. AIME*, 1963, **227,** 14–17.
22. T. Wahlberg and L. Fredholm: *Jernkont. Ann.*, 1953, **137,** 1–26.
23. G. Hägg: Allmän och oorganisk kemi, 1963, Uppsala, Almqvist och Wiksell.
24. H. Widmark: *Jernkont. Ann.*, 1959, **143,** 426–449.
25. R. C. Evans: 'An introduction to crystal chemistry', 2nd ed., 1964, 157, Cambridge University Press.
26. A. Hultgren: *Jernkont. Ann.*, 1945, **129,** 633–671.
27. O. W. Flörke: *Ber. Dt. Keram. Ges.*, 1961, **38,** 89–97.
28. H. Pettersson: *Jernkont. Ann.*, 1946, **130,** 653–663.
29. R. Ward: *Progr. Inorg. Chem.*, 1959, (1), Chapt.III, 479–488.
30. S. Aramaki and R. Roy: *J. Am. Ceram. Soc.*, 1962, **45,** 229–242.
31. F. Harders and S. Kienov: 'Feuerfestkunde', 1960, Berlin, Springer.
32. C. W. Burnham: *Z. Krist.*, 1963, **118,** 127–148.
33. N. V. Belov: 'Crystal chemistry of large-cation silicates', Transl., 1963, 20–22, New York, Consultants Bureau.
34. F. P. Glasser: *Am. J. Sci.*, 1958, **256,** 398–412.
35. International tables for X-ray crystallography, Vol.I, 1952, Birmingham, Kynoch Press.
36. J. White: *JISI*, 1943, **148,** 586.
37. J. Bruch: *Arch. Eisenh.* to be published.

38. G. R. Rigby and A. T. Green: 'The thin-section mineralogy of ceramic materials', 2 edn 1953, 168, Stoke-on-Trent, Brit. Ceram. Res. Ass.
39. P. P. Ewald *et al.*: 'Strukturbericht I', 1913–1938, Leipzig, Akad. Verlagsgesellschaft, photo-lithoprint reprod., Ann Arbor, Michigan, Edwards Bros.
40. W. A. Fischer and H. J. Fleischer: *Arch. Eisenh.*, 1961, **32,** 305–313.
41. T. Malkiewicz and S. Rudnik: *JISI*, 1963, **201,** 33–38.
42. H. Opitz *et al.*: *Arch. Eisenh.*, 1962, **33,** 841–851.
42a. E. M. Trent: *JISI*, 1963, **201,** 1001–1015.
43. L. S. Darken and R. W. Gurry: *J. Amer. Chem. Soc.*, 1946, **68,** 798–816.
44. F. Trojer: *Radex-Rund.*, 1952, 132–136.
45. F. B. Pickering: *Iron and Steel*, 1957, **30,** 3–9.
46. A. C. Turnock and H. P. Eugester: *J. Petrology*, 1962, **3,** 544.
47. N. L. Bowen and J. F. Schairer: *Am. J. Sci.*, 1935, **29,** 151–217.
48. W. Koch *et al.*: *Arch. Eisenh.*, 1960, **31,** 279–286.
49. H. J. Yearian *et al.*: *Corrosion*, 1956, **12,** 515–525.
50. S. Modin: *Met. Treat.*, 1962, March, 89–95.

Appendix 1

Names and ideal formulae of inclusion phases

		Section	Page
almandine	$3FeO.Al_2O_3.3SiO_2$	B 6	76
anorthite	$CaO.Al_2O_3.2SiO_2$	(A 12)	55
chromite	$FeO.Cr_2O_3$	C 4	87
cordierite	$2MgO.2Al_2O_3.5SiO_2$	(A 13)	57
corundum	Al_2O_3	A 2	7
Cr-galaxite	$MnO.Cr_2O_3$	C 3	83
cristobalite	SiO_2	A 3	17
escolaite	Cr_2O_3	C 2	80
fayalite	$2FeO.SiO_2$	B 3	71
Fe-cordierite	$2FeO.2Al_2O_3.5SiO_2$	B 7	78
galaxite	$MnO.Al_2O_3$	A 7	29
grossularite	$3CaO.Al_2O_3.3SiO_2$	(A 11)	52
grunerite	'$FeO.SiO_2$'	B 4	73
jacobite	$MnO.Fe_2O_3$	(A 7)	31
haematite	Fe_2O_3	D 4	94
hausmannite	$MnO.Mn_2O_3$	(A 7)	31
hercynite	$FeO.Al_2O_3$	B 5	74
knebelite	$2Fe_xMn_{1-x}O.SiO_2$	(B 3)	71
Mg-ferrite	$MgO.Fe_2O_3$	(A 7)	31
magnetite	$FeO.Fe_2O_3$	D 3	93
manganosite	MnO	A 6	26
Mn-anorthite	$MnO.Al_2O_3.2SiO_2$	A 12	55
Mn-cordierite	$2MnO.2Al_2O_3.5SiO_2$	A 13	57
mullite	$3Al_2O_3.2SiO_2$	A 8	37
picrochromite	$MgO.Cr_2O_3$	(A 7)	31
pyrope	$3MgO.Al_2O_3.3SiO_2$	(A 11)	52
quartz	SiO_2	A 5	23
rhodonite	$MnO.SiO_2$	A 9	44
spessartite	$3MnO.Al_2O_3.3SiO_2$	A 11	52
spinel	$MgO.Al_2O_3$	(A 7)	31
tephroite	$2MnO.SiO_2$	A 10	48
tridymite	SiO_2	A 4	22
wüstite	'FeO'	B 2, D 2	67

Appendix 2

Metal content of the inclusion phases in their ideal composition

	Name		Section	Page
Mg		wt-% Mg		
$Mg_2Al_4Si_5O_{18}$	cordierite	8	(A 13)	57
$MgFe_2O_4$	Mg-ferrite	12	(A 7)	31
$MgCr_2O_4$	picrochromite	13	(A 7)	31
$MgAl_2O_4$	spinel	17	(A 7)	31
$Mg_3Al_2Si_3O_{12}$	pyrope	18	(A 11)	52
Al		wt-% Al		
$Mn_3Al_2Si_3O_{12}$	spessartite	11	A 11	52
$Fe_3Al_2Si_3O_{12}$	almandine	11	B 6	76
$Ca_3Al_2Si_3O_{12}$	grossularite	12	(A 11)	52
$Mg_3Al_2Si_3O_{12}$	pyrope	13	(A 11)	52
$Mn_2Al_4Si_5O_{18}$	Mn-cordierite	17	A 13	57
$Fe_2Al_4Si_5O_{18}$	Fe-cordierite	17	B 7	78
$MnAl_2Si_2O_8$	Mn-anorthite	18	A 12	55
$Mg_2Al_4Si_5O_{18}$	cordierite	18	(A 13)	57
$CaAl_2Si_2O_8$	anorthite	19	(A 12)	55
$MnAl_2O_4$	galaxite	31	A 7	29
$FeAl_2O_4$	hercynite	31	B 5	74
$Al_6Si_2O_{13}$	mullite	38	A 8	37
$MgAl_2O_4$	spinel	38	(A 7)	31
Al_2O_3	corundum	53	A 2	7
Si		wt-% Si		
$Al_6Si_2O_{13}$	mullite	13	A 8	37
Mn_2SiO_4	tephroite	14	A 10	48
Fe_2SiO_4	fayalite	14	B 3	71
$Fe_3Al_2Si_3O_{12}$	almandine	17	B 6	76
$Mn_3Al_2Si_3O_{12}$	spessartite	17	A 11	52
$MnAl_2Si_2O_8$	Mn-anorthite	19	A 12	55
$Ca_3Al_2Si_3O_{12}$	grossularite	19	(A 11)	52
$CaAl_2Si_2O_8$	anorthite	20	(A 12)	55
$FeSiO_3$	grunerite	21	B 4	73
$MnSiO_3$	rhodonite	21	A 9	44
$Mg_3Al_2Si_3O_{12}$	pyrope	21	(A 11)	52
$Fe_2Al_4Si_5O_{18}$	Fe-cordierite	22	B 7	78
$Mn_2Al_4Si_5O_{18}$	Mn-cordierite	22	A 13	57
$Mg_2Al_4Si_5O_{18}$	cordierite	24	(A 13)	57
SiO_2	cristobalite	47	A 3	17

SiO_2	tridymite	47	A 4	22
SiO_2	quartz	47	A 5	23
Ca		wt-% Ca		
$CaAl_2Si_2O_8$	anorthite	14	(A 12)	55
$Ca_3Al_2Si_3O_{12}$	grossularite	27	(A 11)	52
Cr		wt-% Cr		
$FeCr_2O_4$	chromite	46	C 4	87
$MnCr_2O_4$	Cr-galaxite	47	C 3	83
$MgCr_2O_4$	picrochromite	54	(A 7)	31
Cr_2O_3	escolaite	68	C 2	80
Mn		wt-% Mn		
$Mn_2Al_4Si_5O_{18}$	Mn-cordierite	17	A 13	57
$MnAl_2Si_2O_8$	Mn-anorthite	19	A 12	55
$MnFe_2O_4$	jacobite	24	(A 7)	31
$MnCr_2O_4$	Cr-galaxite	25	C 3	83
$MnAl_2O_4$	galaxite	32	A 7	29
$Mn_3Al_2Si_3O_{12}$	spessartite	33	A 11	52
$MnSiO_3$	rhodonite	42	A 9	44
Mn_2SiO_4	tephroite	54	A 10	48
Mn_3O_4	hausmannite	72	(A 7)	31
MnO	manganosite	77	A 6	26
Fe		wt-% Fe		
$Fe_2Al_4Si_5O_{18}$	Fe-cordierite	17	B 7	78
$FeCr_2O_4$	chromite	25	C 4	87
$FeAl_2O_4$	hercynite	32	B 5	74
$Fe_3Al_2Si_3O_{12}$	almandine	34	B 6	76
$FeSiO_3$	grunerite	42	B 4	73
$MnFe_2O_4$	jacobite	49	(A 7)	31
Fe_2SiO_4	fayalite	55	B 3	71
$MgFe_2O_4$	Mg-ferrite	56	(A 7)	31
Fe_2O_3	haematite	70	D 4	94
Fe_3O_4	magnetite	72	D 3	93
FeO	wüstite	78	B 2, D 2	67

Part II

Inclusions belonging to the systems $MgO–SiO_2–Al_2O_3$, $CaO–SiO_2–Al_2O_3$ and related oxide systems. Sulphide inclusions

Foreword

This monograph is the second part of the work begun in Special Report 90 (1964) of The Iron and Steel Institute 'Non-metallic inclusions in steel'.[51] *In part I special attention was given to the system* $MnO–SiO_2–Al_2O_3$. *This system is of fundamental importance to the formation of all oxide inclusions in modern steels and serves as a convenient base for the discussion of most other oxide systems of interest in steelmaking. It is the main system to which indigenous inclusions are related. In the same work inclusions belonging to the systems* $FeO–SiO_2–Al_2O_3$, $MnO–SiO_2–Cr_2O_3$, *and* $FeO–SiO_2–Cr_2O_3$ *were also discussed, together with the iron oxides.*

The present work deals with inclusions belonging to the systems $MgO–SiO_2–Al_2O_3$ *and* $CaO–SiO_2–Al_2O_3$, *as well as the sulphide inclusions. In respect of the morphology and crystal chemistry of the inclusions, these oxide systems, particularly that with MgO, are closely related to the MnO system. In contrast, the mechanisms of inclusion formation differ between the MnO and MgO or CaO systems, MgO and CaO being especially related to exogenous inclusions. However, the difference between indigenous and exogenous inclusions should not be overemphasized since most oxide inclusions have formed from exogenous nuclei which have grown indigenously.*

The metallography of oxides from the two large systems based on MgO and CaO, as they occur in steelmaking, is not well understood. The aim of the present work has been to include as much metallographic information about these complex systems as possible. The general arrangement of part I has been preserved in presenting information on the phases, but since the systems considered in part II are more complex and less information is available elsewhere, the general discussion sections are longer.

The presentation of the sulphide systems differs in some respects from that of the oxides and the main emphasis has been on the sulphides FeS, MnS and (MnMe)S, which are of fundamental importance in steelmaking. However, several sulphide systems which have as yet received little attention have also been reviewed. These are included because the conditions of inclusion formation are likely to be modified by the development of new processes, such as vacuum treatment, and the controlled addition of rarer elements, as in the micro-alloyed steels. Under these conditions the range of inclusion phases encountered is likely to be extended. The incomplete knowledge of several of the sulphide systems has meant that much of the information given has been based on older research techniques.

To achieve as full a coverage as possible, extensive use has been made of tables and summaries to cover the various systems. Although this tends to give a compressed text, the authors feel that this is the best way to make the information easily accessible.

The experimental methods are the same as those used for the inclusion investigations in part I to which the reader is referred for further information.

In many cases excellent photomicrographs of different inclusions have been shown in the literature, which it has not been possible to reproduce. The authors have therefore added specific references to plates and pages for these illustrations. Two books of general interest for the metallography of inclusions have appeared since part I was published. These are the monograph by Bragg and Claringbull[52] *on the crystal structure of minerals and that by Muan and Osborn*[53] *on oxide systems of interest in steelmaking; frequent reference to these books has been made. In the list of references in part I of older literature, dealing with inclusions, the important work by Portevin and Castro was omitted. This classical work should not be forgotten and reference is made to it on several occasions in the present part.*[115,149,153] *In general, the references have been selected with due regard to experimental techniques used by the original authors.*

The primary aim of part II, as of part I, is to give the ferrous metallurgist as useful a tool as possible for the identification of inclusions and to enable metallographers to use their existing facilities, especially the optical microscope, to the full. In many cases, with an adequate background knowledge of steel and inclusions, it is not necessary to use the modern sophisticated physical methods of investigation in order to identify the inclusions, provided that a thorough optical investigation is carried out. If the metallographer is familiar with the chemistry and physical properties of the principal inclusion systems, considerable economy and precision in the use of experimental methods may be achieved.

It is hoped that the present book, together with part I, will be of use not only to the well-equipped laboratories with full facilities, but also that it will encourage a full use of the optical microscope.

In the concluding chapter (M) the authors have included an identification scheme for the inclusions covered in parts I and II. The aim of this scheme is to help metallurgists who are studying inclusions by ordinary metallographic methods to find those sections in the two parts of the report where thorough information and micrographs of the appropriate inclusion phases are to be found. The scheme is based on information about the steel type and steel analysis. Although incomplete, it may serve as a rough guide and it also should be helpful when deciding if the more expensive methods, like electron probe analysis, should be used for a complete identification of the inclusion phases.

It is also hoped that the more extensive discussions in this part as compared with part I will help the metallographer not only to identify the phases in the inclusions, but also to specify their origin. While a knowledge of the origin often is the answer to problems associated with inclusions, it is obvious that the overall importance of the quantity and nature of inclusions in a steel product cannot be assessed on this basis. Therefore, it is essential that further work is carried out to reveal more about the detailed formation of inclusions and also the important behaviour of inclusions during the working of the steel to the final product. The general attitude to regard an inclusion as a piece of dirt in the steel must be changed. Inclusions in steel are complex units and their influence on steel properties depends on the application of the steel as well as on the detailed physical properties of the different inclusion phases. They are sometimes disastrous, sometimes even advantageous, all depending on the specific use of the steel. The freedom of the steel from inclusions is therefore an important technical as well as an economic question. It is hoped that the present work will stimulate further research in this important field.

Professor Roland Kiessling is Director of the Swedish Institute for Metal Research, where Mr Nils Lange is research assistant. The authors are indebted to Mrs Connie Westman (née *Helin) and Mr Tore Malmberg for valuable help with X-ray diffraction methods and the preparation of synthetic slags. Mr Sten Bergh (of the Research Centre, Stora Kopparbergs Bergslags AB, Domnarvet, Sweden) has contributed much valuable criticism. Dr D. Dulieu again assisted with the English text.*

As was the case for part I, part II is based on a series of communications during 1965 and 1966 to the Swedish journal

Jernkontorets Annaler. *The printing of the English edition has been greatly facilitated by the loan of the original blocks from* Jernkontorets Annaler, *and the help of the editor of this journal is gratefully acknowledged.*

Stockholm, May 1966

F. The MgO-SiO_2-Al_2O_3 system

F 1 General

Magnesium compounds are constituents of many of the refractories used in steelmaking. These ceramic materials are the most important source of magnesium in non-metallic inclusions in steel. A summary of analyses typical of the refractories used in steelmaking is given in Table IX and, as this table shows, the MgO content is high for many

TABLE IX **Typical analyses for different refractories, used in steelmaking, wt-%. Most of the values given have been taken from the summary of Pettersson[57] on refractories, used in Swedish steelmaking**

	Refractory type	CaO	MgO	Al_2O_3	Cr_2O_3	SiO_2	Others (Fe_2O_3, TiO_2, ZrO_2, alk.)
Acid	silicate brick	1·5–2·2	...	0·5–2	...	94–97	1–4
	semi-silicate brick	1·5–2·2	...	7–28	...	70–90	1–4
	chamotte brick	+	...	30–45	...	49–64	1–6
Neutral	forsterite brick (olivine brick)	1	50–63	1–2	+	30–38	4–8
Basic	magnesite brick	0·5–3	>85	0·5–4	+	1–3	2–8
	chrome–magnesite brick	0·5–2	35–50	10–16	25–30	3–5	8–14
	magnesite–chromium brick	0·5–2	55–80	4–7	6–15	3–5	4–7
	dolomite brick	50–60	36–40	0·5–3	...	0·5–5	0·5–3
	acid mass	...	...	...	...	+	+
	basic mass	...	+	...	...	...	+
	aluminate cement	20–35	...	60–70	...	2–6	+

of the different brick compositions. Thus the basic and neutral bricks of magnesite, chrome magnesite, dolomite and forsterite may contain from about 35 wt-% up to more than 85 wt-% of MgO, which is also an important constituent of basic furnace cements.

Non-metallic inclusion nuclei with MgO as one component may be formed as a product of reactions between furnace refractories and the furnace or ladle slags, or with the molten steel itself. These reactions have been extensively studied and have been considered in the Special Reports Nos.74 (1962) and 87 (1964) of The Iron and Steel Institute.[54,55] Table X, based on information given in a paper

TABLE X Phases which have been identified in different types of refractories after service for some time in steelmaking furnaces. The values have been collected from The Iron and Steel Institute's Special Report 87, paper by Duke and Lakin[56]

Section	Phase	Tar-dolomite	Dolomite bricks	Magnesite bricks (unfired)	Silicate bricks	Alumina bricks	Magnesite bricks
F 2	MgO	+	+	+	...	...	+
F 3	$MgO.Al_2O_3$	...	...	+	...	...	+
(F 3)	$MgO.Fe_2O_3$	...	...	+	...	...	+
G 2	CaO	+	+	...	...	...	...
J	$CaO.MgO.SiO_2$	...	...	+	...	...	+
J	$3CaO.MgO.2SiO_2$	...	...	...	...	...	+
G 4	$2CaO.SiO_2$	...	...	+	...	...	+
G 4	$3CaO.SiO_2$	+	+	...	...	...	...
G 3	$CaO.6Al_2O_3$	...	...	...	...	+	...
G 3	$12CaO.7Al_2O_3$	...	+	...	...	+	...
G 3	$3CaO.Al_2O_3$	+	+	...	...	...	...
(G)J	$2CaO.Fe_2O_3$	+	+	+	...	...	...
J	$3CaO.Fe_2O_3.3SiO_2$	...	...	...	+	...	...
B 5	$FeO.Al_2O_3$	...	...	...	...	+	...
D 3	$FeO.Fe_2O_3$	+	...	...	...	...	...
A 2	Al_2O_3	...	...	...	...	+	...
A 3	SiO_2(cristob.)	...	...	...	+	...	...
A 4	SiO_2(trid.)	...	...	...	+	...	...
A 5	SiO_2(quartz)	...	...	...	+	...	...
...	$Ca(OH)_2$	+	...	...	...	...	...
A 8	$3Al_2O_3.2SiO_2$	...	...	...	...	+	...

by Duke and Lakin,[56] summarizes those phases identified in various refractories after service in basic arc furnaces. These phases consist of both the original constituents of the bricks and products resulting from reactions with the liquid phases in the bath. These nuclei may be found as inclusions in the steel, either with their original composition or, more frequently, changed by indigenous growth but retaining a substantial MgO content.

The basic furnace slags form a second source of inclusions containing magnesium. However, when these slags are considered as the origin of inclusions, it is their calcium content which is the most important factor, the behaviour of calcium being discussed further in section G. From Table XIV of that section it will be seen that the MgO content of inclusions originating from certain slags may be up to 5 wt-%.

Magnesium has a very low solubility in iron and is not, therefore, precipitated as an oxide from solid solution. Despite the low free energy of formation of the oxide (Fig.80), magnesium is not usually

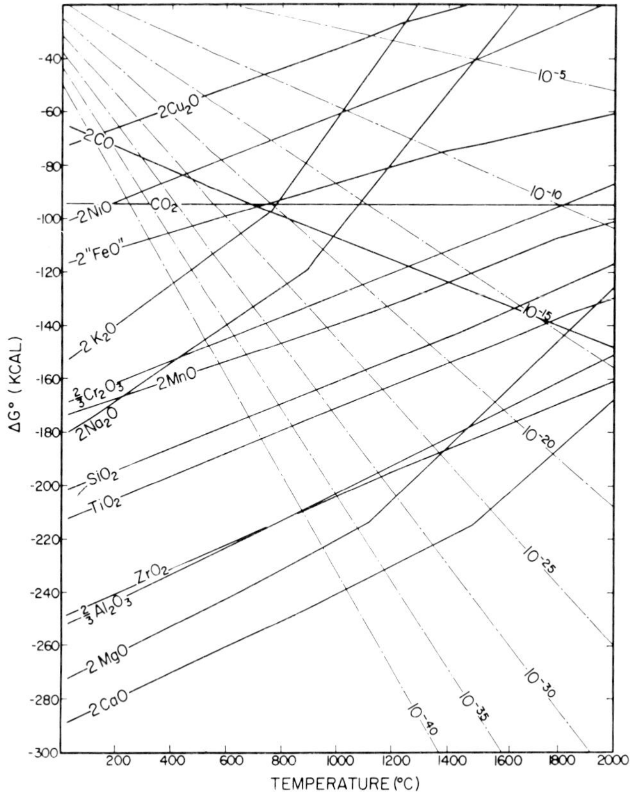

80 Standard free energies of formation $\Delta G°$ (kcal) for various oxides of the elements as a function of temperature. Light dash-dot curves are lines of equal oxygen pressure, (atm) of the gas phase, as labelled on each curve. From Muan and Osborn,[53] Fig.1, p.4

employed as a deoxidant in steel. This would appear to be due to the high volatility of the metal, which evaporates readily at about 1100°C, with a $d(\Delta G^\circ)/dT$ value which increases rapidly with temperature (for further discussion, see section G 1). Metallic magnesium is, however, an additive to nodular cast irons. It is also used in the refining of pig-iron, due to the low free energy of formation of the sulphide, MgS (section K). It is thus possible that metallic magnesium can occasionally be the origin of the magnesium content of inclusions, an example of this is given in Fig.95.

The pure $MgO–SiO_2–Al_2O_3$ system is similar to both the $MnO–SiO_2–Al_2O_3$ and $FeO–SiO_2–Al_2O_3$ systems discussed in parts A and B of the earlier volume. From the structural standpoint the inclusion phases possible with MgO are closely related to those considered previously. Figure 81 indicates schematically the different

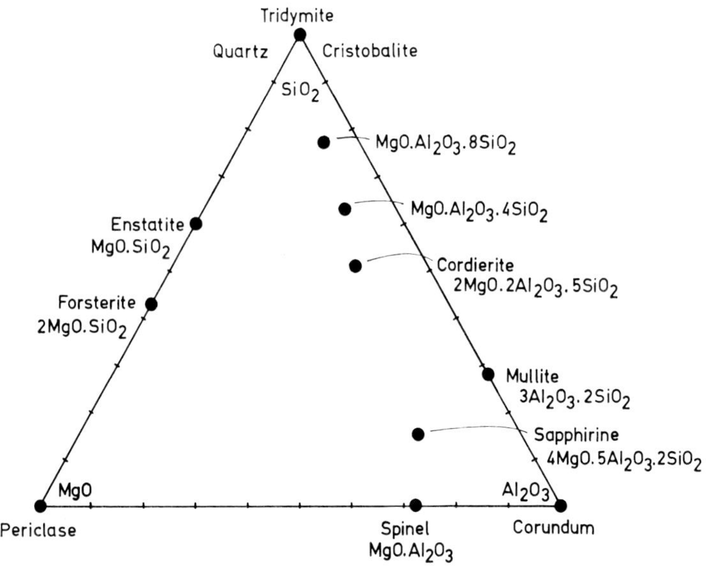

81 Schematic representation of the pseudoternary system $MgO–SiO_2–Al_2O_3$, showing the stoichiometric composition of the different phases which have been reported

phases reported to exist in the system with MgO and this diagram should be compared with Figs.1 and 57 of part I. The compositions of these phases, according to their reported stoichiometric formulae, are listed in Table XI. A projection of the liquidus surface of the ternary system $MgO–SiO_2–Al_2O_3$ is given in Fig.82, based on the summary by Muan and Osborn.[53]

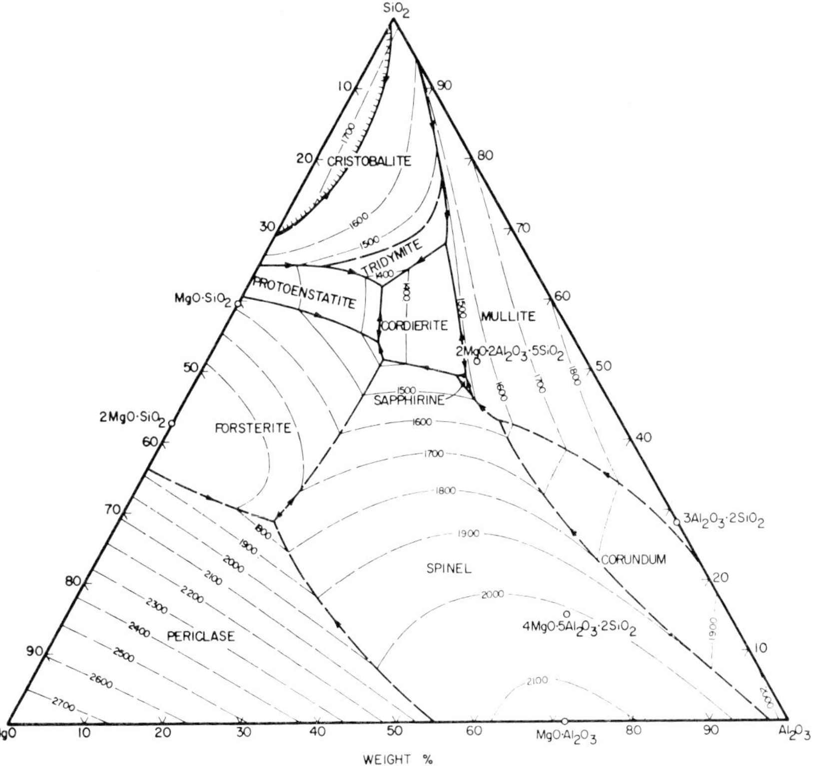

82 Phase diagram for the system $MgO–SiO_2–Al_2O_3$. Projection of the liquidus surface of the system. From Muan and Osborn,[53] Fig.81, p.96

The following survey of magnesium-containing phases is not to be regarded as a general study of the system, the main emphasis being on those phases which are of interest as nuclei or constituents of inclusions. Of the phases listed in Table XI some exist in different modifications or only within limited temperature ranges, while others have only been reported as minerals of unknown purity.

In summary, inclusions containing MgO are characterized by an exogenous origin, the main sources being refractories and the furnace or ladle slags. To consider the system adequately, therefore, it has been necessary to examine and analyse slag and refractory samples containing inclusion phases in addition to the inclusions themselves.

TABLE XI Phases reported to exist in the $MgO–SiO_2–Al_2O_3$ system

Name	Chemical formula	Stoichiometric comp. wt-% MgO	SiO_2	Al_2O_3	Section
Corundum	Al_2O_3	...	...	100	A 2
Cristobalite	SiO_2	...	100	...	A 3
Tridymite	SiO_2	...	100	...	A 4
Quartz	SiO_2	...	100	...	A 5
Periclase	MgO	100	...	...	F 2
Spinel	$MgO.Al_2O_3$	28	...	72	F 3
Mullite	$3Al_2O_3.2SiO_2$	...	28	72	A 8
Enstatite (Protoenstatite, Clinoenstatite)	$MgO.SiO_2$	40	60	...	F 4
Forsterite	$2MgO.SiO_2$	57	43	...	F 5
Cordierite (α, β, μ)	$2MgO.2Al_2O_3.5SiO_2$	14	51	35	F 6
Pyrope	$3MgO.Al_2O_3.3SiO_2$	30	25	45	F 7
Sapphirine	$4MgO.5Al_2O_3.2SiO_2$	20	15	65	F 8
(Osumilite)	$MgO.Al_2O_3.4SiO_2$	10	63	27	F 9
(Petalite)	$MgO.Al_2O_3.8SiO_2$	7	77	16	F 10

F 2 $MgO–SiO_2–Al_2O_3$: Periclase

Chemical formula MgO

Physical properties

Melting point: 2800°C

Density: 3·581

Microhardness: 1000 kp/mm^2

Structure type

Cubic; space group Fm3m

a=4·213 Å, ASTM data card 4–0829

NaCl-type. Wyckoff I, chap. III, text p.1, table p.13

Isostructural with manganosite (A 6), wüstite (B 2 and D 2) and CaO (G 2)

Monochromatic X-ray powder photograph of manganosite and wüstite, *see* Figs.19 and 75 (part I)

Solid solubility and equilibria

A wide range of solid solubility exists between periclase and several Me(II)-oxides. Of special importance are the systems MgO–FeO, MgO–MnO and MgO–CaO. In the MgO–FeO and MgO–MnO systems a complete solid solubility exists, whereas CaO is insoluble in MgO.[8] At temperatures above 1 000°C, small amounts of the oxides Al_2O_3, Cr_2O_3 and Fe_2O_3 are soluble in MgO, the solid solubility increasing with temperature.[47] These solid solutions of ternary oxides are not stable at lower temperatures, and the ternary oxides are precipitated, usually as double oxides with MgO. A detailed study of the system MgO–Al_2O_3 has been published by Stubican and Roy,[58] according to whom about 4 mol% of Al_2O_3 are soluble in MgO. Al_2O_3 is precipitated from the solid solution as spinel (F 3) at lower temperatures, the precipitated spinel phase being localized at grain boundaries and intergranular pores.

Phase equilibria in basic refractories have been studied by White,[59] who used the seven-component system CaO–MgO–FeO–Fe_2O_3–Al_2O_3–Cr_2O_3–SiO_2 as a base for investigation. White found that the equilibrium between the readily identifiable periclase and other phases in the system was governed mainly by the ratio CaO:SiO_2. If this ratio is 2:1 then there is an equilibrium between periclase and the phases 2CaO·SiO_2 (G 4) and MgO·Fe_2O_3 (A 7, J). If the ratio exceeds 2:1 then there is apparently formation of the low melting-point phase 2CaO·Fe_2O_3 (J), whereas if the ratio is less than 2:1 all the Fe_2O_3 appears to form a double oxide of the spinel type without the formation of the low melting-point phase. The influence of the other components in the system on these equilibria are further discussed in White's paper.

Microscopy and electron probe analysis

Periclase (MgO) is an important structural component of several basic refractories. In magnesite bricks it is usually visible as rather

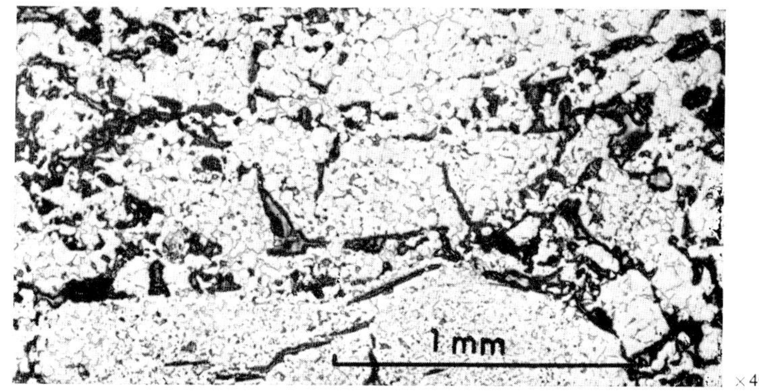

×40

Magnesite is an impure carbonate, $MgCO_3$. During the firing of the carbonate, carbon dioxide is liberated, followed by a sintering and recrystallization of the periclase phase

83 Periclase in burnt magnesite (sintermagnesite) at a low magnification (cf. Fig.84). Light, rounded periclase grains constitute the main part of the structure, in which several dark pores are also present

a 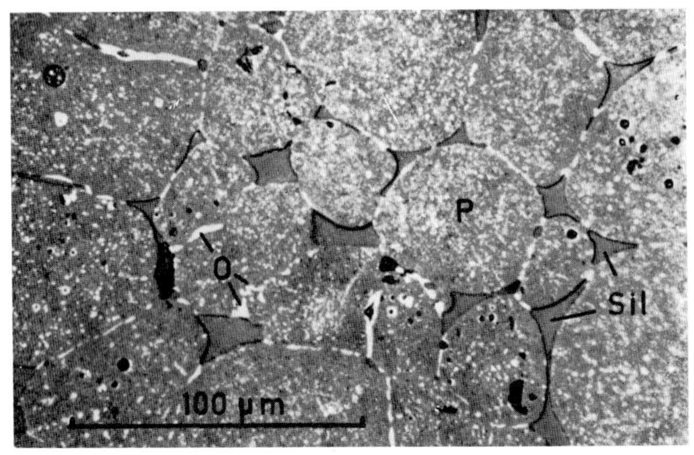

×400

b 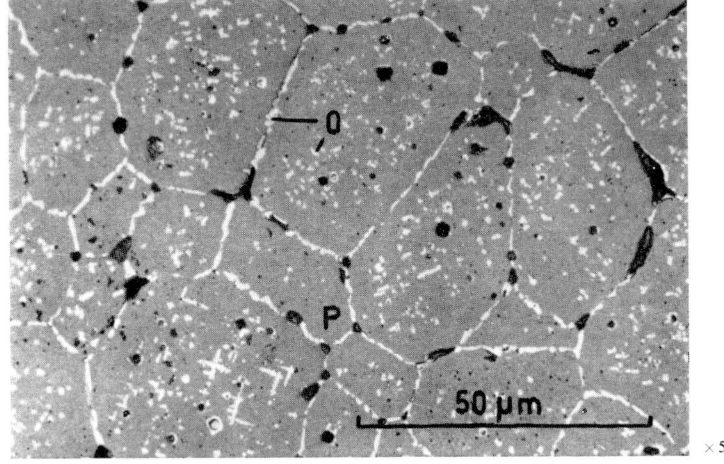

×500

84*a* Most of the magnesium ferrite is visible as precipitates inside the periclase grains, which have partly sintered. Many of the interstices between the periclase grains are filled with glassy silicate matrix.
Typical analysis for magnesite, % > 85MgO, 0·5–3CaO, 0·5–4Al_2O_3, 1–3SiO_2, 2–8Fe_2O_3

84*b* The microstructure of this brick differs from that shown in Fig.84*a*. (The brick shown in Fig.84*b* had a lower Si content and was fired differently.) The periclase grains are smaller and the magnesium ferrite phase is more crystallized and is found not only inside the periclase grains but also to a considerable extent in the grain boundaries. The porosity is greater due to the absence of silicate phase

84a Periclase (P), magnesium ferrite (O) and glassy silicate matrix (Sil) in magnesite bricks containing iron.

84b Periclase (P) and magnesium ferrite (O) in iron-containing magnesite bricks

large, rounded grains (Figs.83 and 84), but the grain size depends upon firing temperature and brick composition and the periclase structure may be finer than that shown in the two examples. Magnesite usually contains iron impurities and during the firing process, when periclase is formed from magnesite ($MgCO_3$), this iron is taken into solid solution in the periclase phase and precipitated as magnesium ferrite, the double oxide $MgO{\cdot}Fe_2O_3$. These precipitates are visible inside the periclase grains in Fig.84. In addition the ferrite phase is sometimes present as a binding phase between the boundaries of the periclase grains, as shown in Fig.84*b*.

Periclase is also a structural component of burnt dolomite which is shown in Figs.85 and 104. The periclase phase is visible as white –

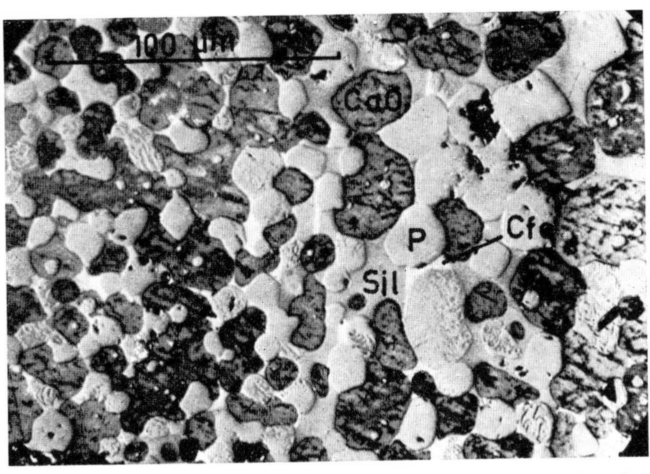

×400

Dolomite is a double carbonate, $CaMg(CO_3)_2$ with small amounts of different oxides present. During the firing of the carbonate, carbon dioxide is liberated and periclase and calcia (lime) are formed; these are insoluble in each other. The two oxides form a sintered structure, and the sintering is facilitated by the other oxides present as well as by the silicate phase.

The CaO-phase shows a number of surface nuclei of $Ca(OH)_2$, formed by the action of moisture in the air (*see* Fig.104). Calcium ferrite is present as a white phase in several grain boundaries and also as a fine precipitate inside most of the periclase grains. (The photomicrograph is reproduced from the paper by Brunner[60])

85 Periclase (P), calcia (lime, CaO), calcium ferrite (Cfe) and silicate phase (Sil) in burnt dolomite

grey, rounded grains with an even surface. They remain unchanged in moist atmospheres, unlike CaO, the other main structural component, which is attacked. An example of periclase with a dendritic morphology has been given by Brunner. [60a] Such dendritic periclase structures can be obtained upon cooling in refractories containing

MgO which have been in contact with liquid CaO–SiO_2 slags at high temperatures. A liquid slag rich in MgO forms on the surface of the refractory, dendritic periclase crystallizing from this slag on cooling.

F 3 MgO–SiO_2–Al_2O_3: Spinel

Chemical formula $MgO{\cdot}Al_2O_3$

Physical properties

Melting point: 2135°C
Density: 3·58
Microhardness: 2100–2400 kp/mm^2
Fluorescence: Spinel shows a green fluorescence in the electron probe analyser

Structure type

Cubic; space group Fd3m
a=8·080 Å. ASTM data card 5–0 672
Spinel type. Wyckoff II, chap. VIII, text p.16, table p.41
Isostructural with galaxite (A 7), hercynite (B 5), chromium galaxite (C 3), chromite (C 4) and magnetite (D 3)
Monochromatic X-ray powder photographs of galaxite, hercynite and magnetite, *see* Figs.24, 63 and 75

Solid solubility

The spinel phase $MgO{\cdot}Al_2O_3$ has given its name to a whole family of double oxides with the general formula $AO{\cdot}B_2O_3$, with the result that the nomenclature may sometimes be confusing. Mg may be replaced completely by Fe(II) and Mn(II). These double oxides of the spinel type have been considered earlier in section A 7. The compositions of some substituted magnesium spinels found in slag inclusions are given in Table XII, which should be compared with the double oxides listed in Table IV of section A 7.

The ternary oxide of the spinel phase, Al_2O_3, may also be substituted and the substitution of chromium for aluminium has been observed in non-metallic inclusions. The resulting substituted spinel has the general formula $MgO{\cdot}Al_{2-x}Cr_xO_3$. The lattice parameter of the phase increases continuously from about 8·08 Å for $x=0$ to about 8·13 Å for $x=2$.[61] Complete substitution of Fe(III) for Al is also possible, the end member is then magnesium ferrite, $MgO{\cdot}Fe_2O_3$, a common precipitate in magnesite refractories (F 2).

Microscopy and electron probe analysis

Spinel usually forms idiomorphic, hard, characteristic crystals, easily identifiable in the microstructure by their regular sections. The

TABLE XII Composition, wt-%, of the spinel phase in different inclusions with MgO (cf. Table IV, section A 7)

Inclusion type	MgO	MnO	FeO	CaO	Al_2O_3	Cr_2O_3	Steel type	Fig.
Spinel	30	...	...	...	70	...	rolled stainless	89
Spinel in Ca-aluminate	33	...	...	...	68	...	low alloyed	91
Spinel in Ca-aluminate	31	...	...	...	67	...	low alloyed	...
Spinel in Ca-aluminate	31	+	+	2	68	...	low alloyed	88
Spinel and forsterite in silicate	25	7	...	...	70	...	vacuum steel	99
Spinel in Ca-Al-silicate	26	6	...	...	67	...	low alloyed	...
Spinel in silicate	25	12	...	2	62	...	vacuum steel	86
Spinel	12	23	...	...	59	...	low alloyed	90
Spinel in silicate	5	34	1	...	54	...	rolled carbon steel	25
Spinel in silicate	5	25	1	...	46	12	rolled stainless	...
Spinel in silicate	3	34	2	...	61	...	electric steel	27

phase is cubic, and is inactive in polarized light unlike the anisotropic phases corundum (A 2) and escolaite (C 2) which are sometimes similar in appearance. Several examples of the spinel phase in multiphase inclusions may be seen in Figs.86–89. These inclusions have been chosen from a variety of steels to illustrate the frequent occurrence of spinels. Figure 90 shows spinel present in an inclusion together with corundum. The two phases are similar in appearance but corundum is the harder, is active under polarized light and may be distinguished in the electron probe by its reddish fluorescence, in contrast to the spinel phase which fluoresces green. The inclusions in Fig.91 illustrate spinel in a rolled steel, together with one of the calcium aluminates (G 3). In the ingot the inclusion was of the spherical, two-phase type shown in Figs.87 and 88. The heavy working of the steel matrix during the rolling operation has separated the two phases, which appear as more or less isolated crystals in elongated strings. Like the spinel phase the calcium aluminate (here the phase $CaO{\cdot}2Al_2O_3$, *see* G 3) is also idiomorphic with a regular section, it is distinguished from spinel being softer and giving a blue fluorescence under the electron beam. The response of this type of inclusion to deformation of the steel matrix differs markedly from

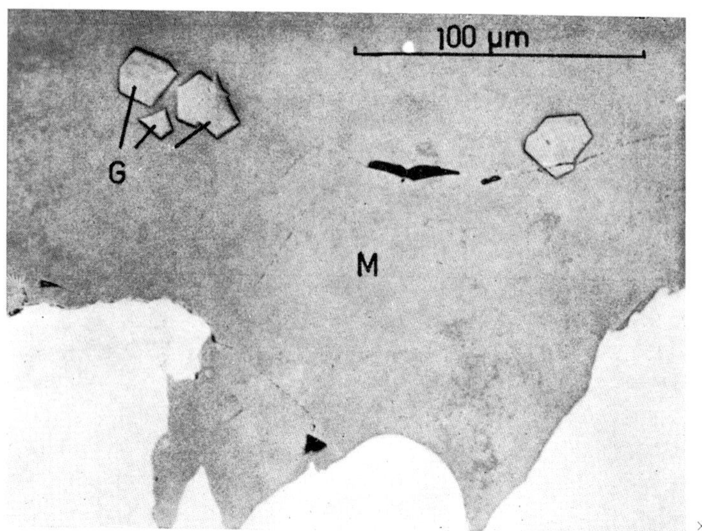

×400

STEEL TYPE Silicon-killed electrosteel, vacuum treated. From a defective railway rail

STEEL ANALYSIS, % 0·62 C, 0·47 Si, 1·78 Mn, 0·034 P, 0·018 S, 0·007 N

INCLUSION, % (G) Spinel with 25 MgO, 12 MnO, 2 CaO, 62 Al_2O_3. (M) Matrix with 13 MnO, 7 MgO, 23 CaO, 41 SiO_2, 20 Al_2O_3. Mean analysis % 13 MnO, 7 MgO, 23 CaO, 41 SiO_2, 20 Al_2O_3

86 Oxide inclusion in steel: spinel in glassy silicate matrix

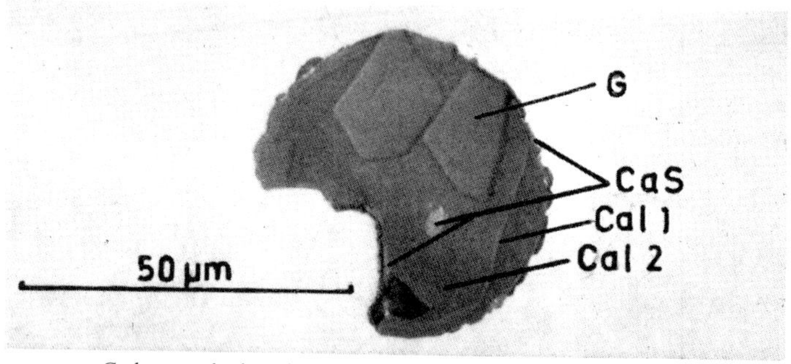

×800

STEEL TYPE Carbon steel, deoxidized with CaSi

STEEL ANALYSIS, % 0·16 C, 0·37 Si, 1·44 Mn, 0·018 P, 0·009 S, 0·031 Al, 0·009 N

INCLUSION, % (G) Spinel with 30 MgO, 70 Al_2O_3 (Cal 1) Calcium aluminate with 16 CaO, 76 Al_2O_3 (G 3) (Cal 2) Calcium aluminate with 3 MgO, 31 CaO, 60 Al_2O_3 (G 3) (CaS) Calcium sulphide with 48 Ca (K 11). Mean analysis % 10 MgO, 23 CaO, 67 Al_2O_3

COMMENT The calcium sulphide phase is mainly situated as a scale around the inclusion in the border between the inclusion and the steel phase. The sulphide is soluble in water and therefore the preparation of the microsection has to be done with great care in order to show the sulphide phase

87 Oxide inclusion in steel: spinel in calcium aluminate, partly surrounded by a scale of calcium sulphide

the deformation of high-silica inclusions (*see*, for instance, Figs.28 and 29). Further examples are given in section G 3.

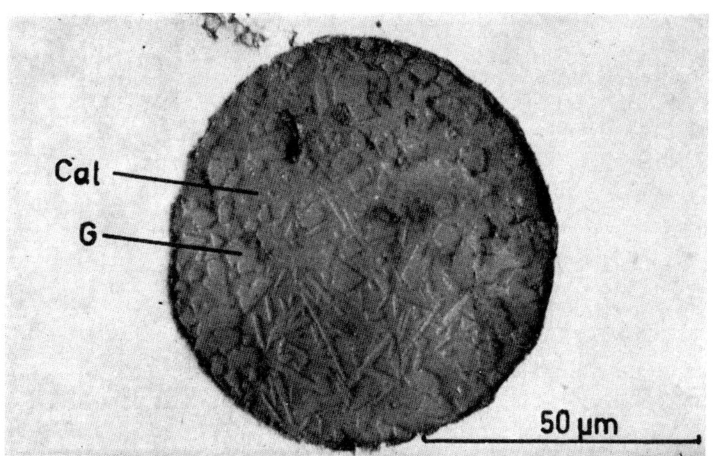

×800

STEEL TYPE Low-alloyed chromium steel, heat-treated 925°/12h → 830°/9·5h → oil 50°. From a broken fatigue test sample
STEEL ANALYSIS, % 0·15 C, 0·29 Si, 0·75 Mn, 0·86 Cr, 1·45 Ni, 0·08 Mo

INCLUSION, % (G) Spinel with 68 Al_2O_3, 31 MgO, 2 CaO, trace of FeO and MnO. (Cal) Matrix 71 Al_2O_3, 29 CaO, trace of FeO (G 3). Mean analysis 64 Al_2O_3 13 MgO, 18 CaO

88 Oxide inclusion in steel: spinel in calcium aluminate

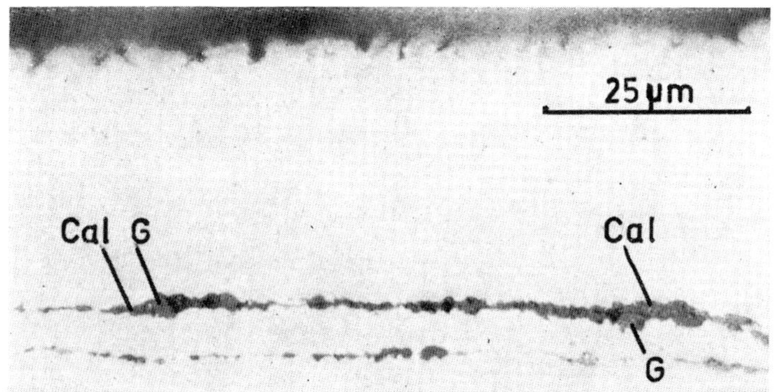

×1000

STEEL TYPE Austenitic stainless, cold-rolled steel. Surface defects, appearing as blisters on the steel surface
STEEL ANALYSIS, % 0·037 C, 0·64 Si, 1·22 Mn, 0·025 P, 0·009 S, 18·1 Cr, 9·0 Ni, 0·20 Mo, 0·0048 O, 0·024 N
INCLUSION, % (G) Spinel with 30 MgO, 70 Al_2O_3 (Cal) Calcium aluminate with 20 CaO, 80 Al_2O_3 (G 3)
COMMENT This inclusion was of the spherical type in the ingot as shown for instance in Figs.87 and 88, but as a result of deformation of the steel these spheres have been elongated to strings in the rolling direction. The harder spinel phase has been partly separated from the calcium aluminate, which has been crushed, and the two phases appear as mixed constituents of the strings, two of which are visible in this microsection (*see* also Fig.91). This figure should be compared with Figs.27 and 29, part I, showing deformed spherical inclusions with galaxite in a silicate matrix. The galaxite phase, which is of the spinel type and has similar physical properties to spinel, remains undeformed during rolling of the steel phase and separates partly from the matrix. The silicate matrix, however, is more plastic than the calcium aluminates and is therefore deformed without being crushed

89 Oxide inclusion in steel: spinel and calcium aluminate

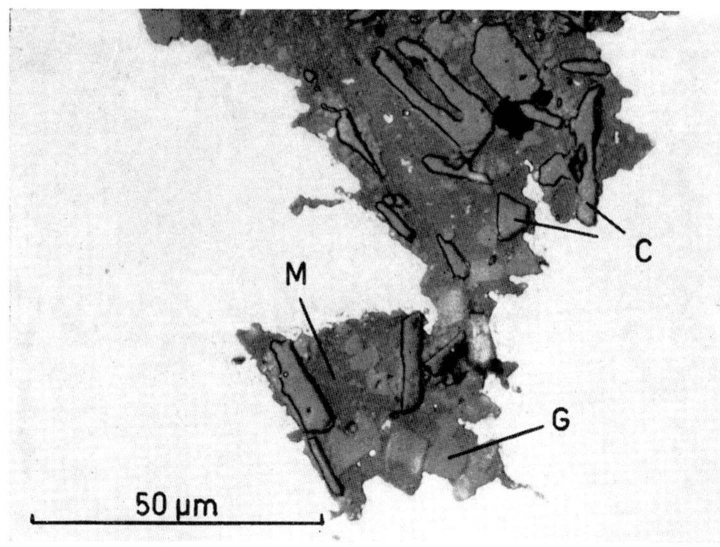

× 600

STEEL TYPE Low-alloyed
STEEL ANALYSIS, % 0·44 C, 0·28 Si, 0·72 Mn, 0·011 P, 0·051 S, 0·95 Cr, 0·11 Ni, 0·19 Mo, 0·14 Cu
INCLUSIONS, % (G) Spinel with 12 MgO, 23 MnO, 59 Al_2O_3, (C) Corundum with 100 Al_2O_3, (M) Matrix with 3 MnO, 14 CaO, 44 SiO_2, 40 Al_2O_3. Mean analysis 4 MnO, 1 MgO, 10 CaO, 32 SiO_2, 53 Al_2O_3
COMMENT It has not been possible to determine if the matrix (M) is glassy calcia-alumina-silica or a very fine crystalline precipitate of anorthite (G 5)

90 Oxide inclusions in steel: spinel and corundum in calcia-alumina-silica

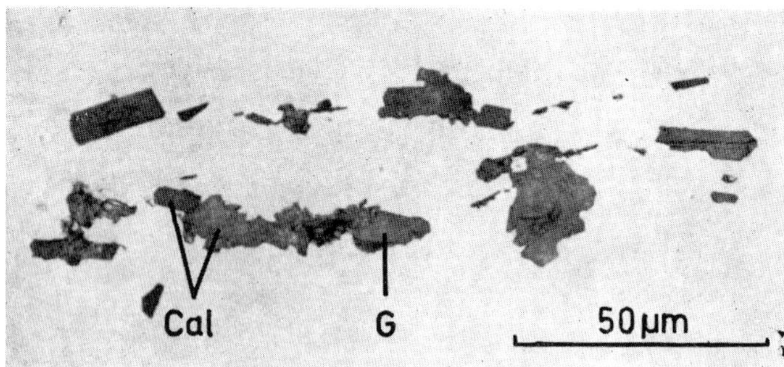

× 800

STEEL TYPE Arc-remelted electric steel. From rolled bar
STEEL ANALYSIS, % 0·35 C, 0·25 Si, 0·40 Mn, 1·4 Cr, 3·0 Ni, 0·50 Mo
INCLUSIONS, % (G) Spinel with MgO as only binary metal, (Cal) Calcium aluminate with 3 MnO, 14 CaO and 78 Al_2O_3. This phase is probably the aluminate $CaO{\cdot}2Al_2O_3$ (CA_2, section G 3) and has partly crystallized as idiomorphic crystals with a rectangular section

91 Oxide inclusions in steel: spinel and calcium aluminate, which is partly glassy, partly crystalline

F 4 $MgO-SiO_2-Al_2O_3$: Enstatite, protoenstatite, clinoenstatite

Chemical formula $MgO \cdot SiO_2$

Modifications and their relation to inclusions

Three modifications of $MgO \cdot SiO_2$ are known: enstatite, protoenstatite, and clinoenstatite. The transformations are influenced by

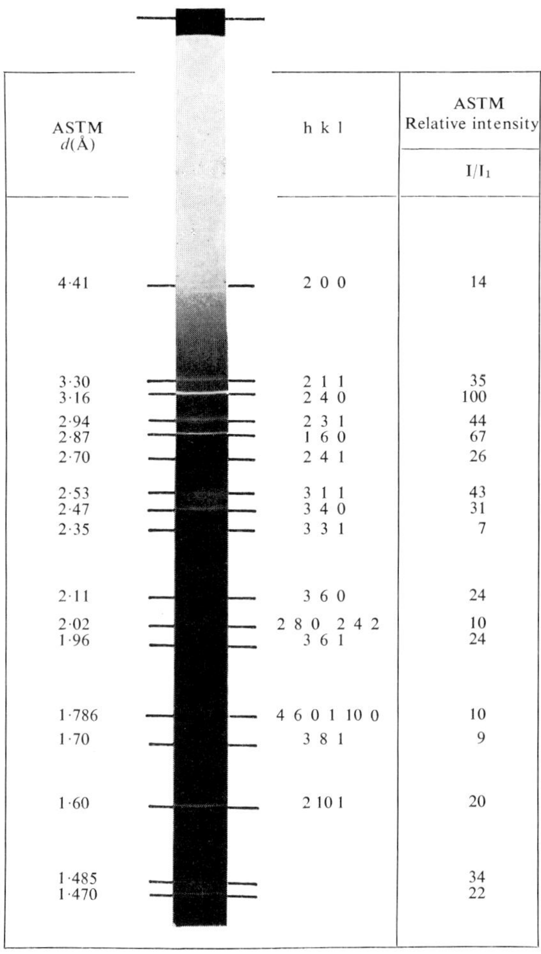

ASTM d(Å)	h k l	ASTM Relative intensity I/I_1
4·41	2 0 0	14
3·30	2 1 1	35
3·16	2 4 0	100
2·94	2 3 1	44
2·87	1 6 0	67
2·70	2 4 1	26
2·53	3 1 1	43
2·47	3 4 0	31
2·35	3 3 1	7
2·11	3 6 0	24
2·02	2 8 0 2 4 2	10
1·96	3 6 1	24
1·786	4 6 0 1 10 0	10
1·70	3 8 1	9
1·60	2 10 1	20
1·485		34
1·470		22

d-values, hkl-values and relative intensities for the different reflections are quoted from the ASTM index, data card 7–216

92 X-ray powder photograph of enstatite; monochromatic FeKα-radiation; Guinier camera

temperature and high pressure.[62] Enstatite is the stable low-temperature form and it transforms at 1 042°C to protoenstatite, the stable high-temperature phase.[63,64] However, with the usual cooling rate for steel, protoenstatite transforms to a third modification clinoenstatite, upon cooling, and this phase can exist in a metastable condition at room temperature. The clinoenstatite form usually transforms to the stable enstatite form after a short holding time in the temperature range between 865° and 1 042°C, but no transformation occurs at normal pressure if the temperature is below 865°C.[9] Due to these transformations, which are all possible within the normal temperature range for handling steel, it is difficult to establish definitely which of the modifications of $MgO \cdot SiO_2$ is present in an actual inclusion without a direct phase analysis. It should be noted that the nomenclature for the modifications differs in the standard reference literature (for example between Trojer,[9] Harders and Kienov[31], and Bragg and Claringbull[52]). During the present investigation it was possible to establish directly the modification of $MgO \cdot SiO_2$ by X-ray diffraction methods. For the inclusion shown in Fig.93, the phase could be definitely identified

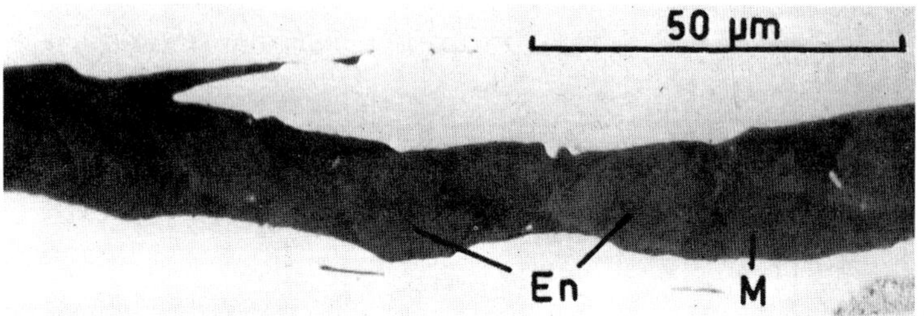

STEEL TYPE Silicon-killed electrosteel for railway rails
STEEL ANALYSIS, % 0·59 C, 0·39 Si, 1·53 Mn, 0·023 P, 0·18 S, 0·008 N
INCLUSIONS, % (En) Enstatite with 26 MgO, 34 MnO, 39 SiO_2, (M) Matrix with 4 MgO, 27 MnO, 7 CaO, 45 SiO_2, 14 Al_2O_3. Mean analysis 14 MgO, 31 MnO, 4 CaO, 43 SiO_2, 8 Al_2O_3

93 Oxide inclusions in steel: enstatite in glassy silicate matrix

as the enstatite modification. Consequently the authors have named the $MgO \cdot SiO_2$ phase in inclusions as enstatite, this, however, does not exclude the possibility of its being clinoenstatite in some of the inclusion examples discussed.

Enstatite is a possible reaction product between different refractories and slags, but its formation depends on the basicity of the slag (*see* Harders and Kienov,[31] p.676). It may not be stable together with periclase (F 2) under equilibrium conditions. The most common direct reaction product between magnesite bricks and acid slags is probably the silica-poor magnesium-silicate forsterite (F 5) and enstatite is first formed when all the periclase phase has reacted with the slag. In olivine and forsterite bricks enstatite may exist as a separate phase, as shown by the example in Fig.98.

Physical properties

Melting point: 1557°C
Density: 3·19
Microhardness: 1000 kp/mm^2
Some data for the two other polymorphs of $MgO{\cdot}SiO_2$, clino- and protoenstatite, are given in Trojer,[9] pp.334–335

Structure type

Orthorhombic; space group Pbca
a=18·20 Å, b=8·860 Å, c=5·204 Å. ASTM data card 7–216
Pyroxene type. Wyckoff II, chap. XII, text p.46
Monochromatic X-ray powder photograph, Fig. 92
Isostructural with 'grunerite', $FeO{\cdot}SiO_2$ (B 4)

The two modifications clinoenstatite and protoenstatite have ASTM data card respectively 3–0696 and 3–0523. A discussion of their structural relation to enstatite is given by Bragg and Claringbull.[52]

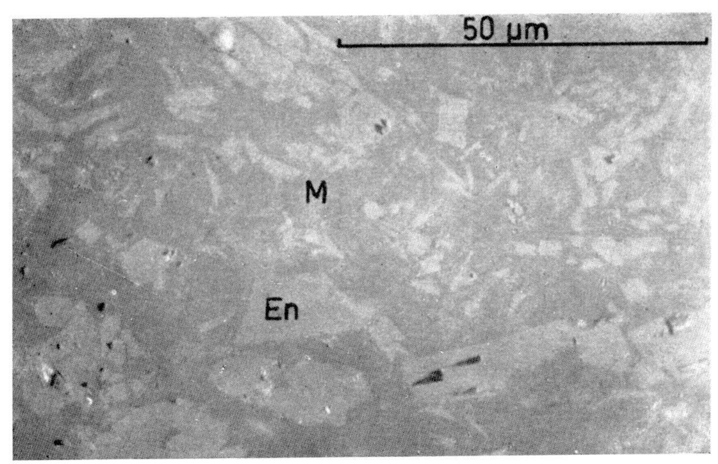

×1000

STEEL TYPE A broken roll
STEEL ANALYSIS, % 0·6 C, 0·3 Si, 0·6 Mn, 0·6 Cr
INCLUSION, % (En) Enstatite with 40 MgO, 58–64 SiO_2, (M) Matrix with 50–75 SiO_2, 20–33 Al_2O_3, trace MgO. Mean analysis 14 MgO, 66 SiO_2, 20 Al_2O_3

94 Oxide inclusions in steel: enstatite in glassy silicate matrix

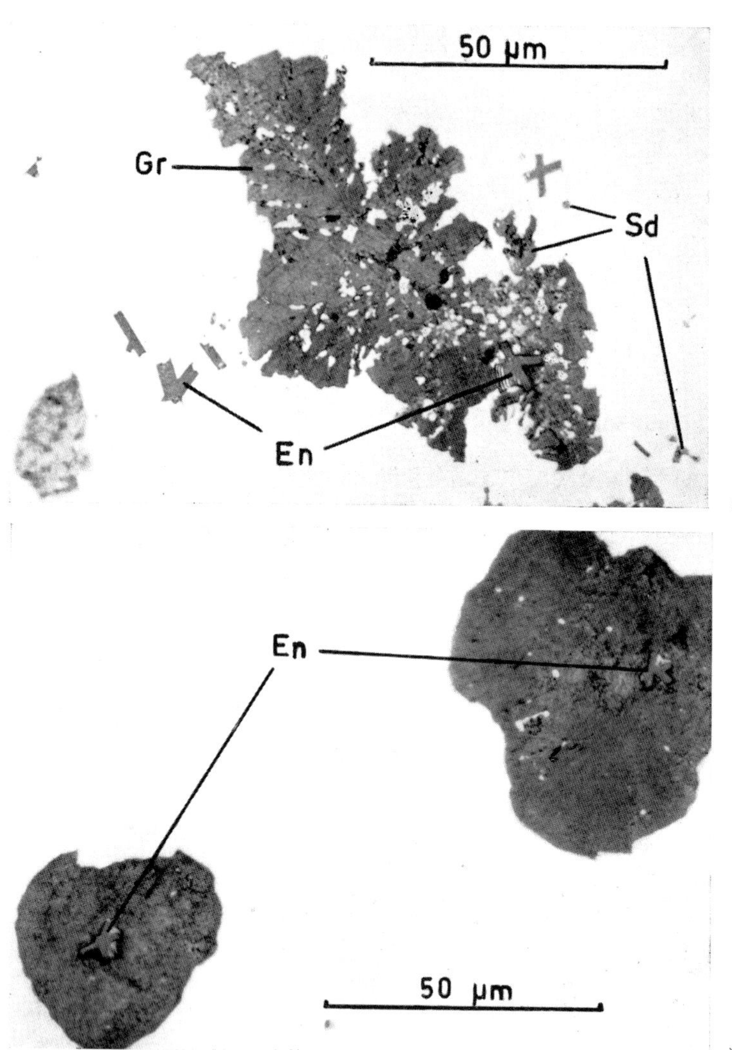

STEEL TYPE Nodular cast iron, inoculated with FeSi
STEEL ANALYSIS, % 3·53 C, 1·91 Si, 0·86 Mn, 0·011 P, 0·008 S, 0·02 Cr, 1·91 Ni
INCLUSIONS, % (En) Enstatite with 27 MgO, 67 SiO_2, 10 Al_2O_3, (Gr) Graphite nodules, (Sd) Ca-Mg-sulphide (K 11)

COMMENT The enstatite phase is situated directly in the iron phase as well crystallized rods, where part of the crystals seem to have acted as nuclei for the graphite nodule. Similar nodules with enstatite are also shown in Fig.95*b*

95a Oxide inclusions in nodular cast iron. Separate enstatite crystals in the iron phase and enstatite as nuclei (?) within the graphite nodules

95b Oxide inclusions in nodular cast iron: graphite nodules with nuclei of enstatite phase in their centres

Solid solubility
MgO in MgO·SiO_2 can be completely substituted by FeO, but pure FeO·SiO_2 is not stable at normal pressures (B 4). MgO can also be substituted to about 50 wt-% by CaO. The end member of this solid solutions range, MgO·CaO·2SiO_2, has the name diopside. MgO in MgO·SiO_2 may also be partly replaced by MnO. This is relevant to the formation of non-metallic inclusions since, according to Trojer,[9] MnO stabilizes the high-temperature modification protoenstatite down to room temperature (*see also* Tables XX and XXI, section G 4).

Microscopy and electron probe analysis
In microsections of inclusions containing enstatite the phase usually forms characteristic sections with some irregular grain boundaries. Such multiphase inclusions are shown in Figs.93 and 94. An example of monophase inclusions of enstatite is given in Fig.95, showing enstatite crystals in nodular cast iron. The enstatite particles are situated both within the iron phase, forming monophase inclusions of characteristic microsection, and also at the centres of the graphite nodules. This observation indicates that the enstatite crystals may have acted as nuclei for the nodules and thus one effect of magnesium additions on the formation of nodular cast iron is to produce compounds which nucleate favourable carbon growth. (Other mechanisms of nodule formation are also possible. It has been reported that the centres of nodules are increased in magnesium content without there being a separate magnesium phase.) In Fig.98 enstatite is shown together with the other magnesium silicate forsterite (F 5) in an olivine brick.

F 5 MgO–SiO_2–Al_2O_3: Forsterite

Chemical formula 2MgO·SiO_2

Physical properties
Melting point: 1 890°C
Density: 3·22
Microhardness: 1 200 kp/mm^2

Structure type
Orthorhombic; space group Pbnm (*see* note p.48, part I); a=4·76 Å, b=10·20 Å, c=5·99 Å. ASTM data card 4–0768
Chrysoberyl type. Wyckoff II, chap. VIII, text p.20 and chap. XII, text p.6.
Monochromatic X-ray powder photograph, Fig.96
Isostructural with tephroite (A 10), fayalite (B 3) and γ–Ca_2SiO_4 (G 4)

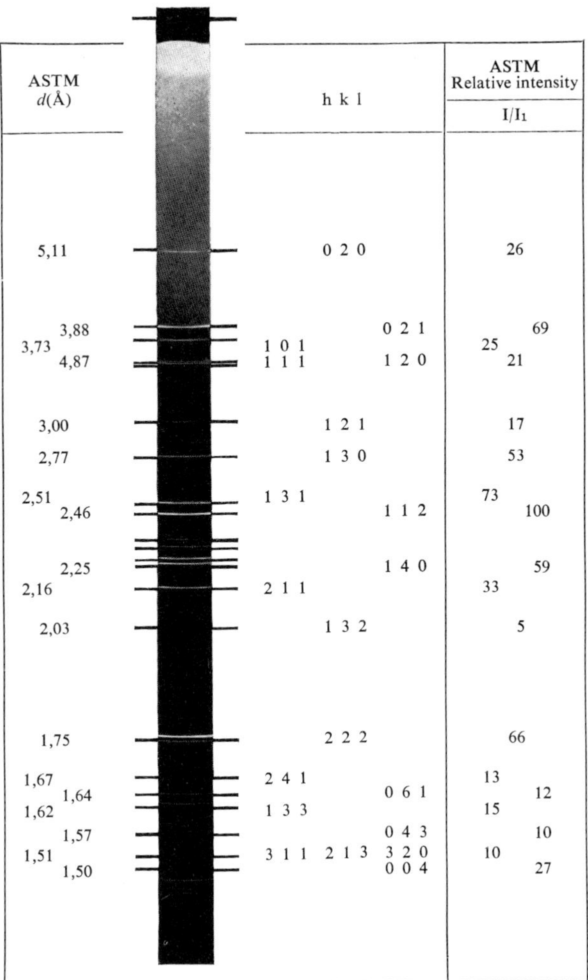

ASTM
d(Å)
h k l
ASTM
Relative intensity
I/I1
5,11
0 2 0
26
3,88
0 2 1
69
3,73
1 0 1
25
4,87
1 1 1
1 2 0
21
3,00
1 2 1
17
2,77
1 3 0
53
2,51
1 3 1
73
2,46
1 1 2
100
2,25
1 4 0
59
2,16
2 1 1
33
2,03
1 3 2
5
1,75
2 2 2
66
1,67
2 4 1
13
1,64
0 6 1
12
1,62
1 3 3
15
1,57
0 4 3
10
1,51
3 1 1
2 1 3
3 2 0
10
1,50
0 0 4
27

d-values, hkl-values and relative intensities for the different reflection are quoted from the ASTM index, data card 4–0768

96 X-ray powder photograph of forsterite; monochromatic FeKα-radiation; Guinier camera

Solid solubility

MgO in forsterite can be completely substituted by FeO (fayalite, B 3) and MnO (tephroite, A 10). Varying amounts of several other oxides are also soluble in forsterite. Of especial interest as a constituent of inclusions is CaO. Thus the γ-modification of $2CaO{\cdot}SiO_2$ is isostructural with forsterite.[65] In addition there exists an intermediary compound $MgO\cdot CaO\cdot SiO_2$, monticellite, which is a structural component of certain slags (G 4, J). Wide solid solubility of CaO in forsterite is possible, therefore, limited however by the temperature dependence of the structure of $2CaO\cdot SiO_2$ (the modifications α, α', β, and γ are discussed further in section G 4). All the substituted silicates belong to the olivine group but the name is usually retained for the silicates of the type $2(Mg,Fe)O\cdot SiO_2$.

Microscopy and electron probe analysis

Forsterite is the main structural component in forsterite and olivine refractories and is also often used as a bonding agent for periclase in magnesite bricks.[57] It is a common constituent in the reaction zone between MgO-refractories and acid slags. The forsterite phase in forsterite and olivine bricks fired in an oxidizing atmosphere is shown in Figs.97 and 98. The main structural component is forsterite with a white, fine-grained precipitate in the grain boundaries. This precipitate is rich in iron and chromium and is probably the double oxide chromite (C 4). At high magnifications enstatite, the magnesium silicate of higher silica content (F 4) may often also be found in the grain boundaries (Fig.98). In Fig.99 an

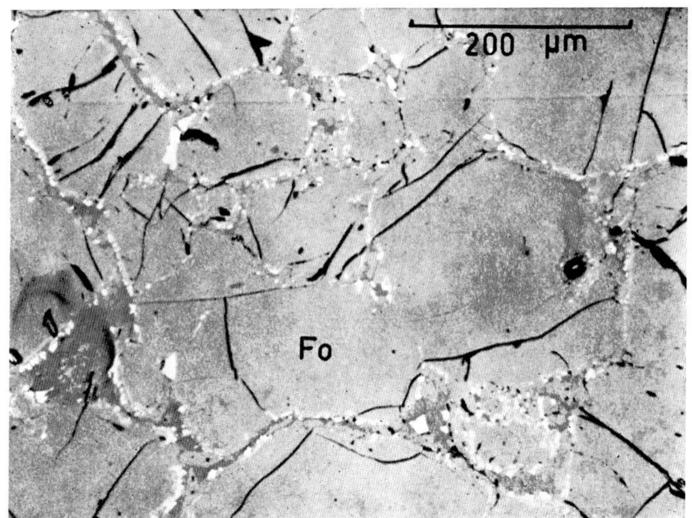

×150

(Fo) forsterite. Between the forsterite grains, a glassy silicate matrix is visible, which also contains enstatite (*see* Fig.98). The white phase along the boundaries of the forsterite grains is chromite (*see* C 4). Typical analysis for olivine bricks % 50–63 MgO, 1 CaO, 1–2 Al_2O_3 + Cr_2O_3, 30–38 SiO_2, 4–8 Fe_2O_3

97 Forsterite in an olivine brick, which has been fired in an oxidizing atmosphere inclusion from a vacuum degassed steel is shown, which consists of large, rounded forsterite crystals in a glassy silicate matrix. This inclusion was of exogenous origin and shows a striking resemblance to the brick structures of Figs.97 and 98. A hypothesis for its

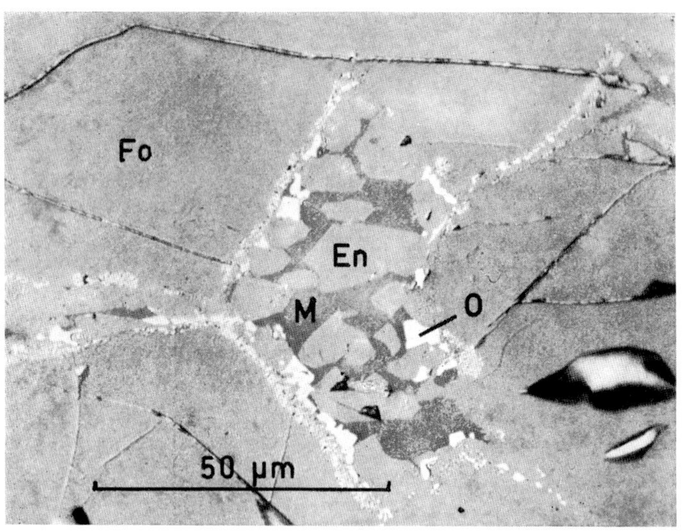

×800

STRUCTURE COMPONENTS, % (Fo) Forsterite with 48 MgO, 5 FeO, 44 SiO_2, (En) Enstatite with 29 MgO, 8 FeO, 60 SiO_2, (O) Chromite with 55–70 FeO, 5 Al_2O_3, 21 Cr_2O_3, (M) Matrix with 6 FeO, 2 MgO, 4 CaO, 72 SiO_2, 15 Al_2O_3

98 Forsterite, enstatite and chromite in olivine bricks which have been fired in an oxidizing atmosphere. This is the same sample, which was shown in Fig.97, showing a grain boundary section in a higher magnification

formation is that a small part of a forsterite brick was carried into the steel and this acted as an exogenous nucleus for further indigenous precipitation of Al_2O_3 and SiO_2. The forsterite grains have been partly dissolved due to the corresponding change in composition of the inclusion. Alternatively, the particle may have formed from an exogenous nucleus rich in magnesia, such as a particle or droplet of a dispersed MgO-rich silica slag. This nucleus may then have gained alumina and silica and come into the range of composition from which forsterite could have crystallized. This latter hypothesis is supported by the occurrence of a steel particle within the inclusion.

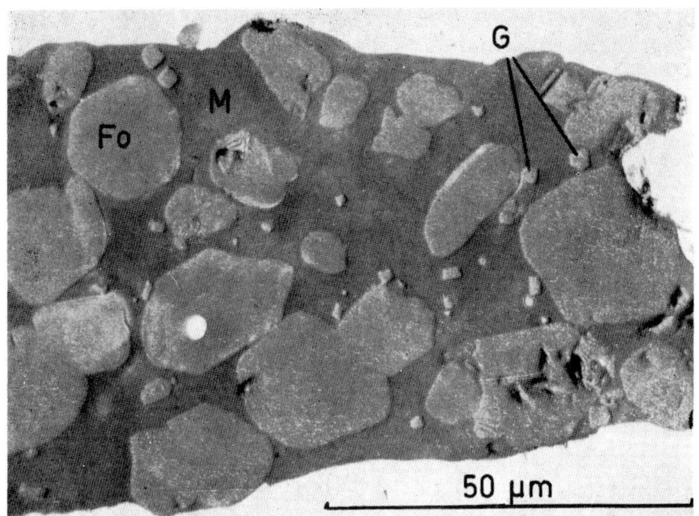

× 1 000

STEEL TYPE Silicon-killed carbon steel, which has been vacuum treated. Defect in the web of a railway rail
STEEL ANALYSIS, % 0·59 C, 0·39 Si, 1·53 Mn, 0·023 P, 0·018 S, 0·008 N
INCLUSIONS, % (Fo) Forsterite with 55 MgO, 8 MnO, 3 FeO, 37 SiO_2, (G) Spinel with 25 MgO, 7 MnO, 70 Al_2O_3, (M) Matrix with 7 MnO, 3 CaO, 6 MgO, 57 SiO_2, 22 Al_2O_3. White, round particle is steel. Mean analysis 1 FeO, 8 MnO, 34 MgO, 1 CaO, 10 Al_2O_3, 46 SiO_2
COMMENT The origin of this inclusion is probably a piece of MgO-rich refractory, which through the action of ladle slag has been dispersed in the steel and acted as an exogenous nucleus. The nucleus has grown indigenously and the MnO-, SiO_2 and Al_2O_3 contents have increased as compared with the original refractory

99 Oxide inclusions in steel: forsterite and spinel in a glassy silicate matrix

F 6 MgO–SiO_2–Al_2O_3: Cordierite

Chemical formula $2\,MgO{\cdot}2Al_2O_3{\cdot}5SiO_2$
Polymorphs
Three modifications of cordierite are known, designated α, β, and μ. α-cordierite is the stable high-temperature form above 830°C. It can easily be super-cooled. β-cordierite is the stable low-temperature polymorph and μ-cordierite is a metastable low-temperature polymorph. Little is known about β- and μ-cordierite, and only the properties of α-cordierite are mentioned in this report.

Physical properties
Melting point: 1460°C
Density: 2·55

Structure type

Orthorhombic; space group Cccm

a=9·67 Å, b=17·03 Å, c=9·35 Å. ASTM data card 9–326

Monochromatic X-ray powder photograph, Fig.100

Isostructural with Mn-cordierite (A 13), Fe-cordierite (B 7) and Ca-cordierite (G 5)

Microscopy and electron probe analysis

No example of cordierite in non-metallic inclusions in steel has been found in the material investigated or in literature references.

F 7 MgO–SiO$_2$–Al$_2$O$_3$: Pyrope

Chemical formula $3MgO{\cdot}Al_2O_3{\cdot}3SiO_2$

Structure type

Cubic; space group Ia3d

a=11·525 Å. ASTM data card 2–1008

Granate type. Wyckoff I, chap. XII, text p.3

Isostructural with spessartite (A 11), almandine (B 6) and grossularite (G 5)

Monochromatic X-ray powder photograph of spessartite, *see* Fig.46, part I

Solid solubility

Several isostructural compounds are known, where MgO is substituted by MnO (spessartite), FeO (almandine) and CaO (grossularite), and a wide solid solubility exists between these different compounds.

Microscopy and electron probe analysis

No example of pure pyrope in steel inclusions has been found in the material available or in the literature. However, several inclusions with the feather-like, branched structure of spessartite (A 11) as an inclusion phase have been observed with considerable amounts of MgO in solid solution together with MnO and CaO. An example is shown in Fig.133, section G 5, and further examples have been given in Table XXIII. (The observation by the authors in section A 11, part I, that the spessartite phase in inclusions only holds small amounts of MnO, FeO or CaO in solid solution was therefore premature.) Like the isostructural phases spessartite and grossularite, pyrope probably only crystallizes slowly from the matrix after long heat-treatment of the inclusions.

ASTM d(Å)	h k l	ASTM Relative intensity I/I_1
8·54	1 1 0 2 0 0	100
4·91	3 1 0 0 2 9	50
4·66	0 0 2	30
4·09	1 1 2 2 0 2	80
3·37	3 1 2 0 2 2	80
3·13	2 2 2 4 0 2	80
3·07	5 1 1	80
2·64	5 1 2 4 2 2	50
2·42	6 2 0 0 4 0	50
2·34	0 0 4	50
2·10		50
1·94		40
1·87		50
1·80		50
1·69		65
1·59		50
1·46		30

d-values, hkl-values and relative intensities for the different reflections are quoted from the ASTM index, data card 9–326

100 X-ray powder photograph of α-cordierite; monochromatic FeKα-radiation; Guinier camera

F 8 MgO–SiO_2–Al_2O_3: Sapphirine

Chemical formula $4MgO{\cdot}5Al_2O_3{\cdot}2SiO_2$

Physical properties

Melting point: 1482°C, forming spinel and mullite

Density: 3·45

Structure type
Monoclinic
$a=9{\cdot}72$ Å, $b=14{\cdot}58$ Å, $c=10{\cdot}07$ Å, $\beta=100°13'$
ASTM data card 11–607

Comments
Sapphirine has not been observed in steel inclusions. It has been reported as a structural component of certain types of porcelain.

F 9 $MgO–SiO_2–Al_2O_3$: Phase of osumilite type

Chemical formula $MgO{\cdot}Al_2O_3{\cdot}4SiO_2$
Comments
According to Trojer[9] this phase is metastable but can be formed in the temperature range of 1 050°C to 1 250°C. It has not been observed in non-metallic steel inclusions. The X-ray data for a complicated isostructural mineral are given in ASTM index, data card 10–413. The structure of this phase is also mentioned in Wyckoff I, table 23, where it is reported to have a hexagonal symmetry.

F 10 $MgO–SiO_2–Al_2O_3$: Phase of petalite type

Chemical formula $MgO{\cdot}Al_2O_3{\cdot}8SiO_2$
Comments
According to Trojer,[9] this phase is metastable but can be formed in the temperature range of 900°C to 1 000°C.

It has not been observed in non-metallic steel inclusions. The X-ray data for an isostructural lithium mineral are given in ASTM index, data card 7–60. The structure of this phase is also mentioned in Wyckoff II, table p.23, where it is reported to have a monoclinic symmetry.

G. The CaO-SiO_2-Al_2O_3 system

G 1 General

Calcium has been regarded as a metal, which from a technical point of view is insoluble in iron. The general assumption is, therefore, that endogenous inclusions with CaO are not found and this must be regarded as a general rule valid for most inclusions. Recently, however, a slight solubility for calcium of liquid iron has been reported by Sponseller and Flinn,[66] who at 1600°C found a content of 0·032 wt-% Ca in pure iron. An increase in solubility was observed if also aluminium, carbon, nickel and silicon were present, and for a silicon content of 10·5 wt-% Si they found a solubility in the liquid iron phase of 0·36 wt-% Ca.

Even if exogenous sources are without doubt the main origin for inclusions containing CaO, the observations by Sponseller and Flinn may have important implications for such metallurgical operations as the deoxidation of steel with calcium alloys. Several alloys of calcium with iron, aluminium and titanium are used as deoxidants, the compositions of some are given in Table XIII. The free energy

TABLE XIII The composition of different types of deoxidation alloys with calcium (from Koch[68])

Type of alloy	Composition, wt-% Ca	Si	Fe	Al	Ti	C
75/2	2·2	73·6	21·6	1·2	0·15	...
75/5	5·8	75·4	16·1	1·4	0·10	...
75/10	11·0	81·7	4·8	1·1	0·08	...
55/14 Ti Al	13·9	56·1	16·85	1·97	8·91	0·38
55/17 Ti	17·3	54·6	17·2	0·2	7·44	0·64
75/20	17·9	77·8	2·1	0·66	0·06	...
60/30	31·1	60·2	5·7	1·4	...	...

of formation of CaO is the lowest among the common elements, as shown by the diagrams in Fig.80. Calcium therefore has the highest affinity for oxygen and should be the most effective deoxidant, but in practice the deoxidation process is complicated.

Calcium metal has a low boiling point (810°C), which makes its addition to molten steel difficult. The change of entropy for the formation of CaO from the elements is given by the relation

$$\frac{d(\Delta G^\circ)}{dT} = -\Delta S^\circ$$

According to Fig.80, the ΔG° curve has a positive derivative with relation to temperature and the entropy change ΔS° will therefore be negative for the formation of CaO. This means that deoxidation should be carried out at as low a temperature as possible. The (ΔG°, T) curve also has a steepening slope above the boiling point of calcium. For these reasons deoxidation with calcium in the temperature range of the molten steel is not so favourable as the free energy diagrams of Fig.80 might indicate.

On the other hand, the vapour pressure of calcium-silicon alloys at the temperature of deoxidation is so high that they partly volatilize during the deoxidation operation. The vapour pressure of silicon is therefore increased which means that its activity and deoxidizing effect increase. The detailed mechanism for the deoxidation of steel with calcium alloys is still uncertain and it is not clear whether calcium acts directly as a deoxidizer or if it takes part in the deoxidizing process only by increasing the vapour pressure of silicon.[67,68] Steel inclusions with a CaO-component may, however, be formed as a result of the deoxidizing process.

CaO is a component of most refractories used in steel practice, dolomite and aluminate cement (Table IX) are particularly high in calcia. Such materials are therefore possible sources for exogenous inclusions containing calcium, which may result from reactions between the slag or the molten steel and the refractories. As shown in Table X a great number of phases with CaO as one component have been identified in used refractories of different types.

CaO is usually called lime in commercial practice but calcia conforms to the systematic nomenclature for oxides and has been preferred for use in the present context. It is the principal slag former in basic steel processes and is an important component of both acid and basic slags. As a guide to possible inclusion sources, Table XIV gives the analyses of slags typical of the more important steelmaking processes. (These analyses must be regarded as only giving the relative distributions of the major phases. For further details references 69–71, from which the table was compiled, should be consulted). The most important equilibrium diagrams for different slag systems have been summarized by Muan and Osborn.[53] These authors have considered the thermodynamic basis of the diagrams for several important systems and Rein and Chipman[72] have contributed a more thorough discussion of this aspect. Reactions between slags and different basic refractories have been summarized in the two ISI Special Reports, Nos.74 and 87[54,55] and some reactions involving dolomite and magnesite have been studied in detail by Brunner.[60] It should be emphasized that slags and their reaction products with

TABLE XIV **Typical analyses, wt-%, for different slag types used in steelmaking. The values have mainly been taken from the Sandviken reference book[69]**

Steel type	CaO	MgO	Al_2O_3	Cr_2O_3	SiO_2	FeO	MnO	Fe_2O_3	P_2O	S	Others
Acid OH	5		5	...	50	25	15		+		
Basic OH	48	4	4	...	20	15	7		2		
Bessemer slag	1	0·5	2–4	...	48–60	6–10	30–40		0·01		
Thomas slag	45–55	2–4	1–2	...	7–10	7–12	4–6	5–6	16–20		
Electric arc, slag I	47	...	...	...	19	18	6		0·15	0·12	
Electric arc, slag II	68	...	...	...	17	1	0·1		0·01	0·31	
LD-slag[70]	45–46	5	1·5–2·5	...	15–16	20	6		1·1–1·5		
Kaldo[72]	54	5	...	...	9	8	3		17		4

refractories are the principal exogenous sources of calcium-containing inclusions.

In the following sections different types of oxide inclusions with a calcium component are illustrated and discussed. These are both inclusions with an entirely exogenous origin and those which have an exogenous nucleus which has grown indigenously in the steel bath. The former type has the main components of CaO, MgO and Al_2O_3 (sometimes also a small amount of SiO_2), whereas the latter contain in addition considerable amounts of MnO and SiO_2.

Oxide inclusions with calcia can also be formed as a result of the welding of steel. These are located mainly in the fusion line of the welds. They may either have been present in the steel before the welding and merely increased in size due to coalescence during the welding operation or they may have been formed directly as a result of the welding process. For instance, it is possible to form inclusions by a reaction between calcium-metasilicate flux and manganese in the steel.[73] Therefore a thorough determination of the composition of such oxide inclusions with calcia is necessary in order to establish their origin.

A convenient basis for a discussion of oxide inclusions with calcia is the ternary system $CaO–SiO_2–Al_2O_3$. This system is more complicated than the systems $MnO–SiO_2–Al_2O_3$, $FeO–SiO_2–Al_2O_3$, $MgO–SiO_2–Al_2O_3$ and corresponding chromia systems, which have been discussed earlier in this report. This is evident from Table XV, which gives a summary of the stoichiometric composition and the names of the different phases existing in the $CaO–SiO_2–Al_2O_3$ system. The great number of phases and the many modifications are a characteristic feature of the system. The stoichiometric composition of the phases has been schematically summarized in Fig.101 which should be compared with the corresponding summaries in Figs.1, 57, 66, 67 and 81 for the other oxide systems relevant to inclusion formation. Fig.102 shows a ternary diagram of the system with a projection of the different liquidus surfaces. Some observations of phases belonging to the more general system (Ca, Mg, Mn, Fe)O–SiO_2–(Al, Cr, Fe)$_2O_3$ have also been included in the discussion. The $CaO–SiO_2–Al_2O_3$ system gives only a simplified view of the true compositions of the oxide phases with calcia but is none the less fundamental.

In the present discussion of the $CaO–SiO_2–Al_2O_3$ system, it has not been possible to discuss all the phases separately, as has been done in previous sections. Many of them have been discussed in groups, always related to their appearance in non-metallic steel inclusions. Most of the examples have been chosen from the

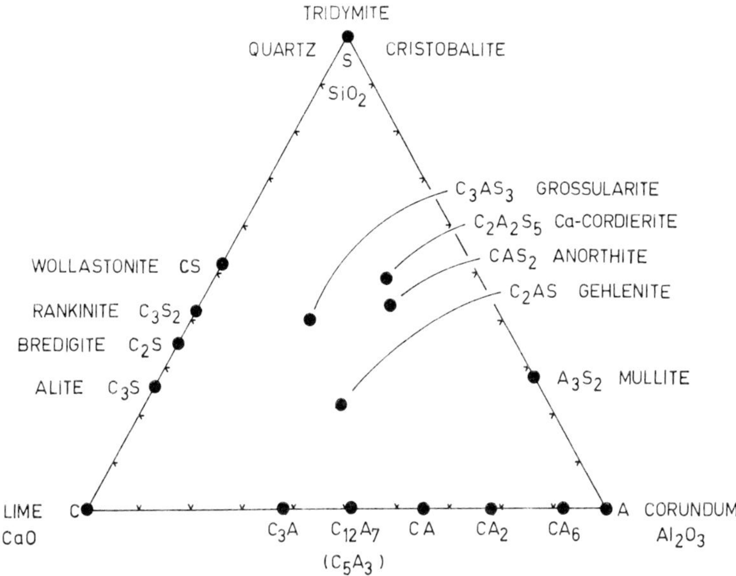

101 Schematic representation of the pseudoternary system $CaO–SiO_2–Al_2O_3$, showing the stoichiometric composition of the different phases which have been reported, $CaO = C$, $SiO_2 = S$, $Al_2O_3 = A$

authors' sources of steel samples, but for some of the phases reference has also been made to good photomicrographs in the literature.

Inclusions rich in calcia are more difficult to study experimentally both in the microscope and in the electron probe analyser than the other types of oxide inclusions. There are several reasons for this. For instance, the optical properties are similar between the different phases, which are not easily distinguished in the microscope. Moreover, phases with CaO are often glassy and only crystallize with difficulty. They are also often sensitive to water which makes the preparation of the microsections more difficult than for other inclusions and polishing and grinding must be done with waterfree liquids, e.g. ethanol.

In spite of these difficulties, it is essential to establish whether inclusions contain any calcium and if so to make also as complete an investigation as possible. As already mentioned, the sources for such inclusions are exogenous and with a detailed knowledge of their structure and composition it is often possible to localize their origin. This is discussed further in section G 6.

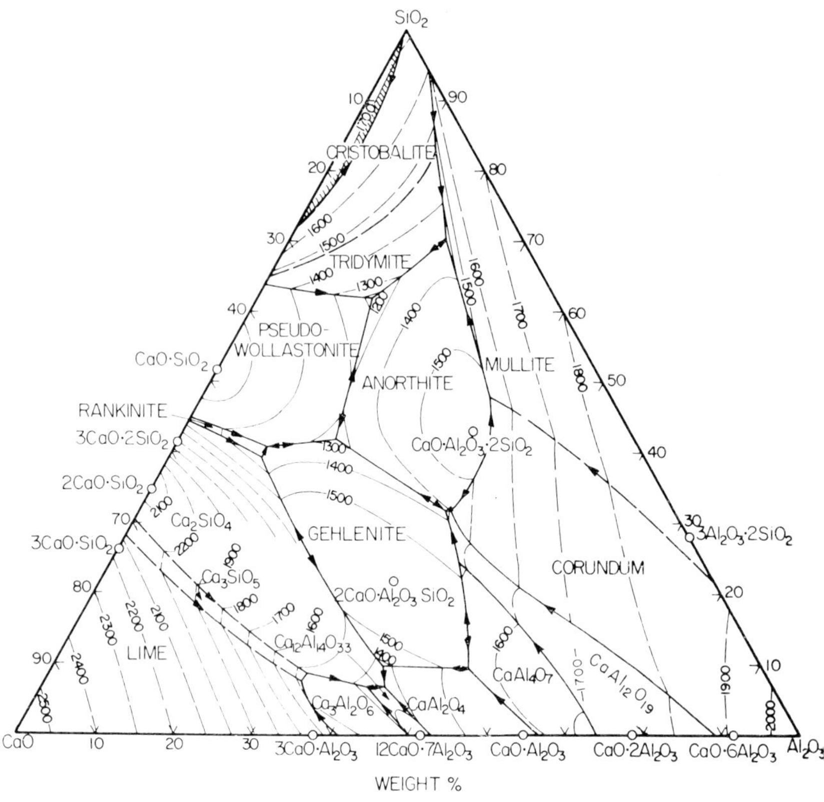

102 Phase diagram for the system $CaO–SiO_2–Al_2O_3$. Projection of the liquidus surface of the system. From Muan and Osborn,[53] Fig.80, p.95

G 2 $CaO–SiO_2–Al_2O_3$: CaO

Chemical formula CaO

Physical properties

Impure CaO, lime, is produced by the burning of different types of calcium carbonate. This oxide varies in purity and is hygroscopic. Recently remelting of lime on a commercial scale has been possible,[74] resulting in a purer oxide with a purity of 99·75 wt-% CaO. This oxide is also much less hygroscopic than the normal product. The physical properties given below are those of the pure oxide (calcia).

Melting point: 2·570°C
Density: 3·341
Microhardness: 400 kp/mm^2

Structure type

Cubic; space group Fm3m

$a=4{\cdot}8075\pm0{\cdot}0007$ Å, ASTM data card 4–0777

NaCl-type. Wyckoff I, chap. III, text p.1, table p.12

Isostructural with manganosite (A 6), wüstite (B 2, D 2) and periclase (F 2)

Monochromatic X-ray powder photographs of the isostructural compounds manganosite and wüstite have been given in Figs.19 and 75 respectively (part I)

Solid solubility

A complete solid solubility exists in the CaO–MnO system. CaO also dissolves about 10 wt-% of FeO at 1 150°C and probably considerable amounts of SrO.[9] The solid solubility for MgO, however, is small.

Microscopy and electron probe analysis

Dolomite is an important raw material containing CaO used for refractories in steel practice. It is a mineral with the ideal composition $CaMg(CO_3)_2$ but dolomite usually also contains varying amounts of FeO, Al_2O_3 and SiO_2. The microstructure of dolomite is shown in Fig.103 and consists of rather large, homogeneous

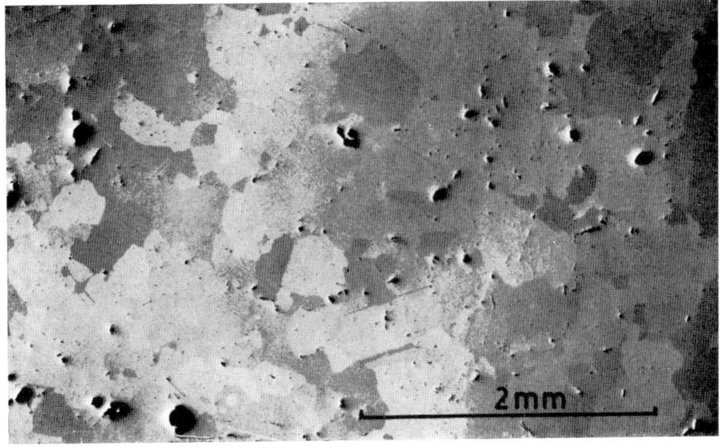

×20

103 Microsection of the mineral dolomite in a low magnification. Dolomite is a double carbonate with the approximate composition $CaMg(CO_3)_2$. The two carbonates of calcium and magnesium have a complete solid solubility. The dolomite grains therefore consist of one phase only and they are homogeneous. Small amounts of FeO, Al_2O_3, SiO_2 and sometimes also other oxides are usually present in solid solution

grains of calcium magnesium carbonate (the two phases $CaCO_3$ and $MgCO_3$ show a complete solid solubility). Several types of refractories are fabricated from dolomite according to different technical processes (for an outline see Harders and Kienov[31]). In all these procedures the dolomite phase is burned. This results in decomposition of the monophase carbonate to gaseous CO_2 and a fine-grained, two-phase mixture of the oxides CaO and MgO, which are only slightly soluble in each other. Burnt dolomite is shown at different magnifications in Fig.104 and also in Fig.85. The structure is slightly porous with small, grey, hard periclase grains and larger, darker grains of calcia. In the grain boundaries and at interstices between these two main structural components two more phases are visible, white calcium ferrite (J) and a greyish calcium silicate (G 4). The calcia phase is hygroscopic and is attacked by water vapour after some hours exposure in air, resulting in the formation of $Ca(OH)_2$ nuclei on the calcia surface. This is shown in Fig.104*c*, and it is therefore easy to identify the calcia phase.

As already mentioned, a purer form of calcia has been prepared by remelting of CaO in an electric arc. Colour photographs, showing the appearance of the remelted, non-hygroscopic calcia have been published by Fischer and Hoffman.[74]

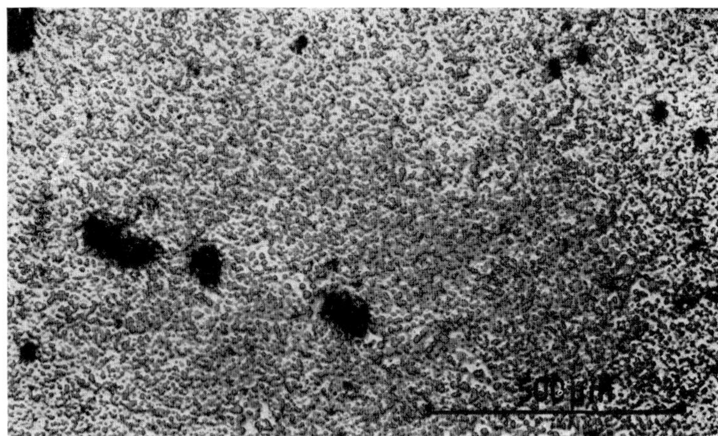

×70

104a The microstructure of burnt dolomite in a low magnification. During the firing of dolomite, carbon dioxide is liberated and the monophase, homogeneous, coarse carbonate structure is transformed to a more fine-grained, partly sintered, two-phase structure of calcia (lime, CaO) and periclase (MgO, F 2) which are insoluble in each other. The structure is also slightly porous. Typical analysis for burnt dolomite bricks, % 50–60 CaO, 36–40 MgO, 0·5–3 Al_2O_3, 0·5–5 SiO_2, 0·5–3 Fe_2O_3

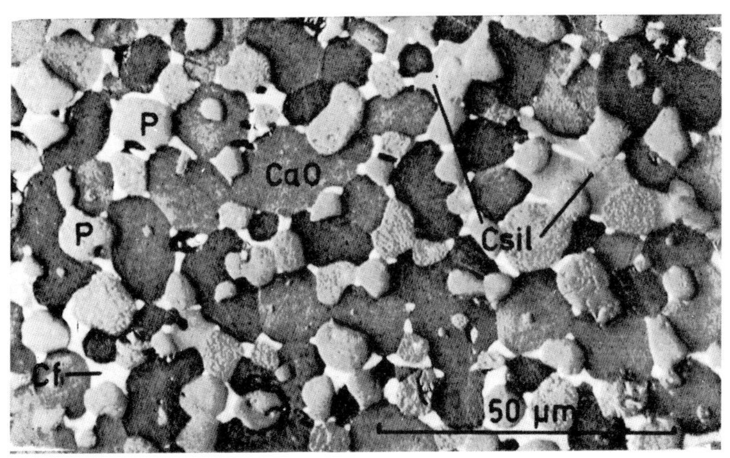

×800

104b Burnt dolomite in a high magnification, photographed immediately after the preparation of the microsection. Four different phases are visible. The two main phases are the rounded, light-grey, hard periclase grains (P) and the larger, darker irregular calcia (lime) grains (CaO). In some grain boundaries a white phase, calcium ferrite (Cf) is visible and in interstices between the periclase and calcia grains also a grey, softer phase appears, which is a calcium silicate, probably C_3S (C sil, G 4). The calcia phase is hygroscopic

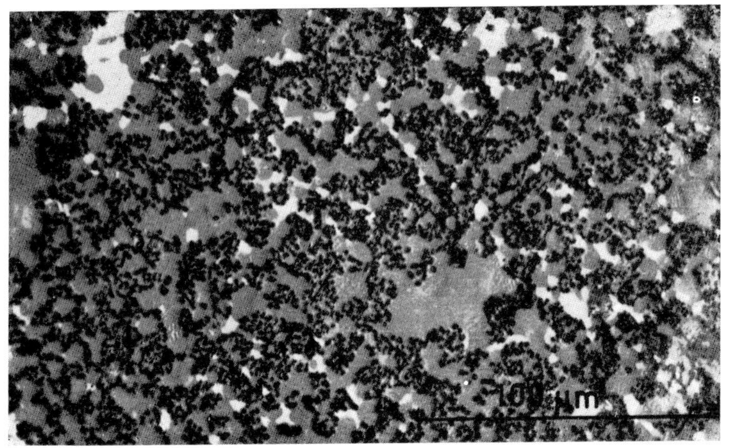

×800

104c Burnt dolomite, microsection photographed after 5 h in air. The hygroscopic calcia (lime) phase has been attacked by the moisture of the air, resulting in a great number of nuclei of $Ca(OH)_2$ all over the calcia phase. The periclase, calcium ferrite and calcium silicate phases have not been attacked

Dolomite bricks are often stabilized against moisture by a thin tar layer (tar-bonded dolomite bricks). Also chemically stabilized types exist, in which most of the calcia phase has been transformed to a calcium silicate, $3CaO{\cdot}SiO_2$ (G 4). This is achieved by means of an addition of magnesium silicate to the raw dolomite before burning. The different calcium silicates influence the properties of burnt dolomite due to their polymorphic changes; this is discussed in section G 4.

CaO is thus a common phase in refractories used in steelmaking, but it is not usually found as a separate phase in steel inclusions. The reason for this seems to be that exogenous inclusions from these refractories change their composition during their further indigenous growth in the steel bath. MnO, SiO_2, and Al_2O_3 are precipitated on the exogenous nucleus, or alternatively the CaO-phase reacts with other components in slag, refractory or steel forming other CaO-rich phases. This is also the reason why lime, which is added to steel slags as a separate phase, is not found as a phase in steel inclusions. It reacts readily in the molten slag with other components forming calcia compounds such as silicates or aluminates.

G 3 CaO–SiO_2–Al_2O_3: The calcium aluminates

Chemical formulae

Six intermediate phases have been reported to exist in the binary system CaO–Al_2O_3 (Table XV). Their stoichiometric compositions are usually given as

$3CaO{\cdot}Al_2O_3$ (C_3A)*
$12CaO{\cdot}7Al_2O_3$ ($C_{12}A_7$)
$5CaO{\cdot}3Al_2O_3$ (C_5A_3)
$CaO{\cdot}Al_2O_3$ (CA)
$CaO{\cdot}2Al_2O_3$ (CA_2)
$CaO{\cdot}6Al_2O_3$ (CA_6)

It is doubtful if the pure compound C_5A_3 exists as a separate phase, it may be a ternary oxide compound containing iron[9] and is not included in recent equilibrium diagrams. The pseudobinary equilibrium diagram for the system CaO–Al_2O_3 is given in Fig.105.

Physical properties

The melting points, densities and microhardness values for the different calcium aluminates have been summarized in Table XVI.

* For brevity the symbols *C*, *A* and *S* have been used to denote the phases CaO Al_2O_3 and SiO_2.

TABLE XV **Phases reported to exist in the $CaO–SiO_2–Al_2O_3$–system. The different polymorphs and the mineral names of the different phases have also been given**

Name	Chemical formula	Notation	Stoichiometric comp., wt-% CaO	SiO₂	Al₂O₃	Section
Corundum	Al_2O_3	...	...	...	100	A 2
Cristobalite	SiO_2	...	...	100	...	A 3
Tridymite	SiO_2	...	...	100	...	A 4
Quartz	SiO_2	...	...	100	...	A 5
Calcium oxide	CaO	...	100	...	...	G 2
Mullite	$3Al_2O_3.2SiO_2$	A_3S_2	...	28	72	A 8
...	$3CaO.Al_2O_3$	C_3A	62	...	38	G 3
...	$12CaO.7Al_2O_3$	$C_{12}A_7$	48	...	52	G 3
...	$5CaO.3Al_2O_3$	C_5A_3	48	...	52	G 3
...	$CaO.Al_2O_3$	CA	35	...	65	G 3
...	$CaO.2Al_2O_3$	CA_2	22	...	78	G 3
...	$CaO.6Al_2O_3$	CA_6	8	...	92	G 3
Pseudo-wollastonite	α-$CaO.SiO_2$	CS	48	52	...	G 4
Wollastonite	β-$CaO.SiO_2$	CS	48	52	...	G 4
Para-wollastonite	β'-$CaO.SiO_2$	CS	48	52	...	G 4
Rankinite	$3CaO.2SiO_2$	C_3S_2	58	42	...	G 4
...	α-$2CaO.SiO_2$	C_2S	65	35	...	G 4
Bredigite	α'-$2CaO.SiO_2$	C_2S	65	35	...	G 4
Larnite	β-$2CaO.SiO_2$	C_2S	65	35	...	G 4
Shannonite	γ-$2CaO.SiO_2$	C_2S	65	35	...	G 4
Alite	$3CaO.SiO_2$	C_3S	74	26	...	G 4
Anorthite	$CaO.Al_2O_3.2SiO_2$	CAS_2	20	43	37	G 5
Gehlenite	$2CaO.Al_2O_3.SiO_2$	C_2AS	41	22	37	G 5
Grossularite	$3CaO.Al_2O_3.3SiO_2$	C_3AS_3	37	40	23	G 5
Ca-cordierite	$2CaO.2Al_2O_3.5SiO_2$	$C_2A_2S_5$	18	49	33	G 5

TABLE XVI **Physical properties of the different calcium aluminates**

Phase	Melting point °C	Density	Microhardness kp/mm²
C_3A	1535	3·04	...
$C_{12}A_7$	1455	2·83	...
CA	1605	2·98	930
CA_2	~1750	2·91	1100
CA_6	~1850	3·38	2200

The information has been collected from the books of Trojer[9] and Muan and Osborn[53] except for the microhardness values, which have been determined by the authors.

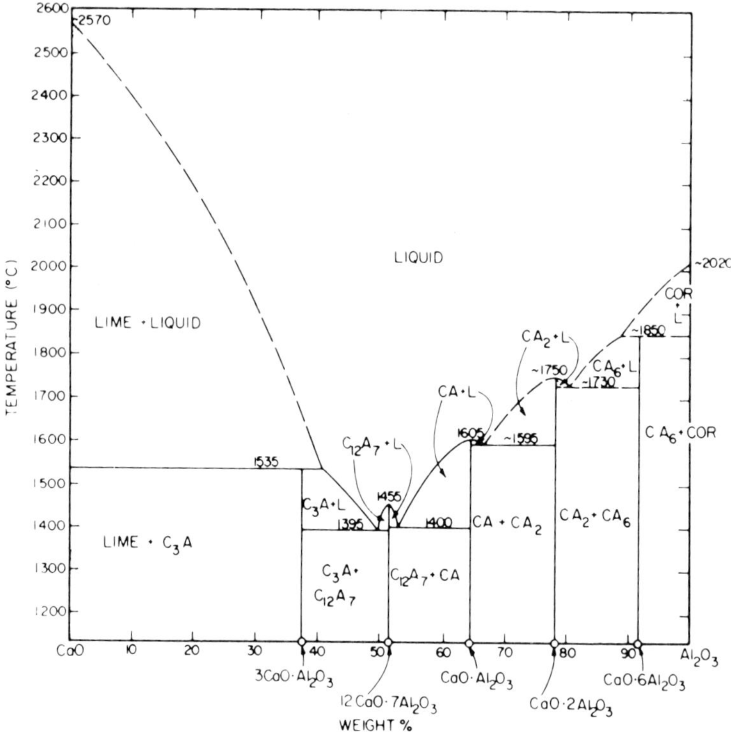

105 The equilibrium diagram for the pseudobinary system $CaO–Al_2O_3$. From Muan and Osborn,[53] Fig.26, p.43

TABLE XVII Crystallographic characteristics of the different calcium aluminates

Phase	System	Space group	Parameter (A)	Reference W	ASTM
C_3A	cubic	Pm3m	7·639	XI, text p. 48	8–6
$C_{12}A_7$	cubic	Id3d	11·95	XI, text p.49	9–413
CA	unknown	...	...	...	1–0888
CA_2	monoclinic, ev. also tetrag. modif.	C2/c (?)	...	XI, table p.26	7–82
CA_6	hexag.		$a = 5·54$ $c = 21·82$		7–85

Structure type

Available information about the crystal chemistry of the different calcium aluminates has been summarized in Table XVII.

Monochromatic X-ray powder photographs of C_3A, $C_{12}A_7$, CA and CA_2, *see* Figs.106–109.

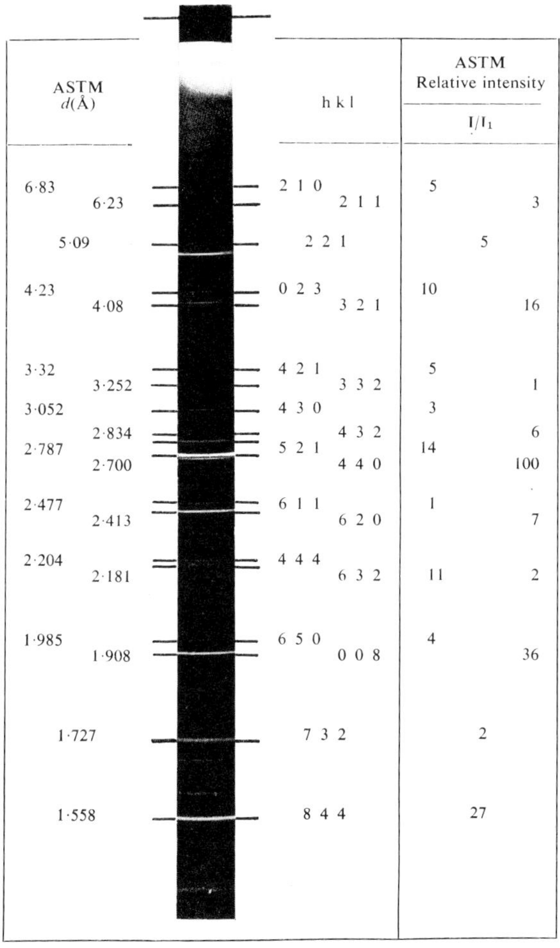

d-values, hkl-values and relative intensities for the different reflections are quoted from the ASTM index, data card 8–6

106 X-ray powder photograph of $3CaO.Al_2O_3$ (C_3A); monochromatic FeKα-radiation; Guinier camera

ASTM d(Å)	h k l	ASTM Relative intensity I/I_1
4·89	2 1 1	95
4·24	2 2 0	5
3·79	3 1 0	15
3·204	3 2 1	25
2·998	4 0 0	45
2·680	4 2 0	100
2·556	3 3 2	17
2·446	4 2 2	50
2·350	5 1 0	9
2·189	5 2 1	40
2·054	5 3 0	9
1·945	6 1 1	30
1·850	5 4 1	7
1·767	6 3 1	5
1·730	4 4 4	11
1·695	7 1 0	7
1·662	6 4 0	30
1·630	7 2 1	9
1·601	6 4 2	30
1·522	7 3 2	5
1·498	8 0 0	5
1·475	8 1 1	7

d-values, hkl-values and relative intensities for the different reflections are quoted from the ASTM index data card 9–413

107 X-ray powder photograph of 12CaO.7Al$_2$O$_3$ ($C_{12}A_7$); monochromatic FeKα-radiation; Guinier camera

Solid solubility

In the different calcium aluminate phases in steel inclusions, both CaO and Al$_2$O$_3$ can be substituted to a varying extent. Thus MnO, FeO and MgO have been found partly to replace CaO and Cr$_2$O$_3$ and Fe$_2$O$_3$ may partly substitute Al$_2$O$_3$. These different oxides are

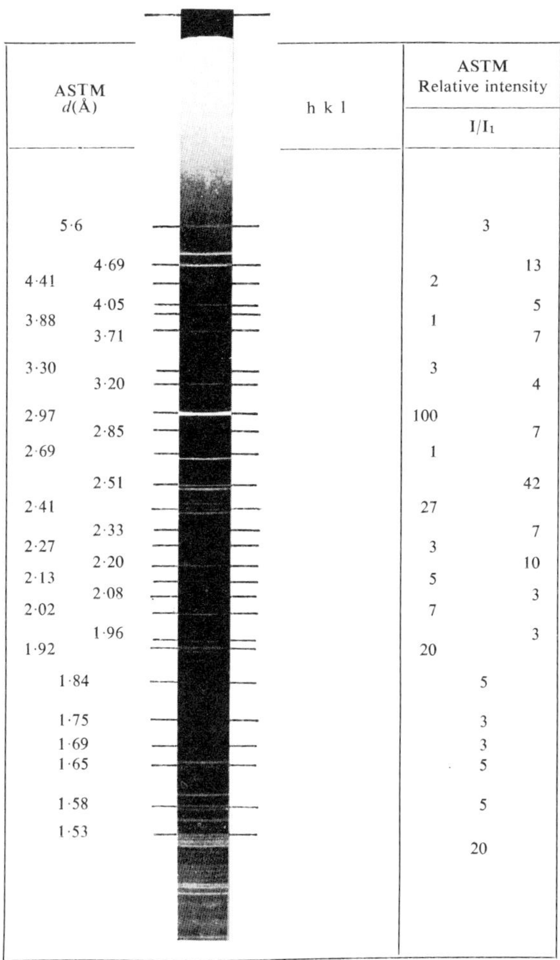

ASTM d(Å)	h k l	ASTM Relative intensity I/I₁
5·6		3
4·69		13
4·41		2
4·05		5
3·88		1
3·71		7
3·30		3
3·20		4
2·97		100
2·85		7
2·69		1
2·51		42
2·41		27
2·33		7
2·27		3
2·20		10
2·13		5
2·08		3
2·02		7
1·96		3
1·92		20
1·84		5
1·75		3
1·69		3
1·65		5
1·58		5
1·53		20

d-values and relative intensities for the different reflections are quoted from the ASTM index, data card 1–0888

108 X-ray powder photograph of $CaO.Al_2O_3$ (*CA*); monochromatic FeKα-radiation; Guinier camera

often found dissolved in the aluminate phases in steel inclusions, which may also dissolve oxides of titanium, zirconium, vanadium and other metals. Most of the limits for solid solubility are not accurately known.

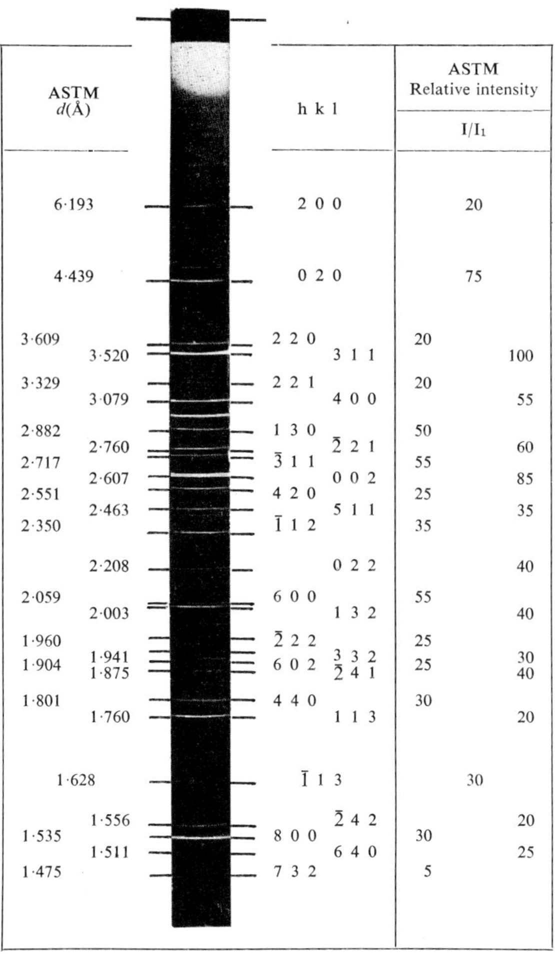

ASTM d(Å)	h k l	ASTM Relative intensity I/I_1
6·193	2 0 0	20
4·439	0 2 0	75
3·609	2 2 0	20
3·520	3 1 1	100
3·329	2 2 1	20
3·079	4 0 0	55
2·882	1 3 0	50
2·760	$\bar{2}$ 2 1	60
2·717	$\bar{3}$ 1 1	55
2·607	0 0 2	85
2·551	4 2 0	25
2·463	5 1 1	35
2·350	$\bar{1}$ 1 2	35
2·208	0 2 2	40
2·059	6 0 0	55
2·003	1 3 2	40
1·960	$\bar{2}$ 2 2	25
1·941	3 3 2	30
1·904	6 0 2	25
1·875	$\bar{2}$ 4 1	40
1·801	4 4 0	30
1·760	1 1 3	20
1·628	$\bar{1}$ 1 3	30
1·556	$\bar{2}$ 4 2	20
1·535	8 0 0	30
1·511	6 4 0	25
1·475	7 3 2	5

d-values, hkl-values and relative intensities for the different reflections are quoted from the ASTM index, data card 7–82

109 X-ray powder photograph of $CaO.2Al_2O_3$ (*CA_2*); monochromatic FeKα-radiation; Guinier camera

Microscopy and electron probe analysis

Oxide inclusions with calcium aluminates in steel ingots usually have a spherical shape. The inclusions have been spherical droplets in the steel bath, which then solidify. Typical inclusions of this type

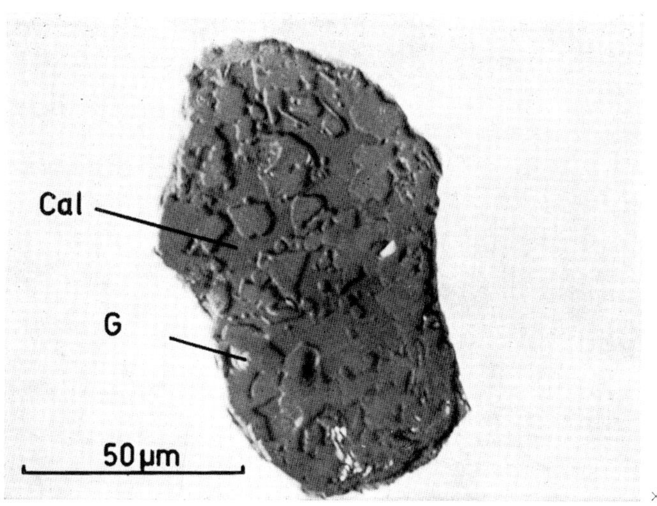

×600

STEEL TYPE Carbon steel, CaSi deoxidized. From ingot
STEEL ANALYSIS, % 0·26 C, 0·48 Si, 0·84 Mn, 0·022 P, 0·011 S, 0·038 Al, 47 ppm O
INCLUSIONS, % (Cal) Glassy calcium aluminate with 19 CaO, 76 Al_2O_3, trace of FeO, (G) Spinel with 21 MgO, 70 Al_2O_3, trace of FeO and MnO. Mean analysis 5 CaO, 17 MgO, 78 Al_2O_3

110 Oxide inclusions in steel: glassy calcium aluminate matrix (appr. CA_2) with spinel

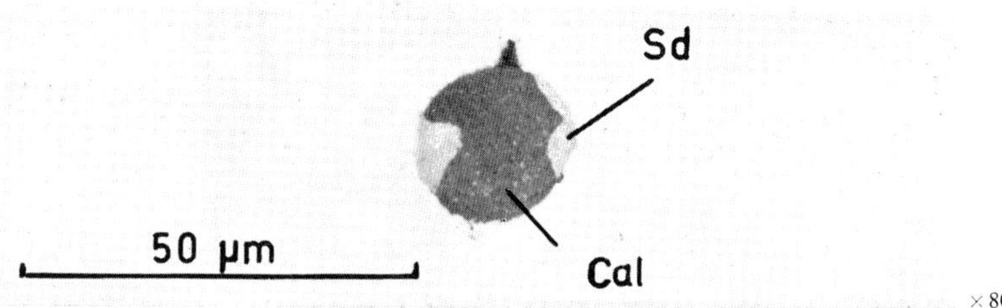

×800

STEEL TYPE Electrosteel, CaSi deoxidized, 6 ton ingot rolled to 50 mm sheet
STEEL ANALYSIS, % 0·17 C, 0·40 Si, 1·50 Mn, 0·016 P, 0·015 S, 0·011 N, 0·035 Nb, 0·020 Al (soluble), 0·016 Al (indissoluble)
INCLUSIONS, % (Cal) Finely crystallized calcium aluminate with 30 CaO, 70 Al_2O_3 (CA ?), (Sd) Calcium-manganesesulphide with 25 Mn, 35 Ca, 41 S
COMMENT The ratio between manganese and calcium in the sulphide phase varies for different inclusions within the limits 22–54% Mn and 5–35% Ca

111 Oxide inclusions in steel: finely crystallized calcium aluminate matrix (appr. CA) with calcium-manganese sulphide

are shown in Figs.87, 88, 110–113. If they are free of MgO and contain only small amounts of MnO and SiO_2, it is probable that they originate from the deoxidation process.[75,76] However, when MgO can be detected in an inclusion, this indicates that it has

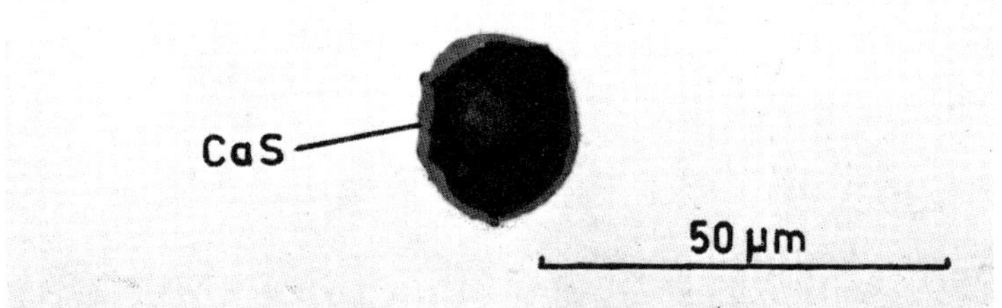

STEEL TYPE Carbon steel. From rolled plate
STEEL ANALYSIS, % 0·16 C, 0·37 Si, 1·44 Mn, 0·018 P, 0·009 S, 0·09 Cu, 0·02 Cr, 0·07 Ni, 0·031 Al (soluble), 0·009 N
INCLUSIONS, % Matrix with 26–39 CaO, 74–65 Al_2O_3 (measured in six different inclusions), rim phase CaS

COMMENT The sulphide shows a yellow-brown to yellow-green fluorescence in the electron probe analyser. The hard aluminate inclusion has maintained its spherical shape in spite of the deformation of the steel phase at the rolling operation. Compare the comment to Fig.114

112 Oxide inclusions in steel: calcium aluminate surrounded by a scale of calcium sulphide

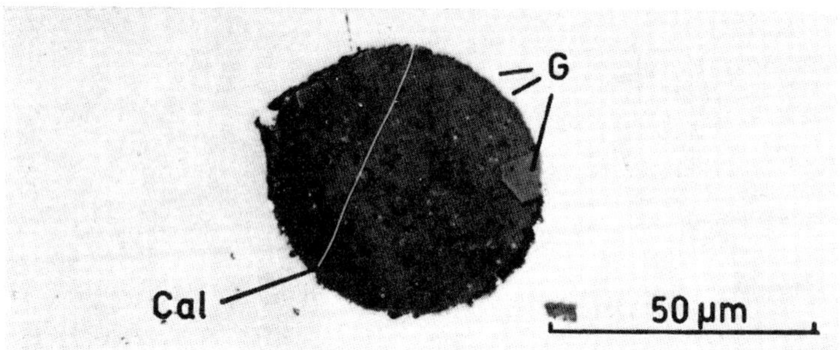

STEEL TYPE Roller bearing steel, *not* deoxidized with Ca
STEEL ANALYSIS, % 0·27 C, 0·25 Si, 0·70 Mn, 1·10 Cr, 0·20 Mo
INCLUSIONS, % (Cal) Calcium aluminate with 37 CaO, 5 SiO_2, 57 Al_2O_3, trace of MgO, (CA), (G) Spinel with 31 MgO, 67 Al_2O_3. Mean analysis 36 CaO, 1 MgO, 5 SiO_2, 58 Al_2O_3

COMMENT This inclusion cannot be a deoxidation product as the steel was not Ca-deoxidized. It has a high MgO-content, indicating that it originated from the refractories (*see* G 6)

113 Oxide inclusions in steel: calcium aluminate (*CA*) with spinel

originated from other sources. Deoxidation alloys are generally free of magnesium (Table XIII), whereas many refractories have a high MgO content (Table IX), and several furnace and ladle slags also contain magnesia (Table XIV). This is an example where electron probe analysis is of special value since the optical appearance of such calcia inclusions is often independent of the magnesia content. A thorough microscopic investigation may indicate the presence of magnesia, especially if the idiomorphic spinel phase appears as in Fig.87. However if the inclusions are glassy, electron probe analysis is always recommended for a thorough investigation.

Calcium aluminate inclusions, which are free of SiO_2, are comparatively hard. They often keep their spherical shape even after a deformation of the steel matrix. If the steel is more heavily deformed especially at higher temperatures, they split into strings of aluminates and other non-silica phases present, e.g. spinels. These strings are orientated in the deformation direction. Fragments of such spherical inclusions are shown in Figs.89, 91, and 114. The deformation behaviour of these silicate-free inclusions differs from that of silicate-rich inclusions, where the silicate phase deforms plastically (*see* Figs.27 and 29, part I).

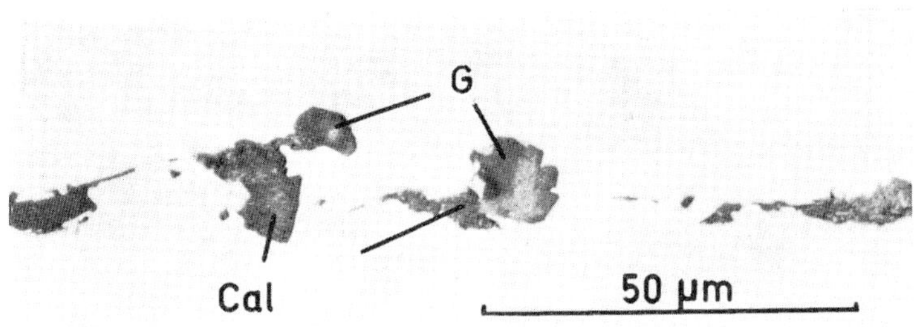

×800

STEEL TYPE Reduced electric steel, CaSi-deoxidized 1·2 ton ingot. Sample from rolled bar, 150 mm diam, reduction 1:5
STEEL ANALYSIS, % 0·35 C, 0·25 Si, 0·40 Mn, 1·4 Cr, 3·0 Ni, 0·50 Mo
INCLUSIONS, % (Cal) Calcium aluminate with 14 CaO, 3 MnO, 78 Al_2O_3 (CA_2), (G) Spinel, $MgO{\cdot}Al_2O_3$
COMMENT Spherical calcium aluminate inclusions of the type shown in Figs.110, 111, 112 and 113 do not deform readily (Fig.112). Only if the steel is heavily reduced, for instance, during rolling, do the inclusions change their shape. They are usually crushed to strings of the hard inclusion components in the rolling direction. Inclusions with the softer silicate matrix are more readily deformed plastically (*see* Figs.27, 29 and 72)

114 Oxide inclusions in steel: calcium aluminate (CA_2) and spinel in heavily rolled steel

The calcium aluminates do not crystallize readily under the conditions found in steel practice and therefore CaO and Al_2O_3 in inclusions are often found as components of the glassy inclusion matrix. However, the composition of the matrix often corresponds to one of the calcium aluminates. The early stages of crystallization are sometimes visible in calcia inclusions, the glassy matrix may be opalescent or have a very fine grain structure. The inclusion in Fig.110 has a glossy calcium aluminate matrix of an approximate composition corresponding to CA_2, in which spinel (F 3) has crystallized. The inclusion in Fig.88 is of a similar kind. The inclusions in Figs.111 and 87 have a matrix in which crystallization has started. The composition of matrix for both these inclusions is close to CA. They are both multiphase inclusions, the former with a second phase of calcium sulphide, the latter with spinel, sulphide and CA_2 crystallizing in the matrix. The inclusion of Fig.112 also has a matrix on the verge of crystallization, its appearance is typical for calcium aluminate inclusions in steels deoxidized with CaSi. (The inclusion also has a sulphide rim, which is discussed further in section K 11.)

According to the present investigation the most common calcium aluminate in steel inclusions seems to be the phase $CaO \cdot 2Al_2O_3$ (CA_2). This is in agreement with the findings of Bruch *et al.* from studies of macroinclusions.[76] In microsections of steel inclusions CA_2 is usually found as idiomorphic, platelike crystals, often with a rectangular section. Typical crystals are present in the inclusions shown in Figs.91 and 114–117. The CA_2 sections are sometimes also needle-like as in Fig.87 (Cal 1). Koch has identified the CA_2-phase in inclusion isolates and pictures of such isolated CA_2-crystals have been published in his monograph on metal analysis,[77] p.375.

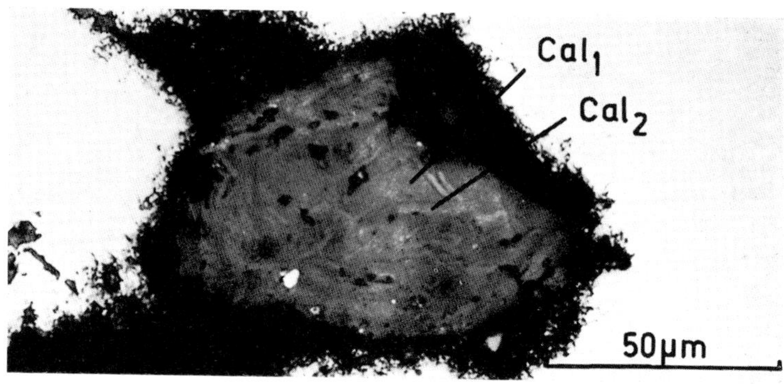

×600

STEEL TYPE Reduced electric steel, Ca-deoxidized. From forged bar. The bar was forged from the lower part of a 3·5 ton ingot, reduction 1:10, diameter 200 mm
STEEL ANALYSIS Compare Fig.114

INCLUSIONS, % (Cal_1) 20 CaO, 70 Al_2O_3 (CA), (Cal_2) 15 CaO, 80 Al_2O_3 (CA_2). Mean analysis 20 CaO, 80 Al_2O_3
COMMENT Compare Fig.116

115 Oxide inclusions in steel: the calcium aluminates CA and CA_2

The calcium aluminate $CaO \cdot Al_2O_3$ (CA) has also been identified as an inclusion phase in the present investigation. It usually appears as a fine-grain precipitate in the inclusion matrix (*see* Figs.87, 111 and 113).

The authors have thus only been able to identify the two calcium aluminates CA and CA_2 in steel inclusions. The other aluminates, C_3A, $C_{12}A_7$ and CA_6 have not been found and no reference to their appearance in non-metallic steel inclusions seems to exist. The phases have been observed, however, in reaction products between

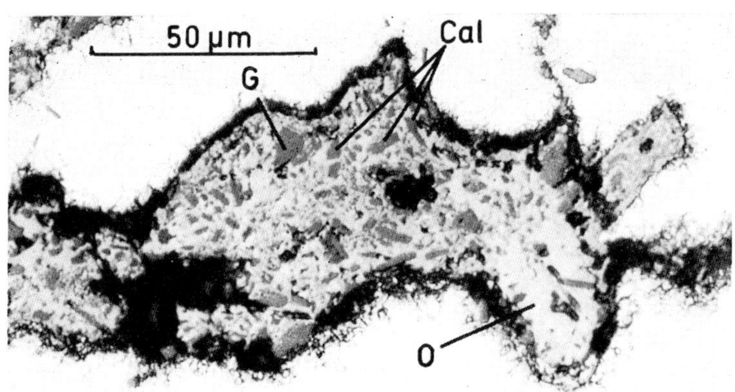

×600

STEEL TYPE Compare Fig.115
STEEL ANALYSIS Compare Fig.114
INCLUSIONS, % (Cal) Calcium aluminate with 15–20 CaO, 70–80 Al_2O_3 (CA_2), (G) Spinel, $MgO{\cdot}Al_2O_3$, (O) iron oxide, probably FeO
COMMENT This inclusion was probably initially similar to the inclusion shown in Fig.115. The heating during the forging operation has caused oxidation of the steel around the inclusion which as a result has taken up a considerable amount of iron oxide

116 Oxide inclusions in steel: calcium aluminate (CA_2), spinel and iron oxide

×400

STEEL TYPE Low-alloyed steel, normalized at 880°C quenched and annealed at 650°C
STEEL ANALYSIS, % 0·39 C, 0·35 Si, 0·44 Mn, 0·007 P, 0·006 S, 1·39 Cr, 2·88 Ni, 0·61 Mo, 0·003 N
INCLUSIONS, % (CA_2) 12 CaO, 79 Al_2O_3, trace of TiO_2, (G) Spinel with 25 MgO, 65 Al_2O_3, trace of CaO and TiO_2, (M) Matrix with 16 CaO, 40 SiO_2, 42 Al_2O_3. Mean analysis 12 CaO, 7 MgO, 27 SiO_2, 54 Al_2O_3

117 Oxide inclusions in steel: calcium aluminate (CA_2) and spinel in a silicate matrix

steel slags and refractories. For example, the C_3A-phase has been identified in worn chrome-magnesite bricks[59] and dolomite blocks (*see* Table X). The $C_{12}A_7$-phase has been found in used dolomite blocks and in high-alumina roof bricks (*see* Table X) and the CA_6-phase after slag attack on high-alumina arc-furnace roof refractories. In the paper by Hayhurst and Webster,[78] excellent microsections of the reaction zones in these alumina refractories have been given. They show a characteristic crystallization pattern of the CA_6-phase, which forms lath-shaped crystals, distinctly different from the corundum crystals of the alumina refractory.

The authors have also prepared a series of synthetic calcium aluminates from the pure oxides CaO and Al_2O_3. Different mixtures of these oxides have been melted in an electron beam furnace and then heat-treated for 2 h at 1 150°C. Microsections of some of these melts are shown in Fig. 118*a–c*. They have been chosen to illustrate the microscopic crystallization pattern of the two-phase ranges CA_6+CA_2, CA_2+CA and $CA+C_{12}A_7$, and all show primary crystals of the former phase in a matrix of the latter (*see* the equilibrium diagram Fig.105).

G 4 $CaO–SiO_2–Al_2O_3$: The calcium silicates

Chemical formulae

Four intermediate phases have been reported to exist in the binary system $CaO–SiO_2$ (Table XV). Several modifications of the calcium

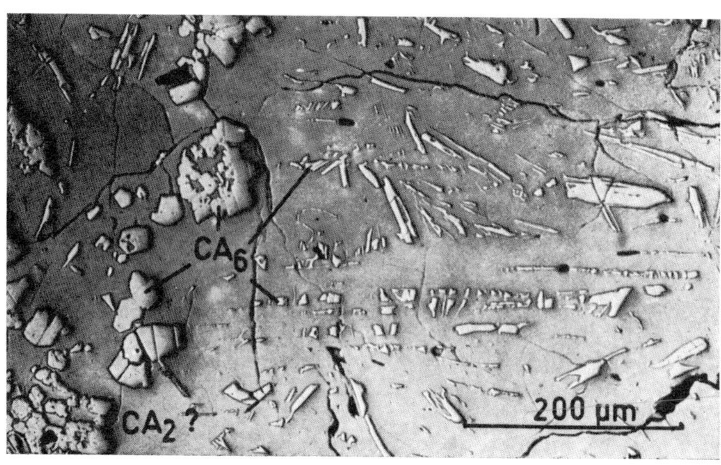

Slag analysis, % (CA_6) 9 CaO, 93 Al_2O_3, (CA_2) 19 CaO, 81 Al_2O_3

118a CA_6 in matrix of CA_2

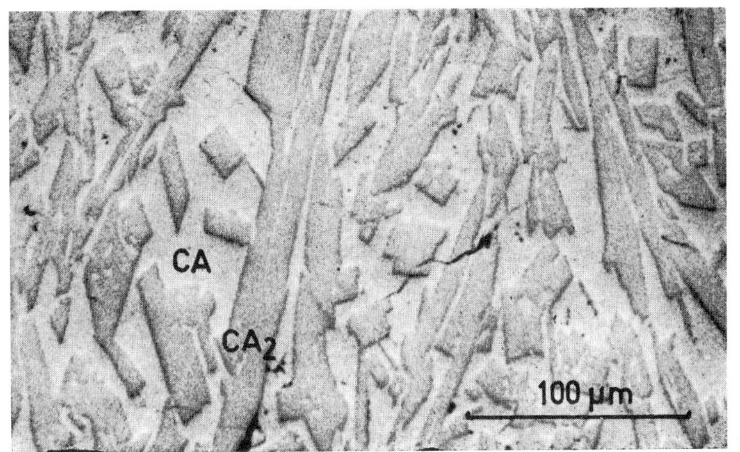

×300

Slag analysis, % (CA_2) 19 CaO, 81 Al_2O_3, (CA) 31CaO, 70 Al_2O_3

118b CA_2 in matrix of CA

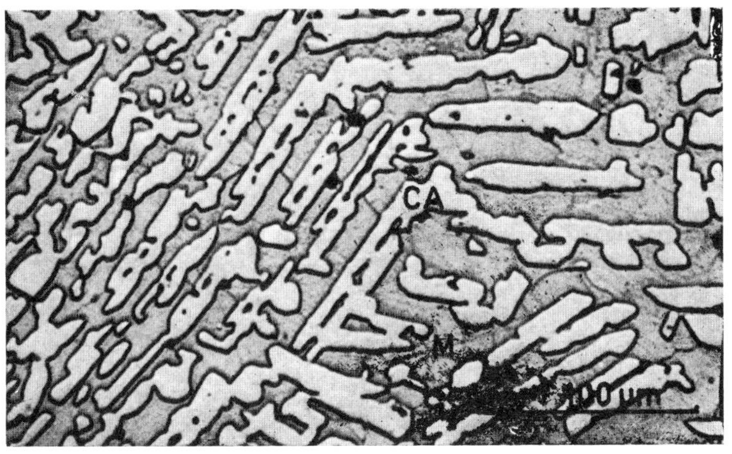

×300

Slag analysis, % (CA) 34 CaO, 63 Al_2O_3, ($C_{12}A_7$?) 46 CaO, 53 Al_2O_3

118c CA in matrix of $C_{12}A_7$ (?)

118 Synthetic slag, prepared from calcia-alumina mixture by argon arc melting at about 2 500°C. Compare the equilibrium diagram for the CaO–Al_2O_3 system, Fig.105

silicates are also known. The stoichiometric composition of the different calcium silicates are usually given as follows.

$CaO \cdot SiO_2$ (*CS*). Three modifications are known
α-*CS* (pseudo-wollastonite), β-*CS* (wollastonite) and
β'-*CS* (para-wollastonite)
$3CaO \cdot 2SiO_2$ (C_3S_2, rankinite)
$2CaO \cdot SiO_2$ (C_2S). Four modifications are known
α-C_2S, α'-C_2S (bredigite), β-C_2S (larnite) and γ-C_2S (shannonite)
$3CaO \cdot SiO_2$ (C_3S, alite). Three modifications might exist, but the available information is incomplete.

The pseudobinary equilibrium diagram for the system $CaO–SiO_2$ is given in Fig.119

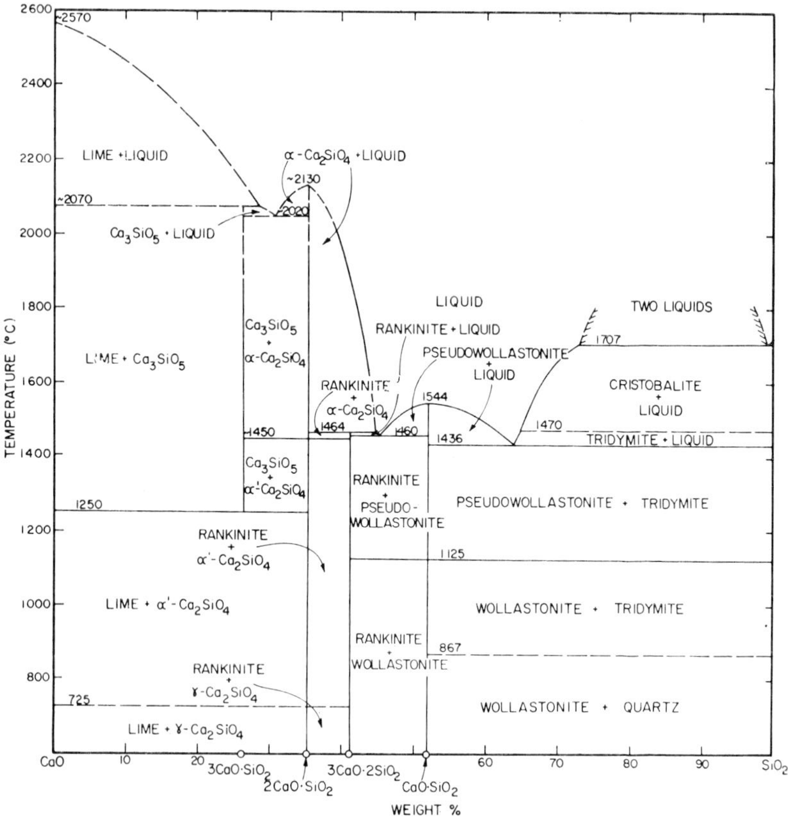

119 The equilibrium diagram for the system $CaO–SiO_2$. From Muan and Osborn[53] Fig.18, p.36

Physical properties

The melting points, stability ranges of the different modifications, densities and microhardness values for the different calcium silicates have been summarized in Table XVIII. The information has been

TABLE XVIII Physical properties of the different calcium silicates

Modification	Melting point °C	Stability range °C	Density	Microhardness kp/mm^2
α-CS	...	<1125	2·90	1000 (*CS*)
β-CS	1544	1125–1544	2·92	...
β′-CS	...	...	2·92	...
C_3S_2	1475	...	...	...
α-C_2S	~2130	~1450–2130	3·70	980 (C_2S)
α′-C_2S	...	~725–1450	3·31	...
β-C_2S	...	metastable	3·38	...
γ-C_2S	...	<725	2·97	...
C_3S(rombohedr)	~2070	~970–2070	3·22	
C_3S (mon.)	...	~920–970	...	...
C_3S (tricl.)	...	<~920	...	...

TABLE XIX Crystallographic characteristics of the different calcium silicates

Modification	System	Space group	Parameter (Å)	Reference W	Reference ASTM
α-CS	triclinic	...	...	XII, table p.24	10–486
β-CS	triclinic	P1	...	...	10–487
β′-CS	monoclinic	$P2_1/a$	...	XII, text p.38	10–489
C_3S_2	monoclinic	$P2_1/a$	...	XII, table p.24	11–317
α-C_2S	hexagonal	$C\bar{3}m$	$a = 5·45$ $c = 7·18$	...	...
α′-C_2S	rhombic	Pmnm	$a = 10·93$ $b = 18·41$ $c = 6·75$	VIII, text p.20	11–585
β-C_2S	monoclinic	$P2_1/n$	...	XII, text p.6	9–351
γ-C_2S	rhombic	Pmcn	$a = 5·06$ $b = 11·28$ $c = 6·78$	...	9–369
C_3S	rhombohedr.	R3m	$a = 7·0$ $c = 25·0$	XII, text p.26	9–352
C_3S	monoclinic	...	...	...	...
C_3S	triclinic	...	...	...	...

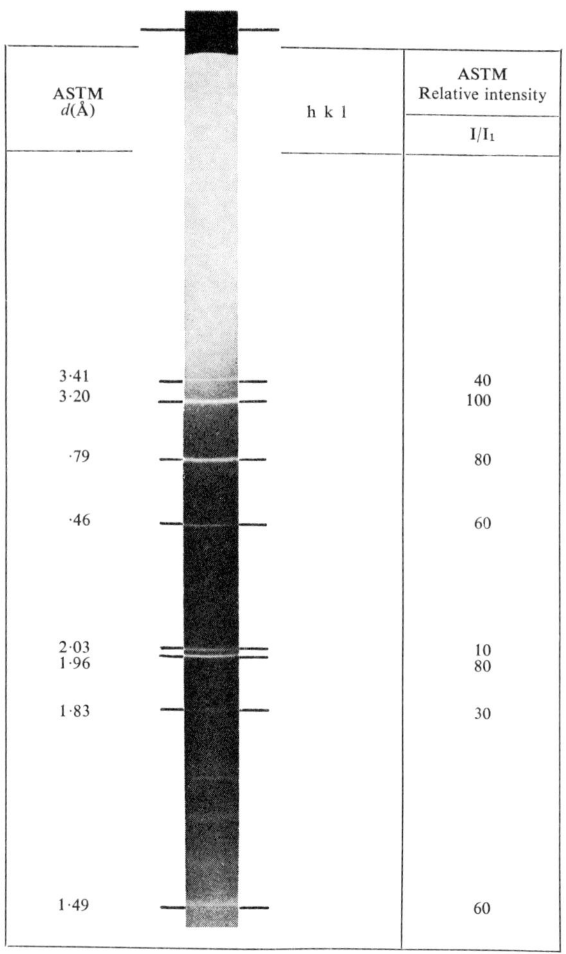

d-values and relative intensities are quoted from the ASTM index, data card 10–486

120 X-ray powder photograph of α-$CaO{\cdot}SiO_2$ (α-*CS*), pseudo-wollastonite; monochromatic FeKα-radiation; Guinier camera

collected from the books of Trojer[9] and Muan and Osborn[53] except for the microhardness values, which have been determined by the authors.

Structure type

Available information about the crystal chemistry of the different calcium silicates has been summarized in Table XIX.

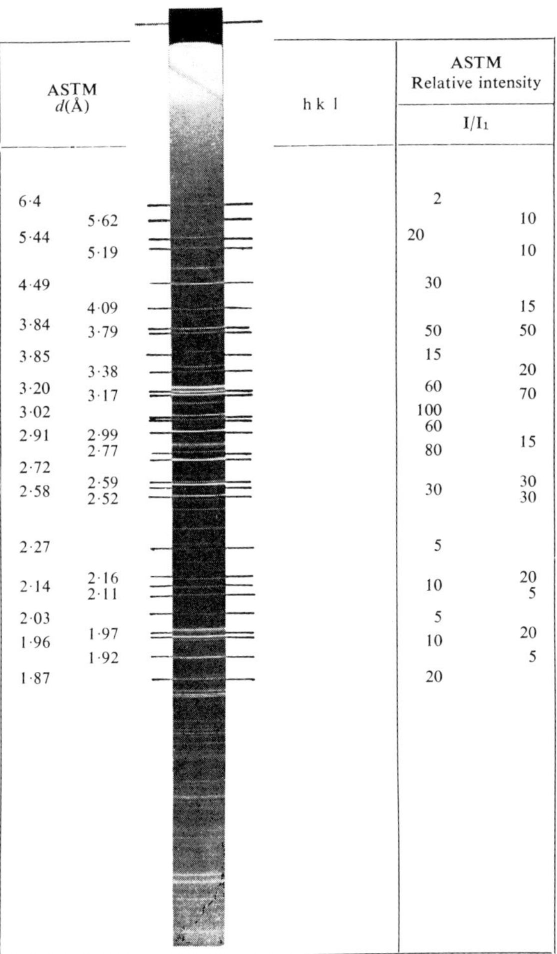

ASTM d(Å)		hkl	ASTM Relative intensity I/I₁	
6·4			2	
	5·62			10
5·44			20	
	5·19			10
4·49			30	
	4·09			15
3·84	3·79		50	50
3·85			15	
	3·38			20
3·20	3·17		60	70
3·02			100	
2·91	2·99		60	15
	2·77		80	
2·72				
2·58	2·59		30	30
	2·52			30
2·27			5	
2·14	2·16		10	20
	2·11			5
2·03			5	
1·96	1·97		10	20
	1·92			5
1·87			20	

d-values and relative intensities are quoted from the ASTM index, data card 11–317

121 X-ray powder photograph of $3CaO{\cdot}2SiO_2$ (C_3S_2), rankinite; monochromatic FeKα-radiation; Guinier camera

Monochromatic X-ray powder photographs of α-CS, C_3S_2, β-C_2S and γ-C_2S, *see* Figs.120–123.

Transformations

The transformations between the different polymorphic forms of the calcium silicates are complicated and not fully known, and they are influenced by impurities in the silicate.

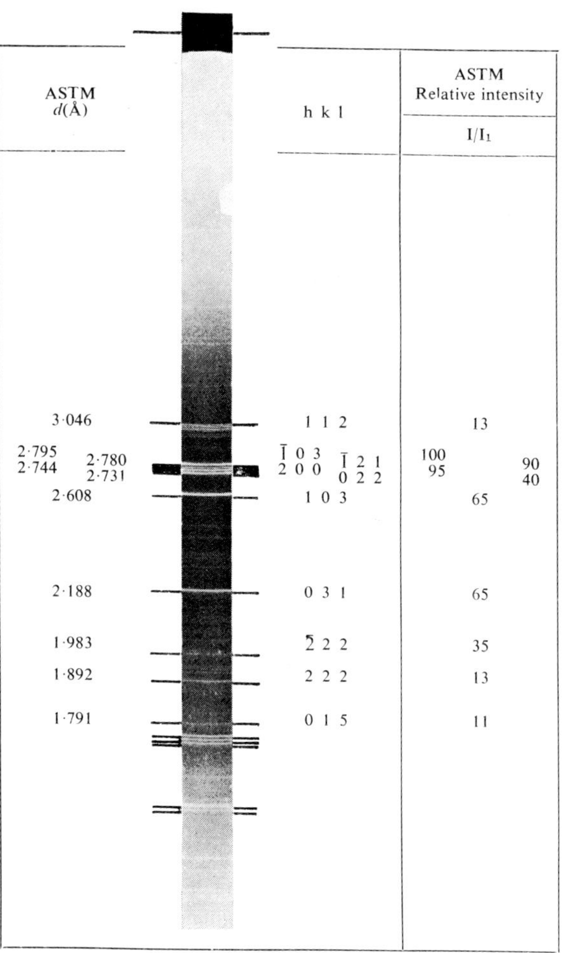

d-values, hkl-values and relative intensities are quoted from the ASTM index, data card 9–351

122 X-ray powder photograph of β-$2CaO{\cdot}SiO_2$ (β-C_2S), larnite; monochromatic FeKα-radiation; Guinier camera

α-*CS*, pseudo-wollastonite, is a low-temperature polymorph, which transforms to β-*CS*, wollastonite, at 1125°C. This transformation is reversible. β′-*CS*, para-wollastonite, is a metastable polymorph without metallurgical interest.

The transformations between the different crystalline modifications of dicalcium silicate (C_2S) are of especial interest as this silicate is often a constituent of refractories used in steelmaking systems. The

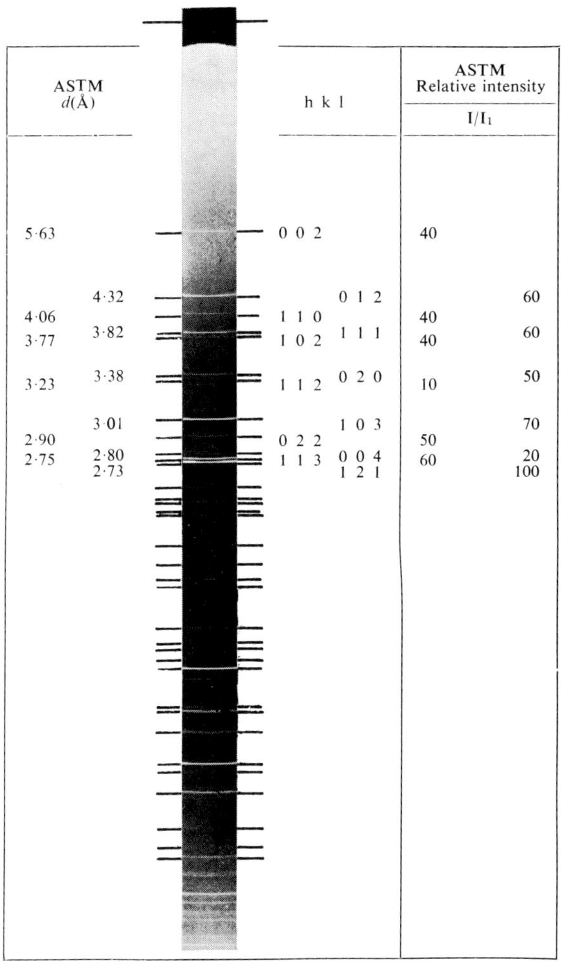

d-values, hkl-values and relative intensities are quoted from the ASTM index, data card 9–369

123 X-ray powder photograph of γ-2CaO·SiO$_2$ (γ-C_2S), shannonite; monochromatic FeKα-radiation; Guinier camera

relative stabilities of the various C_2S polymorphs at different temperatures are indicated in Fig.124. (The diagram has been taken from the monograph by Muan and Osborn,[53] whose discussion of the transformations is summarized here.)

C_2S is known in four polymorphic forms. The three forms known as α-, α'- and γ-C_2S exist as stable phases over the temperature

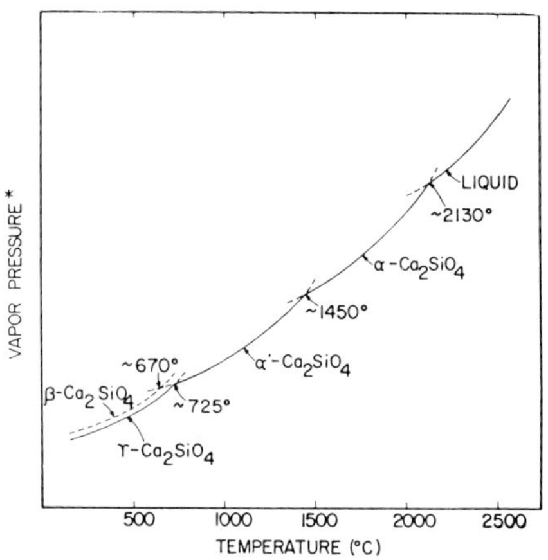

124 Schematic representation of the relative stability of the different Ca_2SiO_4 polymorphs. The vertical vapour pressure scale is only relative. From Muan and Osborn,[53] Fig.10, p.21

ranges indicated by the solid curves in Fig.124. The β-form is considered to be metastable at atmospheric pressure in the entire temperature range of the diagram. α-C_2S, the high temperature form, transforms on cooling to α'-C_2S, bredigite, in a rapid and reversible manner at about 1 450°C. On further cooling the structural changes may follow one of four paths, as indicated in Fig.124.

1. α'-C_2S may persist to room temperature
2. α'-C_2S may transform to β-C_2S, larnite, which then persists to room temperature
3. α'-C_2S may transform to γ-C_2S, shannonite, the modification stable at room temperature
4. α'-C_2S may transform to β-C_2S, which in turn transforms to γ-C_2S

These transformations are of fundamental interest for the cement industry and detailed studies of the different variables which may have an influence, as well as discussions of their significance for the fabrication of cement, are to be found in the literature.[79,80] The transformations also have some interest for metallurgists as they influence the stability of several of the refractories used in steelmaking. The density of the γ-C_2S modification is considerably lower than the density of the other C_2S modifications (Table XVIII) at corresponding temperatures. Volume changes during transformations involving α, α' and β are comparatively small and have no influence on the mechanical properties of the refractories. In contrast the transformation to γ results in a large volume increase ($\sim$10%) of the silicate phase and may cause disintegration or crack formation in refractory bodies ('dusting'). Delayed disintegration of used magnesite bricks due to transformation to γ-C_2S in the silicate phase has been reported in literature[81] and various means of stabilizing the α'- and β-forms and prevent the γ-formation are used in practice. Usually stabilized refractories have an addition of various inhibitors, such as Cr_2O_3, B_2O_3 and P_2O_5, which are effective in inhibiting the transformation to the γ-form.

The C_3S phase, the main component in cement, is also known to exist in several polymorphic forms.[82] It has a rhombohedral high-temperature modification, which inverts to a monoclinic form at about 970°C. The monoclinic modification transforms at about 920°C to a triclinic form, stable down to room temperature. The transformations have not been studied in detail and it is a matter of discussion whether they occur in pure C_3S or if they are caused by different impurities. Such impurities also influence the transformation temperatures. It is believed that the monoclinic and triclinic modifications of C_3S are only formed if traces of MgO and Al_2O_3 have been dissolved in the silicate phase.

The pure C_3S-phase is reported to be stable only at temperatures higher than 1 200°C,[83] but forms C_2S at lower temperatures according to the reaction

$$3CaO \cdot SiO_2 \rightarrow 2CaO \cdot SiO_2 + CaO$$

This reaction is of interest in steelmaking as both the C_3S and C_2S phases are sometimes present in burnt dolomite blocks (Table X and Fig.104*b*). During the steelmaking process, a temperature zone in these blocks may be within the critical temperature range, resulting in crack formation.[84]

Solid solubility

Reliable values for the solid solubility limit in the *CS*-polymorphs of the oxides MnO, FeO, MgO, Al_2O_3, Cr_2O_3, and Fe_2O_3 are difficult to find and are sometimes contradictory. This probably results from the complicated structural building of the different $MeO \cdot SiO_2$ phases. The calcium silicates of composition $CaO \cdot SiO_2$, the modifications of $MgO \cdot SiO_2$ (enstatite, proto- and clinoenstatite, F 4), $MnO \cdot SiO_2$ (rhodonite, A 9), $FeO \cdot SiO_2$ (grunerite, B 4) as well as $CaO \cdot MgO \cdot 2SiO_2$ (diopside, J) and $CaO \cdot MnO \cdot 2SiO_2$ (bustamite, johannsenite, J) belong to a group known as the chain silicates. They are all structurally characterized by chains of SiO_4-tetrahedra, but with slight differences in the grouping of the chains.[52] Those chain silicates which are of interest as possible inclusion phases have been summarized in Table XX and divided into two subgroups, which are closely related but differ in their arrangement of the SiO_4 chains. $MgO{\cdot}SiO_2$ and $CaO{\cdot}SiO_2$ belong to different silicate groups.

TABLE XX $MeO.SiO_2$ Silicates with chains of SiO_4-tetrahedrons

Type	Ideal formula	Name	Section
Pyroxene	$MgO.SiO_2$	Enstatite (*a*)	F 4
	$FeO.SiO_2$	Grunerite	B 4
	$(Mg, Fe)O.SiO_2$	Hyperstenite	...
	$CaO.MgO.2SiO_2$	Diopside	J
	$(Ca, Mg, Fe)O.SiO_2$	Pigeonite	...
	$CaO.FeO.2SiO_2$	Hedenbergite	J
	β-$CaO.MnO.2SiO_2$	Johannsenite	...
Others	$CaO.SiO_2$	Wollastonite (*a*)	G 4
	$MnO.SiO_2$	Rhodonite	A 9
	α-$CaO.MnO.2SiO_2$	Bustamite	J
	$(Mn, Fe, Ca, Mg)O.SiO_2$	Pyroxmangite	...

(*a*) Diff. modifications

The analytical results for $MeO{\cdot}SiO_2$ phases in non-metallic steel inclusions have been summarized in Table XXI. The results show that $CaO{\cdot}SiO_2$ may dissolve varying amounts of MnO, FeO and Al_2O_3 but *not* MgO. $MgO{\cdot}SiO_2$ may dissolve varying amounts of

TABLE XXI The composition of the $MeO.SiO_2$ phase in different inclusion examples. It should be noted that MgO and CaO never have been found to be present in the same phase in silicates of the composition $MeO.SiO_2$

$MeO.SiO_2$-type	Composition of the inclusion phase (wt-%)						Fig.
	MnO	FeO	MgO	CaO	SiO_2	Al_2O_3	
$MgO.SiO_2$	34	...	26	...	39	...	93
	...	...	40	...	60	...	94
	...	...	27	...	67	10	95*a*
	...	8	29	...	60	...	98
$(Ca,Mn)O.SiO_2$	60	2	...	...	37	2	40
	52	+	...	...	43	...	42
	...	15	...	30	50	...	126*a*
	7	30	...	12	43	...	125*c*
	6	20	...	18	52	...	125*c*
	19	+	...	23	51	7	125*b*
	4	...	...	39	34	12	125*a*

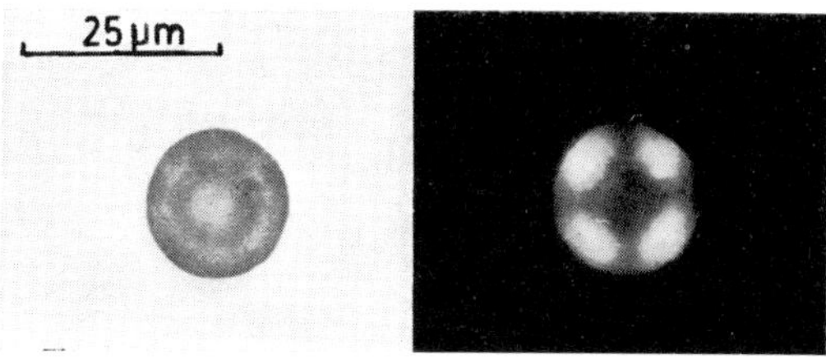

a ×700

125a Glassy *CS*-phase with only limited CaO-substitution but with a high Al_2O_3 content

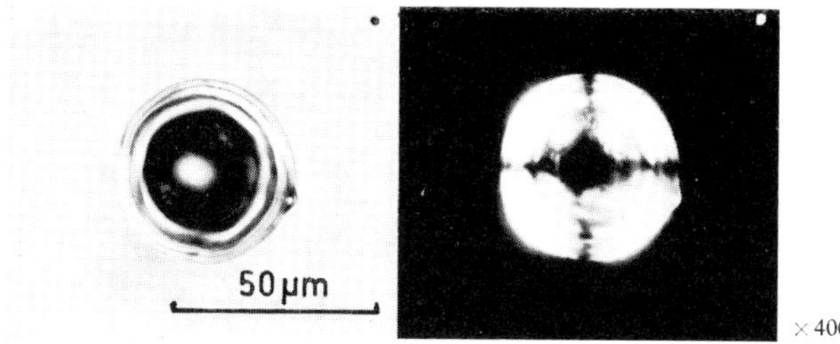

b ×400

125b Glassy *CS*-phase with MnO partly substituting for CaO

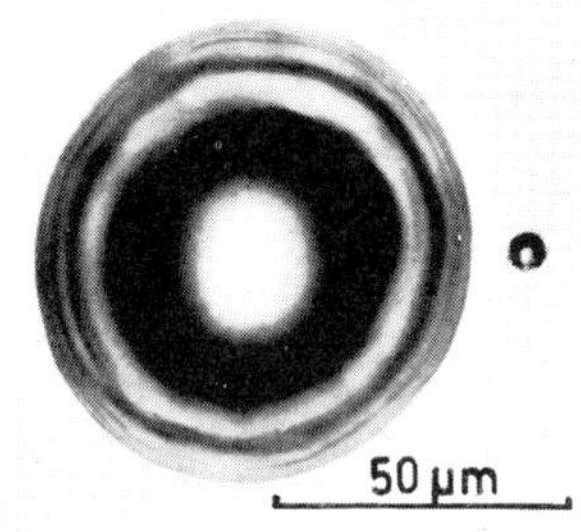

c ×400

125c Glassy *CS*-phase with FeO partly substituting for CaO

a STEEL TYPE Austenitic stainless, CaSi-deoxidized. From a scoop sample taken from the mould
STEEL ANALYSIS, % 0·043 C, 0·40 Si, 1·34 Mn, 0·020 P, 0·005 S, 18·5 Cr, 10·1 Ni, 0·50 Mo, 0·029 N
INCLUSIONS, % Glassy CS-phase with 4 MnO, 39 CaO, 3 MgO, 34 SiO_2, 12 Al_2O_3, trace of TiO_2 and Cr_2O_3
COMMENT Electron probe analysis showed the presence of a small amount of MgO. This oxide, according to the authors' opinion, is not present in solid solution in the CS-phase, but as a very fine precipitate of spinel ($MgO{\cdot}Al_2O_3$, F 3) which has clouded the glassy silicate phase

b STEEL TYPE Carbon steel, CaSi-deoxidized
STEEL ANALYSIS, % 0·18 C, 0·27 Si, 0·69 Mn, 0·009 P, 0·021 S, 0·009 N
INCLUSIONS, % Glassy CS-phase with 19 MnO, 23 CaO, 51 SiO_2, 7 Al_2O_3, trace of FeO. Different inclusions showed analysis values 19–22 MnO, 10–23 CaO, 43–51 SiO_2, 7–22 Al_2O_3
COMMENT Compare rhodonite, $MnO{\cdot}SiO_2$. Fig.40

c STEEL TYPE Scoop sample from Kaldo furnace at a low C-content. The steel was cast into a small sample mould with 5 g CaSi
STEEL ANALYSIS, Unknown
INCLUSION, % Glassy CS-phase with 7 MnO, 30 FeO, 12 CaO, 43 SiO_2. Different inclusions showed analysis values 6–7 MnO, 20–30 FeO, 12–18 CaO, 43–52 SiO_2

125 Oxide inclusions in steel: glassy calcium silicate of CS-type but with CaO partly substituted by other metal oxides (*see* Table XXI). The microphotographs to the right have been taken with polarized light

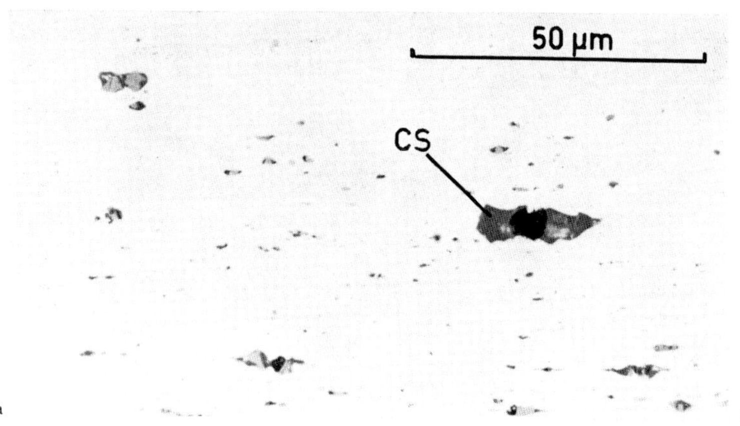

a ×80

MATERIAL TYPE Welding electrode
MATERIAL ANALYSIS Unknown

INCLUSIONS, % Crystalline CS-phase with 15 FeO, 30 CaO and 50 SiO_2

126a Oxide inclusions in a welding electrode: *CS*-phase

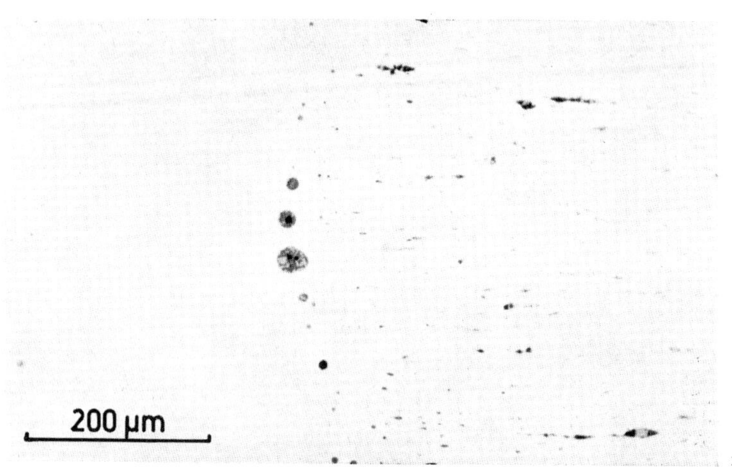

b ×1

126b Inclusions, rich in calcium silicates in the fusion line between the steel and the welding electrode, and in a section through the welding electrode. The larger inclusions in the fusion line mainly consist of CaO and SiO_2, but they also hold MnO, TiO_2 and V_2O_5. They have a spherical shape as they were liquid during the welding operation. Their origins are the calcium silicate inclusions in the electrode. These have formed larger droplets due to coalescence which have then gained TiO_2 and V_2O_5 from the electrode coating and MnO and SiO_2 indigenously from the steel phase

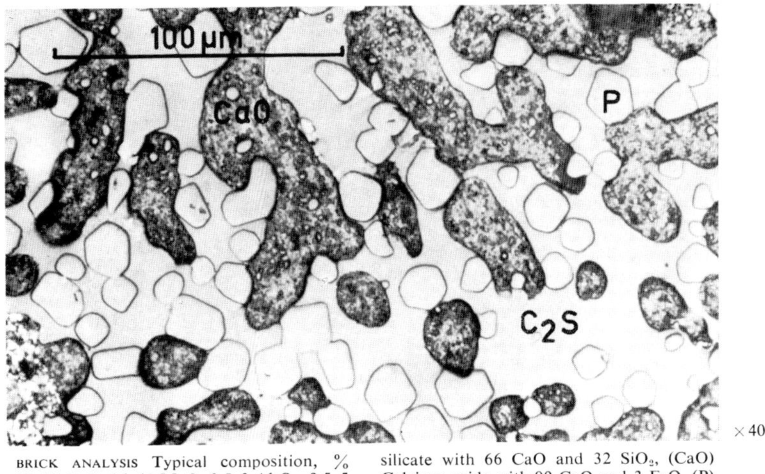

×400

BRICK ANALYSIS Typical composition, % 50–60 CaO, 36–40 MgO, 0·5–3 Al_2O_3, 0·5–5 SiO_2, others 3
STRUCTURE COMPONENTS, % (C_2S) Calcium silicate with 66 CaO and 32 SiO_2, (CaO) Calcium oxide with 90 CaO and 3 FeO, (P) Periclase with 100 MgO

127 The calcium silicate C_2S in burnt dolomite bricks

MnO, FeO and Al_2O_3 but *not* CaO, even in such inclusions as that shown in Fig.93, where enstatite has been precipitated in a CaO-rich matrix. This result is thus consistent with what would be expected from the structural building of the silicate phases.

No information concerning the solid solubility of different oxides in the C_3S_2-phase has been found.

For the C_2S-modifications the information available is also scanty. It seems that the α- and γ-phases have a higher solid solubility for the binary oxides MnO, FeO, MgO than for the ternary oxides Al_2O_3, Cr_2O_3 and Fe_2O_3. For the α'- and β-phases the reverse solubility rule seems to hold with a higher solid solubility for Al_2O_3, Cr_2O_3 and Fe_2O_3 than for MnO, FeO and MgO.

According to Trojer,[9] the C_3S-phase has only a small solid solubility for MnO. No other information about solid solubility of oxides of interest as inclusion components has been found.

Microscopy and electron probe analysis

The different calcium silicates are all somewhat difficult to identify microscopically and they are often glassy. Etching reactions for their identification in ceramics are known and Trojer's monograph[9] should be consulted for etching many of the silicate phases. In the present summary it has not been possible to examine samples of all the different calcium silicates, but some typical inclusions and refrac-

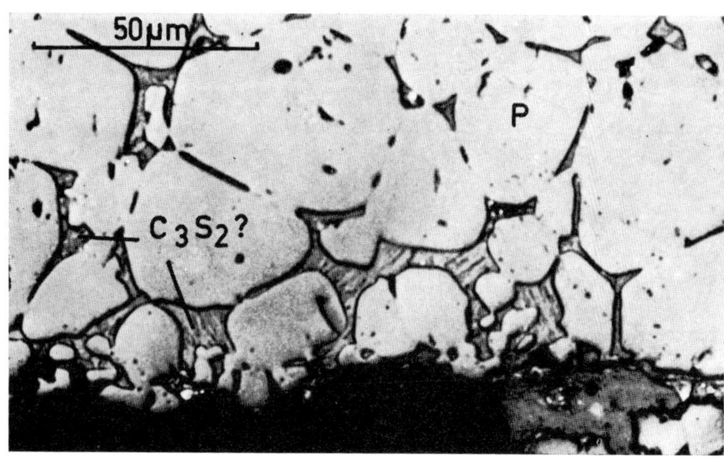

×600

BRICK ANALYSIS Typical composition, % 0·5–3 CaO > 85 MgO, 0·5–4 Al_2O_3, 1–3 SiO_2, rest 2–8, (C_3S_2) Calcium-silicate with 54 CaO, 4 MgO and 40 SiO_2, (P) Periclase with 100 MgO

COMMENT This phase may also be one of the C_2S-modifications but it is microscopically similar to synthetic rankinite (Fig.129). It is also, like rankinite, sensitive to moisture

128 The calcium silicate C_3S_2 (?) in magnesite bricks

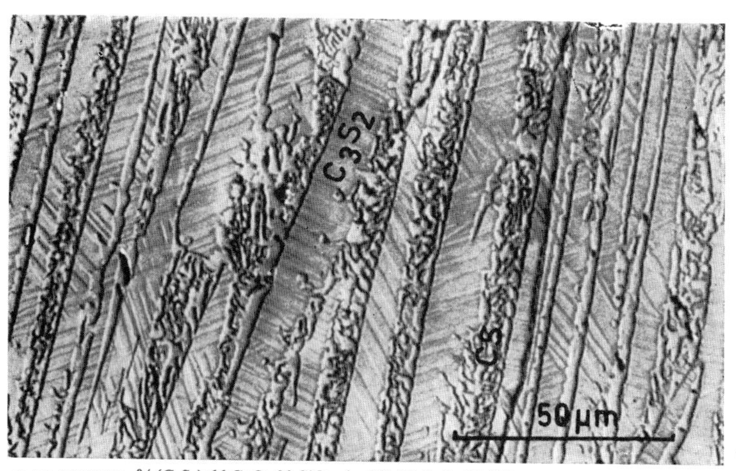

×600

SLAG ANALYSIS, % (C_3S_2) 66 CaO, 30 SiO_2, (α-CS) 56 CaO, 39 SiO_2

129 The calcium silicates C_3S_2, rankinite, and α-CS, pseudo-wollastonite, in a synthetic CaO–SiO_2 slag. The slag was prepared from calcia-silica mixture by argon arc melting at about 2 000°C, followed by a heat-treatment 3 h at 1 150°C. The phases were identified by X-ray diffraction analysis

tories with these phases have been analysed and are illustrated in Figs.125–130. It is essential for the ferrous metallurgist to know if an inclusion phase with calcium is an aluminate or a silicate, whereas the exact phase identification is often of less importance. This is discussed in more detail in section G 6, which deals with general conclusions on the origin of inclusions with calcium.

Monophase, glassy inclusions with a composition corresponding to $CaO{\cdot}SiO_2$ constitute a common inclusion type in steel deoxidized with CaSi. The inclusions are usually spherical in undeformed steel and are sometimes cloudy due to very fine crystallization. Part of the calcia is often substituted by MnO or FeO and the ratio CaO:MnO:FeO often varies considerably even in the inclusions within the same steel ingot. MgO is usually absent, as mentioned

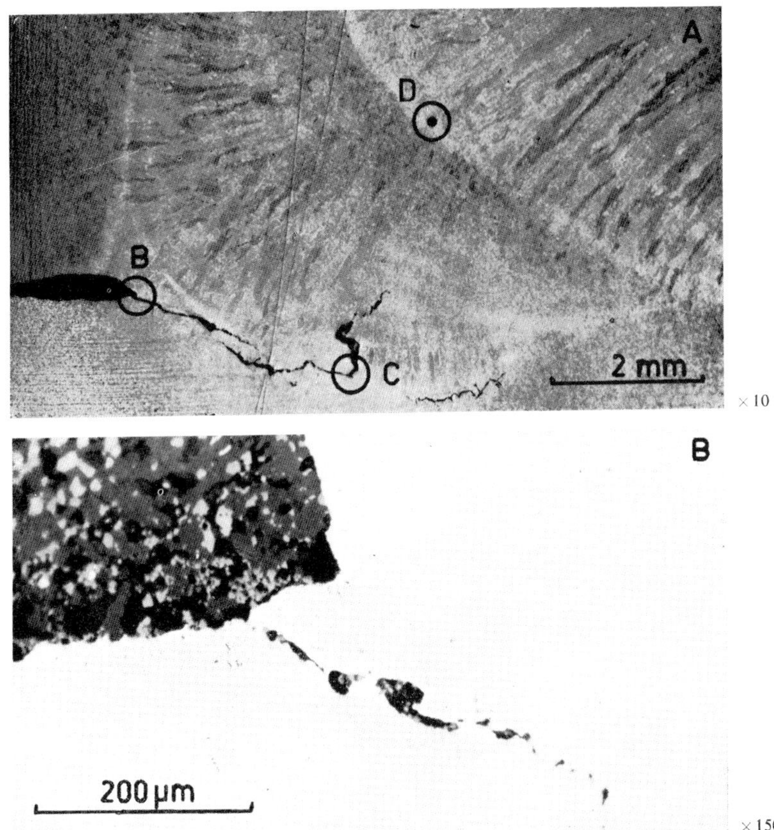

130 *A*, *B* (*C*, *D* overleaf)

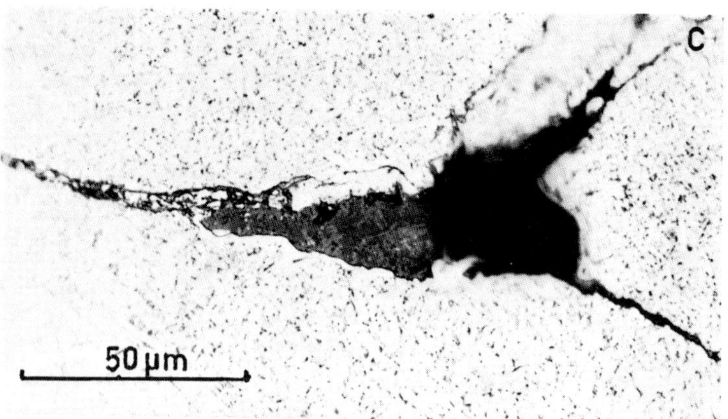

×600

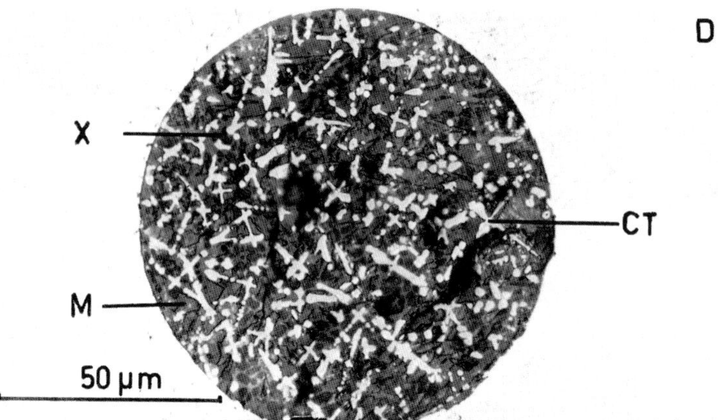

×600

STEEL TYPE Carbon steel. Welded cross test piece with cracks in the welding zone

STEEL ANALYSIS, % 0·15 C, 0·36 Si, 1·43 Mn

INCLUSIONS, % (X) Calcium silicate with 52 CaO, 13 SiO_2, 4 TiO_2 (analysis uncertain, not complete), (CT) Calcium-titanium oxide with 35–40 CaO and 44–59 TiO_2, (M) Matrix with 2–10 MnO, 48–55 CaO, 38–45 SiO_2, 3–16 TiO_2

COMMENT The electrode coating was rich in TiO_2. The Ti-content of the inclusion in the border between the two fusion beads (*see* area D) is a clear indication of its relation to the electrode coating

130*A* Microsection of the weld in a low magnification. The areas B, C and D are shown in a higher magnification in Figs.130*B-D*

130*B* The bottom of the spall and the nearest part of the crack, filled with electrode slag (area B)

130*C* Inclusion in the fusion line, which has acted as a starting point for a secondary crack (area C)

130*D* Inclusion in the border between the two fusion beads (area D)

130 Oxide inclusions in steel: electrode slag in the fusion line after welding

above. This inclusion type may also hold up to about 10 wt-% of Al_2O_3 or Cr_2O_3. Examples are given in Table XXI and microsections of similar inclusions with different composition are shown in

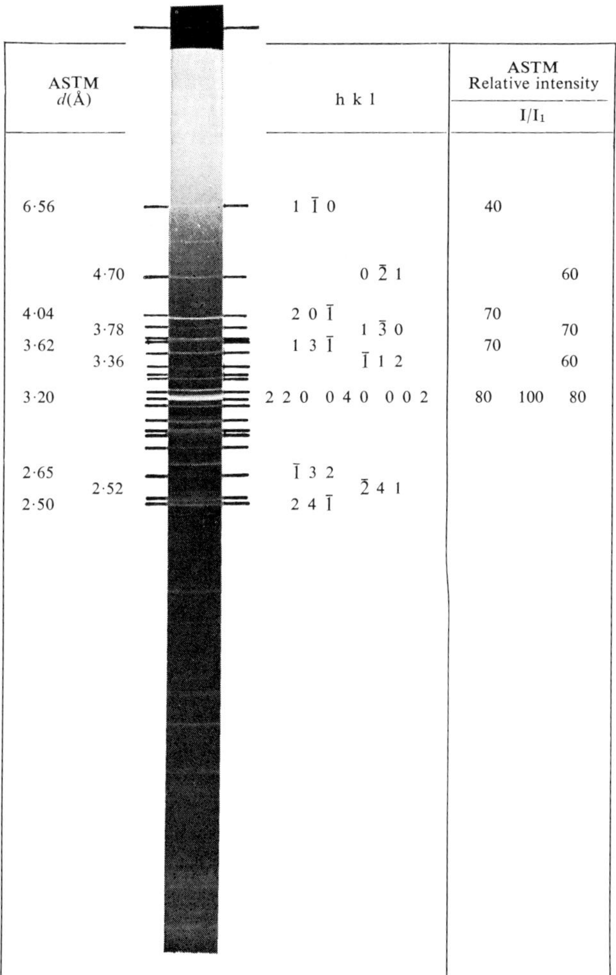

d-values, hkl-values and relative intensities are quoted from the ASTM index, data card 9–464. Only some of the more important diffraction lines have been indexed

131 X-ray powder photograph of δ-$CaO \cdot Al_2O_3 \cdot 2SiO_2$ (δ-CAS_2), anorthite; monochromatic FeKα-radiation; Guinier camera

Fig.125*a–c* and in Fig.134. The glassy rhodonite inclusion in Fig.40 (section A 9, part I) also belongs to this inclusion type with all the calcia substituted by MnO. It is essential to analyse this kind of inclusion by the electron probe method in order to determine the metals if the origin of the inclusions is to be traced. Inclusions of this

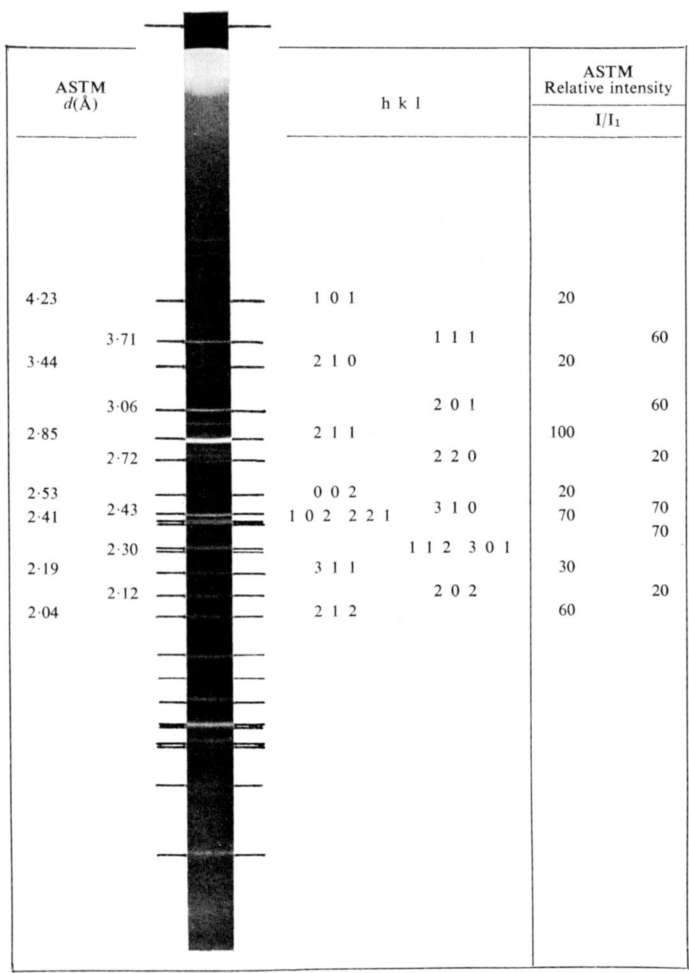

d-values, hkl-values and relative intensities are quoted from the ASTM index, data card 4–0690. Only some of the more important diffraction lines have been indexed

132 X-ray powder photograph of $2CaO{\cdot}Al_2O_3{\cdot}SiO_2$ (C_2AS), gehlenite, monochromatic FeKα-radiation; Guinier camera

type are often referred to as silica inclusions in the older literature, which is not in agreement with the present results.

If the steel matrix is deformed, for instance by rolling, forging or drawing, the spherical calcium silicate inclusions will be deformed

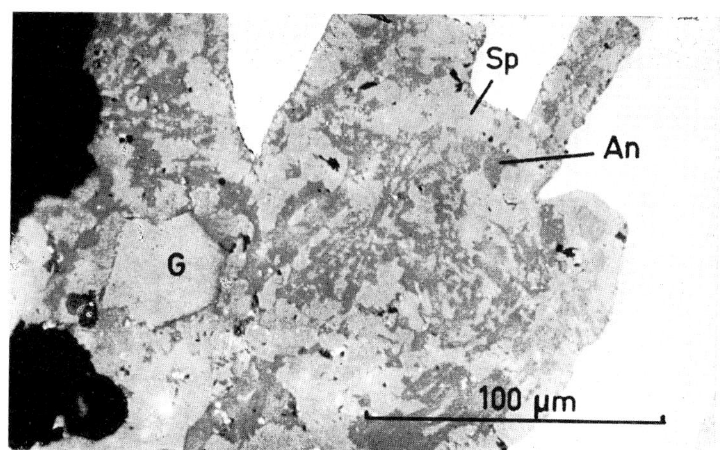

×400

STEEL TYPE Si-killed electric steel, vacuum treated. The steel sample was heat-treated for 2 h at 1 100°C
STEEL ANALYSIS Compare Fig.86
INCLUSIONS, % (An) Anorthite with 15 CaO, 2 MgO, 40 SiO_2, 36 Al_2O_3, (Sp) Spessartite with 9 MnO, 17 CaO, 13 MgO, 40 SiO_2, 17 Al_2O_3, (G) Spinel with 12 MnO, 25 MgO, 60 Al_2O_3
COMMENT The steel sample from Fig.86 was heat-treated for 2 h at 1 100°C and slowly cooled. This resulted in precipitation of crystalline anorthite and grossularite (spessartite) from the glassy matrix of the inclusion in Fig.86. The grossularite phase in this inclusion holds considerable amounts of MnO and MgO in solid solution. The phase could therefore be regarded as C_3AS_3 with MnO and MgO substituting for part of CaO, or as spessartite ($3MnO \cdot Al_2O_3 \cdot 3SiO_2$, A 11) with MgO and CaO in solid solution, or as pyrope ($3MgO \cdot Al_2O_3 \cdot 3SiO_2$, F 7) with MnO and CaO in solid solution. All these phases are isostructural and probably show a complete range of mutual substitutional solid solubility

133 Oxide inclusion in steel: anorthite, spinel and grossularite (spessartite)

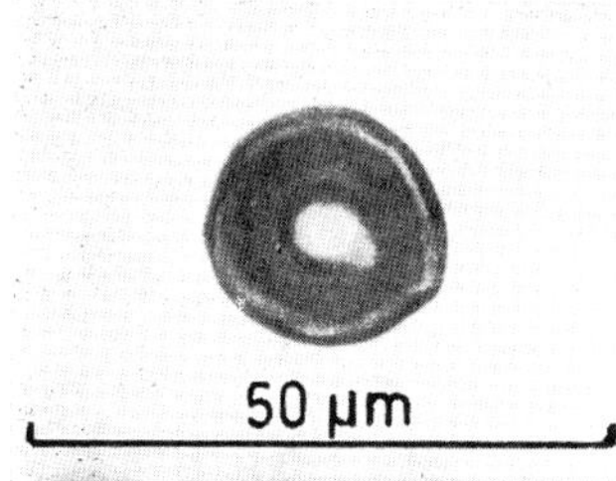

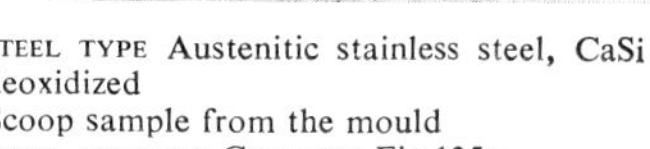

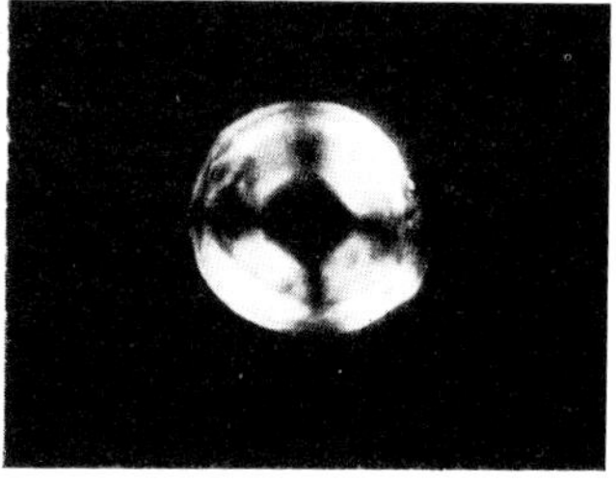

×800

STEEL TYPE Austenitic stainless steel, CaSi deoxidized
Scoop sample from the mould
STEEL ANALYSIS Compare Fig.125*a*
INCLUSIONS, % Glassy *C-A-S* phase with 30 MnO, 6 CaO, 2 MgO, 35 SiO_2, 10 Al_2O_3, 7 Cr_2O_3, trace of TiO_2
COMMENT Compare Figs.40, 125*a-c* and 135

134 Oxide inclusion in steel: glassy inclusion in the calcia-alumina-silica range but with CaO as well as Al_2O_3 partly substituted in the direction of the deformation of the steel phase. They deform plastically in a similar manner to the manganese silicates (*see* Figs.27, 29 and 72, part I). However, the deformation pattern is different from that of the hard calcium aluminates, which break up to strings of crystals in the deformation direction of the steel phase (Figs.89 and 114). Calcium silicate inclusions usually remain in a glassy state after deformation of the steel phase and give interference colours in the microscope, whereas calcium aluminates have a higher crystallization tendency. It has not been possible to obtain any good microsections of pure, deformed calcium silicate inclusions, but in section G 5, Fig.135, such a deformed silicate inclusion is shown

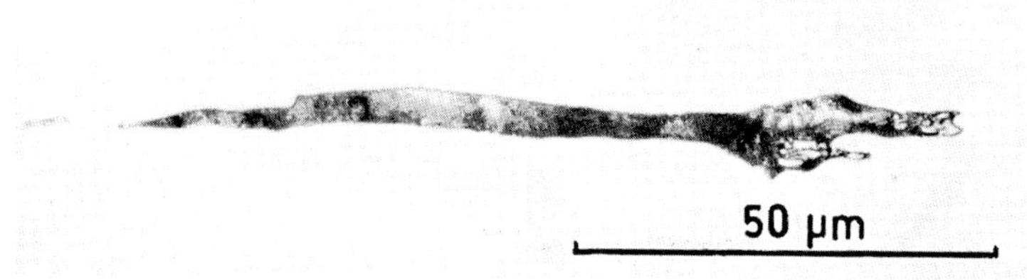

STEEL TYPE Rolled carbon steel
STEEL ANALYSIS Unknown
INCLUSIONS, % *C-A-S* phase with 9 MnO, 15 CaO, 39 SiO_2, 29 Al_2O_3, 3 TiO_2
COMMENT Spherical, glassy monophase inclusions of the types shown in Figs.125 and 134 deform plastically if the steel phase is deformed, but they usually remain glassy. Compare the deformation pattern of calcium aluminate inclusions (*see* Figs.89 and 114) and multiphase silicate inclusions (*see* Figs. 27, 29 and 72)

135 Oxide inclusions in steel: glassy inclusion in the calcia-alumina-silica range in a rolled steel plate

with some Al_2O_3 in solid solution. It has not been possible to decide if the inclusion shown is to be regarded as a glassy silicate with Al_2O_3 in solid solution, or as a glassy ternary $CaO–SiO_2–Al_2O_3$ phase.

Crystallized calcium silicate inclusions of the *CS*-type have been observed in certain types of welding electrodes (Fig.126*a*). They have an irregular shape, and part of the calcia is substituted by FeO. In the fusion line of the weld, such inclusions often increase in size due to coalescence. They may also take up MnO from the steel. Due to the comparatively high welding temperature, these inclusions form molten droplets in the weld pool. Normally they separate to the weld surface, but they may be trapped in the weld border forming spherical inclusions. A microsection of such a fusion line with inclusions is shown in Fig.126*b*. These inclusions often hold small amounts of oxides like TiO_2 or V_2O_5, originating from the coating of the welding electrodes.

TABLE XXII Crystallographic characteristics of the ternary CaO–SiO_2–Al_2O_3 compounds

Modification	System	Space group	Parameter (Å)	Reference W	Reference ASTM	Isostructural phases
α-CAS_2	hexag.	C6/mmm	a=5·10 c=14·72	XII, text p.97	10–379	...
β-CAS_2	rhombic	...	a=8.224 b=4·836 c=8·606	XII, text p.97	...	...
γ-CAS_2	triclinic					
δ-CAS_2	triclinic	$P\bar{1}$	...	XII, text p.95	9–464	Mn-anorthite, A12
C_2AS	hexagonal	...	a=7·690 c=5·067	XII, text p.32	4–0690	...
$C_2A_2S_5$	rhombic	Cccm	...	...	...	Cordierite, F6(A13, B7)
C_3AS_3	cubic	Ia3d	a=11·851	...	3–0826	Spessartite, A11

The other calcium silicate phases are not normally found as inclusions in steel phases but they sometimes appear as structural components in refractories, steelmaking slags or as reaction products between slags and refractories. The silicate phase in the burnt dolomite of Figs.85 and 104, as well as the corresponding phase in the dolomite lining of Fig.127, is probably the C_2S-phase. The phase penetrates the main structure of calcia and periclase grains, and is usually unevenly distributed in the grain interstices where it acts as a binder. The corresponding calcium silicate phase in magnesite bricks is shown in Fig.84*a*, where it has a glassy appearance. In used magnesite bricks the silicate sometimes crystallizes, as in Fig.128. The phase in this example has a fine, ribbon-like structure and a composition close to 3CaO·$2SiO_2$, rankinite. Its structural appearance is similar to rankinite in synthetic slags, as shown in Fig.129. (The rankinite phase in the last figure has been definitely identified by X-ray methods.) The other calcium silicate phase in Fig.129 was found to be the α-form of CaO·SiO_2, pseudo-wollastonite. It is often difficult to establish which of the different calcium silicates, C_3S_2, C_2S or C_3S may be present in a refractory, and as mentioned above the last two phases are also closely related. The main problem in steelmaking practice is, however, to know which of the steel refractories holds silicate phase, to determine its location

TABLE XXIII **The composition, wt-%, of ternary CaO–SiO_2–Al_2O_3 phases in different inclusion examples from different material types. Many of the C–A–S phases are glassy, and their classification into different ternary crystalline C–A–S phases on the basis of composition is therefore often only formal**

Phase type	Comp. from electron probe analysis						Material type	Fig.
	CaO	MnO	MgO	SiO_2	Al_2O_3	Cr_2O_3		
CAS_2 theor.	20	...	...	43	37	...	...	...
(anorthite)	22	...	...	40	40	...	low all. steel	137
	24	...	...	42	32	2	ferrochrome	138
	14	3	...	44	40	...	low all. steel	136, 90
	22	...	...	44	42	...	synthetic slag	144
	16	...	...	47	38	...	slag-ladle lining	143
	16	...	...	40	42	...	low all. steel	117
	+	14	...	48	39	...	low all. steel	142*b*
	15	...	2	40	36	...	electro steel	133
C_2AS theor. (gehlenite)	41	...	...	22	37	...	...	...
C_3AS_3 theor.	37	...	...	40	23	...	...	...
(grossularite)	23	13	7	41	20	...	carbon steel	86
	37	...	+	36	27	...	slag-dolomite	141
	39	4	3	34	12	+	18–11 stainless steel	125*a*
	16	15	...	50	8	...	ladle slag	143
	6	30	2	35	10	7	18–11 stainless steel	134
	17	9	13	40	17	...	electro steel	133
$C_2A_2S_5$ theor.	18	...	...	49	33	...	...	...
	13	3	...	49	28	...	high speed steel	139
	12	5	...	58	23	...	carbon steel	140
	3	7	6	57	22	...	carbon steel	99

and its approximate amount. The greyish silicate ‘islands’ between the calcia and magnesia grains in dolomite and magnesite refractories are relatively characteristic. They are usually easy to distinguish from the white, well crystallized calcium and magnesium ferrites, which are also localized between the calcia and magnesia grains (*see* Figs.84, 85 and 104).

Several of the calcium silicates are also found in solidified basic slags. Examples of the microstructure of C_2S in such slag samples have been given by Pettersson and Eketorp,[85] and of C_3S in CaO–SiO_2 slag by Brunner.[60b] Ferrero, Tino and Creton[86] have identified some of the C_2S-modifications by X-ray methods in electrolytically isolated steel inclusions and so far this seems to be the only positive identification of C_2S-phase in inclusions.

TABLE XXIV The mean composition, wt-%, of different SiO_2-free calcium aluminate inclusions from different steel types

Inclusion type	CaO	MgO	MnO	FeO	Al_2O_3	Fig.
Glassy aluminate with spinel	5	17	+	+	78	110
Aluminate and sulphide	30	...	...	...	70	111
Aluminate and sulphide	26–39	...	...	...	74–65	112
Aluminate with spinel	36	1	...	...	58	113
Aluminate with spinel	14	...	3	...	78	114
Aluminate	20	...	...	...	80	115
Aluminate, spinel and sulphide	23	10	...	...	67	87
Aluminate with spinel	18	13	...	...	64	88
Aluminate with spinel	10	15	...	...	75	89

TABLE XXV Mean composition, wt-%, of welding electrode slags from different electrode types (according to Eriksson[93])

Electrode type	K_2O	Na_2O	CaO	MgO	TiO_2	SiO_2	FeO	MnO	Cr_2O_3	Al_2O_3
Fining	0·7	1·0	0·7	5·2	0·3	20·6	53·2	9·6	...	8·3
Acid	2·7	0·1	0·9	3·3	0·8	43·0	22·6	22·5	...	3·8
Acid	2·6	1·2	2·6	2·5	2·0	30·4	23·9	20·3	...	10·9
Acid	2·5	0·3	3·8	4·5	16·0	32·9	17·9	15·1	...	6·3
Rutile	2·5	0·8	0·4	1·9	49·5	17·4	12·0	11·0	...	3·5
Rutile	3·3	0·9	0·7	2·2	45·0	21·8	10·0	10·6	...	4·7
Rutile	3·0	0·2	0·9	3·4	38·8	17·7	13·7	13·7	...	4·4
Rutile	2·6	0·4	7·5	4·6	25·5	27·5	14·0	12·5	...	5·4
Rutile, cellulose	0·8	1·0	0·6	6·5	42·8	21·8	17·3	9·2	...	1·6
Basic*	0·8	1·0	31·0	1·1	4·3	24·0	6·0	4·0	...	6·0
Rutile, Mo-alloyed	2·8	0·2	6·4	4·0	21·5	29·1	12·3	16·7	...	5·8
Rutile Cr-Mo-alloyed	2·8	0·3	6·0	3·9	20·8	28·3	13·5	13·2	1·5	6·1
Rutile, penetrating	2·4	1·8	0·4	1·7	28·3	28·3	10·2	17·2	...	5·1

* With CaF_2

For welding purposes, electrodes with a coating rich in calcium silicates are often used. This coating forms an electrode slag, which may occasionally be trapped in the fusion line of the weld, as already mentioned above. Such trapped slag inclusions may be starting points for crack formation, and an example is shown in Fig.130*a–d*. Both the spherical inclusion D in the border between the two weld beads (Fig.130*a* and *d*) and the large triangular inclusion in the fusion line (Fig.130*c*) originate from the slag layer B (Fig.130*b*). They are rich in calcia, and one of the phases in D is a calcium silicate with titanium in solid solution. The latter element in this example is a constituent of the electrode coating and definitely establishes the origin of the inclusion. Titanium is often a valuable inactive tracer element for inclusions in welding technology (*see* H 2, Table XXV).

G 5 $CaO–SiO_2–Al_2O_3$: Ternary oxide phases

Chemical formulae

Four intermediate ternary phases, one with four polymorphic forms, have been found to exist in the ternary system $CaO–SiO_2–Al_2O_3$ (Table XV). Their stoichiometric compositions are usually given as:

$CaO \cdot Al_2O_3 \cdot 2SiO_2$ (CAS_2) (four modifications are known)
α-CAS_2, β-CAS_2, γ-CAS_2 (high anorthite) and δ-CAS_2 (low-anorthite)
$2CaO \cdot Al_2O_3 \cdot SiO_2$ (C_2AS, gehlenite)
$2CaO \cdot 2Al_2O_3 \cdot 5SiO_2$ ($C_2A_2S_5$, Ca-cordierite)
$3CaO \cdot Al_2O_3 \cdot 3SiO_2$ (C_3AS_3, grossularite)

Physical properties

Melting points: CAS_2 1 550°C; C_2AS 1 590°C
Densities: δ-CAS_2 2·77; C_2AS 3·04; C_3AS_3 3·5

Structure type

Available information about the crystal chemistry of the different ternary calcium-aluminium-silicates has been summarized in Table XXII.

Monochromatic X-ray powder photographs of δ-CAS_2 and C_2AS have been reproduced in Figs.131 and 132.

Transformations

CAS_2 belongs structurally to the feldspar group of silicates and several polymorphic forms are known. They are structurally closely related but differ in the arrangement of SiO_4-tetrahedra, and the polymorphic transformations are therefore of the same type as for the SiO_2-transformations (A 3–5) and for the calcium silicate transformations (G 4). The α-CAS_2 and β-CAS_2 modifications are metastable forms, which transform to γ-CAS_2 at about 1 200°C. γ-CAS_2, high-anorthite, and δ-CAS_2, low-anorthite, are the two stable high- and low-temperature modifications of anorthite which are of metallurgical interest. The inversion temperature and the effect of impurities on the transformation are not fully known.

Solid solubility

The ternary phases of the CaO–SiO_2–Al_2O_3 system in inclusions have been found to dissolve not only MnO and FeO but also MgO. They therefore differ from the binary *C-S* phases, which do not show any solid solubility for MgO (G 4). Pickering, for instance, has reported that inclusion phases with a composition in the range C_2AS–C_2MS_2 are common in steels, deoxidized with CaSi.

C_3AS_3 is isostructural with spessartite (A 11), almandine (B 6) and pyrope (F 7) and a complete range of solid solubility seems to exist with possible substitution of CaO by MnO, FeO and MgO.

Microscopy and electron probe analysis

The ternary phases in the *C-A-S* system are often glassy and only crystallize with difficulty. They are therefore usually glassy in oxide inclusions in steel ingots, which have cooled past the critical crystallization temperature range too rapidly for crystallization to occur. Calcium inclusions with a glassy matrix corresponding to one of the ternary phases in the *C-A-S* system are relatively common in different steel types with partial substitution of MnO, FeO or MgO for CaO and Cr_2O_3 for Al_2O_3. The relation between the matrix, which is to be regarded as the glassy residual melt of the inclusions, and one of the ternary phases is often only formal and the phase which may crystallize from the inclusion matrix often has a different composition. This is shown by the inclusion in Fig.133 from a Si-killed electro-steel. It is the same inclusion shown in Fig.86, but the steel sample was heat-treated for 2 h at 1 100°C and slowly cooled. The matrix had a composition within the range of C_3AS_3 but the precipitates are CAS_2 with MnO and MgO in solid solution and C_3AS_3 but with a composition differing from the matrix.

A summary of the analytical results for ternary *C-A-S* phases found in inclusions from different steels with reference to the different photomicrographs is given in Table XXIII.

Glassy, monophase inclusions with a composition within the *C-A-S* range are common in steel. Such an inclusion is shown in

Fig.134 and it belongs to the same inclusion type as the glassy *C-S* inclusions in Fig.125*a–d* and the glassy rhodonite inclusion in Fig.40. Such inclusions are often referred to as SiO_2-inclusions, but, as already mentioned in section G 4, this is usually not the case and electron probe analysis should be used in order to establish their composition.

If a steel containing this spherical inclusion type is deformed the *C-A-S* inclusions plastically deform in the same direction as the steel, but the inclusions usually remain glassy, with visible interference fringes on their surface. A deformed inclusion with a

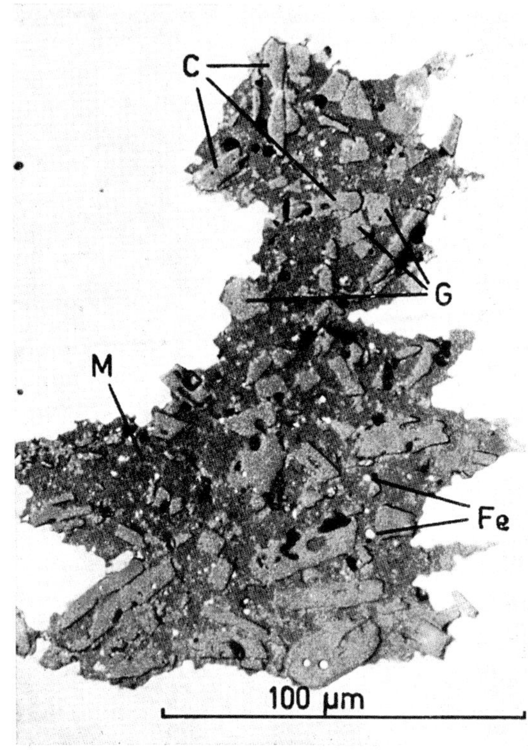

×500

STEEL TYPE Low alloy
STEEL ANALYSIS, % 0·44 C, 0·28 Si, 0·72 Mn, 0·011 P, 0·051 S, 0·95 Cr, 0·11 Ni, 0·19 Mo, 0·14 Cu
INCLUSIONS, % (M) *C-A-S* matrix with 14 CaO, 3 MnO, 44 SiO_2, 40 Al_2O_3 (C) Corundum with 100 Al_2O_3, (G) Galaxite with 23 MnO, 12 MgO, 59 Al_2O_3. Mean analysis 4 MnO, 1 MgO, 10 CaO, 32 SiO_2, 53 Al_2O_3
COMMENT Matrix is here crystalline with very fine-grained anorthite

136 Oxide inclusions in steel: calcia-alumina-silica matrix in the CAS_2-range with corundum and galaxite

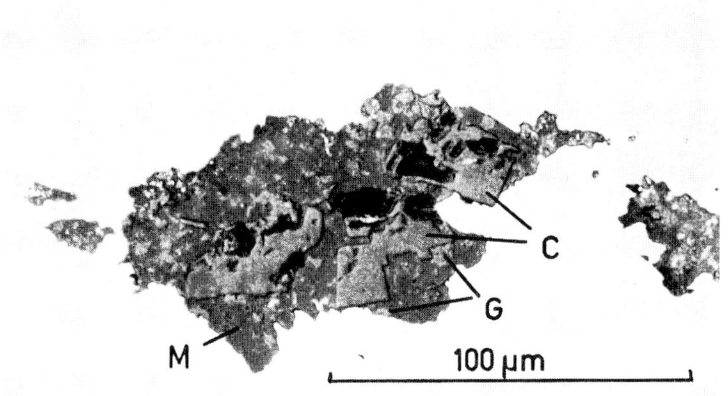

×500

STEEL TYPE Low-alloyed electric steel, top cast. Heat-treated at 850°C, quenched in oil and annealed at 590°C in air
STEEL ANALYSIS, % 0·37 C, 0·23 Si, 0·74 Mn, 0·022 P, 0·045 S, 1·07 Cr, 0·09 Ni, 0·01 Mo, 0·12 Cu, 0·01 V
INCLUSIONS, % (M) *C-A-S*-matrix with 22 CaO, 40 SiO_2, 40 Al_2O_3 (approx. composition), (C) Corundum with 100 Al_2O_3, (G) Galaxite, $MnO{\cdot}Al_2O_3$. Mean analysis 1 MnO, 14 CaO, 28 SiO_2, 57 Al_2O_3
COMMENT Matrix may be glassy or may be a very fine structure of crystalline anorthite

137 Oxide inclusions in steel: calcia-alumina-silica matrix in the CAS_2-range with corundum and galaxite

composition in the anorthite-Ca-cordierite range is shown in Fig.135. Its deformation mode is completely different from that of the brittle, SiO_2-free calcium aluminate inclusions, which only change their spherical shape after heavy working of the steel phase. They do not deform plastically but break down to strings of crystalline aluminates and the other inclusion phases, e.g. spinel, in the direction of steel deformation. Such inclusions in a rolled steel are shown in Figs.89 and 114.

Microsections of multiphase calcium inclusions with a matrix composition within the range of the different *C-A-S* phases are shown in Figs.136–138 (CAS_2), Figs.139–140 ($C_2A_2S_5$) and Fig.141 (C_3AS_3). The matrix is usually glassy, but early stages of crystallization are sometimes visible. As already mentioned the referring of the glassy matrix composition to different crystalline phases is only formal, and phases with a composition different from matrix are often precipitated. The matrix is the part of the inclusion with lowest melting point and is to be regarded as the residual melt. The inclusion examples have been observed in quite different steel types and materials such as carbon steel, alloyed steel, high-speed tool steel,

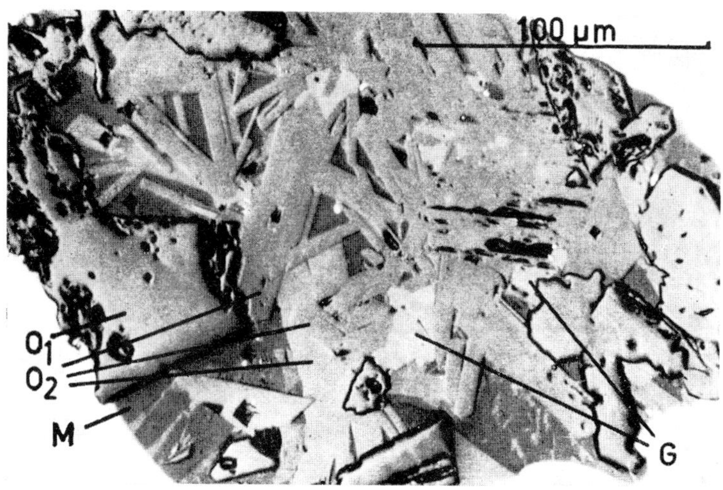

× 400

MATERIAL TYPE Ferrochromium suraffiné
MATERIAL ANALYSIS, % 0·03 C, 0·08 Si, 73 Cr, rest Fe
INCLUSIONS, % (M) *C-A-S* matrix with 24 CaO, 42 SiO_2, 32 Al_2O_3, trace of Cr_2O_3, (O_1) Corundum-escolaite with 78 Al_2O_3, 23 Cr_2O_3, (G) Chromium galaxite with 18 MnO, 26 Al_2O_3, 54 Cr_2O_3, (O_2) Double oxide with 10–20 CaO, 20–24 MnO, 25–35 Al_2O_3, 28–33 Cr_2O_3
COMMENT The phase O_2 may be a calcium aluminate with MnO and Cr_2O_3 in substitutional solid solution or a double oxide of the spinel type. The microscopic structure is more closely similar to that of the calcium aluminates

138 Oxide inclusions in ferro alloys: calcia-alumina-silica matrix in the CAS_2-range with several oxide phases

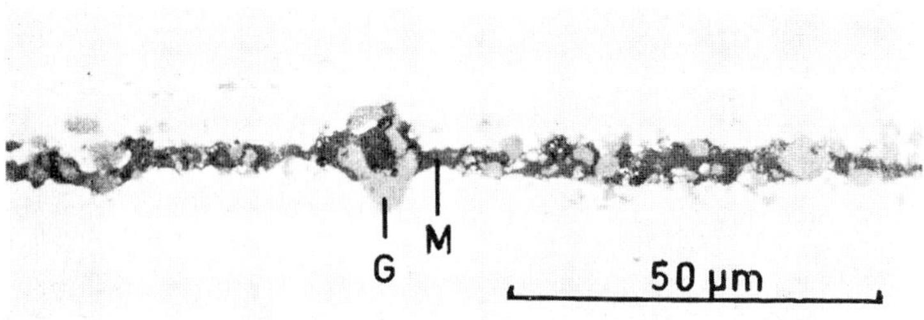

× 800

STEEL TYPE High speed steel, from the centre of a rolled bar
STEEL ANALYSIS, % 0·9 C, 0·2 Si, 0·3 Mn, 4·3 Cr, 3·3 Mo, 6·4 W, 2·1 V
INCLUSIONS, % (M) *C-A-S* matrix with 13 CaO, 3 MnO, 49 SiO_2, 28 Al_2O_3, (G) Double oxide with 28 MnO, 33 V_2O_3, 14 Al_2O_3, 23 Cr_2O_3
COMMENT The composition of the matrix is such that it may also be situated within the CAS_2-range

139 Oxide inclusions in high speed steel: calcia-alumina-silica matrix in the $C_2A_2S_5$ range with double oxides of the spinel type

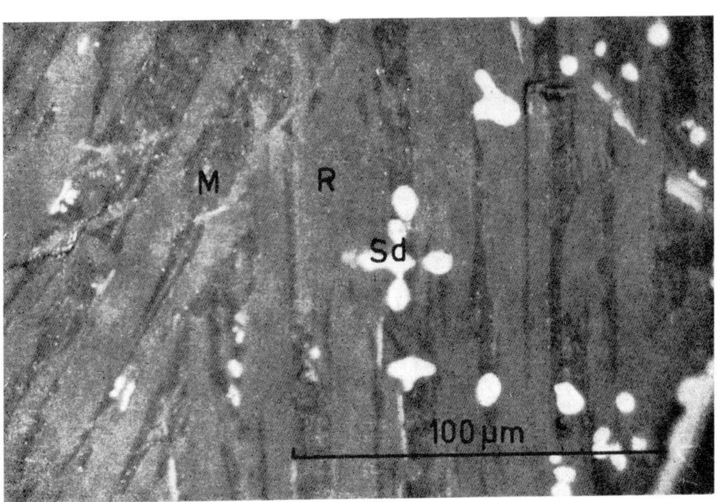

× 500

STEEL TYPE Basic electric Si-killed; inclusion isolated from the lower central part of top-cast 6-ton ingot
STEEL ANALYSIS, % 0·16 C, 0·35 Si, 0·67 Mn, 0·015 P, 0·017 S, 0·009 N, 100 ppm O; Al not detectable
INCLUSIONS, % (M) *C-A-S* matrix with 12 CaO, 5 MnO, 58 SiO_2, 23 Al_2O_3, (R) Rhodonite with 39 MnO, 6 MgO, 43 SiO_2, 8 TiO_2, trace of CaO, (Sd) Sulphide with 59 Mn, 38 S, trace of Fe. Mean analysis 40 MnO, 48 SiO_2, 12 Al_2O_3
COMMENT The matrix of this inclusion is finely dispersed and therefore the limits of error for the analysis are rather large

140 Oxide inclusions in steel: calcia-alumina-silica matrix in the $C_2A_2S_5$ range with rhodonite and sulphide

ferrochromium and in the reaction zone between dolomite and slag. This clearly shows that the details of steelmaking practice have a much greater influence on the composition of the inclusions than the steel analysis itself for inclusions belonging to the CaO–SiO_2–Al_2O_3 system. This conclusion is in agreement with the results from the investigation on inclusions from the MnO–SiO_2–Al_2O_3 and related systems, reported in section E, Table VIII, part I.

Crystalline ternary phases from the CaO–SiO_2–Al_2O_3 system in steel inclusions are less common than glassy phases. The most common of the crystalline phases is anorthite, CAS_2, which in microsections appears as dark, branched crystals. It usually precipitates only after heat-treatment of the steel and is therefore not generally found in ingot inclusions. An inclusion with anorthite phase together with galaxite and grossularite is shown in Fig.133. The steel had been Si-killed and heat-treated. In Fig.142*a*, characteristic anorthite crystals are shown in a glassy silicate matrix. The morphology is similar to that of mullite (A 8), but anorthite is darker than the silicate whereas mullite is lighter. In Fig.142*b*, an

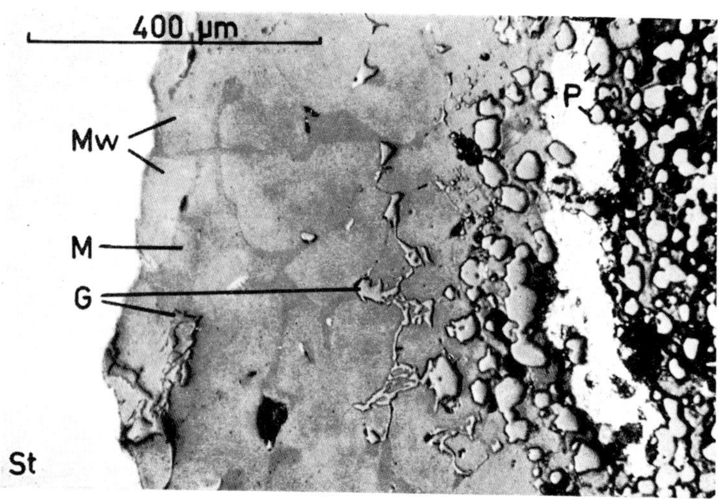

× 100

MATERIAL TYPE Reaction zone steel-slag-burnt dolomite
SLAG PHASES, % (M) *C-A-S* matrix with 37 CaO, 36 SiO_2, 27 Al_2O_3, trace of MgO, (G) Spinel with 30 MgO, 3 FeO, 54 Al_2O_3, (Mw) Merwinite with 46 CaO, 11 MgO, 46 SiO_2 (see section J), (St) Steel
COMMENT The burnt dolomite (to the right) has reacted with the liquid slag phase, which has dissolved calcia (lime) and periclase. The galaxite phase has precipitated during the cooling of the slag layer

141 Calcia-alumina-silica matrix in the C_3AS_3 range with merwinite and galaxite

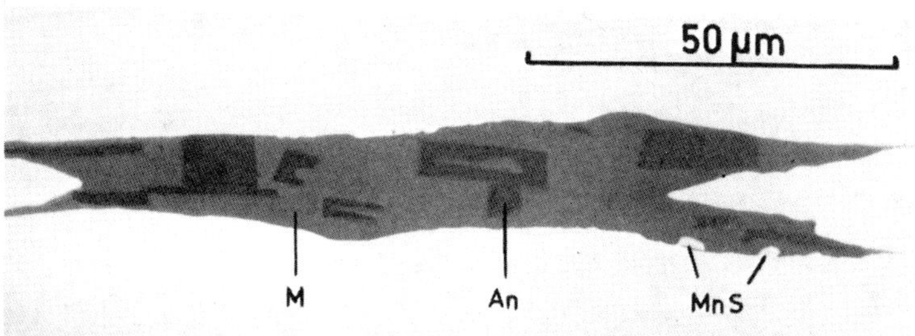

STEEL TYPE Kaldo steel, CaSi-killed in ladle. From rolled plate, longitudinal section
STEEL ANALYSIS, % 0·12 C, 0·29 Si, 1·01 Mn, 0·014 P, 0·025 S
INCLUSIONS (An) Anorthite with 18 CaO, 43 SiO_2, 42 Al_2O_3, (M) Matrix with 12 MnO, 24 CaO, 44 SiO_2, 21 Al_2O_3, (MnS) Manganese sulphide

142a Oxide inclusions in steel: anorthite in silicate matrix

inclusion with the isostructural compound Mn-anorthite is to be seen. The microscopic appearance of the Mn phase is the same as that of anorthite. The crystalline phase was precipitated during heat-treatment of the steel sample, and it contains traces of CaO in solid solution. As mentioned in section A 12, the Mn-anorthite

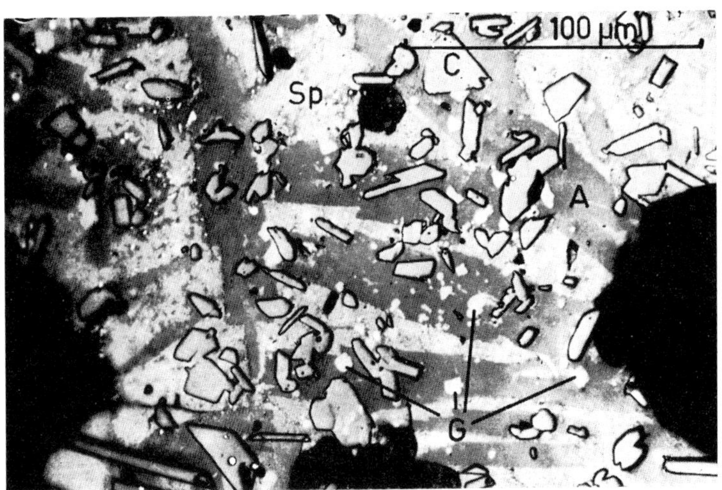

×400

STEEL TYPE Basic open-hearth steel, ferro-silicon-killed before tapping and aluminium-killed in the ladle. From forged ring
STEEL ANALYSIS, % 0·33 C, 0·34 Si, 0·60 Mn, 0·013 P, 0·018 S, 2·43 Cr, 0·12 Ni, 0·49 Mo, 0·05 Cu
INCLUSIONS, % (A) Mn-anorthite with 14 MnO, 48 SiO_2, 39 Al_2O_3, trace of CaO, (C) Corundum with 100 Al_2O_3, (Sp) Spessartite with 37 MnO, 40 SiO_2, 24 Al_2O_3, (G) Galaxite with 38 MnO, 42 Al_2O_3, 19 Cr_2O_3. Mean analysis 27 MnO, 39 SiO_2, 33 Al_2O_3
COMMENT This inclusion initially had a glassy matrix, but Mn-anorthite and spessartite (A 11) were precipitated during the heat-treatment for 2 h at 1 100°C of the steel sample. The Mn-anorthite contains traces of calcia in solid solution, this oxide seems to be necessary in order to stabilize this phase (*see* A 12)

142b Oxide inclusions in steel: Mn-anorthite, corundum, galaxite and spessartite

probably only exists if stabilized by calcia. Anorthite may also be formed as a reaction product between certain slags and refractories, if the cooling conditions are favourable for its crystallization. An example of this is shown in Fig.143, where the border zone between ladle slag and ladle refractories has been remelted at 2 500°C and heat-treated at 1 150°C. Further examples of anorthite as a reaction product from slag-refractory reactions have been given by Hayhurst and Webster,[78] who show an example of anorthite as a reaction product between high-alumina refractories and an iron and lime-rich slag. Characteristic anorthite crystals in a synthetic *C-A-S* melt are shown in Fig.144.

Crystalline, ternary phases of the C_2AS, C_3AS_3 and $C_2A_2S_5$ type are not common inclusion phases, but some examples are found in the literature. Inclusions with gehlenite, C_2AS, have been recorded by Rait and Pinder,[166] who also have identified the phase in slag from the firebrick portion of the nozzle. As shown in their figures, the phase has a featherlike structure, similar to anorthite. Opitz

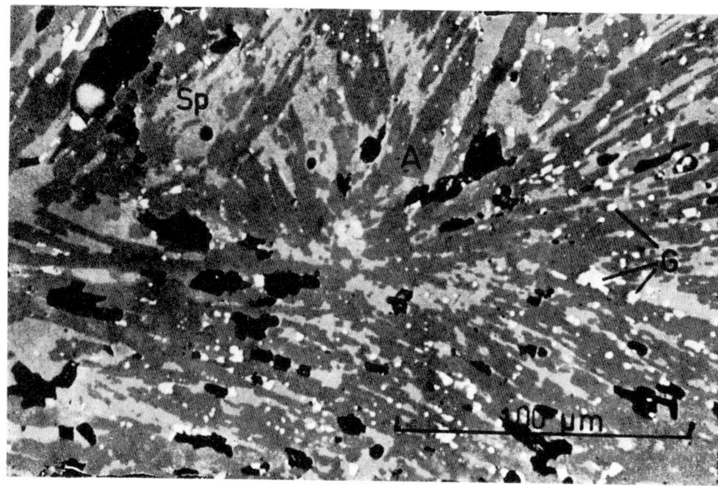

MATERIAL TYPE Sample from the border zone between ladle slag and ladle refractories, remelted in electron beam melting furnace at 2 500°C and heat-treated for 2 h at 1 150°C
SLAG PHASES, % (A) Anorthite with 16 CaO, 47 SiO_2, 38 Al_2O_3, (Sp) Spessartite with 15 MnO, 16 CaO, 50 SiO_2, 8 Al_2O_3 (G) Galaxite, $MnO{\cdot}Al_2O_3$
COMMENT Spessartite with CaO in substitutional solid solution could also be regarded as grossularite, C_3AS_3, with MnO in substitutional solid solution, *see* Fig.133

143 Crystalline anorthite (CAS_2), spessartite (grossularite C_3AS_3) and galaxite from the slag lining of the ladle

et al.[87] have identified the phase by X-ray crystallographic methods on the surface of a hard metal tool tip. It was not possible to decide, however, if the phase was originally present in the steel inclusions or if it was formed from $CaO–SiO_2–Al_2O_3$ inclusion components on the tool surface during the machining of the steel. The phase is sometimes to be found in solidified slag samples, and a discussion of the thermodynamics of its formation is given by Rein and Chipman.[72] Hayhurst and Webster[78] have indicated that gehlenite, which has been formed as a reaction product between slag and high-alumina roof refractories, may be washed into the molten steel with the liquid reaction products. Macroinclusions from ladle slag with gehlenite phase have been identified by Bruch.[76]

Crystalline grossularite, C_3AS_3, is similar in appearance to the isostructural spessartite phase (A 11). It has a featherlike, branched, dendritic structure, and in inclusions it usually contains some MnO in solid solution. It may also hold FeO (B 6) or MgO (F 7) and, as mentioned, all these ternary phases have a wide range of mutual solid solubility. Crystalline grossularite solid solutions in inclusions are shown in Figs.133 and 143.

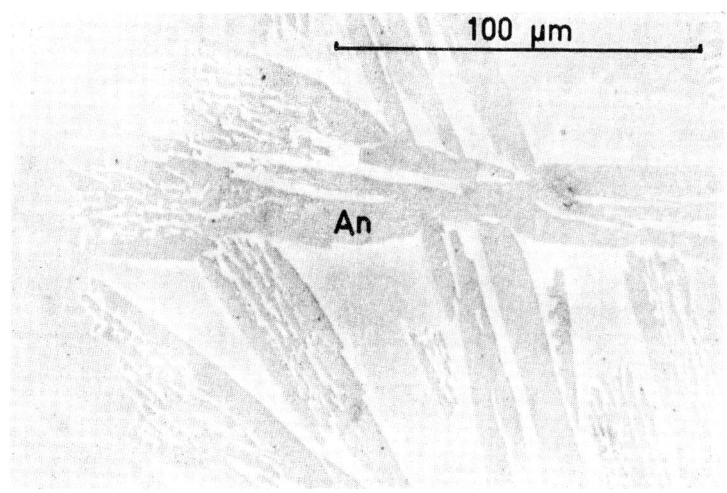

×500

MATERIAL TYPE Synthetic slag, prepared by melting a calcia-alumina-silica mixture in an electron beam melting furnace, followed by a heat-treatment for 2 h at 1 150°C and slowly cooled

SLAG ANALYSIS, % (An) Anorthite with 22 CaO, 44 SiO_2, 42 Al_2O_3. Matrix not analysed

144 Anorthite, δ·CAS_2, in a synthetic calcia-alumina-silica slag

A phase with the composition $2CaO{\cdot}2Al_2O_3{\cdot}5SiO_2$ ($C_2A_2S_5$) has not been reported previously, but evidence of its existence has been found in the present work. X-ray diffraction studies of synthetic $CaO–SiO_2–Al_2O_3$ melts with a composition close to $C_2A_2S_5$ gave evidence of a phase with a powder diffraction pattern identical with that of cordierite ($2MgO{\cdot}2Al_2O_3{\cdot}5SiO_2$, F 6). An isostructural calcium compound, probably with the stoichiometric formula $2CaO{\cdot}2Al_2O_3{\cdot}5SiO_2$, therefore apparently exists and the authors have called it Ca-cordierite. The phase could not be identified microscopically and no evidence for its formation in steel inclusions was found.

G 6 $CaO–SiO_2–Al_2O_3$: Summary and discussion

Calcium can be used as an inactive tracer element in non-metallic inclusions to determine their origin. The oxide CaO has the lowest free energy of formation among the elements usually found in steel-making (Fig.80). As a result the CaO component of slag droplets or refractories is not reduced and is unchanged in the molten steel, remaining as a component of inclusions in the final product. It is of importance, however, to know not only the calcium content but also the amounts of the elements Mg, Al, Si, Mn, and Ti present in the inclusions. If the concentrations of all these elements are known approximately, then much valuable information concerning the origin of the calcia inclusions may be obtained. This section comprises a recapitulation and summary of the occurrence and properties of calcia inclusions. Emphasis has been placed on the important problem of locating the source of an inclusion, often the practical object of inclusion analysis. Although the summary is intended to be as generally applicable as possible, it is, of course, based largely on the present authors' own observations and will be liable to modification in the light of more extensive investigations.

1. There are three main sources of calcium inclusions: the furnace and ladle refractories (Table IX), the furnace and ladle slags (Table XIV), and the deoxidants containing calcium (Table XIII). The ternary MeO–SiO_2–Al_2O_3 diagram of Fig.145 summarizes the mean compositions of all the inclusions containing calcium which have been studied in the present investigation. The binary oxide MeO is mainly calcia, but it often contains varying amounts of the oxides MnO, FeO and MgO in solid solution. For many steels the mean composition of the calcia inclusions varies even though they are taken from a small volume of a steel sample. However, the values given in the diagram may be considered as representative analyses of the inclusions.

2. Exogenous particles or drops rich in calcia act as nuclei for inclusions, but they grow and change their composition in the steel bath by the indigenous addition of oxides such as MnO and SiO_2 as long as the steel remains liquid. As a result the mean composition of calcia inclusions is usually different from the mean composition of the exogenous nuclei. Support for this view of the growth of calcia inclusions is given by a comparison between Fig.145 and the corresponding summary of the mean inclusion compositions for the MnO–SiO_2–Al_2O_3 system (Fig.56, part I). Two main concentration areas for inclusion composition are visible in the CaO–SiO_2–Al_2O_3 diagram of Fig.145. One is situated on the CaO–Al_2O_3 axis of the diagram, representing silica-free calcium aluminate inclusions with an Al_2O_3 content of between 65 and 80 wt-%. These inclusions will be further discussed below, point 4*b*. The other main concentration area is limited by the lines L_1 and L_2, representing maximum and minimum *Me*O:SiO_2 ratios. The position of these lines is nearly identical with the corresponding lines in the MnO–SiO_2–Al_2O_3 system. The same relation between the *Me*:Si ratio of the steel and and *Me*O:SiO_2 ratio of the inclusions therefore also seems to be valid for inclusions with calcia partly substituted for MnO. This is

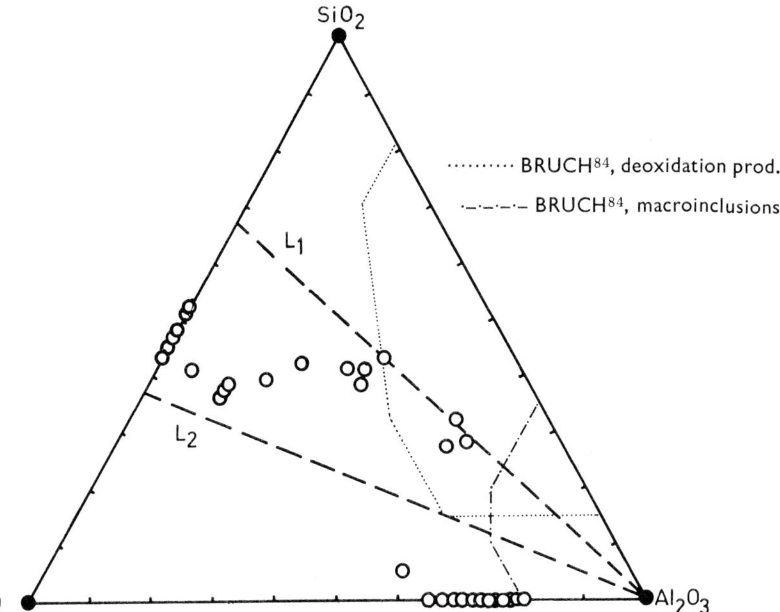

145 Summary of the mean analyses, wt-%, for the different calcia inclusions in the CaO–SiO_2–Al_2O_3 system which were studied during the present investigation
The inclusions were found in several different steel types. In several inclusions CaO is partly substituted by MnO and MgO and Al_2O_3 by Cr_2O_3. The findings of Bruch *et al*, [76,88] regarding the composition areas for deoxidation products in the CaO–SiO_2–Al_2O_3 system as well as for macro-inclusions in the same system, have been indicated. The former area is along the Al_2O_3–SiO_2 line of the diagram, the latter against the Al_2O_3-corner
Many inclusions are silica-free and are situated between about 65 and 80 wt-% Al_2O_3 on the CaO–Al_2O_3 line. (Comments are given in section G 6, point 4*b*.) Several inclusions are also situated on the CaO–SiO_2 line. (They are also discussed in section G 6, point 5.) All the other inclusions except one are situated in the central part of the ternary composition diagram, between the line L_1 and L_2. This composition area coincides largely with the low-melting parts of the system and the lines L_1 and L_2 run very close to the lines L_1 and L_2 in the MnO–SiO_2–Al_2O_3 system (*see* Fig.56). This is discussed in section G 6, point 2

an indication of a similar formation mechanism, e.g. an indigenous growth mechanism for the exogenous calcia nuclei.

3. Calcia inclusions high in MgO, containing about equal amounts of the two oxides, most probably originate from the refractories. Several compositions are high in magnesia as well as

calcia (Table IX) and may contribute to inclusions either directly or by reaction with the slags. The alternative sources of calcia–magnesia inclusions, namely entrapped furnace slag and deoxidation products, are less probable. Most of the furnace slags are low in magnesia (Table XIV) and therefore the ratio CaO:MgO in inclusions nucleated by furnace slags must be considerably greater than unity. The only exception among the more common slag types are those found in the acid open hearth, but these usually have a low total CaO+MgO content. The common deoxidants containing calcium are free from magnesium (Table XIII) and thus cannot be a source of inclusions containing MgO. These generalizations will, of course, break down where there are unusual combinations of refractory, slag and deoxidation practice. However, the general principle of the importance of the relation between the MgO and CaO contents of both the inclusions and their possible exogenous sources remains valid.

4. Inclusions rich in CaO often contain varying amounts of Al_2O_3. There are two types of such CaO–Al_2O_3 inclusions, those with SiO_2 and those free of SiO_2.

(*a*) The most common type among those with SiO_2 are monophase, glassy, inclusions with a variable composition in the *C-A-S* range. A typical example is shown in Fig.134. They should be analysed with the electron probe method since their composition may vary within wide limits without any detectable difference in their optical appearance (Figs.40, 125 and 134).

(*b*) Many inclusions containing CaO have been found to be free of SiO_2 and have an Al_2O_3 content of 65–80 wt-%. They often also contain varying amounts of MgO (Fig.145, Table XXIV). A possible mode of formation would be reduction of the SiO_2-component of silicates in the ladle slag to Si by Al during the Al-deoxidation of the steel in the ladle. The ladle slag often holds calcium silicates as well as varying amounts of magnesium silicates. In the border zone between the steel and the slag phases, as well as around drops of liquid ladle slag, dispersed in the molten steel, the following reactions are thermodynamically possible ($\Delta G°$ for $Al_2O_3 < SiO_2$, Fig.80). $CS + \underline{Al} \rightarrow CA + \underline{Si}$ resp. $MS + \underline{Al} \rightarrow MA + \underline{Si}$. The solid reaction products are the aluminates of calcium and magnesium, which may be dispersed in the steel phase and thus appear as inclusions.

Another hypothesis would be formation of this inclusion type during a reaction between Al_2O_3 and dolomite. Experiments have shown that if burnt dolomite is melted in an alumina crucible, a reaction starts between the acid Al_2O_3 and the basic dolomite, the resulting reaction zone being microscopically very similar to this type of inclusions (Fig.146).

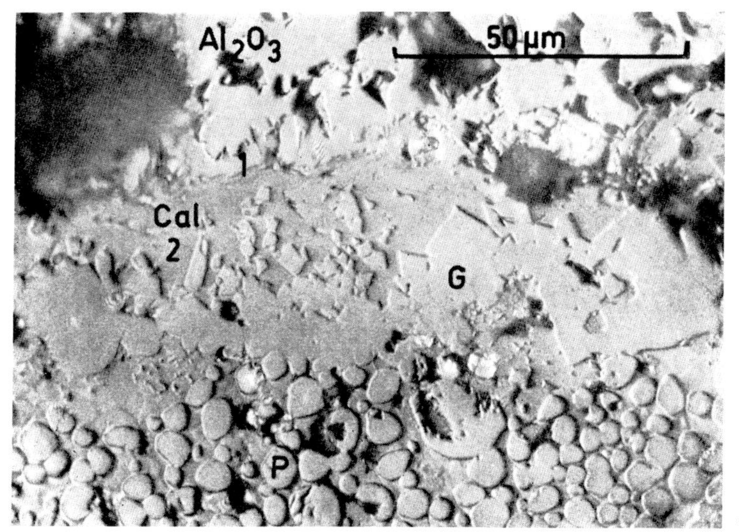

×800

MATERIAL TYPE Laboratory melt of burnt dolomite in a pure alumina crucible. The crucible was heated in air to about 1 000°C
STRUCTURE COMPONENTS, % (Cal 1) CA_6, (Cal 2) CA_2 (probably), (G) Spinel with 18 MgO, 78 Al_2O_3, (P) Periclase
COMMENT Compare the appearance of silica-free calcium aluminate inclusions in section G 3, for instance Fig.110, and the discussion in section G 6, point 4*b*

146 The reaction zone between burnt dolomite and Al_2O_3

(*c*) The calcium-containing inclusions discussed above have a characteristic response to deformation of the steel matrix. They are only broken up by severe hot-working of the steel, forming strings of hard particles elongated in the direction of working. This response to deformation is distinct from that of the plastically deformable silicate inclusions (compare, for example, Figs.113–114 with Figs.134–135 or Figs.28–29).

5. Calcium inclusions rich in SiO_2 may also be free from Al_2O_3. If such inclusions are found in the formerly molten parts of the weld, they may not be inclusions in the parent steel but originate from the silicate flux or the oxide coating of the electrode. TiO_2 is often a component of coatings and therefore electron probe analysis of such inclusions is of value when deciding if they originate from the steel or from the welding process (Table XXV). An example of inclusions formed during the welding process is shown in Fig.130.

6. Bruch *et al.*[76,88] have studied the composition of calcium inclusion isolates after CaSi deoxidation of the steel. The mean

composition of these calcia inclusions was found to be situated within the dotted area of Fig.145. They also studied a number of macroinclusions with calcia, which all had a composition near the Al_2O_3-corner of the CaO–SiO_2–Al_2O_3 diagram (Fig.145). These results are therefore different from those of the present work. The reason for this difference in composition of the calcia inclusions is very likely that the main refractory material in the studies of Bruch was chamotte, which is an aluminium-silicate refractory. In contrast the refractories for nearly all the steels investigated by the present authors were neutral or basic bricks of different kinds (Table IX). The difference in the results as well as the lack of correlation between steel analysis and inclusion analysis (*see* E) is a further indication of the influence of the many details of the steel practice on the inclusion composition. A knowledge of the furnace and ladle slag composition, the composition of furnace, ladle, and runner refractories as well as the timing of the deoxidation procedure is usually necessary to trace the origin of the calcia inclusions, but as their origin is exogenous this is often possible.

7. Salmon Cox and Charles[126] have pointed out that potassium compounds are often used as additions to refractories for different purposes. This potassium behaves similarly to calcium and can therefore be used to trace the origin of calcia inclusions. They have used electron probe analysis, which is often sensitive enough to indicate the presence of traces of potassium in the inclusions ($>$0·1 wt-% have been detected).

8. Finally, a phase has been observed in melts of CaO–SiO_2–Al_2O_3 which has not been previously described. Its powder diffraction diagram is analogous to that of cordierite (F 6), and its composition is close to that of $2CaO{\cdot}2Al_2O_3{\cdot}5SiO_2$ (*cf.* cordierite, $2MgO{\cdot}2Al_2O_3{\cdot}5SiO_2$). It was therefore called Ca-cordierite.

H. Oxide inclusions with the Group III, IV and V transition metals

H 1 Oxide inclusions with Group III transition metals (the rare earth elements)

General

In Group III, only the rare earth elements are of any practical interest in steelmaking. They all have a strong oxygen affinity and have been tried as deoxidizers for steel. An alloy of about 45–55 wt-% of cerium, 22–30 wt-% of lanthanum, 15–18 wt-% of neodymium and 5 wt-% of praseodymium is known as mischmetall and is occasionally used in steelmaking. Not much information is available about oxide inclusions with these metals, but LaO_2 and CeO_2 have been identified in steel inclusions, as well as oxysulphides of the rare earth metals.[89–92] They also have a strong sulphur affinity and they are in fact more related to sulphide inclusions than to oxide inclusion (*see* section K).

H 2 Oxide inclusions with Group IV transition metals (Ti, Zr, Hf)

General

Titanium is an alloying element in certain steels, and alloys with titanium as well as zirconium are of some interest as deoxidizers for steel. The free energy of formation is low for TiO_2 and ZrO_2 (Fig.80), and titanium is also present in several calcium–silicon deoxidation alloys (Table XIII), as well as in some ferroalloys. Oxide inclusions with titanium or zirconium oxides may therefore be formed from these different sources.

TiO_2 and ZrO_2 are also present in some refractories, used in steelmaking. The content of these oxides is usually low (Table IX), but oxide inclusions with different amounts of titanium and zirconium oxides may result from slag or steel reactions with these refractories.

The coatings of welding electrodes are often rich in TiO_2 and this coating forms a slag during welding, which is often rich in titanium oxides. The analytical composition of such slags, formed by different electrode types, has been experimentally determined[93] and a summary of analyses for the principal phases is given in Table XXV. The table shows that all electrode slags have a titanium content, the amount of which, however, may vary from traces only up to about 50 wt-%, depending on the type of electrode.

A study of the deoxidation mechanism of iron by titanium has been carried out by Evans and Sloman,[94] who also investigated the composition of the resulting inclusions. A whole series of different titanium–iron–oxygen compounds was formed, which could be identified by X-ray diffraction methods. If these phases are arranged according to their increasing oxygen content, the following sequence of phases was found:

$TiO(?) \rightarrow Ti_2O_3 \rightarrow Ti_3O_5 \rightarrow TiO_2$(rutile) $\rightarrow FeO{\cdot}TiO_2$(ilmenite) $\rightarrow$ Fe_2TiO_4 (inverse spinel) and an unknown phase called X-phase.

The inclusions were further studied by Pickering,[95] who identified microscopically all the phases except the Ti_3O_5 and X-phases. However, he questioned the identity of the TiO-phase and suggested that this could also be TiN.

These pure titanium phases are not usually present in inclusions, but some comments about their structure may still be helpful as an indication of the solubility and phase relations obtained with titanium in oxide steel inclusions.

TiO. Cubic; space group Fm3m, a=4·18 Å. The lattice parameter varies due to a varying oxygen content and due to a wide range of solid solubility between TiO, TiC and TiN. ASTM data card 8–117, NaCl-type. Wyckoff I, chap. III, text p.1, table p.13. Isostructural with manganosite (A 6), wüstite (B 2 and D 2), periclase (F 2), calcia (G 2), TiC and TiN.

Monochromatic X-ray powder photograph of manganosite and wüstite, *see* Figs.19 and 75, part I.

It is questionable whether TiO appears as a deoxidation product, however, this oxide is of metallurgical importance since oxygen, nitrogen and carbon all are soluble in the fcc metal lattice of TiX with NaCl structure. TiN appears as a precipitate in several steels containing titanium, where it forms idiomorphic, regular, yellowish crystals usually in the grain boundaries.[96] These precipitates often hold both carbon and oxygen in solid solution. The TiC-phase in sintered carbides also often holds oxygen in solid solution.

Ti_2O_3 A high- and low-temperature modification are known. Only the low-temperature form, α-Ti_2O_3 is of metallurgical interest. It is hexagonal and isostructural with corundum (A 2), escolaite (C 2) and haematite (D 4), but the solid solubility for these Me_2O_3-oxides in α-Ti_2O_3 is limited. In microsections of iron containing α-Ti_2O_3 inclusions the phase is globular, chocolate-coloured, often with a banded, twin-like substructure. Photomicrographs of the phase are to be found in the monographs of Trojer,[9] and Allmand.[2] According to these authors the banded structure is a result of the transformation of high to low Ti_2O_3.

Ti_3O_5, anosovite The phase is monoclinic and only stable above 1 200°C, but it may be supercooled down to room temperature. The phase has been identified by X-ray diffraction methods in inclusion isolates from iron deoxidized with titanium.[94] Pickering,[95] however, could not identify it *in situ* in the same material and the nature of appearance in inclusions is unknown.

TiO_2, rutile Tetragonal, space group P4/mnm. a=4·58 Å, c=2·95 Å. ASTM data card 4–0551. Cassiterite type. Wyckoff I:IV, table p.15, text p.5. Bragg and Claringbull,[52] p.107. Two other modifications of TiO_2 are known, brookite and anatase. TiO_2 as rutile is a component of several refractories (Table IX) and also, in varying amounts, of different types of welding electrodes (Table XXV). The metals Cr, V and Nb are soluble in the rutile phase but not Fe, Al and Mn.

$FeO{\cdot}TiO_2$, ilmenite The phase is hexagonal and structurally similar to haematite (Fe_2O_3, D 4). At higher temperatures a complete range of solid solubility exists, but an immiscibility gap forms at lower temperatures. The structural similarity comes from the possibility of substituting $2Fe^{3+}$ for $Fe^{2+}+Ti^{4+}$ in the hexagonal lattice. The same type of substitution is also possible between ilmenite and corundum (Al_2O_3, A 2) and escolaite (Cr_2O_3, C 2), a factor of importance for inclusion formation. The structure of ilmenite is discussed by Bragg and Claringbull.[52] ASTM data card 3–0781.

Fe_2TiO_4 Cubic; space group Fd3m. a=8·50 Å. Inverted spinel type. Wyckoff II:VIII, table p.72. The inverted spinels are closely related in structure to the double oxides $AO{\cdot}B_2O_3$ of the spinel type (A 7, F 3), and this iron titanate should be written $FeO{\cdot}(Fe^{2+}Ti^{4+})O_3$ to indicate that $Fe^{2+}+Ti^{4+}$ are structurally equivalent to the $2B^{3+}$ atoms in the spinel lattice. These relations have been discussed by Bragg and Claringbull.[52] Within inclusions the phase may therefore have other double oxides of spinel type in solid solution and, conversely, titanium is often to be found as a solid solution in the spinel inclusion phases.

Several other phases with titanium are of interest as possible constituents of inclusions. For example, $MnO{\cdot}TiO_2$ and $MgO{\cdot}TiO_2$

are both isostructural with ilmenite, $FeO{\cdot}TiO_2$, with a range of solid solubility. Although $CaO{\cdot}TiO_2$ is also known its structure is different, belonging to a crystallographically important class of double oxides, the perovskites. For further discussion of the crystal structures, *see* Bragg and Claringbull. Several other Me–Ti–O compounds are known, but most of them are of no particular interest for steel inclusions. Some ternary $MeO–SiO_2–TiO_2$ compounds exist, of which the mineral $CaO{\cdot}SiO_2{\cdot}TiO_2$, called sphene (Bragg and Claringbull[52]) is the most important for inclusion formation. In the concluding Tables XXVI to XXIX, which summarize the oxide inclusion phases, more relevant titanium-oxygen compounds are mentioned.

Many of the titanium-oxygen phases are therefore possible as oxide components of inclusions. Titanium is usually found in solid

TABLE XXVI Summary of phases of the types AO, SiO_2 and B_2O_3 in the different $AO–SiO_2–B_2O_3$ inclusion systems discussed in the present report. Some Ti-, Zr- and V-phases have also been included. The amount of each element, wt-%, in the different phases has been calculated

Type	Ideal formula	*Me*	O	Name	Reference
*A*O	MnO	77	23	Manganosite	A 6
	FeO	78	22	Wüstite	B 2, D 2
	MgO	60	40	Periclase	F 2
	CaO	72	28	Calcia (lime)	G 2
	TiO (?)	75	25	...	H 2, ref.95, p.148 ev. TiN
B_2O_3	Al_2O_3	53	47	Corundum	A 2
	Cr_2O_3	68	32	Escolaite	C 2
	Fe_2O_3	70	30	Haematite	D 4
	Ti_2O_3	67	33	...	H 2, ref.95, p.148–149
	V_2O_3*	68	32	...	H 3, ref.9, p.68
MeO_2	TiO_2	60	40	Rutile (Anatase, Brookite)	H 2, ref.95, p.148 Ref.77, p.232
	ZrO_2	74	26	Baddeleyite	H 2, ref.77, p.232, ref.9, p.8
	VO_2*	61	39	...	H 3, ref.9, p.86
Me_2O_3	V_2O_3*	56	44	...	H 3, ref.9, p.68
SiO_2	SiO_2	47	53	Cristobalite	A 3
				Tridymite	A 4
				Quartz	A 5

* Phases marked with an asterisk have not been found in inclusions, slags or refractories met in steelmaking.

TABLE XXVII Summary of phases of the types AO–SiO_2 in the different AO–SiO_2–B_2O_3 inclusion systems discussed in the present report. The amount of each element, wt-%, in the different phases has been calculated according to the ideal formula of the compounds

Type	Ideal formula	A_1	A_2	Si	O	Name	Reference
$AO.SiO_2$	$MnO.SiO_2$	42	...	21	37	Rhodonite	A 9
	$FeO.SiO_2$	42	...	21	37	Grunerite′	B 4
	$MgO.SiO_2$	24	...	28	48	Enstatite (Clino- and Protoenstatite)	F 4
	$CaO.SiO_2$	35	...	24	41	Wollastonite (Pseudo- and Para-wollastonite)	G 4
	$CaO.MnO.2SiO_2$	16	22	23	39	Bustamite	Fig.125*b*
	$CaO.FeO.2SiO_2$	16	22	23	39	Hedenbergite	Figs.125*c* and 126*a*
	$CaO.MgO.2SiO_2$	19	11	26	44	Diopside	ref.85, p.672
	$MnO.MgO.2SiO_2$	24	11	24	41	...	Fig.93
$3AO.2SiO_2$	$3CaO.2SiO_2$	42	...	19	39	Rankinite	G 4
	$2CaO.FeO.2SiO_2$*	27	18	18	37	Ferroackermannite	ref.9, p.356
	$2CaO.MgO.2SiO_2$	29	9	21	41	Ackermannite	ref.9, p.355
$2AO.SiO_2$	$2MnO.SiO_2$	54	...	14	32	Tephroite	A 10
	$2FeO.SiO_2$	55	...	14	31	Fayalite	B 3
	$2MgO.SiO_2$	35	...	20	45	Forsterite	F 5
	$2CaO.SiO_2$	47	...	16	37	Bredigite (Larnite, Shannonite)	G 4
	$CaO.MnO.SiO_2$*	22	29	15	34	Glaukochroite*	ref.9, p.346
	$CaO.FeO.SiO_2$	21	30	15	34	Kirschsteinite*	ref.9, p.347
	$CaO.MgO.SiO_2$	26	15	18	41	Monticellite	ref.60, pp.495 and 496
	$3CaO.MgO.2SiO_2$	37	7	17	39	Merwinite	Fig.141, ref.60, p.490
$3AO.SiO_2$	$3CaO.SiO_2$	53	...	12	35	Alite	G 4
$(3B_2O_3.2SiO_2)$	$(3Al_2O_3.2SiO_2)$	38	...	13	49	(Mullite)	A 8

* Phases marked with an asterisk have not been found in inclusions, slags or refractories met in steelmaking.

TABLE XXVIII **Summary of phases of the types $AO–B_2O_3$ in the different $AO–SiO_2–B_2O_3$ inclusion systems discussed in the present report. Some SiO_2-free Ti- and V-phases have also been included. The amount of each element, wt-%, in the different phases has been calculated according to the ideal formula of the compound**

Type	Ideal formula	A	B	O	Name	Reference
$3AO.B_2O_3$	$3CaO.Al_2O_3$	45	20	35	...	G 3
$2AO.B_2O_3$	$2CaO.Fe_2O_3$	30	41	29	...	ref.98, pp.156 and 168
	$4CaO.Al_2O_3.Fe_2O_3$				Brownmillerite	ref.9, p.263
$12AO.7B_2O_3$	$12CaO.7Al_2O_3$	33	34	33		G 3
$AO.B_2O_3$	$MnO.Al_2O_3$	32	31	37	Galaxite	A 7
	$FeO.Al_2O_3$	32	31	37	Hercynite	B 5
	$MgO.Al_2O_3$	17	38	45	Spinel	F 3
	$CaO.Al_2O_3$	25	34	41	...	G 3
	$MnO.Cr_2O_3$	24	47	29	Cr-galaxite	C 3
	$FeO.Cr_2O_3$	25	46	29	Chromite	C 4
	$MgO.Cr_2O_3$	13	54	33	Picrochromite	ref.9, p.241
	$CaO.Cr_2O_3$	19	50	31	...	ref.9, p.273
	$MnO.Fe_2O_3$	24	48	28	Jakobite	ref.9, p.254
	$FeO.Fe_2O_3$	24	48	28	Magnetite	D 3
	$MgO.Fe_2O_3$	12	56	32	...	Fig.84
	$CaO.Fe_2O_3$	19	52	29	...	Figs.85 and 104*b*
	$MnO.Ti_2O_3$	25	45	30	...	ref.9, p.203
	$FeO.(Fe^{2+}Ti)O_3$	50	21	29	...	H 2. ref.95, pp.148, 149
	$MgO.(Mg,Ti)O_3$					ref.9, p.201
	$CaO.Ti_2O_3$	20	48	32	...	ref.9, p.199
	$MnO.V_2O_3$	25	46	29	...	H 3, Figs.15 and 139
	$MnO.(MnV)O_3$	49	23	28	...	H 3
	$FeO.V_2O_3$	25	46	29	...	H 3. ref.9, p.167
	$MgO.V_2O_3$	13	54	33	...	H 3. ref.9, p.167
	$MgO.(MgV)O_3$	30	31	39	...	H 3. ref.9, p.169
	$CaO.V_2O_3$	19	50	31	...	H 3. ref.9, p.167
$AO.2B_2O_3$	$CaO.2Al_2O_3$	15	42	43	...	G 3
	$CaO.2Fe_2O_3$	11	59	30	...	ref.98, pp.171 and 177
$AO.6B_2O_3$	$CaO.6Al_2O_3$	6	49	45	...	G 3
$AO.TiO_2$	$MnO.TiO_2$	36	32	32	Pyrophanite	H 2. ref.9, p.203
	$FeO.TiO_2$	37	31	32	Ilmenite	H 2. ref.95, pp.148–149
	$MgO.TiO_2$	20	40	40	Geikielite	H 2. ref.9, p.199
	$CaO.TiO_2$	30	35	35	Perovskite	H 2. Fig.130
$AO.2TiO_2$	$FeO.2TiO_2$	24	41	35	...	ref.9, p.202

TABLE XXIX **Summary of ternary phases of the types $AO–B_2O_3–SiO_2$ in the different $AO–SiO_2–B_2O_3$ inclusion systems discussed in the present report. The amount of each element, wt-%, in the different phases has been calculated according to the ideal formula of the compound**

Type	Ideal formula	A	B	Si	O	Name	Reference
$2\textit{A}O.B_2O_3.SiO_2$	$2CaO.Al_2O_3.SiO_2$	29	20	10	41	Gehlenite	G 5
	$2CaO.(AlFe^{3+})O_2.SiO_2$	27	9(Al) 18(Fe)	9	37	Fe-gehlenite	ref.9, p.356
$3\textit{A}O.\textit{B}_2O_3.3SiO_2$	$3MnO.Al_2O_3.3SiO_2$	33	11	17	39	Spessartite	A 11
	$3FeO.Al_2O_3.3SiO_2$	34	11	17	38	Almandine	B 6
	$3MgO.Al_2O_3.3SiO_2$	18	13	21	48	Pyrope	F 7
	$3CaO.Al_2O_3.3SiO_2$	27	12	19	42	Grossularite	G 5
	$3MgO.Cr_2O_3.3SiO_2$	16	23	19	42	Hanleite	ref.9, p.363
	$3CaO.Cr_2O_3.3SiO_2$	24	21	17	38	Uwarowite	ref.9, p.362
	$3MnO.Fe_2O_3.3SiO_2$	30	20	15	35	Calderite	ref.9, p.363
	$3CaO.Fe_2O_3.3SiO_2$	24	22	16	38	Andradite	ref.9, p.363
$\textit{A}O.\textit{B}_2O_3.2SiO_2$	$MnO.Al_2O_3.2SiO_2$	19	18	19	44	Mn-anortite	A 12, G 5, Fig.142*b*
	$CaO.Al_2O_3.2SiO_2$	15	19	20	46	Anorthite (high-anorthite)	G 5
$2\textit{A}O.2\textit{B}_2O_3.5SiO_2$	$2MnO.2Al_2O_3.5SiO_2$	17	17	22	44	Mn-cordierite	A 13. ref.76, p.213
	$2FeO.2Al_2O_3.5SiO_2$	17	17	22	44	Fe-cordierite	B 7
	$2MgO.2Al_2O_3.5SiO_2$	8	19	24	49	Cordierite	F 6
	$2CaO.2Al_2O_3.5SiO_2$	13	17	23	47	Ca-cordierite	G 5
$4\textit{A}O.5\textit{B}_2O_3.2SiO_2$	$4MgO.5Al_2O_3.2SiO_2$	12	34	7	47	Sapphirine	F 8
$\textit{A}O.\textit{B}_2O_3.4SiO_2$	$MgO.Al_2O_3.4SiO_2$	6	14	30	50	Osumilite-type	F 9
$\textit{A}O.\textit{B}_2O_3.8SiO_2$	$MgO.Al_2O_3.8SiO_2$	4	9	36	51	Petalite-type	F 10

solution in the phases described in sections A-G, and several examples of this will be mentioned below.

Zirconium is closely related to titanium both from a crystallographic and a metallurgical point of view. Some information about different oxide compounds with zirconium is given by Trojer.[9]

Hafnium is also very similar to titanium, but little is known about its oxide compounds.

Microscopy and electron probe analysis

Pure titanium-oxygen phases of the types discussed above are not very common in steel inclusions, and few examples have been found in literature. Pickering[97] has identified α-Ti_2O_3 in a titanium-stabilized 18:10 steel. TiO_2 (and ZrO_2) were found in inclusion isolates from titanium-alloyed (and zirconium-alloyed) steel reported by Koch.[77] In the BISRA report on the microscopic identification of inclusions in steel by Allmand,[2] some examples of different inclusions with titanium-oxygen phases are shown and their microscopic appearance discussed. They were mostly taken from steels, deoxidized with titanium or ferrotitanium.

Several examples of oxide inclusions with varying amounts of titanium in solid solution in different phases have already been given. Titanium was present in double oxides of the spinel type in the inclusions shown in Figs.28 and 117. Titanium in rhodonite was found in the inclusions in Figs.11, 17, 41, and 140 and in the glassy *C-A-S* matrix in Figs.125*a*, 134, and 135. Titanium in weld inclusions is shown in Figs.126 and 130, the origin for the titanium component of these inclusions probably being the electrode coating. The inclusion in Fig.130*d* has several phases rich in titanium, and in this inclusion an example of the pure perovskite phase, $CaO{\cdot}TiO_2$, was also observed.

H 3 Oxide inclusions with Group V transition metals (V, Nb, Ta)

General

These metals are mainly of interest as alloying additions to the steel, rather than as deoxidizers. Their oxygen compounds are not usually found in refractories but they are sometimes components of welding electrode coatings. Their oxides are therefore not common in steel inclusions but some comments may still be of value.

Vanadium Several compounds V–*Me*–O exist which are of interest as inclusion phases and vanadium oxides with a valency of the metal of 3, 4, and 5 are known.

$\mathbf{V_2O_3}$ is isostructural with the other ternary inclusion oxides Al_2O_3, Cr_2O_3, Fe_2O_3 and α-Ti_2O_3. Double oxides of the spinel type and

with extended homogeneity ranges are known, such as $FeO{\cdot}V_2O_3$, $MnO{\cdot}V_2O_3$, and $MgO{\cdot}V_2O_3$, where V^{3+} can be partly substituted by Al, Cr and Fe^{3+}. A compound with calcia, $CaO{\cdot}V_2O_3$, has also been found to exist, but it is not of the spinel type.
$\mathbf{VO_2}$ is monoclinic. Double oxides of the inverse spinel type (compare ilmenite, H 2) are known, of which $MgO{\cdot}(Mg^{2+}V^{4+})O_3$ and $MnO{\cdot}(Mn^{2+}V^{4+})O_3$ are of interest as possible inclusion phases.
$\mathbf{V_2O_5}$ is orthorhombic. Calcium vanadates and magnesium-vanadates with V^{5+} are known.
Niobium and *tantalum* usually have the valency 5, and several oxide compounds of these metals with Mn, Fe, Mg and Ca are known.

Microscopy and electron probe analysis
No pure vanadium phases have been observed in oxide inclusions during the present investigation, but in several inclusions vanadium was found in solid solution in the spinel phase. These inclusions were all present in steel high in vanadium or, in one example, in a weld. In Fig.15, an inclusion from a chromium steel with 0·25% V is shown. Vanadium was found to be present in only one of the inclusion phases, galaxite (A 7), and this double oxide should probably here be written as $MnO{\cdot}(Cr,V,Al)_2O_3$.

An inclusion in a high-speed tool steel, shown in Fig.139, was found to be high in vanadium. Again, vanadium was only found in the double oxide phase, which held more than 30-wt% of V_2O_3 and was of the same type as that in the chromium steel.

Double oxides of the spinel type with vanadium have also been found in other inclusions in high-speed steels, for instance of the general type $(Mn,Fe)O{\cdot}(Cr,V)_2O_3$. The inclusions usually have several phases but vanadium has only been found to be dissolved in the double oxide phase. Finally, vanadium has also been found in inclusions from the fusion line of a weld (Fig.126*b*). The origin of the vanadium component of these inclusions was most probably the electrode coating, which contained vanadium.

J. Oxide inclusions: Tabular summary of phases

Oxide inclusions in steel can nearly always be referred to one of the oxide systems which have been discussed in sections A–G, i.e. the general system $AO–SiO_2–B_2O_3$. A stands for one or several of the binary metals Mn, Fe^{2+}, Mg and Ca; B for one or several of the ternary metals Al, Cr and Fe^{3+}. Other metals, such as Sr, Ti, Zr, V and Nb are sometimes also present in oxide inclusions, but they are nearly always found in solid solution in the phases discussed in sections A–G.

In the different sections of the present review, only phases from the pure systems $AO–SiO_2–B_2O_3$ have been discussed, i.e. such systems where A and B have been only one of the binary or ternary metals mentioned above. Some comments on solid solubility have also been given, but phases such as $CaO·MgO·SiO_2$, monticellite, have not been systematically discussed. These phases may be components of steelmaking slags but have not been found in non-metallic inclusions (*see* Tables XXVI to XXIX).

In this section a summary, in the form of tables, has been given of all the different oxide phases discussed in the present report with references to the appropriate sections. The tables also include phases with more than one of the binary and ternary metals, which are of interest for oxide inclusions, e.g. monticellite. If possible reference to their microscopic appearance has been given as well as other information relevant to their connexion with oxide inclusions. The gravimetric composition with regard to the different elements present in all phases has also been calculated in order to facilitate the identification of the different phases in oxide inclusions by electron probe analyses.

Table XXVI deals with phases of the types AO, B_2O_3 and SiO_2, Table XXVII with phases in the systems $AO–SiO_2$ or $B_2O_3–SiO_2$, Table XXVIII with phases in the systems $AO–B_2O_3$, and Table XXIX with ternary phases in the systems $AO–SiO_2–B_2O_3$. Some phases with titanium, zirconium, and vanadium have also been included.

K. Sulphide inclusions

K 1 General

Sulphur is soluble in the molten steel phase, but its solubility in the solid steel phase is very low. It is precipitated in the form of metal sulphides during solidification of the steel and the precipitation pattern is influenced by its strong segregation tendency.

The stability of the different metal sulphides relative to the elements is indicated by the curves for the standard free energy of formation against temperature. A graph showing this for a number of sulphides important in steelmaking is shown in Fig.147 (from Muan and Osborn[53]). The curves represent equilibrium values for standard states, i.e. pure substances. The activity for the elements is thus equal to 1. Lines of equal partial sulphur pressure in the gas phase are also given in Fig.147, calculated according to the relation

$$\Delta G^\circ = RT \ln p_{S_2}$$

It can be concluded from Fig.147 that the tendency for sulphide formation of the pure metals increases in the approximate sequence Ni→Fe, Cu→Mn, Ti→Al→Mg, Na, K→Ca. The familiar action of metallic calcium and magnesium as purifiers for sulphur in steel depends on the high stability (low free energy of formation) of the sulphides of these metals relative to the elements. The relative stability of the different sulphides is, however, dependent on temperature and sulphur pressure, as can be seen from the figure.

Figure 147, as well as the corresponding graph for oxides in Fig.80, gives the ΔG° values, i.e. the conditions for the pure elements when the activity is 1. This assumption is usually oversimplified in steelmaking. The molten steel is to be regarded as a dilute solution of a great number of elements in iron, and for many of them the solution is far from saturation and from the point where the activity is unity. Furthermore, activities may be decreased or increased when elements are dissolved in iron. The activity of sulphur is, for instance, strongly and positively influenced by carbon, silicon and phosphorus. The diagrams in Fig.147 (and Fig. 80) should therefore only be used

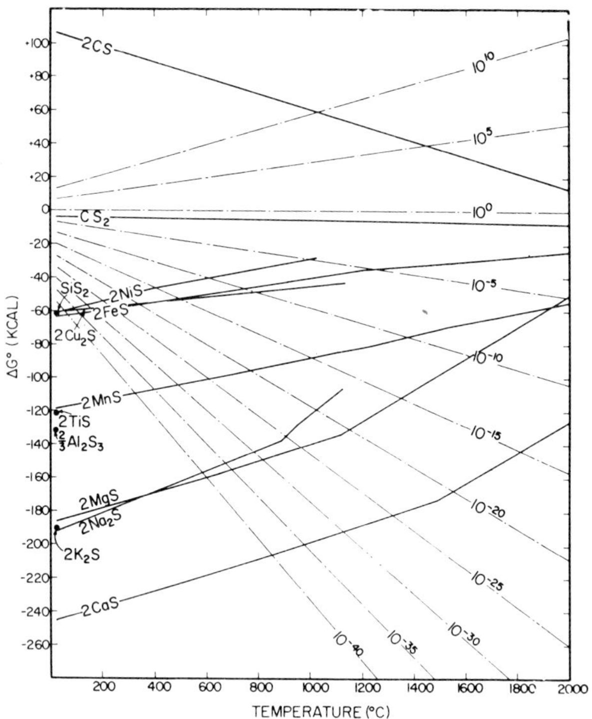

147 Standard free energies of formation $\Delta G°$(kcal) of various sulphides from the elements as a function of temperature. Light dash-dot curves are lines of equal sulphur pressure, atm, of the gas phase, as labelled on each curve. From Muan and Osborn,[53] Fig.3, p.8

with great care. If possible ΔG values should be calculated in each special case from the $\Delta G°$ values with the necessary corrections for the actual contents and activity coefficients.*

The sulphur potential–temperature diagrams for sulphides in iron- and steelmaking processes may also be corrected in a similar way, giving equilibrium lines for contents usual in steel practice. Such a correction has been made by Jorgensen[100] for the reaction

$$2Mn\ (\%) + S_2 \rightarrow 2MnS_{(s)}$$

---

* Eketorp[99] has made such corrections for some oxygen potential–temperature diagrams for oxides in iron- and steelmaking processes. The actual positions of equilibrium lines for contents usual in steelmaking deviate considerably from the $\Delta G°$ diagrams and it is possible to use the corrected diagrams directly for a discussion of the different chemical reactions in steelmaking.

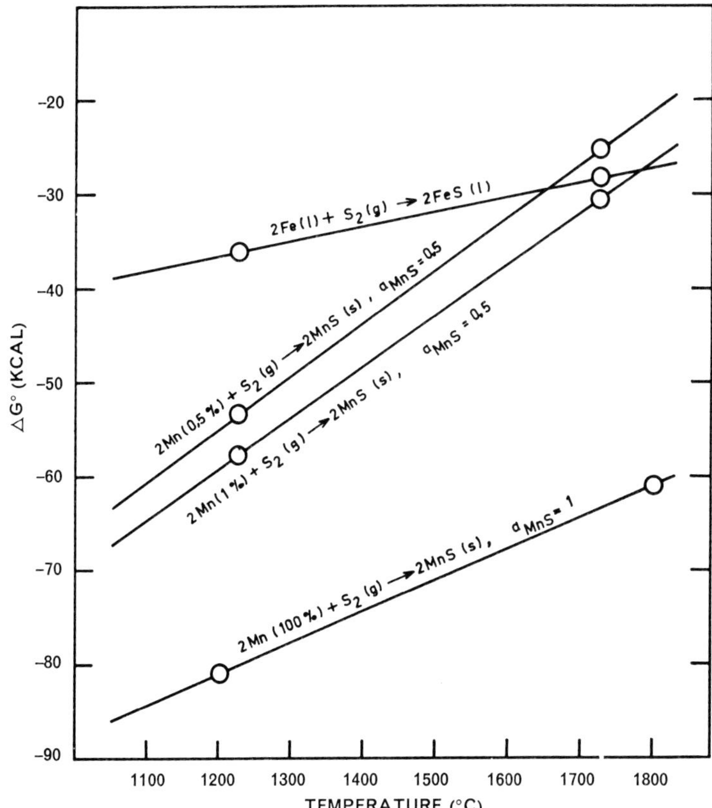

148 Free energy of formation $\Delta G = RT \ln p_{S_2}$ for the reaction $2Mn(\%)+S_2 \rightarrow 2MnS(S)$ as a function of temperature. ΔG has been calculated for a 0·5% and 1% solution of Mn in pure Fe. The activity for MnS has been given a value of 0·5 as MnS often holds Fe in solid solution (*see* K 4). The standard free energies of formation for FeS and MnS according to the reactions $2Fe+S_2 \rightarrow 2FeS$ and $2Mn+S_2 \rightarrow 2MnS$ have also been included, taken from Fig.147

Corrected ΔG diagrams have been calculated for a content of 1% and 0·5% of Mn in Fe. The results are shown in Fig.148, where the ΔG° values for FeS formation from pure Fe have also been shown. It is evident from Fig.148 that the sulphur potential temperature diagrams for MnS under actual steelmaking conditions have been moved due to the lower Mn-activity in the steel as compared with pure Mn and now are closer to the FeS-diagram. The equilibrium diagrams for MnS and FeS at actual steelmaking conditions now intersect in the temperature range of 1 600–1 800°C instead of at temperatures higher than 2 000°C. The implication of this is that the tendency for the formation of FeS and MnS at steelmaking concentrations and temperatures is about the same, but that MnS is the more stable phase at room temperature. In contrast, according to the $\Delta G°$ equilibrium curves (Fig.147), FeS should never be formed except in Mn-free steels. This will be further discussed below.

Elements like Ca, Mg, etc. are usually not present as pure metals but as oxide slags in steelmaking. The desulphurizing power of the slag phase is therefore dependent on the relative stability of the sulphides of these metals relative to those of the oxides of the same element. Muan and Osborn[53] have given curves, showing differences $\Delta(\Delta G°)$ between the standard free energies of formation of sulphides and the corresponding oxides and some of their curves are shown in Fig.149. The diagrams can be used in order to estimate the tendency for the reaction

$$2MeO_{(c)}+S_2 \rightarrow 2MeS_{(c)}+O_2$$

to proceed from left to right, i.e. for one mole of oxygen to exchange position with one mole of sulphur. This tendency will be the higher, the less positive the $\Delta(\Delta G°)$ value is for the reaction. These diagrams should also be used with great care. They are all based on $\Delta G°$

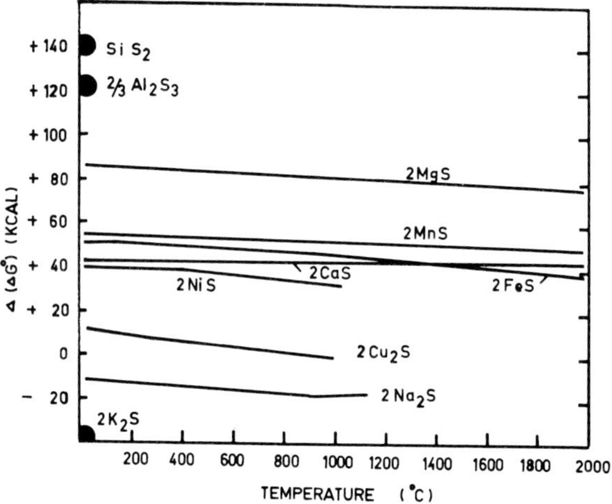

149 Differences in free energies of formation $\Delta(\Delta G°)$ kcal, between sulphides and corresponding oxides at different temperatures. From Muan and Osborn[53] Fig.6, p.11

values for pure elements with unit activity and they do not therefore correctly represent the actual steelmaking conditions. They do at least indicate the general tendency of different elements to form sulphides in the presence of both sulphur and oxygen. This tendency increases in the approximate sequence Si, Al, Ti, Mg, Mn, Fe, Ca, Ni, Cu, Na, K. Following the discussion by Muan and Osborn,[53] (p.7) the p_{S_2}/p_{O_2} ratio of the gas phase in equilibrium with the two condensed phases oxide and sulphide is determined according to the reaction above by the relation

$$\log p_{S_2}\,(\text{eq}) - \log p_{O_2}\,(\text{eq}) = \frac{\Delta(\Delta G^\circ)}{2{\cdot}3RT}$$

As shown by Fig.149, $\Delta(\Delta G^\circ)$ is positive for most oxide-sulphide combinations except for Na and K. The partial pressure of sulphur must therefore exceed that of oxygen in the gas phase in order to form sulphides from the oxides. Vacuum melting of the steel is one method of increasing the ratio between sulphur and oxygen in the gas phase, and may therefore increase the sulphide:oxide inclusion ratio.

Metal–slag reactions are the main means of removing sulphur from the steel, and the free-energy curves are also, as mentioned by Muan and Osborn, a good guide to indicate the relative sulphur-removing capacity of the various constituents of a silicate slag. Strong bases (e.g. CaO) are much better desulphurizers than strong acids (e.g. SiO_2). Even more effective than CaO are the alkaline oxides Na_2O and K_2O, but the high volatility of the latter limits their usefulness in steelmaking processes. They also have a deleterious effect on refractories.

CaS is one of the metallurgically important sulphides which have a low $\Delta(\Delta G^\circ)$ value. This sulphide is also stable in direct contact with the corresponding oxide phase. An example with CaS inside an inclusion with a CaO component is shown in Fig.87. Also MnS and FeS may be stable in oxide inclusions, if the partial pressure of sulphur has been high. Several such inclusions have already been shown, in Figs.16, 17, 30, 41, and 42 different examples of the appearance of MnS precipitates in oxide inclusions have been given, and corresponding inclusions with FeS are shown in Figs.23 and 162*b*. The $\Delta(\Delta G^\circ)$ value for Al_2S_3-formation in the presence of oxygen is unfavourable (K 12).

A general discussion of the formation and distribution of sulphide inclusions in steel ingots is given by Baeyertz.[101] The formation of sulphide inclusions is very similar to the separation of oxide inclusions with two distinctions of degree rather than character of the reactions. In the first place, under most conditions encountered in steelmaking practice, the steel will hold sulphur in solution without the separation of a second phase until the steel has cooled below the usual teeming temperatures. Secondly, temperature has a greater effect on the solubility of the usual sulphides in steel than on most of the oxide inclusions. Sulphides precipitate close to the advancing front of crystallization where the liquid steel phase is rich in sulphur. The precipitation of sulphides from the sulphur-rich liquid phase is facilitated by existing nuclei, for instance oxide inclusions in the steel bath. Oxide inclusions with an outer rim of sulphide are therefore common in those parts of the steel ingot which have been rich in sulphur. Examples of this type of sulphide crystallization are shown in Figs.41 and 167 (MnS) and 112 (CaS). This mode of sulphide formation is different from the precipitation of sulphides inside oxide inclusions, which was discussed above.

The distribution of the sulphide inclusions in the steel ingots depends on the steelmaking process and is usually different in rimmed steel and killed steel. In the rimmed steels, most of the sulphide inclusions are localized in the core of the ingot, often as longitudinal segregations, whereas the sulphides in killed ingots are concentrated more to the outer regions of the ingots. This difference in localization has been discussed by Hayes and Chipman.[102]

In rimming steel ingots, the stirring action caused by the liberation of gas almost entirely defeats the mechanism of entrapment of sulphide inclusions by the advancing front of crystallizing metal, but causes a considerable differentiation in the sulphur content of the solid metal and the remaining liquid which solidifies as the core of the ingot. The sulphide inclusions are therefore concentrated in this part. In a killed steel ingot, however, nearly 100% entrapment occurs in the outer region where the freezing is rapid.

In the following discussion of sulphide inclusions, the main emphasis is on the sulphides FeS and MnS. Some other sulphide systems of interest in steelmaking have also been discussed.

The metal–sulphur systems are often rather complicated with a great number of intermediate phases which often have extended homogeneity ranges. Most of the systems are of little interest in steelmaking but the trend in steel metallurgy (vacuum methods, microalloyed steels, etc.) is to systems or methods where a more detailed knowledge of the sulphide inclusions is essential. A knowledge of sulphide inclusions is also essential for a more detailed understanding of the different deformation processes, governing, for example, the machinability of steel. The shape, composition and distribution of the sulphide inclusions is of great importance for free-machining steels, and a major difference exists between the influence and behaviour of sulphide and oxide inclusions in these steels. Therefore comments on some metal sulphides have been included in the present report, although they are only of limited importance. Some general review articles on different metal sulphide systems are available, giving information about the phases and their crystal and chemical properties. Thus Jellinek[103] has summarized the behaviour of sulphides of the Groups IV, V, and VI transition metals, Samsonov[104] the sulphides and oxysulphides of the rare earth metals and Rosenqvist[105] the iron, cobalt and nickel sulphides. These articles, together with the well-known bibliography by Hansen[106] on binary alloys have been used as general references for the different metal-sulphur systems of interest in steelmaking. The systems discussed in the present report are summarized in Table XXX, and they also include some oxy- and carbosulphides (section

TABLE XXX Summary of sulphide systems, briefly discussed in the present report, with special reference to sulphides of interest in steelmaking

Sulphide system	Group in per. table	Phases encountered in steelmaking (ex.)	Section
Those most important in steelmaking			
Fe–S	VIII	FeS	K 2
Mn–S	VII	MnS	K 3
Mn–Me–S	VII	(Mn, *Me*)S, MnS.Me_2S_3	K 4
With transition metals			
La-elements–S	III	LaS, CeS	K 5
Ti,Zr,Hf–S	IV	TiS	K 6
V,Nb,Ta–S	V	VS	K 7
Cr,Mo,W–S	VI	CrS, MoS_2	K 8
Co,Ni–S	VIII	Ni_3S_2	K 9
With non-transition metals			
Cu,Ag–S	I	Cu_2S	K 10
Mg,Zn,Cd–S	II	MgS	K 11
Ca,Sr–S	II	CaS	K 11
Al–S	III	Al_2S_3	K 12
Sn,Pb–S	IV	...	K 13
With non-metals			
Me–S–*X*	*X*=O,C,N	$Ti_4C_2S_2$, Ce_2O_2S	K 14
Me–Se	...	MnSe	L 1
Me–Te	...	MnTe	L 2

K 14). Also some selenium and tellurium compounds, which are closely related to sulphides and are of interest in steelmaking, have been mentioned in sections L 1–2.

K 2 Sulphide inclusions. The iron–sulphur system

General

Two intermediate phases, FeS and FeS_2, have been reported in the FeS system and minerals with these compositions are also known. Several modifications of FeS exist, crystallographically closely related but differing slightly in composition. The most common modifications of FeS are the pyrrhotite and troilite minerals. FeS_2 exists in two modifications, pyrite and marcasite. A thermodynamic study of the FeS system was made by Rosenqvist,[105] whose equilibrium diagram is shown in Fig.150.

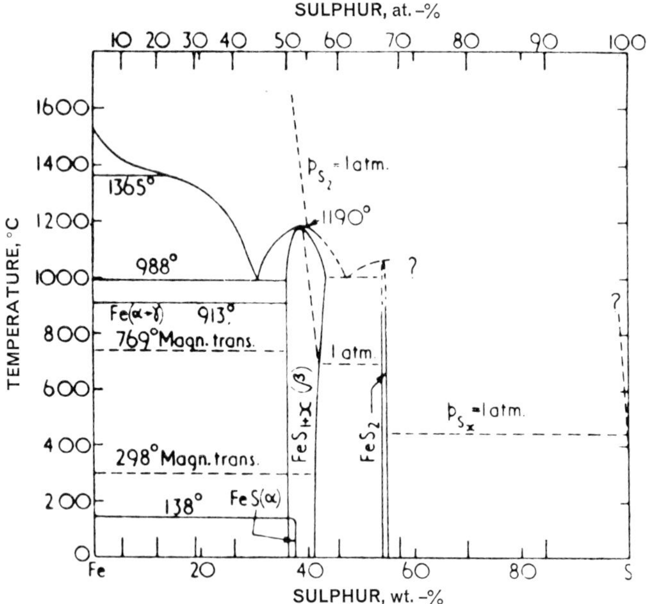

150 The binary system iron-sulphur. From Rosenquist[105], Fig.2, p.40

The main emphasis in this section will be on FeS, troilite, which is the most important of these sulphides in steel inclusions, but some X-ray information about the other sulphides has also been given to facilitate their identification.

Chemical formula $Fe_{1-x}S$

This phase has an extended homogeneity range with a ratio Fe:S$\leqslant$1 and the x-value may vary between about zero and 0·18 (compare wüstite, B 2). Hagg and Sucksdorff[107] have shown that the sulphide has metal vacancies and therefore the phase should be designated $Fe_{1-x}S$ if its exact formula is to be given and not FeS_{1+x}, which has

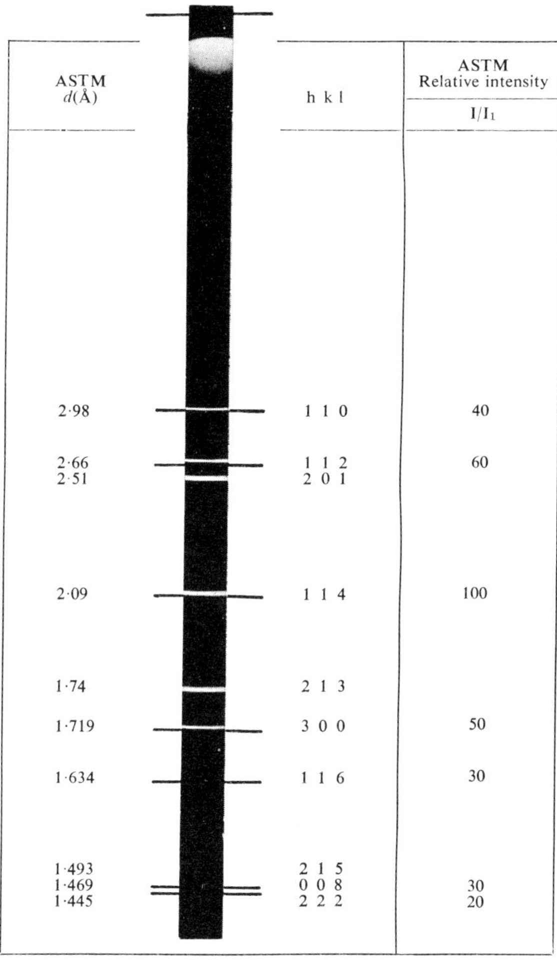

ASTM d(Å)	h k l	ASTM Relative intensity I/I_1
2·98	1 1 0	40
2·66	1 1 2	60
2·51	2 0 1	
2·09	1 1 4	100
1·74	2 1 3	
1·719	3 0 0	50
1·634	1 1 6	30
1·493	2 1 5	
1·469	0 0 8	30
1·445	2 2 2	20

d-values, hkl-values and relative intensities are quoted from the ASTM index, data card 11–151

151 X-ray powder photograph of FeS, troilite; monochromatic FeKα-radiation; Guinier camera

sometimes been the case (for instance in the diagram in Fig.150). In the present report the formula is usually given as FeS irrespective of sulphur content.

Modifications and structure type

Two modifications of FeS exist, these are crystallographically different but still closely related. The structure type depends on the sulphur content of the phase.[107] In addition some modifications are also known (α, β, γ) which have the same crystal structure but differ in magnetic properties.[108]

$Fe_{1-x}S$ with the lowest sulphur content ($x=0$) has a hexagonal lattice and the crystal structure is a superstructure of the NiAs-lattice. This phase as a mineral has the name troilite, and in those cases where FeS inclusions have been isolated from steel and studied by X-ray methods, the FeS phase has always been found to have the structure of troilite. Troilite has a hexagonal symmetry and belongs to space group P6c2 with the parameters $a=5{\cdot}928$ Å and $c=11{\cdot}698$ Å. It is recorded in the ASTM index, data card 11–151. An X-ray diffractograph of troilite has been reproduced in Fig.151.

For higher sulphur contents (increasing x-values) the crystal structure of $Fe_{1-x}S$ changes to the more simple basic structure of NiAs, and the pyrrhotite minerals usually have this crystal structure. This structure type has not been reported, however, for FeS in steel inclusions. It also has a hexagonal symmetry and belongs to space group P6/mmc with parameters $a=3{\cdot}32$ Å and $c=5{\cdot}79$ Å. ASTM data card 4–0832.

The crystal structure of FeS is discussed in Wyckoff I:III, text p.29, table p.38a and in Bragg and Claringbull,[52] p.61.

FeS_2, pyrite, is cubic with the parameter $a=5{\cdot}40$Å and belongs to space group Pa3. ASTM data card 6–0710. The phase is isostructural with MnS_2 (K 3), $MnSe_2$ (L 1) and $MnTe_2$ (L 2). An X-ray diffractograph of pyrite is reproduced in Fig.152.

FeS_2, marcasite, is orthorhombic with the parameters $a=3{\cdot}37$ Å, $b=4{\cdot}44$ Å, and $c=5{\cdot}39$ Å. It belongs to space group Pmnm. ASTM data card 3–0799. An X-ray diffractograph of marcasite is reproduced in Fig. 153.

Physical properties

Melting point: 1190°C

Density: 4·77

Colour: FeS has a yellow ochre colour in the optical microscope. It is anisotropic and therefore is active in polarized light, whereas the cubic MnS is inactive

Microhardness: FeS is reported to be harder than MnS[109a]

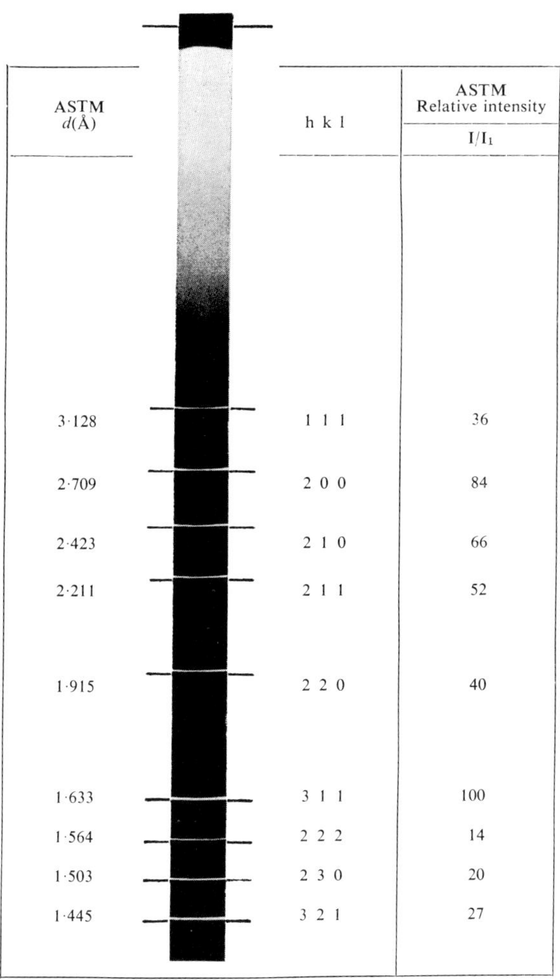

ASTM d(Å)	h k l	ASTM Relative intensity I/I₁
3·128	1 1 1	36
2·709	2 0 0	84
2·423	2 1 0	66
2·211	2 1 1	52
1·915	2 2 0	40
1·633	3 1 1	100
1·564	2 2 2	14
1·503	2 3 0	20
1·445	3 2 1	27

d-values, hkl-values and relative intensities are quoted from the ASTM index, data card 6–0710

152 X-ray powder photograph of FeS_2, pyrite; monochromatic FeKα-radiation; Guinier camera

Solid solubility

FeS-inclusions may have a chromium content of up to 20%, according to electron probe analyses by Philibert *et al.*[110] They have also been found with titanium[111] and vanadium[112] in solid solution,

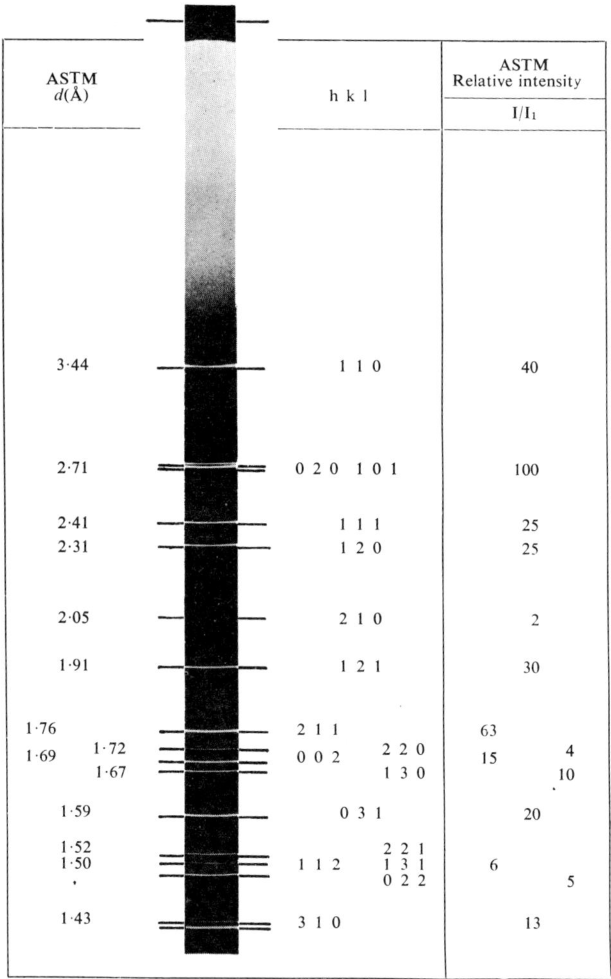

d-values, hkl-values and relative intensities are quoted from the ASTM index, data card 3–0799

153 X-ray powder photograph of FeS_2, marcasite; monochromatic FeKα-radiation; Guinier camera

whereas the solid solubility for manganese is reported to be low.[113] However, the authors have found a solid solubility of about 7 wt-% of manganese in FeS (Fig.159) and Matsubara[114] has also found a rather high solid solubility for manganese.

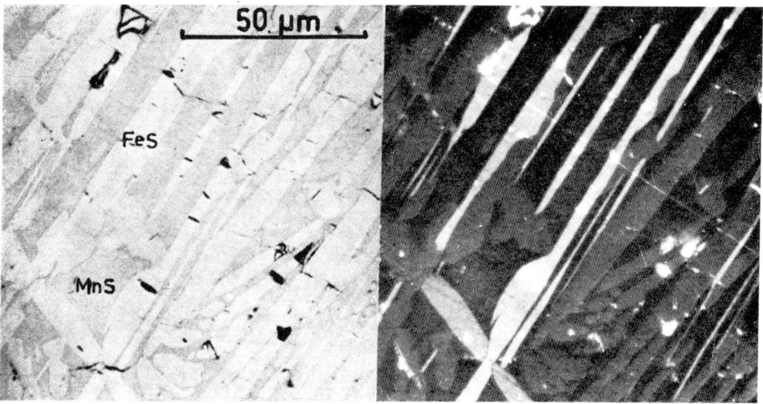

SULPHIDE ANALYSIS, % (FeS) 1 Mn, 58 Fe, balance S; FeS-type (MnS) 24 Mn, 36 Fe, balance S; MnS-type

154 Iron sulphide, precipitated from (Mn,Fe)S in a synthetic sulphide. The solid solution was prepared from Mn, Fe and S-powder and had the approx. composition $Mn_{.2}Fe_{.8}S$. It was heat-treated for 2 h at 1 150°C

The solid solubility of carbon and oxygen in FeS is low. No carbosulphides or oxysulphides of iron have been reported, i.e. compounds, where sulphur is substituted by carbon or oxygen, but two-phase inclusions with FeS and FeO or with FeS and Fe-silicate are common (Figs.23, 155–158). Such inclusions have also been photographed by Portevin and Castro.[115a]

The formation of FeS in steel; hot-shortness
Sulphur is soluble in liquid iron but the solubility is very low in the solid iron phase. When iron solidifies, sulphur therefore precipitates as FeS, either as primary sulphide dendrites or as a eutectic constituent with iron. This eutectic has a low melting point (988°C, Fig.150) and the melting point is decreased by the influence of oxygen. For instance, the FeO–FeS eutectic has a melting point of about 940°C. Sulphur therefore segregates to those parts of the ingot which are the last to solidify and the grain boundaries of these parts will be rich in FeS. The sulphide phase may either form primary FeS (melting point 1 190°C) or be present in different eutectics. If the ingot is reheated to temperatures in the range of 900–1 200°C, a liquid phase may be formed in the grain boundaries. This may cause cracking of the iron, starting in the grain boundaries, if the iron is plastically deformed. This phenomenon is called hot-shortness and was early described by Rinman,[116] who associated this failure

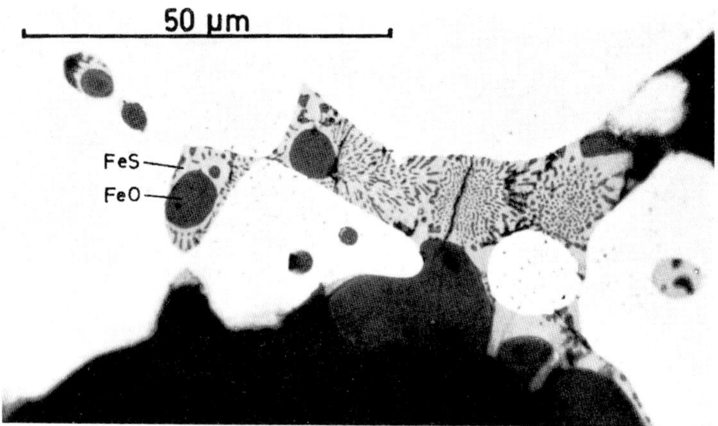

STEEL TYPE Carbon steel, annealed in oil-gas-heated furnace for 32 min at 1 050°C
GAS ANALYSIS, % 1 O_2, 1·7 CO, 14 CO_2, 0·8 H_2. Sulphur content of oil 2·1
STEEL ANALYSIS, % 0·09 C, trace Si, 0·53 Mn, 0·008 P, 0·024 S, 0·13 Cu

SURFACE LAYER (FeS) Iron sulphide, (FeO) Wüstite
COMMENT Such surface layers may accidentally be worked into the steel during rolling or forging, resulting in exogenous steel inclusions. Compare Figs.23 and 159

155 Iron sulphide-wüstite eutectic in oxide scale

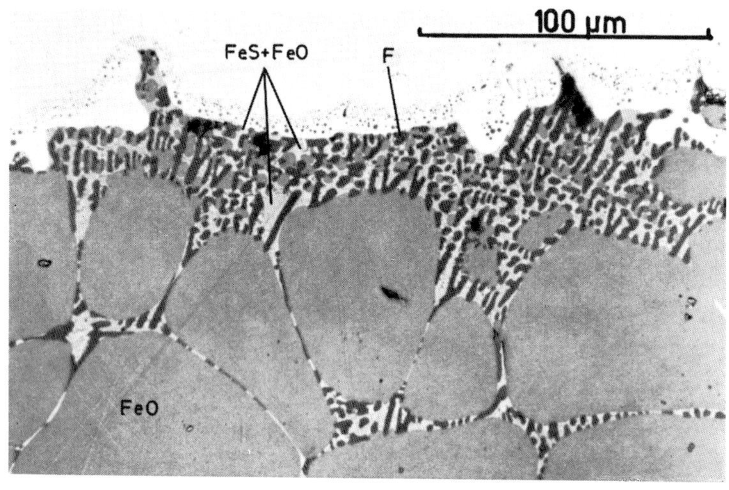

STEEL TYPE Carbon steel, annealed in oil-gas-heated furnace for 8 min at 1 150°C
GAS ANALYSIS, %1 O_2, 1·7 CO, 14 CO_2, 0·8 H_2. Sulphur content of oil 2·1
STEEL ANALYSIS %0·61 C, 1·79 Si, 0·93 Mn, 0·010 P, 0·008 S

SURFACE LAYER (FeS) Iron sulphide, (FeO) Wüstite, (F) Fayalite, $2FeO \cdot SiO_2$
COMMENT *See* Fig.155. A zone in the steel phase below the surface oxide layer consists of small oxide inclusions, formed from 'inner oxidation'

156 Iron sulphide, wüstite and fayalite in oxide scale

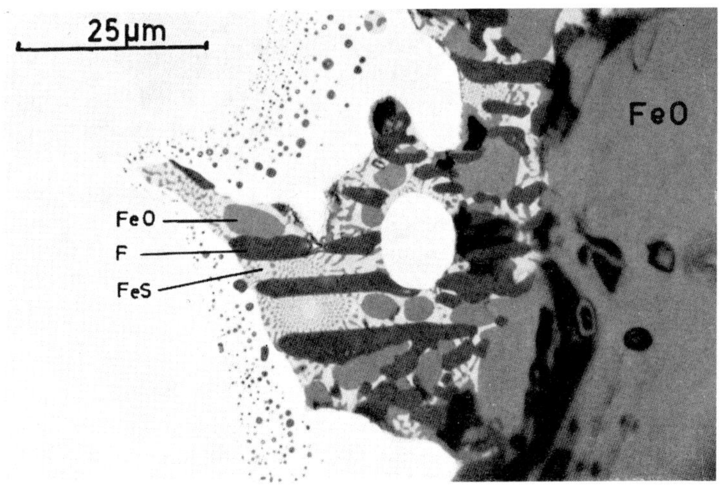

STEEL TYPE *See* Fig.156
STEEL ANALYSIS *See* Fig.156
SURFACE LAYER (FeS) Iron sulphide, (FeO) Wüstite, (F) Fayalite, $2FeO{\cdot}SiO_2$
COMMENT *See* Figs.155 and 156

157 Iron sulphide, wüstite and fayalite in oxide scale

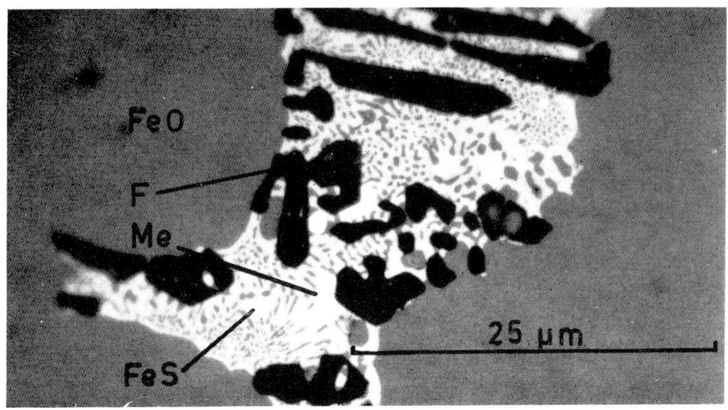

STEEL TYPE *See* Fig.156
STEEL ANALYSIS *See* Fig.156
SURFACE LAYER (FeS) Iron sulphide, (FeO) Wüstite, (F) Fayalite, $2FeO{\cdot}SiO_2$, (Me) Steel
COMMENT This figure shows part of the oxide scale in high magnification, using oil immersion optic. The fayalite is therefore darker than in Figs.156–157

158 Iron sulphide, wüstite and fayalite in oxide scale

of the iron with its sulphur content. Several photomicrographs of FeS in grain boundaries have been given by Portevin and Castro.[115b]

In their study of hot-shortness Hultgren and Herrlander[117] also found evidence of cracking in cases where FeS could not be microscopically observed in the grain boundaries. They therefore concluded that FeS could be present in the grain boundaries as submicroscopic precipitates or that a supersaturated solid solution of sulphur in or near the grain boundaries could also exist.

In modern steels formation of FeS is avoided by manganese additions. As shown in Figs.147 and 148, MnS has a lower free energy of formation than FeS. The sulphur will therefore be bonded to manganese instead of iron, provided that the manganese content has been high enough. In steelmaking practice it has been found that a sulphide of α-MnS type will replace the FeS-type sulphide if the ratio Mn:S is higher than about 4. MnS has a much higher melting point than FeS (1 610°C as compared with 1 190°C for FeS) and also the different possible MnS-eutectics have higher melting points than those with iron sulphide. Therefore the formation of liquid phases in the grain boundaries at hotworking temperatures is avoided.

In exceptional conditions FeS may also be formed in steel with manganese. One possible source is precipitation of FeS from the supersaturated solid solution of Fe in MnS (section K 4), which may be formed at steelmaking temperatures and supercooled to room temperature. A Widmannstätten-like precipitate of FeS inside the MnS-phase is sometimes formed.[109b] but the FeS-precipitate may also be precipitated with other morphological patterns, as shown by the synthetic sulphide of Fig.154. These (Mn,Fe)S solid solutions may quite possibly be formed at those extreme temperature gradients which exist during welding and cause hot-shortness. Hot-shortness in the presence of manganese has also been observed in steels which have been annealed in oil-heated furnaces and exposed to the oxidation products from the oil, which usually have a comparatively high sulphur content.[118] The sulphur content of the oxide scale on the steel surface is increased to such an extent that a low-melting phase with FeS, wüstite, and often silicates can be formed in eutectics with melting points down to 900°C and below. Microstructures of such oxide scales, containing eutectics with FeS in the grain boundaries, are shown in Figs.155–158. They have all been formed by enrichment of the oxide scale with sulphur from oil oxidation products.

Microscopy and electron probe analysis

FeS is not a common phase in modern steels due to their manganese content. The sulphide is easy to identify in the optical microscope. It has a characteristic yellow ochre colour, is active to polarized light and, due to its formation as a surface reaction product, it is

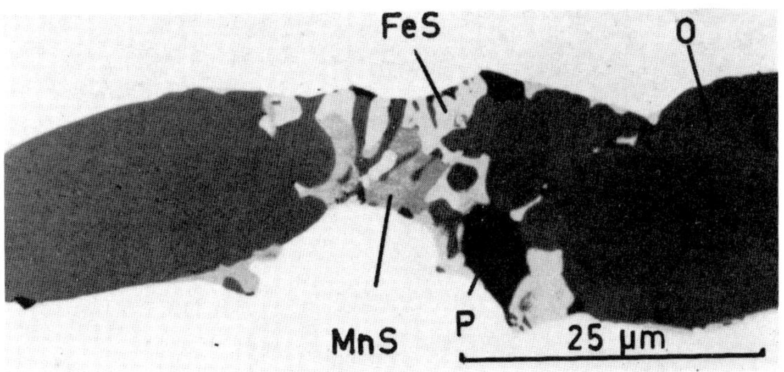

STEEL TYPE Rimmed carbon steel, rolled to 65 mm, longitudinal section
STEEL ANALYSIS, % 0·04 C, 0·14 Mn, 0·012 P, 0·015 S, 0·003 N
INCLUSION (FeS) 55 Fe, 7 Mn, 35 S; FeS-type, (MnS) 37 Fe, 24 Mn, 35 S; MnS-type, (O) 73 FeO, 32 MnO, (P) (Fe, Mn)PO_4

COMMENT Both sulphide phases are solid solutions. This inclusion has probably been formed from an oxide scale of the type shown in Fig.155 and accidentally worked into the steel (the phosphate is a residue from surface treatment)

159 Sulphide inclusions in steel: iron sulphide, manganese sulphide, wüstite and phosphate

usually situated in the austenite grain boundaries at the steel surface. It is easy to distinguish from the light-grey, isotropic MnS phase (Fig.162*b*). The difference in colour and appearance between these two sulphides is shown by the inclusion in Fig.159, which has a eutectic formed by FeS and MnS. A duplex sulphide inclusion with a MnS nucleus and a rim of FeS is shown by Portevin and Castro,[115c] and FeS precipitated within MnS is shown in Fig.154. Morrogh[119] has shown several photomicrographs of the appearance of FeS inclusions in cast steel, often together with other sulphides.

Oxide scales on steel annealed in oil furnaces often have FeS in the grain boundaries. If the silicon content of the steel is low, a eutectic between FeS and wüstite is formed. Such a eutectic in a steel free of Si, in the boundary between the oxide scale and the steel, is shown in Fig.155. For higher silicon contents a liquid Fe–S–O–Si phase may be formed in the boundary between the steel phase and the oxide scale during annealing, from which different phases may be precipitated during cooling. Usually different amounts of fayalite (B 3) will be found together with the FeO–FeS eutectic; several examples from a 1·79 wt-% Si steel are shown in Figs.156–158. If the steel has not been thoroughly freed from such oxide scales, they may be accidentally worked into the steel during rolling or forging, resulting in exogenous steel inclusions containing iron oxides, FeS and often also fayalite. The inclusions shown in Figs.23 and 159 most probably have been formed in this way.

As mentioned above, Mn can be taken up in solid solution by FeS, and in Fig.159 one of the phases in the inclusion is a solid solution of the (Fe,Mn)S type. This solid solution is identical with FeS in the optical microscope and it is necessary to use electron probe analysis in order to establish its existence. This solid solution (Fe,Mn)S of FeS type is quite different in structure and appearance from the solid solutions (Mn,Fe)S of MnS type, which are discussed in section K 4 and of which an example is shown in Fig.159.

K 3 Sulphide inclusions: The manganese–sulphur system

General

Two intermediate phases, MnS and MnS_2, have been reported for the Mn–S system.[120] Three different modifications of MnS are known,[121] α, β, and β′. MnS_2 also forms a mineral, hauerite, and this sulphide phase is isostructural with pyrite. It is not of great interest in steelmaking and will not be discussed further.

Inclusions of the MnS-type are the most frequent and also most important sulphide inclusions in modern steels. The equilibrium diagram for the Mn–MnS system is shown in Fig.160 and it can be

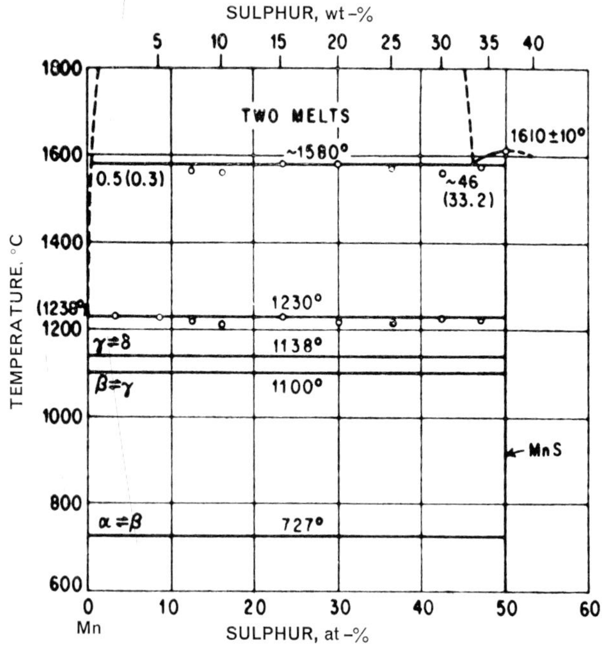

160 The system manganese-sulphur (to MnS) (from Hansen [106]). This system has an immiscibility gap

concluded from this diagram that an immiscibility gap exists for the liquid phase between about 0·3 and 33·5 wt-% S.

Chemical formula MnS
This phase has always been found to have the stoichiometric composition and no extended homogeneity range as for FeS has been reported.

Modifications and structure type
Only α-MnS has been observed in steel inclusions. It has a cubic crystal lattice belonging to space group Fm3m and with the parameter $a=5{\cdot}226$ Å. It is recorded in the ASTM index, data card 6–0518, and is isostructural with manganosite (A 6), wüstite (B 2), periclase (F 2), and CaO (G 2). An X-ray diffractograph of α-MnS has been reproduced in Fig.161.

β-MnS is cubic of zincblende type with the parameter $a=5{\cdot}61$ Å. ASTM data card 3–1065.

β′-MnS (sometimes called γ-MnS) is hexagonal of wurzite type with the parameters $a=3{\cdot}98$ Å. $c=6{\cdot}44$ Å, ASTM data card 3–1062.

MnS_2 is cubic of pyrite type with the parameter $a=6{\cdot}095$ Å, ASTM data card 10–476.

Physical properties
Melting point: 1610°C
Density: 3·99
Microhardness: 170 kp/mm². This sulphide can have large amounts of other metals in solid solution and will then be much harder (Fig.177)
The colour of inclusions of the phase *in situ* is light-grey in the optical microscope; it is isotropic and usually slightly transparent

Solid solubility
Substitutional solid solutions of the α-MnS type are known, where Mn is substituted by other metals. They will be discussed in section K 4. Also S may be substituted, especially by Se, and these solid solutions are mentioned in section L 1, but the solid solubility for Te is low (L 2).

Double sulphides of the spinel type with the formula $MnS{\cdot}B_2S_3$ are known (K 4) which are structurally analogous to double oxides $AO{\cdot}B_2O_3$ of the same type.

At room temperature, the solubility of oxygen in MnS is low,[122,123] but duplex inclusions with the phases MnS+silicate or $MnS+Al_2O_3$ are rather common.[122, 124] Radtke and Schreiber[125]

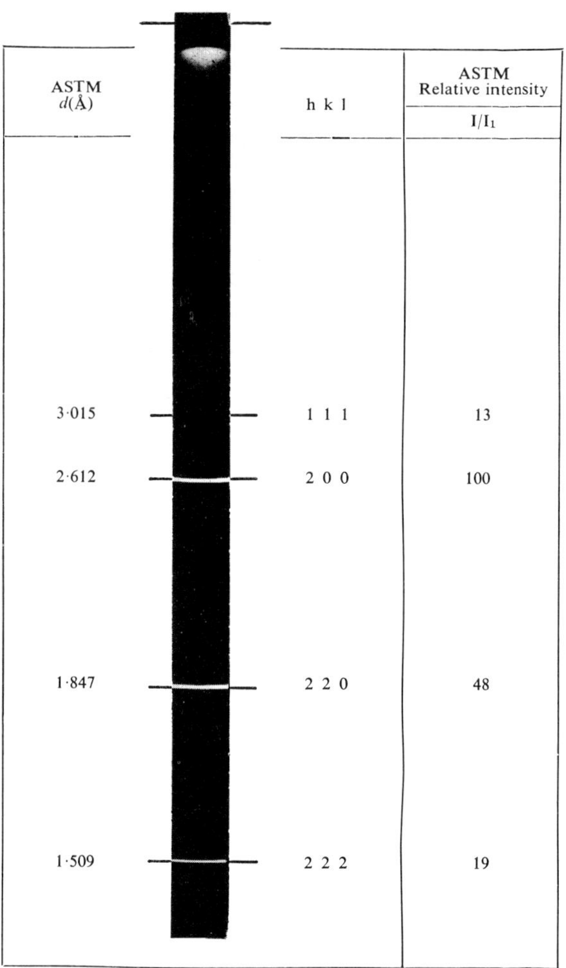

ASTM d(Å)	hkl	ASTM Relative intensity I/I_1
3·015	1 1 1	13
2·612	2 0 0	100
1·847	2 2 0	48
1·509	2 2 2	19

d-values, hkl-values and relative intensities have been quoted from the ASTM index, data card 6–0518

161 X-ray powder photograph of α-MnS; monochromatic FeKα-radiation; Guinier camera

have reported a dark iron-manganese sulphide with more than 30 wt-% O in solid solution. However, according to the present authors' opinion this may not be a single phase but the oxygen may be present as a very fine precipitate of silicate or iron manganese oxide. This dark phase was found in duplex inclusions together with a lighter iron manganese sulphide free from oxygen. Salmon Cox and Charles[126] have found evidence of small amounts of Cu and Ni in solid solution in MnS.

Lead is insoluble in MnS, but the morphology of the MnS-phase seems to be influenced by this metal. MnS inclusions with lead tips have, for instance, been observed in free-machining steels.[127]

Microscopy and electron probe analysis

MnS as an inclusion phase is easily recognized in the optical microscope. It is light-grey, plastic, slightly transparent and inactive to polarized light. According to Gaydos[128] the transparent sulphide is often slightly coloured, green if pure and red if manganese oxides are present.[129] Deformation of the steel usually results in the MnS-phase being elongated in the working direction, forming strings or plates. The characteristic appearance of slightly transparent MnS inclusions in a rolled steel bar is shown in Fig.162. The colour of such inclusions visible in polarized light is due to inner reflection of the light in the inclusion.

The morphology of MnS in steel ingots depends on the steel-making practice. Using the classification given by Sims and Dahle,[130] the sulphide-phase may be described as a type I, type II, or type III sulphide.

Type I (Fig.163) is globular with a wide range of sizes, often duplex with oxygen compounds and with an apparently random distribution. This type is most common in rimmed steel, where Si has been the main deoxidant and is found in carbon as well as in alloyed steels. Duplex inclusions of type I with MnS and silicates

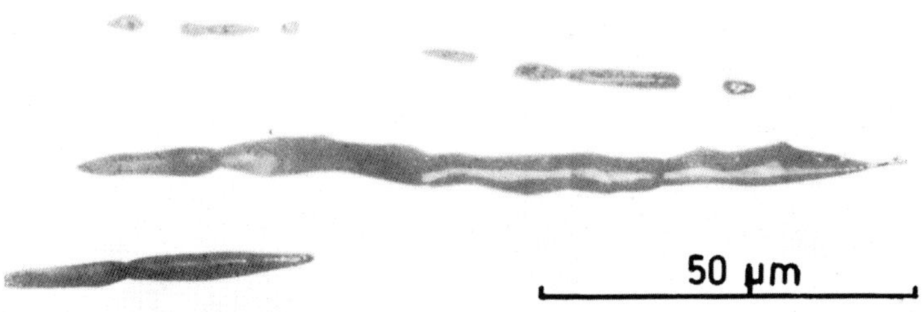

STEEL TYPE Free-cutting steel. Rolled bar
STEEL ANALYSIS, % 0·45 C, 0·18 Si, 0·82 Mn, 0·071 P, 0·150 S
INCLUSION MnS
COMMENT The sulphide is slightly transparent and the light streak in its centre is due to inner reflection from the steel. In polarized light the colour is greenish.[128, 129] No such colour effect is visible for wüstite which phase sometimes is similar in appearance

162a Sulphide inclusions in steel: slightly transparent MnS in steel bar

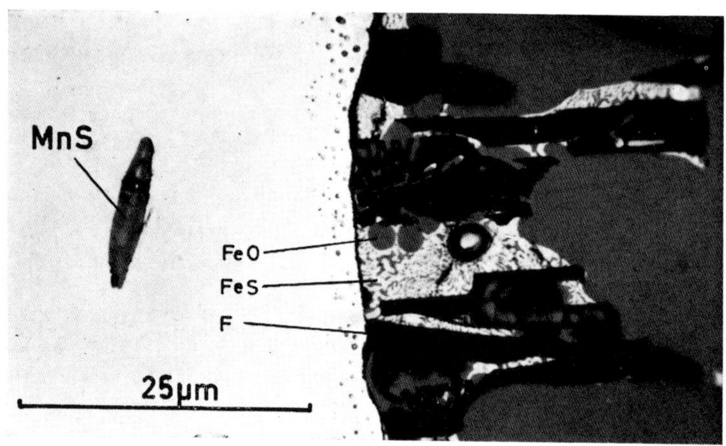

STEEL TYPE Low-alloyed steel, annealed in oil–gas-heated furnace for 30 min at 1 200°C. Gas analysis, % $1O_2$, 1·7 CO, 14 CO_2, 0·8 H_2 Sulphur content of oil 2·1%
STEEL ANALYSIS 0·91 C, 1·34 Si, 0·77 Mn, 0·014 P, 0·013 S, 1·07 Cr, 0·10 Ni
INCLUSION (MnS) Manganese sulphide, (FeS) Iron sulphide, (FeO) Wüstite, (F) Fayalite, $2FeO{\cdot}SiO_2$
COMMENT Small oxide inclusions, formed by inner oxidation of the steel, are visible in the steel phase along the phase boundary. Oil immersion optics (fayalite dark)

162b Sulphide inclusions in steel: MnS inclusion and oxide scale with FeS, wüstite and fayalite

may vary in appearance depending on the sulphur to oxygen ratio and examples of this inclusion type are shown in Figs.164–167.

The MnS phase in this inclusion type often has varying amounts of other elements in solid solution, for instance Cr as in Fig.164*a*

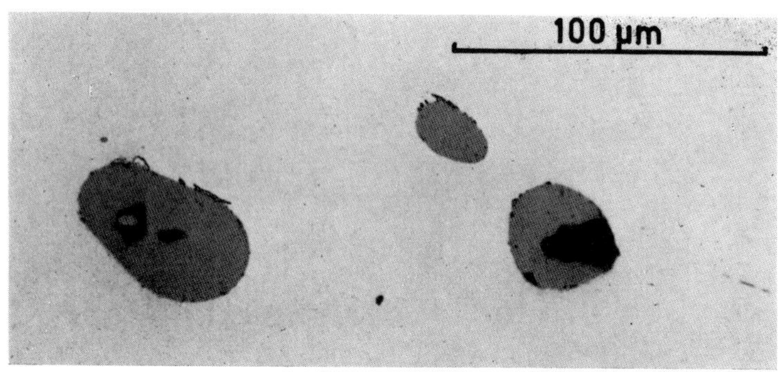

STEEL TYPE Basic electric steel, Si-killed. From ingot
STEEL ANALYSIS, % 0·26 C, 0·48 Si, 0·84 Mn, 0·022 P 0·011 S, 0·001 Al, 74 ppm O
INCLUSION MnS

163 Sulphide inclusions in steel: MnS of type I

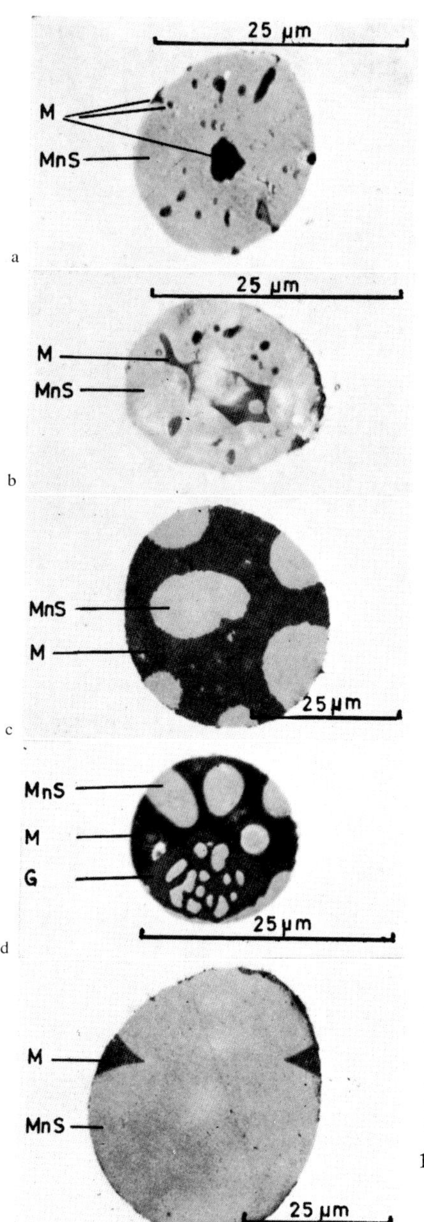

164*a* STEEL TYPE Austenitic stainless
STEEL ANALYSIS, % 0·06 C, 0·5 Si, 1 Mn, 0·02 S, 18 Cr, 8–9 Ni
INCLUSION (MnS) (Mn,Cr)S with 46–56 Mn, 9–12 Cr, 34 S, (M) Silicate

164*b* STEEL TYPE Basic electric steel, Si-killed
STEEL ANALYSIS *See* Fig.163
INCLUSION, % (MnS) 100 MnS, (M) Mn-silicate

164*c* STEEL TYPE Carbon steel ingot
STEEL ANALYSIS, % 0·24 C, 0·30 Si, 1·40 Mn
INCLUSION (MnS) 100 MnS, (M) Mn-silicate

164*d* STEEL TYPE *See* Fig.164*a*
STEEL ANALYSIS, *See* Fig.164*a*
INCLUSION (MnS) (Mn,Cr)S with 59 Mn, 4 Cr, 37 S, (G) Cr-galaxite, MnO·Cr_2O_3, (M) Silicate

164*e* STEEL TYPE *See* Fig.164c
STEEL ANALYSIS *See* Fig.164*c*
INCLUSION, % (MnS) 100 MnS, (M) Mn-silicate

164a–e Sulphide inclusions in steel: MnS of type I in duplex with silicates in different steel types

and *d*. In the inclusion shown in Fig.165, the MnS phase is finely dispersed, and another example of the same type was given in Fig.17*a* and *b*. MnS of type I may also appear as dendrites, as in Fig.166, or as scales on the surface of oxide inclusions, as in Figs.167, 41, and 111. This 'scale' may vary from a very fine rim

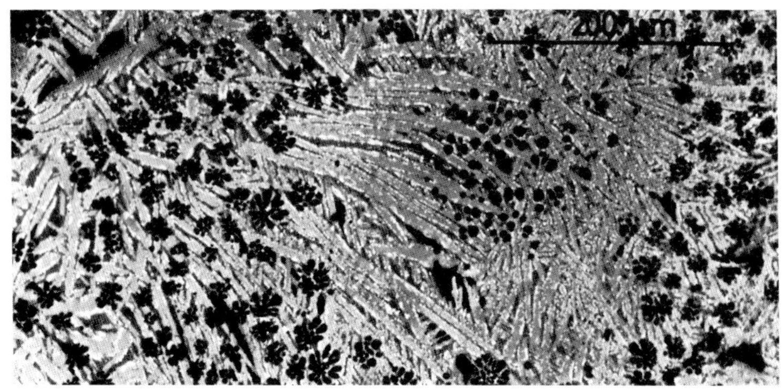

STEEL TYPE Basic electric steel, Si-killed. Inclusion isolated from ingot
STEEL ANALYSIS, % 0·16 C, 0·35 Si, 0·67 Mn, 0·015 P, 0·017 S, 0·039 N, 100 ppm O, no trace of Al
INCLUSION Fine white precipitate of MnS. Dark rosettes of cristobalite. Lathes of rhodonite, $MnO{\cdot}SiO_2$
COMMENT MnS has crystallized from the silicate melt in the inclusion. *See* also Figs.17*a* and *b*

165 Sulphide inclusions in steel: fine crystalline MnS in silicate inclusion with cristobalite

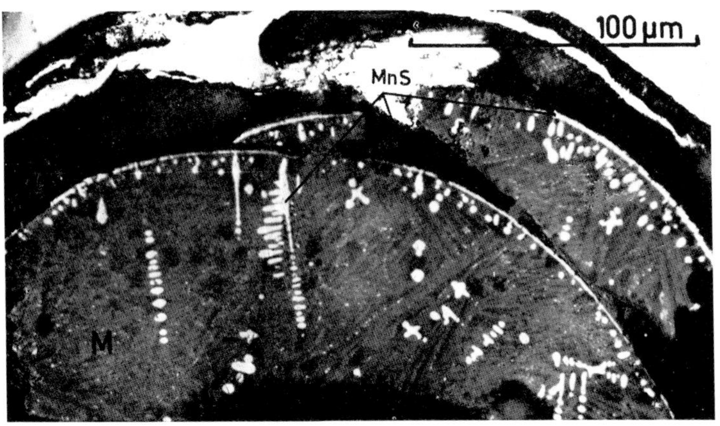

STEEL TYPE *See* Fig.165. Inclusion *in situ*
STEEL ANALYSIS *See* Fig.165
INCLUSIONS, % (MnS) 100 MnS, (M) Mn-silicates with cristobalite

166 Sulphide inclusions in steel: MnS-dendrites, MnS-scale and fine crystallized MnS in silicate inclusions

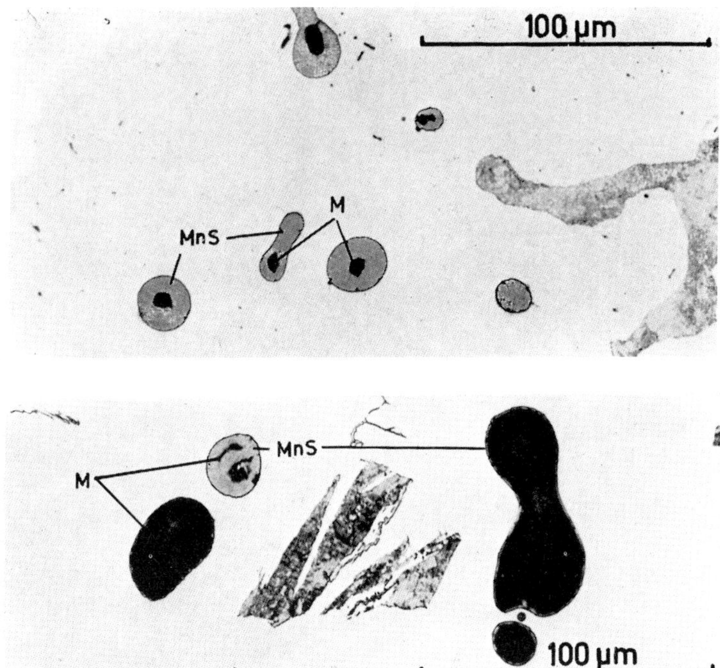

STEEL TYPE *See* Fig.165. Inclusion *in situ*
STEEL ANALYSIS *See* Fig.165
167*a* Sulphide inclusions with nucleus of silicate (thick 'scale' of MnS), (MnS) MnS 100%, (M) Mn-silicate
167*b* Sulphide inclusions with thin 'scale' of MnS, (MnS) MnS, 100%, (M) Mn-silicate with cristobalite

167a-b Sulphide inclusions in steel: different types of 'scales' of MnS around silicate inclusions

to a thick core with only a small silicate nucleus in the centre, as shown by Fig.167*a* and *b*. The MnS phase in the inclusion shown in Fig.111 has calcium in solid solution and was probably formed from a deoxidation nucleus. In alloyed steels those alloying elements which are present in high concentrations in the steel phase may also be present in solid solution in the MnS phase. Such an inclusion from a high chromium steel with Cr in solid solution in MnS is shown in Fig.168. Chromium galaxite (C 3) has been precipitated from the sulphide phase in this inclusion. Solid solutions of the (Mn,*Me*)S type are discussed in section K 4.

MnS of type II (Fig.169) has a dendritic structure and is often called grain-boundary sulphide because it is distributed as chainlike formations or thin precipitates in primary ingot grain boundaries.

Sometimes it seems to have solidified in a eutectic pattern, and it is often found crystallized together with corundum (A 2). This type is found in steels, deoxidized with Al without an excess of the deoxidizing metal.

Type III (Fig.170) is irregular, often angular in shape and randomly distributed in the steel. This type is often similar to type I

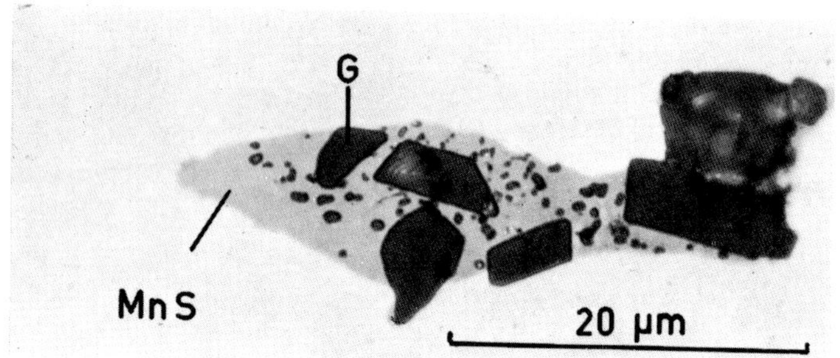

STEEL TYPE Basic electric steel. From billet
STEEL ANALYSIS, % 0·18 Cr, 0·13 Si, 0·39 Mn, 0·023 P, 0·14 S, 13·6 Cr, 0·66 Ni
INCLUSIONS, % (MnS) (Mn,Cr)S with 34 Mn, 24 Cr, 36 S, (G) Cr-galaxite with 30 MnO, 2 FeO, 70 Cr_2O_3

168 Sulphide inclusions in steel: MnS of type I with Cr in solid solution and Cr-galaxite

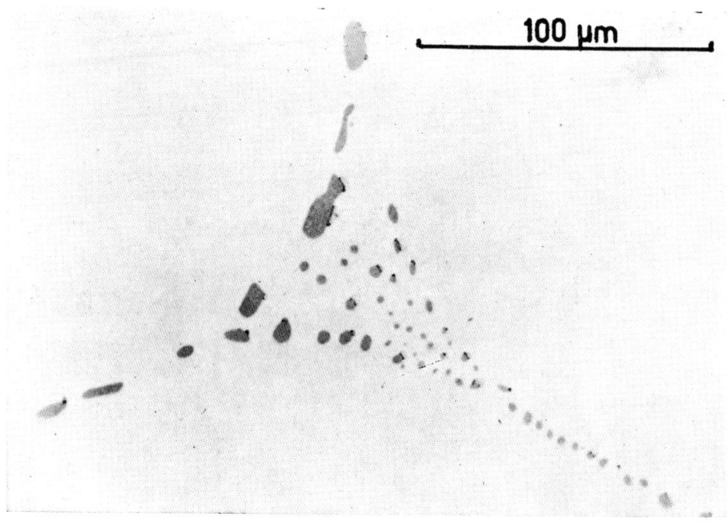

STEEL TYPE Basic electric steel, Al-killed, 1 000 g/ton in ladle. From ingot
STEEL ANALYSIS, % 0·23 C, 0·41 Si, 1·07 Mn, 0·021 P, 0·022 S, 0·007 Al, 40 ppm O
INCLUSION MnS

169 Sulphide inclusions in steel: MnS of type II

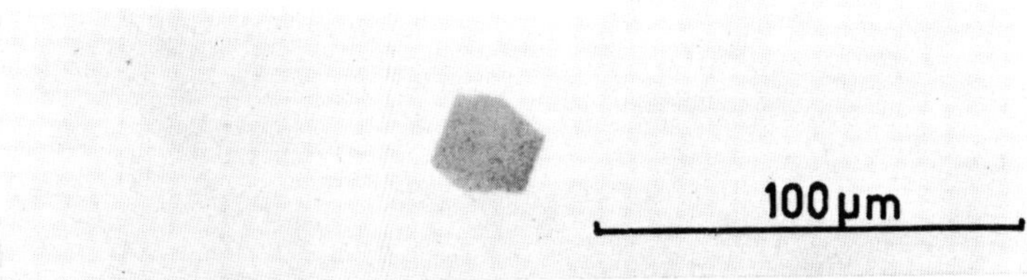

STEEL TYPE Basic electric steel. Al-killed, 700 g/ton in mould
STEEL ANALYSIS, % 0·26 C, 0·48 Si, 0·84 Mn, 0·022 P, 0·011 S, 0·038 Al, 47 ppm O
INCLUSION MnS

170 Sulphide inclusion in steel: MnS of type III

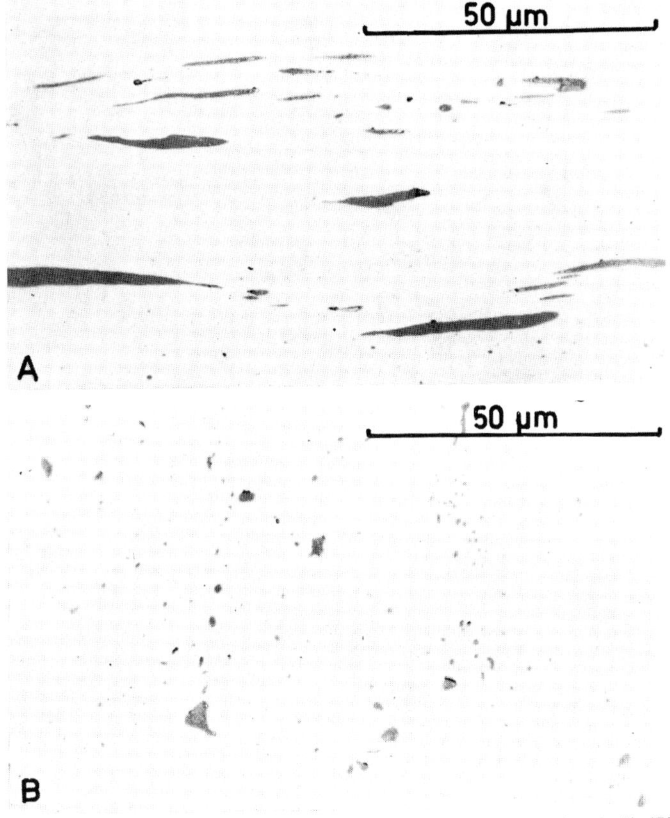

171*a* STEEL TYPE Free-cutting steel. Rolled bar, longitudinal section
STEEL ANALYSIS, % 0·45 C, 0·18 Si, 0·82 Mn, 0·071 P, 0·150 S
INCLUSIONS MnS

171*b* STEEL TYPE Same sample as in Fig.171*a*, transverse section
STEEL ANALYSIS *See* Fig.171*a*
INCLUSION MnS
COMMENT Compare Fig.172*b*

171 Sulphide inclusions in steel: MnS stringers in a rolled bar

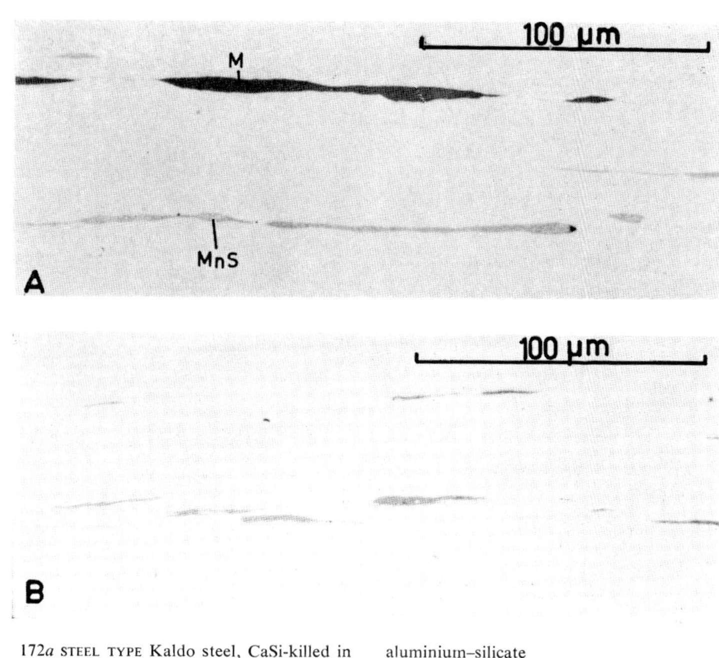

172*a* STEEL TYPE Kaldo steel, CaSi-killed in ladle. From rolled plate, longitudinal section
STEEL ANALYSIS, % 0·12 C, 0·29 Si, 1·01 Mn, 0·014 P, 0·025 S
INCLUSIONS (MnS) 100 MnS, (M) Calcium–aluminium–silicate

172*b* STEEL TYPE Same sample as in Fig.172*a*, transverse section
STEEL ANALYSIS *See* Fig.172*a*
INCLUSION MnS
COMMENT Compare Fig.171b

172 Sulphide inclusions in steel: MnS-plates in a rolled plate

and there is no marked difference between the two types. However, type III always forms monophase inclusions whereas type I usually is present in multiphase inclusions. Type III is most frequently found in steels deoxidized with an excess of Al. The authors have found characteristic examples of the different sulphide types in steels which had the following amounts of Al in solution. Type I, less than 0·001 wt-% Al; type II about 0·007 wt-% Al, added with 1 000 g/ton in the ladle, and type III with 0·038 wt-% Al, added with 700 g/ton in the mould.

The generally accepted explanation for the formation of these three different sulphide types has been given by Sims and Dahle.[130] It is based on the observation that the solubility of sulphur in liquid steel increases with decreasing oxygen solubility. From this general observation the present authors have formed the following conclusions for the formation of the different sulphide types.

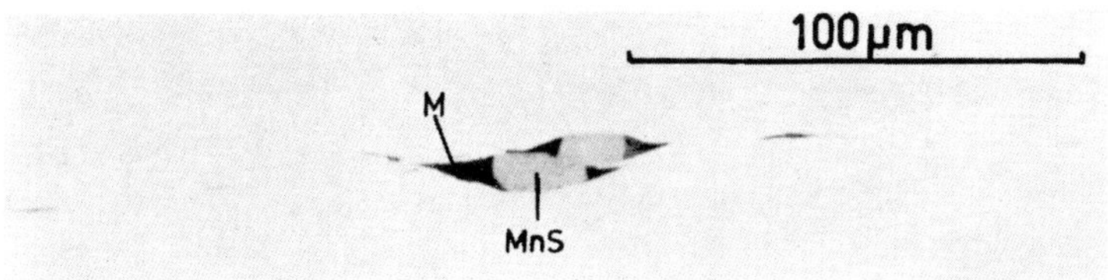

STEEL TYPE *See* Fig.172
STEEL ANALYSIS *See* Fig.172
INCLUSIONS, % (MnS), 100 MnS, (M) Calcium–aluminium–silicate

COMMENT The silicate phase is more plastic than the sulphide phase at working temperature of the steel and has been deformed more. Compare MnS and corundum, Fig.175

173 Sulphide inclusions in steel: MnS and silicate in duplex in rolled steel

MnS of type I is formed in rimmed or semi-killed steel, where the oxygen content of the liquid steel is high and the sulphur solubility low, resulting in a precipitation of sulphide at a comparatively high temperature. The precipitation of sulphide is parallel to the deoxidation process and sulphur and oxygen are precipitated at the same time from the liquid steel. The resulting inclusions are either a result of primary precipitation of both sulphides and oxides from the liquid steel often in duplex (Fig.164) or a sulphur and oxygen-rich silicate melt is formed. In the inclusions, MnS is precipitated from the silicate melt when the temperature falls, followed by crystallization of the oxide component (manganese silicate or aluminate Figs.165–166). In alloyed steel, the high alloy components may be present in the inclusions in solid solution in the sulphide phase and also as precipitated oxide phases, e.g. (Mn,Cr)S and $MnO \cdot Cr_2O_3$ (Fig.168).

Type II is found in killed steels, thoroughly deoxidized with Al but without excess and where the oxygen content is low. As a consequence, these steels have a high sulphur solubility and the sulphide phase precipitates late in the last parts of the steel ingot to solidify. Therefore, the type II sulphide is found in the primary grain boundaries in a dendritic, eutectic pattern. Al_2O_3 is formed by the Al-deoxidant, and therefore corundum often acts as a nucleus for the sulphide phase or is found mixed with the sulphide, but always as a separate phase (Figs.174 and 175). The Al-content in solution in the steel phase is relatively low, due to the moderate use of deoxidants, which are not used in excess for steels with this sulphide phase. The precipitation pattern for type II sulphide depends more on temperature and oxygen content of the steel than for type I sulphides, which are precipitated at higher temperatures and oxygen contents.

Type III is found in steels which have been deoxidized with an

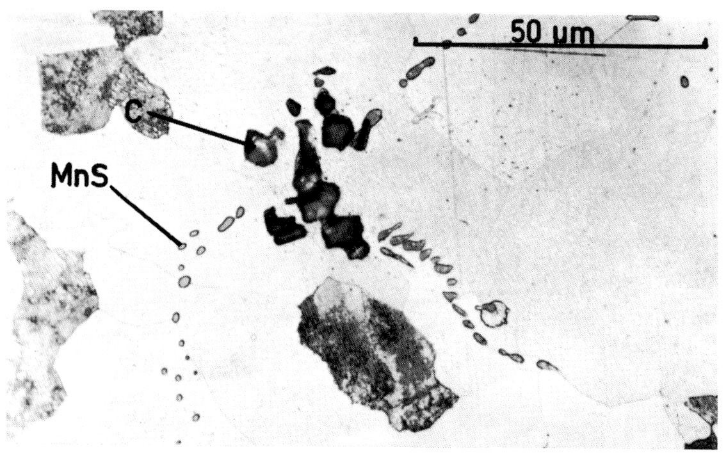

STEEL TYPE Carbon steel. From ingot
STEEL ANALYSIS, % 0·13 C, 0·98 Si, 1·08 Mn, 0·040 P, 0·030 S, 0·011 Cr
INCLUSIONS, % (MnS) 100 MnS, (C) 100 Al_2O_3

174 Sulphide inclusions in steel: MnS and corundum in ingot

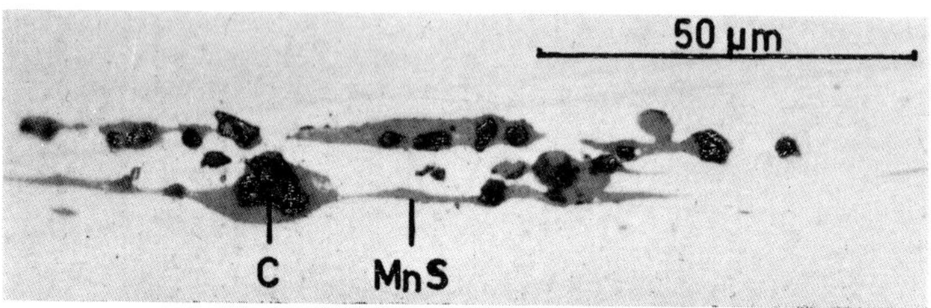

STEEL TYPE Free-cutting steel. From rolled bar
STEEL ANALYSIS *See* Fig.171
INCLUSIONS, % (MnS), 100 MnS, (C) 100 Al_2O_3
COMMENT The corundum phase is harder than the more plastic sulphide phase at working temperature of the steel. Only the sulphide has been deformed. Compare MnS and silicate Fig.173

175 Sulphide inclusions in steel: MnS and corundum in duplex in rolled steel

excess of Al. The oxygen content of the liquid steel is low but the sulphur solubility is also low as compared with steels forming type II sulphides, due to the high Al-content in solution in the liquid steel. This sulphide type is therefore precipitated earlier (at higher temperatures) than the type II sulphides and usually without nuclei. It is more similar to type I sulphide, but usually forms monophase inclusions (there is no precipitation of oxides due to the low oxygen content of the steel), and it is usually free from other metals in solid solution. It was earlier thought that the morphology of type III sulphides was influenced by Al_2S_3 (or Al_2O_3) nuclei. These should be present in solid solution in the MnS-phase, influencing its external form and giving it a shape similar to the hexagonal symmetry of Al_2S_3 (or Al_2O_3). Kiessling, Bergh and Lange,[14] however, have studied the Al-content of the different sulphide types and found that no Al was present in solid solution in any of the sulphides, at least in amounts detectable with electron probe analysis ($\geqslant 0{\cdot}1$ wt-%). The nuclei of Al_2O_3 which are sometimes associated with type II sulphides are always present as a separate phase and have no influence on the outer form of the sulphide.

When the steel matrix is deformed, the plastic MnS-phase usually also changes its shape. The sulphide inclusions will form strings or plates, depending on the mode of deformation of the matrix. Strings of MnS in a rolled steel bar are shown in Fig.171, and plates of MnS in a rolled steel plate in Fig.172. Microsections of MnS inclusions in the steel parallel and perpendicular to the working direction of the steel are shown.

In worked steel with duplex inclusions of sulphide and silicate, the silicate phase deforms more readily than the sulphide phase and is therefore often localized as 'tips' in the elongation of the partly deformed sulphide phase (Fig.173). In duplex inclusions with sulphide and corundum, the hard corundum phase stays undeformed and will be present as hard grains in the deformed sulphide phase (Figs.174–175). Inclusions with sulphides and aluminates are deformed in a similar manner.

Inclusions in leaded steels, where lead has been added in order to improve the free machining properties of the steel, have been studied by Blank and Johnson.[127] They report that 'tips' of lead metal are formed in rolled such steel around the partly deformed sulphide inclusions and that no lead sulphide inclusions were observed (K 13).

Additions of rare earth metals to the steel are reported to change the morphology of the sulphide inclusions in worked steel from elongated stringers to more rounded, shorter sulphide inclusions (K 5). The latter distribution is reported to increase the machinability of the steel. Selenium additions to the steel are also reported to have the same influence on the morphology of the sulphide phase (L 1).

It also seems possible to change the shape of the MnS inclusions by heat-treatment of the steel. A comparatively high temperature is necessary; for instance Sims[131] found spheroidization of the sulphide inclusions after holding for 24 h at 1315°C.

K 4 Sulphide inclusions: Solid solutions of the (Mn, Me)S type and double sulphides of the $MnS{\cdot}B_2S_3$ type

α-MnS can take considerable amounts of other transition metals, as well as of calcium and magnesium, into substitutional solid solution. A systematic study of the solid solubility limit for solid solutions of the (Mn,*Me*)S type, where *Me* is one of the first period transition metals, has been made for synthetic sulphides by Kiessling and Westman.[132] The results have been summarized in Table XXXI and Fig.176. A systematic variation was found for the solid solubility limit, with a maximum solubility of as much as 60–70 wt-% for Cr

TABLE XXXI **Solid solubility of 1st long period transition elements in α-MnS at 1 150°C. Composition, lattice parameter, microhardness and formula at solubility limit are given. The formulae for the Ti, V, and Cr-solutions have been calculated, assuming metal vacancies for the (Mn,Me)S phases**

Solid solution system	*Me*/ Mn+*Me*, at-%	*a*, (Å)	Micro-hardness, kg/mm²	Sulphide formula	Sulphide composition (wt-%) Mn	*Me*	S
(Mn,Ti)S	3	5·209	215	$Mn_{.96}Ti_{.03}S$	61·1	1·7	37·2
(Mn,V)S	30	5·180	340	$Mn_{.61}V_{.26}S$	42·5	16·8	40·7
(Mn,Cr)S	65	5·105	450	$Mn_{.26}Cr_{.49}S$	19·9	35·5	44·6
MnS	100	5·226	170	MnS	63·1	...	36·9
(Mn,Fe)S	65	5·130	300	$Mn_{.35}Fe_{.65}S$	22·0	41·5	36·5
(Mn,Co)S	28	5·205	240	$Mn_{.72}Co_{.28}S$	44·9	18·8	36·3
(Mn,Ni)S	13	5·200	250	$Mn_{.88}Ni_{12}S$	55·3	8·0	36·7

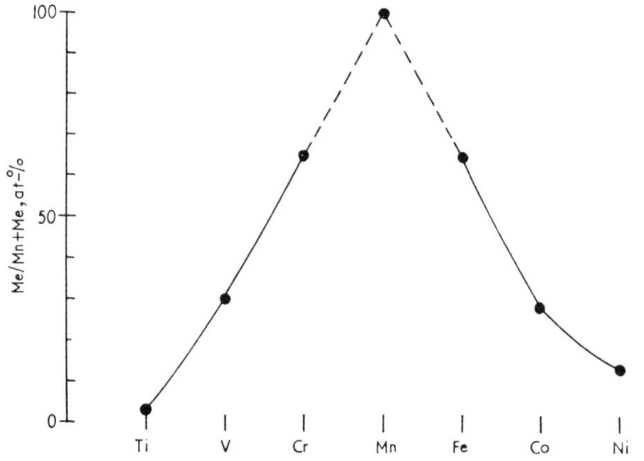

176 Solid solubility limit for the first period transition metals in MnS (at-% of the total metal atoms). Compare MnSe, Fig.181

and Fe in the α-MnS lattice. The microhardness of the solid solutions is reported to be considerably higher than for pure MnS (Fig.177) and also a change in the ratio *Me*:S was found for some of the sulphides. Pure MnS, as well as its solid solutions with Co and Ni, was always found to have a stoichiometric composition, but the solutions with Ti, V, Cr and sometimes also Fe had a higher sulphur content than corresponding to the formula *Me*S. The sulphide lattice of these solid solutions apparently has metal vacancies (Fig.178) which are formed to compensate for the change in valency for the metal ions when Mn(II) is substituted by metals of valency III. Fe can have both valency II and III which explains

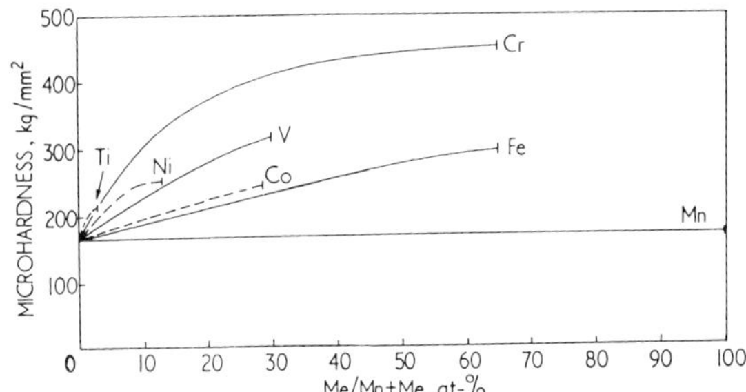

177 Microhardness values for the different (Mn,Me)S solid solutions for different Me-contents (in at-% of the total metal atoms). The solid solutions were quenched at 1 150°C except for (Mn,V)S, where the quenching temperature was 950°C. Compare MnSe, Fig.183

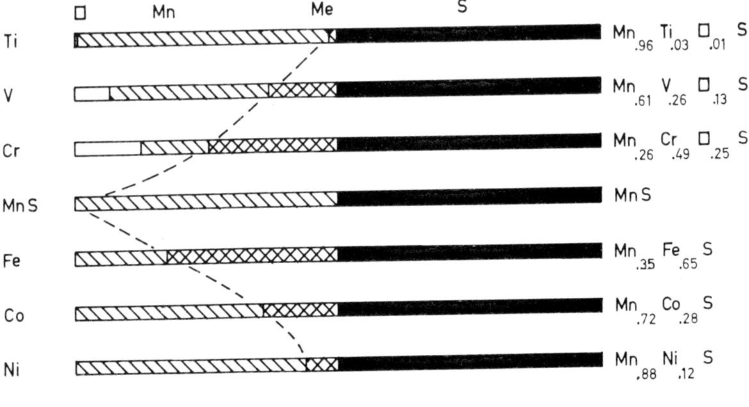

178 Composition of (Mn,Me)S phases at solubility limit for the first period transition metals. Compare MnSe, Fig.182

why some authors have reported a stoichiometric composition for (Mn,Fe)S,[132] and others a metal deficiency.[133]

Sulphide inclusions of MnS-type in alloyed steels may have different amounts of high-alloy elements in solid solution without any change in their optical appearance in the microscope (Fig.179). Some such sulphide inclusions from different steels, where the composition of the sulphide phase was established by electron probe analysis, were investigated earlier by the authors[134] and the results have been summarized in Table XXXII. These solid solutions may

TABLE XXXII Inclusions *in situ* of the (Mn,Me)S type in different steels. Typical electron probe inclusion analyses. A considerable variation in Mn:Me ratio was often observed for inclusions in the same steel sample

Inclusion type	Me-composition of MnS-phase, wt-%				Steel type	Steel analysis			
	Mn	Cr	Fe	Nb		Mn	Cr	Nb	S
(Mn,Cr)S	54	5	3	...	1·5% chromium	0·32	1·52	...	0·017
(Mn,Cr)S	38	25	1	...	13 % chromium				
(Mn,Cr)S	32	26	1	...	18 % chromium	1·47	18·1	...	0·022
(Mn,Fe)S	60	...	3	...	carbon	0·70	...	...	0·034
(Mn,Fe)S	50	...	13	...	carbon				
(Mn,Fe)S	24	...	37	...	carbon	0·14	...	...	0·015

be formed only if the sulphide is precipitated directly from the liquid steel phase. It is concluded that MnS inclusions in alloyed steel can have considerable amounts of Cr and Fe in solid solution. Indications of solid solubility in MnS inclusions, based mainly on the analysis of inclusion isolates have also been given by several authors, for instance Fe,[126,135] Cr,[136] Ni and Mo,[137] and Ti.[138]

The metallurgical significance of these solid solutions in steelmaking should not be overlooked. MnS is regarded as an inner lubricant during the machining of steel due to its plasticity. From Fig.177, however, it is evident that the plastic properties of 'MnS' in alloyed steel may vary, depending on the amount of metal in solid solution, and therefore the influence of the sulphide phase on the machinability of the steel is probably rather complex. The solid (Mn,*Me*)S solutions may also cause microsegregations in the steel. This has been shown to occur for (Mn,Fe)S[114,126] and is in agreement with the thermodynamic information given in Fig.148. If the free energy of formation for FeS and MnS are corrected considering the activities of the different elements in the liquid steel, ΔG for FeS

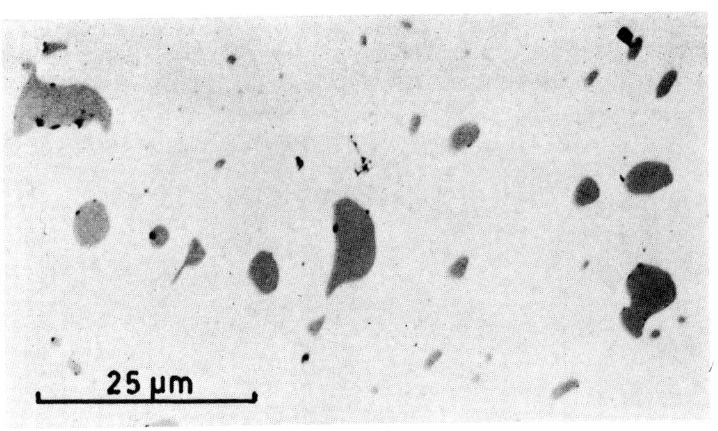

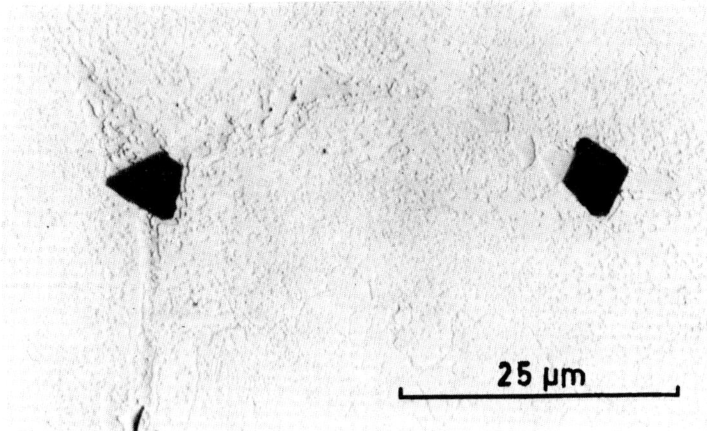

179*a* STEEL TYPE Austenitic stainless. From ingot
STEEL ANALYSIS, % 0·078 C, 0·36 Si, 1·47 Mn, 18 Cr, 9 Ni, 0·22 S
INCLUSIONS, % (Mn,Cr)S of MnS-type with 38–48 Mn, 10–18 Cr, 35–38 S

179*b* STEEL TYPE High-speed steel. From ingot
STEEL ANALYSIS, % 0·83 C, 0·20 Si, 0·26 Mn, 0·020 S, 4·10 Cr, 5·25 Mo, 6·71 W, 0·22 Co, 2·00 V
INCLUSIONS, % (MnS) (Mn,Cr)S of MnS-type with 44 Mn, 16 Cr, 36 S

179 Sulphide inclusions in steel: (Mn,Cr)S in different steel types

and MnS will be of the same order of magnitude at the temperature of the liquid steel, but at lower temperatures ΔG for MnS will be considerably lower than for FeS. This means that if a steel is rich in sulphur and is cooled rapidly, the MnS inclusions may have an iron content which is higher than the equilibrium value and a manganese content which is lower. If such a steel is heat-treated or welded, there may be diffusion of manganese from the steel matrix to the sulphide phase and diffusion of iron in the reverse direction, resulting in Mn-depleted zones around the sulphide inclusions. The existence of such zones has been verified experimentally by Salmon Cox and Charles[126] and Matsubara.[114] The same structural changes should also be expected for (Mn,Cr)S, but

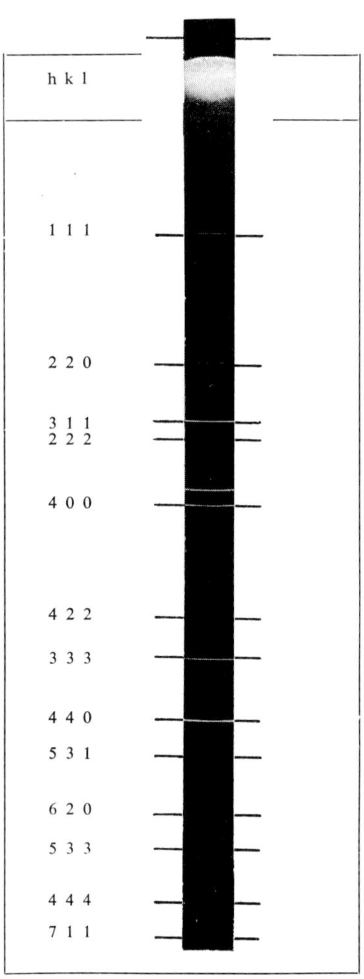

180 X-ray powder photograph of $MnS{\cdot}Cr_2S_3$; monochromatic FeKα-radiation; Guinier camera

no experimental evidence has been published. Other changes in the (Mn,*Me*)S inclusions are also possible after their formation, for instance precipitation of FeS from super-saturated (Mn,Fe)S phase (Fig.154).

Solid solutions of MnS with non-transition metals have also been observed, for instance (Mn,Ca)S and (Mn,Mg)S (Figs.111 and 95*a*).

Double sulphides of the $AS{\cdot}B_2S_3$ type are also known,[139,140] which structurally are of the spinel type ($AO{\cdot}B_2O_3$, A 7). The metal *A* may be Mn, Fe, Co, or Ni and the metal *B* may be Cr. The authors have identified the double sulphide $MnS{\cdot}Cr_2S_3$ by X-ray diffraction methods in synthetic (Mn,Cr)S samples. This phase is cubic with a parameter of $a = 10{\cdot}1$ Å. A monochromatic X-ray diffractograph of this phase is reproduced in Fig.180.

K 5 Sulphide inclusions with transition metals of Group III (the lanthanides)

General

Only the lanthanides (La, Ce . . . Lu) are of interest in steelmaking. They all have a high affinity for oxygen and sulphur. For instance, the free energy of formation for La_2S_3 is $\Delta G^\circ = -315$ cal,* and it is evident that lanthanum is nearly as strong a sulphide former as calcium at corresponding activities. Rare earth metals, added as mischmetal (H 1), have therefore occasionally been used as desulphurizers, mainly for cast steel. The metals influence the sulphur content of the steel and both the morphology and composition of the manganese sulphide inclusions. Two papers on the use of the lanthanides in steelmaking serve as a general introduction to the subject.[90, 91] However, some of the information given in these papers concerning the properties of the lanthanide sulphides is not in agreement with more recent findings. A more complete summary of the crystal chemistry of the different sulphides of the rare earths, together with a summary of their physical properties, is given in the book by Samsonov.[104] Unfortunately this review does not deal with their use in steelmaking. Some information about the metallography of sulphide steel inclusions containing these metals is available,[89–92] but is insufficient to give a real knowledge of their metallurgical behaviour. As the lanthanides are potential additions to steel, a short summary of their properties has been included, based mainly on Samsonov's monograph.

* For comparison, this value must be multiplied by 2:3 to correspond with the ΔG° values given in Fig.147, i.e. $\Delta G^\circ = -210$ cal.

Chemical formulae

Many intermediate phases have been identified in the different Me–S systems of the rare earth metals. Oxysulphides (K 14) are also known. In general, the different systems have intermediate phases with the following ideal compositions:

MeS
Me_5S_4
Me_3S_4
Me_2S_3
MeS_2
Me_2O_2S

The phases of greatest interest for inclusions in steel seem to be MeS, Me_2S_3, and Me_2O_2S, and they are the only ones considered in this section.

Modifications and solid solubility

The phases of MeS-type usually have an extended homogeneity range with metal vacancies, and their formula should therefore be given as $Me_{1-x}S$ (compare $Fe_{1-x}S$, section K 2, and (Mn,*Me*)S, Fig.178, section K 4). For instance $Yb_{1-x}S$ was studied by Domange *et al.*,[141] who found a maximum value for x of about 0·15.

The Me_2S_3-phases exist in several modifications. These have been summarized in Table XXXIII, which also gives an indication of the temperatures of transformation.

TABLE XXXIII **Different modification of the lanthanide sulphides of type Me_2S_3 (from Samsonov[104] p.214, table 45)**

Temperature, °C	Sulphide modification							
	La_2S_3	Ce_2S_3	Pr_2S_3	Nd_2S_3	Sm_2S_3	Gd_2S_3	Dy_2S_3	Y_2S_3
1 500	γ	γ	γ	γ	γ	γ	...	δ
1 400	γ	β	γ	γ	γ	γ	...	...
1 300	γ	β	β	β	γ	γ	...	...
1 100	β	α	β	$\alpha+\beta$	$\alpha+\gamma$	γ	δ	...
1 000	β	α	$\alpha+\beta$	α	α	$\alpha+\gamma$	$\alpha+\delta$	...
850	β	α	α	α	α	α	α	...
800	β	α	α	α	α	α	α	...
650	β	α	α	α	α	α	α	...

The phases of type MeS as well as those of type Me_2S_3 have a solid solubility range for the metals Mn, Fe, Mg, Ca, and Cr. *Vice versa*, the sulphides MnS and FeS have a solid solubility range for most rare earth metals.

Double sulphides of the spinel type are known between MnS and the ternary rare earth metal sulphides (MnS·Me_2S_3, K 4), there are also double sulphides with the ideal composition MnS·2Me_2S_3.

Physical properties

Several of the sulphides of the rare earth metals are very stable compounds with high melting points, for instance CeS melts at about 2450°C, and they are of special interest as possible high-temperature materials. A summary of those physical properties of the compounds *Me*S, Me_2S_3 and Me_2O_2S, which are of interest for inclusions is given in Table XXXIV. (For more information about the physical properties of these phases as well as of other compounds of rare earth metals, the reader is referred to Samsonov's monograph.[104])

TABLE XXXIV Physical properties of some lanthanide sulphides and oxysulphides of interest in steelmaking (from Samsonov[104])

Sulphide phase	Melting point, °C	Density
YS	2060	4·51
LaS	2200	5·75
CeS	2450	5·88
PrS	2230	6·07
NdS	2200	6·24
δ-Y_2S_3	1600	3·87
γ-La_2S_3	2150	4·93
γ-Ce_2S_3	1890	5·18
γ-Pr_2S_3	1795	5·27
γ-Nd_2S_3	2200	5·49
Y_2O_2S	2120	4·89
La_2O_2S	1940	5·77
Ce_2O_2S	1950	5·99
Pr_2O_2S	...	6·16
Nd_2O_2S	1990	6·22

Structure types

The crystallography of the sulphides of the rare earth metals has been summarized in Table XXXV. The phases of *Me*S composition are cubic and isostructural with α-MnS (K 3). The double sulphides of composition MnS·Me_2S_3, MnS·2Me_2S_3 and Me_2O_2S, however, have no structural relations with those sulphides which are of fundamental interest in steelmaking.

TABLE XXXV Crystallographic characteristics of some lanthanide sulphides and oxysulphides of interest in steelmaking (from Samsonov[104])

Sulphide phase	Symmetry	Parameters (Å)	
		a	*c*
YS	cubic	5·466	...
LaS	cubic	5·854	...
CeS	cubic	5·778	...
PrS	cubic	5·747	...
NdS	cubic	5·690	...
δ-Y_2S_3	monoclinic	...	...
γ-La_2S_3	cubic	8·723	...
γ-Ce_2S_3	cubic	8·635	...
γ-Pr_2S_3	cubic	8·611	...
γ-Nd_2S_3	cubic	8·527	...
$MnS.Yb_2S_3$	cubic	10·949	...
$MnS.2Yb_2S_3$	monoclinic	...	...
Y_2O_2S	hexagonal	3·70	6·56
La_2O_2S	hexagonal	4·051	6·943
Ce_2O_2S	hexagonal	4·00	6·82
Pr_2O_2S	hexagonal	3·974	6·825
Nd_2O_2S	hexagonal	3·946	6·790

Microscopy

No information on the microscopic appearance of the lanthanide sulphides as inclusions has been found. The following notes on the changes in the morphology of MnS inclusions in the presence of the lanthanides have been based on the literature.[89–92,142]

Addition of the rare earth metals to steel apparently changes the crystallization pattern of α–MnS from type II and type III (K 3) to type I inclusions. Thus, the long MnS stringers which often appear in rolled steel (Fig.171) are broken up into smaller units by the change in inclusion morphology and it is the globular MnS sulphide dispersion resulting from the lanthanide additions which improve machinability. Since the structure of the lanthanide sulphides are closely related to that of MnS, it is reasonable to suppose, in the absence of experimental evidence, that the sulphide inclusions are solid solutions of the (Mn,*Me*)S type (K 4). It is probable that it is the change in physical properties of the inclusion phase resulting from the substitution of lanthanides for manganese which accounts for the improved machinability of this type of steel.

Microsections of sulphide inclusions in steel, free from rare earth metals as compared with corresponding steels with rare earth metals additions, are given in the references.[91b]

The experimental evidence therefore indicates an effect on the morphology of the important MnS-phase by additions of rare earth metals to the steel but no information about the composition and crystal structure of the sulphide phase after rare earth metals additions to the steel is available according to the author's knowledge. The structure of the sulphides of the rare earth metals is closely related to MnS, however, and it therefore seems reasonable to suppose that rare earth metals are taken up in solid solution by the MnS-phase. These inclusions with solid solutions of the (Mn,*Me*)S type (K 4) should have different properties as compared with pure MnS inclusions. The effect of rare earth metals on different properties of the steel could therefore, at least to some extent, be explained as due to a change in properties of the sulphide phase.

K 6 Sulphide inclusions with transition metals of Group IV (Ti, Zr, Hf)

General

The transition metals of Group IV (Ti, Zr, Hf) are all comparatively strong sulphide formers in the conditions prevailing in steelmaking. The free energies of formation for the titanium sulphides are, for instance, of the same order of magnitude as for MnS (Fig.147). According to Kaneko *et al.*[143] the tendency for sulphide formation of different alloying elements in steel, considering their activities in the molten steel bath, decreases in the approximate order of Zr, Ti, Mn, Nb, V, Al, Mo, W, Fe, Ni, Co, and Si. The crystal chemistry of the group IV sulphides is complicated and the systems Ti–S and Zr–S are not fully known. The available metallographic information about these sulphides is scanty and the composition given for the sulphide inclusions is often not in accordance with recent structural results. The short summary in this section is mainly intended to present those aspects of the present knowledge of titanium and zirconium sulphides which are of interest to the metallographer and it is entirely based on literature references.

Titanium

The system Ti–S is complicated and the information is incomplete. According to the summary by Jellinek,[103] ten intermediate phases have been identified. Those sulphides which have the lowest sulphur content are most likely to occur as inclusions phases in steel. They are as follows:

(*a*) Ti_6S, a hexagonal phase which earlier was thought to be a solid solution of sulphur in titanium metal. It has since been shown that the solid solubility of sulphur in titanium metal is low.

(*b*) TiS_{1-z}. Three phases of this type are known within the composition range of $0{\cdot}3>z>0{\cdot}2$. They have a hexagonal symmetry and the lattice has sulphur vacancies.

(*c*) $Ti_{1-x}S$, with x from about 0·11 to zero. This phase has metal vacancies (compare $Fe_{1-x}S$, K 2) and for $x=0$ it has the hexagonal NiAs-type lattice.

(*d*) *y* (tau-) phase, a phase rich in titanium and sulphur. This phase has been observed in isolates from steels, stabilized with titanium. The composition and nature of the phase has been discussed in a series of papers[144–148] and it was suggested that the phase could be Ti_2S. The present evidence strongly indicates that the phase is not a pure sulphide but a carbosulphide (K 14), stabilized by iron and perhaps also with oxygen and nitrogen in solid solution. It should therefore be written as $(Ti,Fe)_2(S,C)$.

Several studies where titanium sulphides were found as inclusions in steel have been reported in the literature, the sulphide formula usually being given as TiS. With the present knowledge of the Ti–S system, it is difficult to decide which of the titanium sulphides was present as an inclusion phase in the different investigations. In those steels where an X-ray diffraction study of the titanium sulphide in the isolates was possible the sulphide phase was always found to be structurally similar to the NiAs type lattice. This structure type, or structures closely similar, has been found for all three TiS_{1-z} phases as well as for $Ti_{1-x}S$ if $x=0$. 'TiS' could therefore be any of these phases. It seems evident, however, that Ti_6S is not usually an inclusion phase in steel, as this sulphide has a different crystal structure. (An X-ray diffraction photograph of 'TiS' from steel isolates is reproduced in Koch's monograph.[77])

Microsections of titanium sulphide *in situ* are shown by Portevin and Castro[149] in a steel with 0·9% titanium. The phase is described crystallized as long, light-brown lamellae. Tanoue and Ikeda[150] have pointed out that this sulphide is strongly anisotropic. It is hard and does not deform plastically; at large deformations of the steel matrix, the titanium sulphide phase will split up to smaller fragments. Kaneko *et al.*[143] have studied titanium sulphide by optical microscopy, and further information about titanium sulphide inclusions in steel may be found in the Russian literature.[151]

Zirconium

According to Jellinek,[103] the system Zr–S has four intermediate phases. The sulphide with the highest zirconium content is ZrS, which is hexagonal and closely related to the corresponding titanium sulphides. A zirconium sulphide with metal vacancies may exist, with the composition $Zr_{1-x}S$, probably structurally closely related to $Ti_{1-x}S$. X-ray diffraction patterns of both these zirconium sulphides are reproduced by Koch.[77]

Microsections of zirconium sulphide steel inclusions *in situ* are shown by Portevin and Castro,[149b] and metallographic information is also given by Kaneko.[143] The colour of the phase in the optical microscope is said to be dependent of the zirconium content of the sulphide, changing from light yellow for high zirconium contents to blue-grey for lower. For the low zirconium contents the sulphide is similar to MnS and it may be difficult to distinguish between the two phases. The zirconium sulphide is harder, however, and will not deform during deformation of the steel matrix. Kovalenko[151] studied the formation of different sulphides when zirconium metal was added to steel in increasing amounts. MnS (and FeS) gradually disappeared being first replaced by complex 'ZrS'–MnS inclusion. For zirconium additions higher than 0·2 wt-%, separate zirconium sulphide inclusions appeared, which had a hexagonal structure.

Zirconium, like titanium, is known to form oxy- and carbo-sulphides, which are discussed in section K 14.

Hafnium

Little is known about the system Hf–S[103] and no information about hafnium sulphide inclusions is available.

K 7 Sulphide inclusions with transition metals of Group V (V, Nb, Ta)

General

The transition metals of Group V are moderate sulphide formers in steel which, according to Kaneko,[143] should be placed between Mn and Fe (K 6).

Vanadium and niobium may substitute for part of the manganese in MnS, and the usual occurrence of these metals in steel inclusions, if bonded to sulphur, therefore probably is as solid solutions of the (Mn,V)S or (Mn,Nb)S type (K 4).

Vanadium

10 intermediate phases are known in the V–S system and vanadium is unique among the transition metals in its tendency to form poly-sulphides.[103,152] The sulphide with the highest vanadium content is V_3S, but sulphides of the types VS_{1-z} and $V_{1-x}S$ are also known.

Not much is known about vanadium sulphide inclusions in steel, and in the few references to this phase,[143,150] it is reported as 'VS'. The vanadium sulphide is described as a hard, anisotropic phase, which will not be deformed during working of the steel matrix.

Niobium

The Nb–S system is complex and, according to Jellinek,[103] seven intermediate phases exist. The sulphide phase, which is richest in niobium, seems to be of the NbS_{1-z} type with a crystal lattice closely related to NiAs. Its structure, therefore, is similar to the corresponding phases formed by group IV transition metals and also to the corresponding vanadium sulphide. In the few references, where niobium sulphide inclusions are described,[143,150] the phase has been called 'NbS'. It is reported to be a hard, anisotropic phase. The present authors have no evidence of niobium sulphide inclusions in micro-alloyed steels with Nb and an ordinary Mn content.

Tantalum

The system Ta–S is reported to have nine intermediate phases.[103] No inclusions in steel with tantalum sulphides have been reported.

K 8 Sulphide inclusions with transition metals of Group VI (Cr, Mo, W)

The transition metals of Group VI are weak sulphide formers in steel.[143]

Chromium

Chromium has a wide range of substitutional solid solubility in α-MnS, and several examples with sulphide inclusions of the α-MnS type but with considerable amounts of Cr in solid solution have been observed in alloyed steels (Fig.179). The main occurrence of chromium in sulphide inclusions therefore seems to be as (Mn,Cr)S-phase, and inclusions of this type have been discussed in section K 4, where also double sulphides of the type $MnS{\cdot}Cr_2S_3$ have been mentioned. Philibert *et al.*[110] have reported that Cr also may be present in solid solution in inclusions of the FeS type.

The Cr–S system has six intermediate phases[103] and all have rather complicated crystal structures. The sulphide with the highest chromium content has the formula CrS, and has a monoclinic symmetry. Its structure is also related to that of NiAs, the crystal structure type of several of the sulphides of group IV and V transition metals.

Chromium sulphide inclusions in both steel and in cast iron have been mentioned by several authors. Portevin and Castro[153a] have shown several photomicrographs of such inclusions in a 1·5% Cr steel. The phase as shown by them is visible as light-grey, triangular or regular precipitates inside the MnS inclusions.[153b]. The morphology of the precipitates is similar to the corresponding double oxides (A 7). It therefore seems possible that the precipitate is the double oxide (Fig.168). In another plate Portevin and Castro[153c] have shown a chromium sulphide inclusion which is globular with a yellow colour and they have also given an example of a duplex chromium sulphide-chromium oxide inclusion.[153d] The chromium sulphide is reported to be soft and to be plastically deformed during steel working. Morrogh[154] has shown photomicrographs of chromium sulphide inclusions in cast steel. Microsections are also reproduced in the inclusion review of Allman.[2] No reliable determination of the composition of the chromium sulphide inclusions has been reported; the formulae 'CrS' and 'Cr_2S_3' are given but should be used with care.

Molybdenum

Five intermediate phases have been reported in the system Mo–S, Mo_2S_3 being the phase with the highest molybdenum content.[103] The molybdenum sulphides have not been found as steel inclusions. It should be noted, however, that the yellow, hexagonal MoS_2-phase is often present in isolates from steels alloyed with molybdenum, although no inclusions of this phase are found *in situ* in the steel. The reason for this is that the MoS_2-phase will often be formed during the electrolytic isolation procedure as a reaction product between molybdenum in the steel and sulphur compounds in the electrolyte in the presence of reducing agents.

Tungsten

Not much information is available about the W–S system. So far, only the phase WS_2 has been identified, isostructural with one of the MoS_2 modifications.[103] Tungsten sulphide inclusions in steel have not been reported.

K 9 Sulphide inclusions with transition metals of Group VIII (Co, Ni)

The thermodynamics of the systems Co–S and Ni–S were discussed by Rosenqvist[105] (as well as the FeS system, K 2). The stability of the sulphides decreases in the order of iron sulphides, cobalt sulphides, and nickel sulphides, and the most stable sulphides in each system are those with the lowest sulphur content. Kaneko *et al.*,[143] who studied the corresponding sulphides under conditions prevailing in steelmaking, reported a decrease in sulphide formation tendency in the order of iron, nickel, cobalt. Sulphide inclusions with cobalt or nickel sulphides should therefore only be present in steel if they have very high cobalt or nickel contents and only if they are at the same time low in manganese (compare the discussion for FeS, section K 2). However, Melnikov *et al.*[155] have reported that in a steel with 1 wt-% Mn and 1·3 wt-% Ni, most of the sulphide inclusions were of the phase Ni_3S_2 if the steel was quenched at high temperatures, but that only MnS was formed if the steel was quenched from temperatures lower than 600°C.

As mentioned in section K 4, Fig.176, MnS has a range of solid solubility for both Co and Ni. If these metals are bonded to sulphur in steel inclusions, they are therefore probably present as solid solutions in the MnS-inclusions. Examples have been given by Pickering[137] and Salmon Cox and Charles.[126]

It should also be noted that cubic chromium double sulphides of the spinel type are formed both by CoS and NiS.[140] The resulting sulphides, $CoS{\cdot}Cr_2S_3$ and $NiS{\cdot}Cr_2S_3$, are isostructural with the corresponding $MnS{\cdot}Cr_2S_3$ phase (K 4).

K 10 Sulphide inclusions with non-transition metals of Group I (Cu, Ag, Au)

Copper is the only member of this group of interest as a possible sulphide former in steel, but very few references to copper-sulphide inclusions in steel have been found. The most thorough study was made by Morrogh,[119] who investigated sulphide inclusions in cast iron with varying contents of Cu, Mn, and S. He concluded that the phase Cu_2S may be formed in cast iron but only if the Mn content is very low and the Cu and S contents are high. The phase is described as having a blue-brown to dark-brown colour in the optical microscope and it usually appears in duplex inclusions with FeS. Photomicrographs of a phase, which Morrogh considers to be probably Cu_2S or a solid solution in the Cu–Fe–S range, are shown.

Cu_2S has a cubic structure with the parameter a=5·59 Å.

K 11 Sulphide inclusions with non-transition metals of Group II (Mg, Zn, Cd, Hg, Ca, Ba, Sr)

The most important sulphides of this group in steel practice are MgS and CaS. Both these sulphides have a low free energy of formation (Fig.147) and the metals Mg and Ca are used to purify pig iron from sulphur. If oxygen is also present, Ca still has a strong tendency to form sulphides in steelmaking (Fig.148), whereas this tendency is lower for Mg than, for instance, for Mn or Fe.

MgS[156] and CaS[157] are both cubic and isostructural with MnS (K 3) with the parameters a=5·1913 Å and a=5·683 Å respectively. They have a wide range of mutual solid solubility and the solid solubility with MnS probably also extends over the whole solubility range, at least for MgS. CaS has a high melting point, about 2 500°C. Both phases react with water and they therefore often disintegrate during the preparation of microsections in the ordinary way. Ethanol should be used as a substitute for water during grinding and polishing. The CaS-phase is similar in colour to MnS and *not* black, as has sometimes been reported. It darkens rapidly if attacked by water.

CaS is a common inclusion phase. It often appears as a scale round oxide inclusions with CaO, and examples of this have been given (Figs.87, 111, and 112). The sulphide sometimes has Mg and Mn in solid solution. It does not deform as readily as MnS. Church *et al.*[124] have given several examples of CaS inclusions with Mn in solid solution. They also report that CaS inclusions in steel become more frequent if the steel is vacuum-treated.

The authors have not found any evidence of pure MgS inclusions in steel, but inclusions of the (Ca,Mg)S type are known (Fig.95*a*).

No information about steel inclusions with sulphides of other members of group II has been found (ZnS, CdS, HgS, BaS, SrS). Ba and Sr are rather similar to Ca, and therefore sulphides of these metals, probably in solid solution with CaS, should be possible in steel. Zn and Cd have different properties from Ca, and Kröger[158] has shown that these metals are insoluble in solid α-MnS. The sulphides of these metals are of great importance in non-ferrous metallurgy, but this lies outside the scope of the present report.

K 12 Sulphide inclusions with non-transition metals of Group III (Al)

Only Al is of interest in this group, and this metal is a comparatively weak sulphide former in steel practice (Figs.147 and 148). The most important sulphide of aluminium is the phase Al_2S_3, of which three modifications are known.[106,159] The stable low-temperature modification, α-Al_2S_3, is hexagonal with the parameters a=6·423 Å and c=17·83 Å. It transforms at about 1 000°C to a high-temperature modification, γ-Al_2S_3, which is isostructural with corundum (A 2). This form can be easily supercooled to room temperature. Also a third modification, β-Al_2S_3, is known. This phase has a NiAs type of lattice and is metastable over the whole temperature range, but it is said to be stabilized by carbon (K 14). The melting point of Al_2S_3 is 1 130°C.[106]

The possible existence of Al_2S_3 inclusions has been widely discussed. This is due to the fact that the aluminium activity in the steel phase is high during deoxidation and also that γ-Al_2S_3 is isostructural with α-Al_2O_3, corundum, with a possibility for solid solution formation. However, it has usually not been possible to find microscopic evidence for inclusions with Al_2S_3 and it therefore seems that the phase is not very common in steelmaking. In 'Basic open hearth steelmaking' a microsection of a dark phase with a banded structure is shown in a multiphase inclusion with FeS and MnS, which is considered to be Al_2S_3.[70b] The steel melt had a content of 0·97% S, 0·67% Mn, and 0·75% Al and it is concluded that the phase can only be formed if the sulphur and aluminium contents are very high and the manganese content moderate or low. Nor have Kiessling *et al.*[14] found any evidence of Al_2S_3 in MnS inclusions of type III (K 3), which also supports the previous conclusion.

K 13 Sulphide inclusions with non-transition metals of Group IV (Sn, Pb)

Two intermediate phases are reported in the Sn–S system, SnS and SnS_2.[106] SnS has orthorhombic symmetry, whereas SnS_2 is hexagonal with the parameters a=3·65 Å and c=5·88 Å.

In the Pb–S system a cubic, intermediate phase has been found with the composition PbS.[106] This phase is isostructural with MnS, MgS and CaS and its parameter is a=5·936 Å. These intermediate sulphides are, however, not formed in steelmaking where the tendency of these elements to form sulphides is too small compared with the other metals present. Thus metallic lead has been observed in the form of tails around MnS-inclusions where lead has been added in order to increase the machinability of steel.[127] No solid solutions of lead in α-MnS have been observed in spite of the structural identity between PbS and MnS.

In non-ferrous metallurgy, however, the relative tendencies of sulphide formation for tin and lead are often greater than in steelmaking, and sulphide inclusions of these metals are frequently observed, for instance in bearing metals with tin.

K 14 Sulphide inclusions with sulphur partly substituted by the non-metals carbon, oxygen and nitrogen

In this group of inclusions the authors have included those sulphides where sulphur is partly replaced by carbon, oxygen, and nitrogen (the carbo-, oxy-, and nitrosulphides). The sulphur and non-metal are present *in the same phase*. This inclusion type should not be confused with duplex inclusions where the non-metals are present in different phases, for example FeO and FeS. These are considered in the appropriate sections.

Carbosulphides of interest in steelmaking have been observed[147,160] with evidence of the phases $(Ti,Fe)_4C_2S_2$ and $Zr_4C_2S_2$. These phases are sometimes reported as formed in high-carbon, sulphur-rich steels with titanium and zirconium. The 'y-phase' in titanium stabilized steels, which was first thought to be Ti_2S, is probably such a carbosulphide.[147]

The crystal structure of these phases is hexagonal with lattice parameters a=3·20 Å and c=11·18 Å for the titanium phase and a=3·39 Å and c=12·1 Å for the zirconium phase.

It should also be noted that β-Al_2S_3[159] (K 12) is said to be stabilized by carbon, and therefore this sulphide also may be a carbosulphide.

Oxysulphides of interest in steelmaking are formed by lanthanides. The most common type has the composition Me_2O_2S with hexagonal symmetry. They are all very stable compounds with melting points around 2 000°C. (For further details see section K 5, Table XXXIV; more information is given in Samsonov's monograph.[104]) It is not quite clear if these oxysulphides of the lanthanides exist as separate phases also in inclusions, or if duplex inclusions with sulphides and oxides have been observed.

Nitrosulphides of interest in steelmaking have not been described. It is possible, however, that the carbosulphides mentioned above may also sometimes have nitrogen, as well as carbon, partly substituting for sulphur.

L Steel inclusions with selenium and tellurium

L 1 Inclusions with selenium

General

Selenium is sometimes added to martensitic or austenitic steels in order to increase the machinability. Such steels with a selenium content of up to about 0·3% were first developed in the U.S.A.[161] Selenium is chemically similar to sulphur and its main effect seems to be to change the morphology of the MnS inclusions in the steel. α-MnS inclusions in steels high in sulphur but free from selenium are usually deformed to long, narrow stringers during the forging or rolling of the steel (Fig.171). This morphology of the sulphide inclusions is sometimes considered to decrease the free machining properties of the steel and may also influence the directional dependence of the tensile properties. If selenium is added to the steel, the sulphide inclusions are reported to be more globular in shape and the sulphide stringers are broken up to shorter units, presumably enhancing the free machining properties of these steels.[162, 163] It is at present not known if the selenium is taken up in solid solution in the α-MnS phase, or if separate selenides are formed. However, Araki *et al.*[164] have identified the selenide α-MnSe in isolates from steels. A short summary of the crystal chemistry of the manganese selenides is given in this section.

The Mn–Se system

This system is very similar to the MnS system.

A monoselenide, MnSe, as well as a diselenide, $MnSe_2$, are known. Three modifications of MnSe exist (α, β, γ). α-MnSe is cubic of NaCl type with the parameter a=5·46 Å. It is isostructural with α-MnS and the phase has been identified in isolates from a chromium molybdenum steel with selenium additions.[164] β-MnSe is cubic and γ–MnSe hexagonal, and they are isostructural with β-MnS and β′-MnS respectively. $MnSe_2$ is cubic of pyrite type and isostructural with MnS_2.

Solid solutions of the (*Mn*,Me)*Se type*

α-MnSe has a solid-solubility range with several of the metals of interest in steelmaking, forming substitutional solid solutions of the (Mn,*Me*)Se type. A systematic study of these selenides has been made by Kiessling and Westman,[165] who found a great similarity between the selenides and the corresponding substituted manganese sulphides of the (Mn,*Me*)S type (K 4). The solid solubility limit for the first long period transition metals in α-MnSe is shown in Fig.181, which should be compared with the corresponding Fig.176

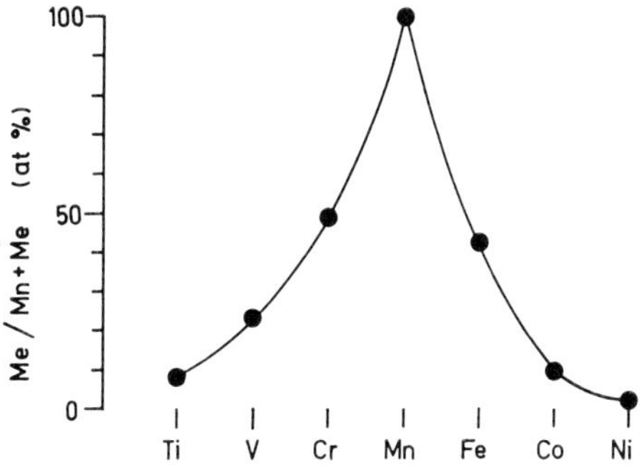

181 Solid solubility limit for the first period transition metals in MnSe (at-% of the total metal atoms). Compare MnS, Fig.176

for the sulphides. It was found that chromium and iron may substitute for as much as about 50 at-% of the manganese atoms in α-MnSe without any change in structure type. The systematic variation in solid solubility limit observed for the sulphides was also found for the selenides, as well as the vacancy formation for metals with valency III (Fig.182). A considerable increase in microhardness was observed for the substituted selenides as compared with α-MnSe (Fig.183) and the selenides of manganese are analogous to the sulphides in this respect also (Fig.177).

Solid solutions of the Mn(*S*,*Se*) *type*

α-MnSe and α-MnS are isostructural and the system MnSe–MnS shows complete solid solubility. The microhardness for α-MnSe was found to be lower than for α–MnS (about 80 kp/mm^2 for MnSe and 170 kp/mm^2 for MnS) and it decreased continuously if selenium was substituted for sulphur (Fig.183).

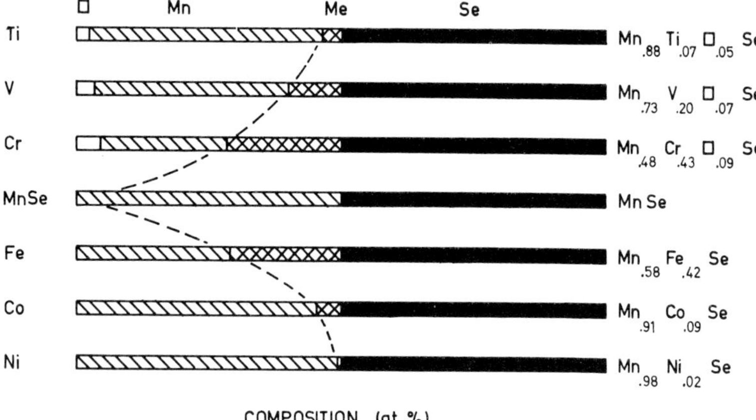

182 Composition of (Mn,*Me*)Se phases at solubility limit for the first period transition metals. Compare MnS, Fig.178

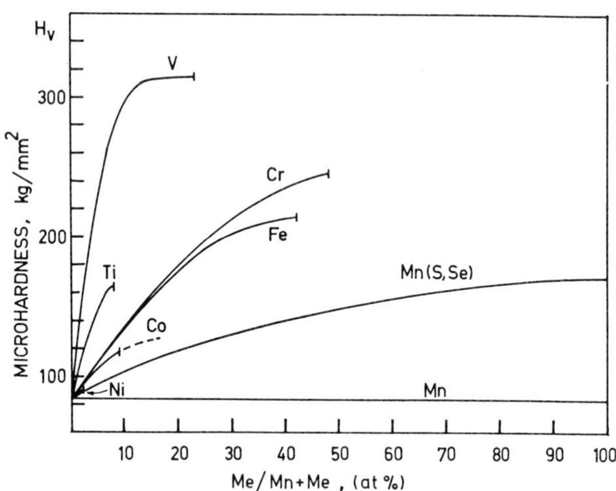

183 Microhardness values for the different (Mn,*Me*)Se solid solutions for different *Me*-contents (in at-% of the total metal atoms). The solid solutions were quenched at 1 150°C. Compare MnS, Fig.177. The microhardness values for Mn(S,Se) are also given

Inclusions

Since there is a wide range of solid solubility between α-MnS and α-MnSe, it is conceivable that the 'MnS' inclusion phase in steels containing selenium is a solid solution of the type (Mn,*Me*)(S,Se).

Therefore, both the change in sulphide inclusion morphology and the improved machinability obtained with selenium are probably related to the difference in physical properties observed between MnS and the solid solution phase.

L 2 Inclusions with tellurium

General

Tellurium belongs to the same group as sulphur and selenium. Tellurium additions to steel have been tried (*cf.* L 1) in order to increase its machinability. No information about successful results has been published, but a short comment on the structural properties of the manganese tellurides will be given in this section.

The Mn–Te system

Two intermediate phases are known, MnTe and $MnTe_2$. MnTe is hexagonal and isostructural with β'-MnS and β-MnSe, but none of the last two phases has been identified in steel inclusions (K 3, L 1). Such inclusions always have phases of the α-MnS or α-MnSe type, but no MnTe-modification of this type is known. $MnTe_2$ is cubic of the pyrite type, being isostructural with MnS_2 and $MnSe_2$.

Solid solutions

Te is only slightly soluble in α-MnS,[165] and wide ranges of solid solubility of the type (Mn,*Me*)Te or Mn(S,Te) or Mn(Se,Te) are not to be expected due to the structural difference between manganese tellurides and manganese sulphides or selenides.

M. Guidance for use

The basic principle for the two parts of the present work on non-metallic inclusions in steel has been to describe systematically the morphology and general properties of as many of the phases present in such inclusions as possible. The authors have considered that the best way to do this is to base the presentation on the composition and structure of the different inclusion phases and then to arrange the phases according to the different ternary oxide diagrams and the different sulphide systems. This arrangement is summarized in the list of contents for the two parts. The photomicrographs have been selected only for the purpose of showing characteristic features of different inclusion phases, independently of steel analysis, steel making practice or steel treatment. The authors think that this is the only possible method to give a systematic presentation of such a complex field as that of non-metallic inclusions in steel. Most of the different possible inclusion phases of modern steels are, according to the authors opinion, included in the present two parts, usually illustrated with adequate photomicrographs.

The authors are aware, however, that a presentation according to such a basic principle also has a serious disadvantage, namely to the metallurgist who wants to identify a certain inclusion based on his knowledge of steel analysis and steel making practice. In most instances, he will find the answer in the present two reports, but he may have difficulties to find his way among the different sections. As a guidance, a schematic indication of the main indigenous inclusion types formed for different steels and for different parts of the steel making practice has been given in Fig.184 and Tables XXXVI and XXXVII, also with an indication of the main origin for the inclusions. The following comments may also be of value.

1. *Indigenous oxygen and sulphur precipitation during solidification and cooling of the steel phase.* The inclusion pattern is mainly dependent on the Mn-, Si- (and soluble Al-) content of the steel, as indicated by the different arrows in the first part of Fig.184. The different

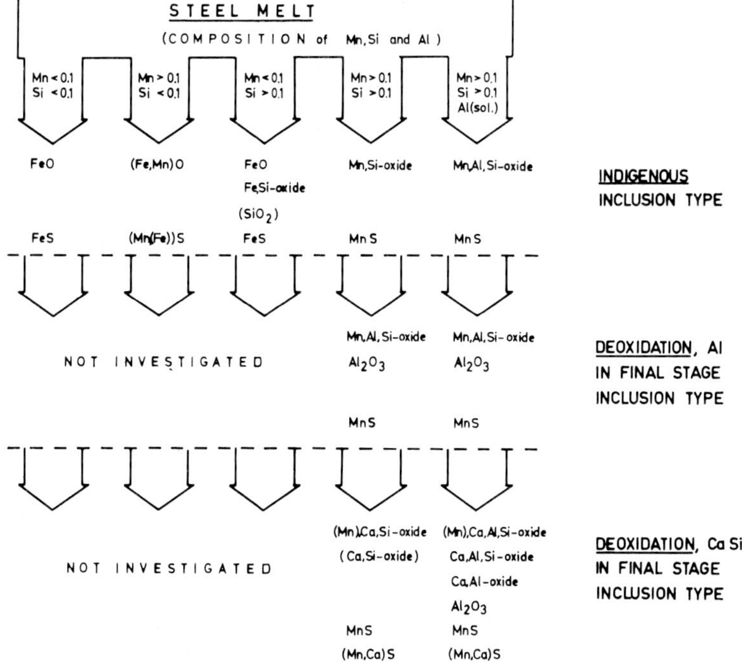

184 Schematic indication of the different indigenous inclusion types which are to be expected for different amounts (wt-%) of the 'base' elements Mn, Si and Al in the steel. The influence of Al and Ca deoxidation is also indicated Reference to the different inclusion phases which may appear in these inclusion types is given in Table XXXVI

inclusion phases which may result, are found in the sections given in Table XXXVI.

2. *Al-deoxidation* will not change the main inclusion pattern, but the oxide inclusions will have an increased Al-content as compared with point 1, which may influence the phases appearing in the inclusions. In addition, Al_2O_3 particles are formed as oxidation products and some of them are trapped in the steel phase. See the second part of Fig.184 and Table XXXVI.

3. *CaSi-deoxidation* will have an additional influence on the pattern and composition of indigenous inclusions resulting in oxide and sulphide inclusions with different amounts of Ca. The Mn-content of the oxide inclusions in such steels is usually lower than for other steel types. The resulting inclusion phases will therefore not only

TABLE XXXVI **The following phases may appear in the different inclusion types mentioned in Fig.184 and more information is found in the appropriate sections**

Inclusion type	*More information in section*																								
(Fig.184)	A											B		D			G				K			M	N
	2	3	4	5	6	7	8	9	10	11	12	2	3	2	3	4	3	4	5	6	2	3	4	7	
FeO												x		x	x	x									
FeMnO					x	x						x		x	x	x									
FeSi-oxide												x	x											x	
SiO_2		x																						x	
Mn, Si-oxide		x						x	x															x	
Mn, Al, Si-oxide	x	x	x			x	x	x	x	x	x													x	x
Al_2O_3	x																								
(Mn)Ca, Si-oxide																		x		x				x	
CaSi-oxide																		x		x				x	
(Mn), Ca, Al, Si-oxide	x					x				x	x						x		x	x				x	
Ca, Al, Si-oxide	x																x		x	x				x	
Ca, Al-oxide	x																x			x					
FeS																					x				
MnS																						x			
(Mn, Fe)S																							x		
(Mn, Ca)S																							x		

TABLE XXXVII Important elements which may be found in inclusions, usually partly substituting for the 'base' elements Fe, Mn, Si and Al

Element	Possible origin (in decreasing order of probability)	*Section* Oxide	Sulphide
Ca	Deoxidation alloys, slags, dolomite refractories	G	K4, K11
Cr	Alloying element, Cr-refractories	C, A7	K4
Mg	Mg-refractories, slags	F, A7	K11
Se	Se-additions	—	L1
Ti	Electrode coatings, alloying element, refractories	H2	K4
Nb	Alloying element	H3	—
V	Electrode coatings, alloying element	H3	K4
Lanthanides	Deoxidation alloys	H1	K5

belong to the $MnO\text{-}SiO_2\text{-}Al_2O_3$ system but also to the more complicated CaO-systems. See the third part of Fig.184 and Table XXXVI.

Inclusions with Ca are also formed from other sources than from the deoxidation alloys. Such sources are discussed in section G6 and indicated in Table XXXVII.

4. *Exogenous inclusions* which have not been much influenced by indigenous precipitation (e.g. parts of refractories, trapped ladle or welding electrode slags, trapped oxide scales, etc.) are usually characterized by their unusual size, shape and microscopic structure. They are often high in such element as Fe, Mg, Ca, Ti and V, and a schematic guidance is given in Table XXXVII.

5. *Most inclusions are formed by indigenous precipitation of Mn, Si (and Al) on exogenous nuclei.* Steel inclusions therefore often have a small amount of other elements than Fe, Mn, Si and Al in solution. Such elements appear in solid solution in different inclusion phases without any change in their morphology. Important such elements are listed in Table XXXVII and they are also usually discussed under 'solid solubility' for the different phases. It is often of value to trace the origin of these exogenous nuclei, as they may influence the general size, shape and number of steel inclusions. Mg is a valuable such trace element as it is insoluble in steel and not used for deoxidation.

6. *Alloying elements of the steel* have an influence on the inclusion pattern only if they are present in high amounts in the steel or added at a late stage of the steel-making process. Cr is the most important of these elements. It usually partly substitutes for Al but has no influence on the main morphology of the inclusion phases, as the

pure inclusion phases formed by these two elements usually are isostructural. Other elements sometimes found in inclusion phases are Ti, V, Nb, lanthanides, and Se. Guidance is given in Table XXXVII.

7. *Glassy inclusions* should be analysed by electron probe analysis, as their composition may vary between wide limits. The main elements to determine for such inclusions are Fe, Mn, Si, Al, Ca, and Mg. The glassy Ca-silicates are more transparent (darker in the microscope) than the glassy Mn-silicates.

8. Precipitation of oxygen and sulphur often results in *complex multiphase oxide-sulphide inclusions.* The sulphide phase usually precipitates in the boundary between the steel phase and the non-metallic inclusion phases. The precipitation mechanism is discussed in section K3.

N. Additions to Part I

The phases quartz (A5) and mullite (A8) in part I were illustrated with Figs.18 and 35–38 respectively, which are not of the highest quality regarding the morphology of these important phases. It is difficult to find good such inclusion examples in steel, the phases are often difficult to photograph. Since part I was written, a steel casting with good examples of these two phases was investigated, and therefore Figs.185–187 have been added.

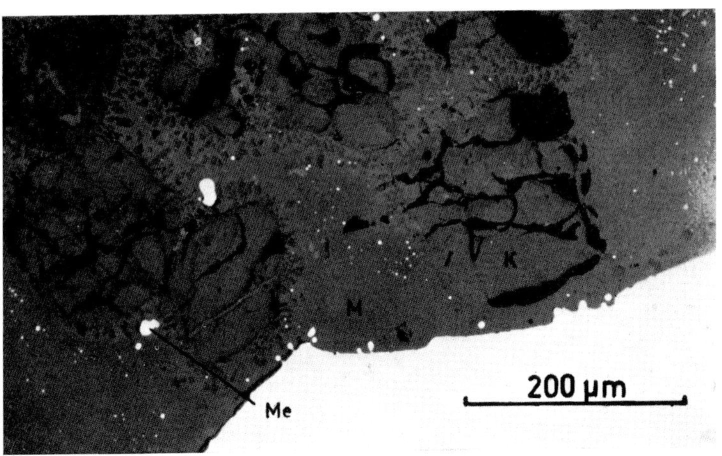

STEEL TYPE Cast steel
STEEL ANALYSIS, % 0·13C, 0·98Si, 1·08 Mn, 0·040 P. 0·030 S, 0·011 Cr
INCLUSION, (K) Quartz with 100 SiO_2 (A5), (Ct) Cristobalite with 100 SiO_2 (A3), (M) Matrix with 10–20 MnO, balance Al_2O_3 and SiO_2, (Me) Steel
COMMENT This inclusion is of exogenous origin (refractories)

185 Oxide inclusion in cast steel: quartz and cristobalite in glassy matrix of manganese oxide-alumina-silica

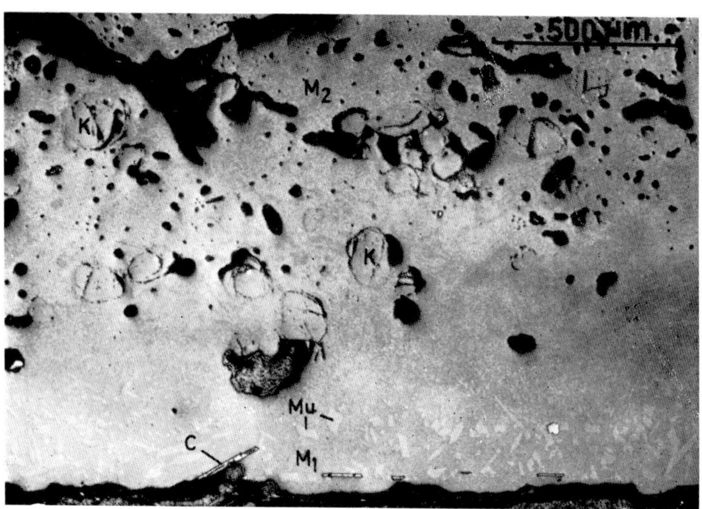

STEEL TYPE Cast steel, Al deoxidized
STEEL ANALYSIS, %0·20 C, 0·28 Si, 0·80 Mn
REACTION ZONE (K) Quartz (A5), (Mu) Mullite (A8), (C) Corundum (A2), (M_1) Matrix with MnO, Al_2O_3 and SiO_2, (M_2) Chamotte
COMMENT The steel surface is along the dark border in the bottom of the figure

186 Reaction zone between molten steel and chamotte: quartz, mullite, corundum, and glassy manganese oxide-alumina-silica

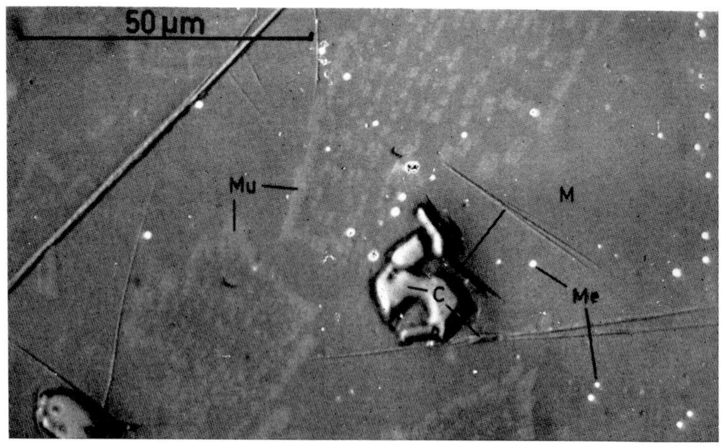

STEEL TYPE *See* Fig.185
STEEL ANALYSIS *See* Fig.185
INCLUSION (Mu) Mullite (A8), (C) Corundum (A2), (M) Matrix with MnO, Al_2O_3 and SiO_2, (Me) Steel

187 Oxide inclusion in cast steel: mullite and corundum in glassy matrix of manganese oxide-alumina-silica

REFERENCES

51. R. Kiessling and N. Lange: 'Non-metallic inclusions in steel', *ISI Spec. Rep. 90*, 1964, London.
52. L. Bragg and G. F. Claringbull: 'Crystal structures of minerals', The Crystalline State IV, 1965, London, G. Bell and Sons Ltd.
53. A. Muan and E. F. Osborn: 'Phase equilibria among oxides in steelmaking', 1965, Reading, Mass., Addison-Wesley.
54. 'Refractories for oxygen steelmaking', *ISI Spec. Rep. 74*, 1962, London.
55. 'Steelmaking in the basic arc furnace', *ISI Spec. Rep. 87*, 1964, London.
56. R. Duke and J. R. Lakin: in *ISI Spec. Rep. 87*, 1964, London, 159–170.
57. H. Pettersson: 'Eldfasta Keramiska material för järn och stålindustrin', 1962, Almqvist och Wiksell, Uppsala, 52.
58. V. S. Stubican and R. Roy, *J. Phys. Chem. Solids*, 1965, **26**, 1293–1297.
59. J. White: in *ISI Spec. Rep. 74*, 1962, London, 9–18.
60. M. Brunner: *Jernkont. Ann*, 1964, **148**, 470–517.

60*a*. *Ibid.*, Fig.14, p.490.

60*b*. *Ibid.*, Fig.31, p.505.

61. R. A. Liutnikov and A. V. Rudneva: *Russian Met. and Mining*, 1963, Nov/Dec, 51.
62. F. R. Boyd, *et al.*: *J. Geophysical Res*, 1964, **69**, 2101–2109.
63. L. Atlas: *J. Geol.*, 1952, **60**, 125–147.
64. J. F. Sarver and F. A. Hummel: *J. Am. Ceram. Soc.*, 1962, **45**, 152–156.
65. D. K. Smith, *et al.*: *Acta Cryst.*, 1965, **18**, 787–795.
66. D. L. Sponseller and R. A. Flinn: *Trans. AIME*, 1964, **230**, 876–888.
67. C. E. Sims: *ibid.*, 1959, **215**, 367–393.
68. W. Koch: *Stahl Eisen*, 1961, **81**, 1592–1598.
69. F. Johansson and N. Winblad: 'Ståltillverkning. Sandviken 1960', Sandviken handbook, D2.

70*a*. 'Basic open hearth steelmaking', 1964, 3 edn. AIME, New York, 981.

70*b*. *ibid.* Fig.12–12, p.472

71. T. O. Dormsjo and D. R. Berg: *Iron Steel Eng.*, 1959, (4), **36**, 11.
72. R. H. Rein and J. Chipman: *Trans. AIME*, 1965, **233**, 415–425.
73. N. Christensen: *Sintef FTR, NTH Report*, March 1964, Trondheim.
74. W. A. Fischer and A. Hoffmann: *Archiv. Eisenh.*, 1964, **35**, 37–38 (errata p.1215).
75. L. O. Uhrus: in 'Clean steel', *ISI Spec. Rep. 77*, 1963, London, 104–109.
76. J. Bruch, *et al.*: *Rheinstahl Technik*, 1965, **2**, 211–222.
77. W. Koch: 'Metallkundliche Analyse', 1965, Stahleisen, Düsseldorf and Chemie, Weinheim, 375.
78. A. Hayhurst and R. Webster: in *ISI Spec. Rep. 87*, 1964, London, 171–176.
79. R. H. Bogue: 'The chemistry of Portland cement', 1955, 2 edn., New York, N.Y., Reinhold Publ. Corp.
80. D. M. Roy: *J. Am. Ceram. Soc.*, 1958, **41**, 293–299.
81. S. Granhed and A. Sander: *Jernkont. Forskning B* 136, 23.6. 1965, 2.
82. G. Yamaguchi and H. Miyabe: *J. Am. Ceram. Soc.*, 1960, **43**, 219–224.
83. E. T. Carlsson: *Nat. Bur. Stand. J. Res.*, 1931, **7**, 893–902.
84. Hiroshi Ohba, *et al.*: in *ISI Spec. Rep. 74*, 1962, London, 59.
85. H. Pettersson and S. Eketorp: *Jernkont. Ann.*, 1946, **130**, 664–677.
86. M. Ferrero, *et al.*: *Mém. Sci. Rev. Mét.*, 1962, **59**, 75–78.
87. H. Opitz, *et al.*: *Arch. Eisenhüttenw.*, 1962, **33**, 841–851.
88. J. Bruch *et al.*: *ibid.*, 1965, **36**, 799–807.
89. Yu. I. Rubenchik *et al.*: *Ind. Lab.*, 1964, **30**, 73–75.

90. W. E. Knapp and W. T. Bolkcom: *Iron age*, 1952, **169**, 129–134 and 140–143.
91*a*. G. A. Lillieqvist and C. G. Mickelson: *J. Met.*, 1952, **4**, 1024–1031.
91*b*. *ibid.* Fig.2, p.1026.
92. Lee Ching-Yuan: *Acta Met. Sinica*, 1963, **6**, 131–154.
93. S. E. Eriksson: 'Elektrisk svetsning. Manuell bågsvetsning', Institutet för Tekniska Kurser, Stockholm, 1965, chap. III:3, p.18.
94. E. L. Evans and H. A. Sloman: *JISI*, 1953, **174**, 318–324.
95. F. B. Pickering: *ibid*, 1955, **181**, 147–149.
96. P. H. Frith: in *ISI Spec. Rep. 50*, 1954, London, Figs.21, 26, and 34.
97. F. B. Pickering: unpublished work.
98. O. Nyquist: *Jernkont. Ann.*, 1962, **146**, 149–194.
99. S. Eketorp: *JISI*, 1966, **204**, 194–202.
100. C. Jörgensen: Inst. for ferrous metallurgy, Royal Inst. of Technology, Stockholm, Internal report.
101. M. Bæyertz: 'Non-metallic inclusions in steel' 1947, Cleveland, Ohio, *ASM*, 72–77.
102. A. Hayes and J. Chipman: *Trans. AIME*, 1939, **135**, 85–125.
103. F. Jellinek: *Arkiv Kemi*, 1963, **20**, 447–480.
104. G. V. Samsonov: 'High-temperature compounds of rare earth metals with non-metals', 1965, New York, Consultants Bureau, 211–280.
105. T. Rosenqvist: *JISI*, 1954, **176**, 37–57.
106. M. Hansen: 'Constitution of binary alloys', 2 edn. 1958, New York/Toronto/London, McGraw-Hill with Elliot, 1st supplement, 1965.
107. G. Hägg and I. Sucksdorff: *Z. Phys. Chem. B*, 1933, **22**, 444–452.
108. H. Haraldsen: *Z. anorg. allg. Chem.*, 1941, **246**, 169–194.
109*a*. H. C. Chao, *et al.*: *Trans. ASM*, 1964, **57**, 885–891.
109*b*. *Ibid*, Fig.6, p.890.
110. J. Philibert *et al.*: *Metallurgia*, 1965, **72**, 203–211.
111. R. Vogel and G. W. Kasten: *Arch. Eisenh.*, 1948, **19**, 65–71.
112. R. Vogel and A. Wüstefeld: *ibid.*, 1938–39, **12**, 261–268.
113. R. Vogel and W. Hotop: *ibid.*, 1937, **11**, 41–54.
114. K. Matsubara: 'The electron microprobe', *Proc. Symp. Electrochem. Soc. Wash. DC*, 1964, Oct., 632–641.
115*a*. A. M. Portevin and R. Castro: *JISI*, 1953, **132**, Figs.5–6, plate XXXIX, p.240.
115*b*. *Ibid.*, Figs.3–4, plate XXXIX, p.240.
115*c*. *Ibid.*, Fig.19, plate XLI, p.240.
116. S. Rinman: 'Försök till järnets historia,' 1782, Stockholm, p.450.
117. A. Hultgren and B. Herrlander: *Trans AIME*, 1947, **172**, 493–509.
118. E. Tholander: *Mekanresultat* 65022, 1965, Aug., 9–12.
119. H. Morrogh: *JISI*, 1946, 154, 399P–408P.
120. S. Furuseth and A. Kjekshus: *Acta Chem. Scand.*, 1965, **19**, 1405–1410.
121. F. Mehmed and H. Haraldsen: *Z. anorg. allg. Chem.* 1937–38, **235**, 193–200.
122. C. A. Müller *et al.*: *Arch. Eisenh.*, 1966, **37**, 27–41.
123. B. S. Ellefson and N. W. Taylor: *J. Chem. Phys.*, 1934, **2**, 58–64.
124. C. P. Church *et al.*: *J. Met.*, 1966, **18**, 62–68.
125. D. Radtke and D. Schreiber: *Stahl Eisen*, 1966, **86**, 89–99.
126. P. H. Salmon Cox and J. A. Charles: *JISI*, 1965, **203**, 493–499.
127. J. R. Blank and W. Johnson: *Steel Times*, 1965, **191**, 110–115, 148–152 and 176–181.
128. R. Gaydos: *J. Met.*, 1964, **16**, 972–977.

129. D. G. Jones and D. A. Melford: Tube Investments Res. Lab., 1966, Hinxton Hall, Cambridge, Techn. Rep. No. 198.
130. C. E. Sims and F. B. Dahle: *Trans. A.F.A.*, 1938, **46**, 65–132.
131. C. E. Sims: *Foundry*, 1951, **79**, 92–97, and 241–253.
132. R. Kiessling and C. Westman: *JISI*, 1966, **204**, 377–379.
133. L. Bäcker *et al.*: *Mem. Sci. Rev. Met.*, 1966, **63**, 319–328.
134. R. Kiessling and N. Lange: *JISI*, 1963, **201**, 761–762.
135. K. Sano and M. Inouye: *J. Iron Steel Inst. Japan*, 1959, **45**, 9.
136. I. S. Brammar and R. W. K. Honeycombe: *JISI*, 1964, **202**, 335–342.
137. F. B. Pickering: in *ISI Spec. Rep. 77*, 1963, London, 18.
138. E. D. Mokhir and Y. G. Gurevich: *Stal in English*, 1964, (8), 643–645.
139. L. Passarini and M. Baccaredda: *Atti. Accad. Lincei*, 1931, **14**, 33–37.
140. F. K. Lotgering: *Philips Res. Rep.*, 1956, **11**, 190–249.
141. L. Domange *et al.*: *Compt. Rend. Acad. Sci.*, 1960, **251**, 1517.
142. E. I. Nikolaev *et al.*: *Chem. Met.*, 1965, **8**, 37–42.
143. H. Kaneko *et al.*: *J. Japan Inst. Met.*, 1963, **27**, 299–304.
144. M. G. Gemmill *et al.*: *JISI*, 1956, **184**, 122–144.
145. J. F. Brown *et al.*: *Metallurgia*, 1957, **56,** 215–223.
146. K. Wetzlar and G. Lennartz: *DEW-Technische Ber.*, 1961, **1**, 15–16.
147. C. Frick *et al.*: *Arch. Eisenh.*, 1960, **31**, 419–422.
148. O. Knop: *Metallurgia*, 1958, **57**, 137–138.
149*a*. A. M. Portevin and R. Castro: *JISI*, 1937, **135**, 223P–244P.
149*b*. *Ibid.*, Fig.24, plate XV, p.240P.
150. T. Tanoue and T. Ikeda: *Tetsu-to-Hagané*, 1964, **50**, 2182–2189.
151. V. S. Kovalenko: *Stal, in English*, 1964, (2), 136–139.
152. H. F. Franzén and S. Westman: *Acta Chem. Scand.*, 1963, **17**, 2353–2354.
153*a*. A. M. Portevin and R. Castro: *JISI*, 1936, **134**, 213P–239P.
153*b*. *Ibid.*, Figs.20–21, plate XXXVII, p.224P.
153*c*. *Ibid.*, Fig.22, plate XXXVIII, p.224P.
153*d*. *Ibid.*, Fig.25, plate XXXVIII, p.224P.
154. H. Morrogh: 'Polarized light in metallography' 1952, Butterworth, Fig.21, p.101.
155. L. M. Mel'nikov: *Fiz-Khim. Osnovy Proizv. Stali*, 1964, 48–56.
156. W. Primak *et al.*: *J. Am. Chem. Soc.*, 1948, **70**, 2043–2046.
157. R. Juza and K. Bünzen: *Z. Phys. Chem.*, 1958, **17**, 82–99.
158. F. A. Kröger: *Z. Krist. A*, 1939–40, **102**, 132–135.
159. J. Flahaut: *Compt. Rend. Acad. Sci.*, 1951, **232**, 334–336 and 2100–2102.
160. H. Kudielka and H. Rohde: *Z. Krist.*, 1960, **114**, 447–456.
161. The Carpenter Steel Comp., *US Pat.* 1,846,140; 2, 009, 713–716.
162. 'Selenium', *Iron Age*, 1952, **170**, 283–284.
163. F. W. Boulger: *Trans. ASM*, 1960, **52**, 698–712.
164. T. Araki *et al.*: *Tetsu-to-Hagané Overseas*, 1965, **5**, 112–122.
165. R. Kiessling and C. Westman: to be published.
166. J. R. Rait and H. W. Pinder: *JISI*, 1946, **157**, Figs.56–58, plate LIV, and Fig.70, plate LV.

Appendix 1

Names and ideal formulae of inclusion phases

		Section	Page
alite	$3CaO \cdot SiO_2$	G 4	50
anatase	TiO_2	H 2	89
anorthite	$CaO \cdot Al_2O_3 \cdot 2SiO_2$	G 5	72
anosovite	Ti_3O_5	H 2	89
bredigite	$2CaO \cdot SiO_2$	G 4	50
brookite	TiO_2	H 2	89
bustamite	$CaO \cdot MnO \cdot 2SiO_2$	G 4, J	57
calcia	CaO	G 2	32
Ca-cordierite	$2CaO \cdot 2Al_2O_3 \cdot 5SiO_2$	G 5	72
clinoenstatite	$MgO \cdot SiO_2$	F 4	15
cordierite	$2MgO \cdot 2Al_2O_3 \cdot 5SiO_2$	F 6	23
diopside	$MgO \cdot CaO \cdot 2SiO_2$	F 4	19
dolomite	$CaMg(CO_3)_2$	G 2	33
enstatite	$MgO \cdot SiO_2$	F 4	15
forsterite	$2MgO \cdot SiO_2$	F 5	19
gehlenite	$2CaO \cdot Al_2O_3 \cdot SiO_2$	G 5	72
grossularite	$3CaO \cdot Al_2O_3 \cdot 3SiO_2$	G 5	72
hauerite	MnS_2	K 3	114
ilmenite	$FeO \cdot TiO_2$	H 2	89
johannsenite	$CaO \cdot MnO \cdot 2SiO_2$	G 4, J	57
larnite	$2CaO \cdot SiO_2$	G 4	50
lime	CaO	G 2	32
magnesite	$MgCO_3$	F 2	9
marcasite	FeS_2	K 2	104
merwinite	$3CaO \cdot MgO \cdot 2SiO_2$	J	78
monticellite	$MgO \cdot CaO \cdot SiO_2$	F 5, J	21
olivine	$2(Mg,Fe)O \cdot SiO_2$	F 5	21
(osumilite-type)	$MgO \cdot Al_2O_3 \cdot 4SiO_2$	F 9	25
para-wollastonite	$CaO \cdot SiO_2$	G 4	50
periclase	MgO	F 2	6
perovskite	$CaO \cdot TiO_2$	H 2	90
(petalite-type)	$MgO \cdot Al_2O_3 \cdot 8SiO_2$	F 10	25
protoenstatite	$MgO \cdot SiO_2$	F 4	15
pseudo-wollastonite	$CaO \cdot SiO_2$	G 4	50
pyrite	FeS_2	K 2	104
pyrope	$3MgO \cdot Al_2O_3 \cdot 3SiO_2$	F 7	24
pyrrhotite	FeS	K 2	104
rankinite	$3CaO \cdot 2SiO_2$	G 4	50
rutile	TiO_2	H 2	89

Appendix 1—*continued*

Names and ideal formulae of inclusion phases

		Section	Page
sapphirine	$4MgO{\cdot}5Al_2O_3{\cdot}2SiO_2$	F 8	25
shannonite	$2CaO{\cdot}SiO_2$	G 4	50
sphene	$CaO{\cdot}SiO_2{\cdot}TiO_2$	H 2	90
spinel	$MgO{\cdot}Al_2O_3$	F 3	10
troilite	FeS	K 2	104
wollastonite	$CaO{\cdot}SiO_2$	G 4	50

Appendix 2

Summary of Tables

Table	*Contents*	*Part*	*Sect.*	*p.*
I	Phases in $MnO–SiO_2–Al_2O_3$ system	I,	A 1,	8
II	Double oxides of spinel type	I,	A 7,	31
III	Solid solubility for double oxides	I,	A 7,	32
IV	Double oxides in different inclusion types	I,	A 7,	33
V	Composition of the spessartite phase in inclusions	I,	A 11,	55
VI	Mn:Si ratio in steel compared with $MnO:SiO_2$ ratio in inclusions	I,	A 15,	64
VII	Phases in $FeO–SiO_2–Al_2O_3$ system	I,	B 1,	69
VIII	Inclusions classified according to steel analysis $MnO–SiO_2–Al_2O_3$ system	I,	E,	98
IX	Analyses for refractories	II,	F 1,	1
X	Phases identified in used refractories	II,	F 1,	2
XI	Phases in $MgO–SiO_2–Al_2O_3$ system	II,	F 1,	6
XII	Composition of the spinel phase in inclusions	II,	F 3,	11
XIII	Composition of deoxidation alloys	II,	G 1,	27
XIV	Analyses for slag types	II,	G 1,	29
XV	Phases in $CaO–SiO_2–Al_2O_3$ system	II,	G 1,	37
XVI	Physical properties of the Ca-aluminates	II,	G 3,	37

Appendix 2—*continued*

Summary of Tables

Table	*Contents*	*Part*	*Sect.*	*p.*
XVII	Crystallographic characteristics of the Ca-aluminates	II,	G 3,	38
XVIII	Physical properties of the Ca-silicates	II,	G 4,	51
XIX	Crystallographic characteristics of the Ca-silicates	II,	G 4,	51
XX	$MeO{\cdot}SiO_2$ chain silicates	II,	G 4,	58
XXI	Composition of $MeO{\cdot}SiO_2$ phase in inclusions	II,	G 4,	58
XXII	Crystallographic characteristics of the ternary $CaO–SiO_2–Al_2O_3$ phases	II,	G 5,	69
XXIII	Composition of ternary $CaO–SiO_2–Al_2O_3$ phases in inclusions	II,	G 5,	70
XXIV	Composition of Ca-aluminate inclusions	II,	G 6,	71
XXV	Composition of welding electrode slags	II,	H 2,	71
XXVI	Phases of types AO, B_2O_3, and SiO_2; summary	II,	J,	90
XXVII	Phases of types $AO–SiO_2$; summary	II,	J,	91
XXVIII	Phases of types $AO–B_2O_3$; summary	II,	J,	92
XXIX	Phases of types $AO–B_2O_3–SiO_2$; summary	II,	J,	93
XXX	Sulphide systems; summary	II,	K 1,	103
XXXI	Solid solubility in α-MnS	II,	K 4,	128
XXXII	Inclusions of (Mn,Me)S type	II,	K 4,	130
XXXIII	Modifications of lanthanide sulphides	II,	K 5,	134
XXXIV	Physical properties of lanthanide sulphides	II,	K 5,	135
XXXV	Crystallographic characteristics of lanthanide sulphides	II,	K 5,	136
XXXVI	Guidance for use, reference to sections	II,	M,	152
XXXVII	Guidance for use, important elements	II,	M,	153

Part III

The origin and behaviour of inclusions and their influence on the properties of steels

Foreword

This monograph is the third part of the work begun in Special Report No.90 (1964) of The Iron and Steel Institute, 'Non-metallic Inclusions in Steel',[167] *which was continued in publication No.100 (1966).*[168] *Several earlier books have appeared on this subject. The first comprehensive work was published in 1930 by Benedicks and Löfquist.*[1] *These authors summarized their views on the role of the metallographer in the study of non-metallic inclusions thus:*

'The importance for practical purposes of the results which may be gained by the aid of microscopic or other metallographic investigations (of inclusions) is to a great extent of a pathological *character. To diagnose different kinds of slag inclusions, to determine their quantity and mode of occurrence, to ascertain the quantity which may be considered as permissible, must of course be questions of great importance as a guide in carrying out metallurgical processes on which the quality of the metal ultimately depends.'*

This could still be used today as a definition of the purpose of metallographic investigations of inclusions. However, the present author feels that the modern metallographer should have as his objective an understanding of how different inclusion types influence steel properties, in addition to the pathological aspects of his studies. This objective may be sought because the gradually increasing knowledge of the parameters governing the mechanical and other properties of steels has been accompanied by the development of powerful research techniques for the metallographer. Benedicks and Löfquist used the term 'pathological', with its implications of disease, and the old concept of regarding all inclusions as being pieces of dirt harmful to the steel remains strong. This must be changed, for while it is true that large exogenous macroinclusions are usually disastrous to the steel and must be avoided, this is only one aspect of the subject of inclusions and the metallographer's work. It should be recognized that the 10^{12}–10^{13} *inclusions present in every ton of steel made are a natural structural component; their presence is sometimes harmful, but sometimes even advantageous. Whatever effect inclusions have, however, they are an inseparable part of the composite product called steel; not only must they be lived with, but they must be exploited and as many of their properties as possible used to the advantage of the final product.*

The smaller microinclusions which make up the majority of the inclusion content of a steel have formation mechanisms, modes of occurrence and an influence on the final properties of the steel which are complicated and difficult to determine. It is all the more important, therefore, for metallographers to devote long-range research effort to these problems, which are of vital importance for the progress of ferrous metallurgy.

In parts I and II of this work the author, with the assistance of Mr N. Lange, made an attempt to summarize in a systematic way the metallography, crystallography and physical properties of the different phases present in non-metallic inclusions found in steel. These two parts together constitute an atlas of inclusions with as much useful information as was possible about the inclusion phases. Since the present volume, part III, forms an inseparable unit of the atlas, the contents of the earlier parts will be briefly summarized.

In Part I special attention was given to the system $MnO–SiO_2–Al_2O_3$ *to which most indigenous oxide inclusions in modern steels are related. This system is of fundamental importance in steelmaking and serves as a convenient basis for the discussion of most of the other oxide systems present in non-metallic inclusions, independent of their origin. The most important of these other systems are* $MgO–SiO_2–Al_2O_3$ *and* $CaO–SiO_2–Al_2O_3$*, both of special relevance to exogenous inclusions in steel. These systems were considered in part II.*

Special attention was given to the oxides FeO, Cr_2O_3 *and* Fe_2O_3*, common constituents of oxide inclusions in steel. The possible inclusion phases which may be formed by these oxides are closely related to phases in the three major systems mentioned above. There are close connections in the morphology, solid solubility, structure and general physical properties between the phases, and the majority of oxide phases in inclusions are those of the limited number of metallic elements Mg, Al, Ca, Fe, and Cr, together with Si. Oxides of metals such as Ti, V, Nb and the lanthanides are sometimes found. However, they do not change the basic pattern of phases in the three major systems and are mostly found in solid solution in the phases from these systems.*

After oxygen, sulphur is the most important non-metallic element in steelmaking practice. Sulphides therefore form a second important group of inclusions. This group was considered in part II, the principal sulphide inclusion types being FeS, MnS and (Mn,Me)S. Other possible sulphide phases of relevance to steelmaking were discussed in general terms, mainly using information published in the literature.

The fundamental principle in parts I and II was to present the inclusion types in a systematic manner based on their chemistry. The illustrations were chosen primarily to illustrate the typical appearance of the phases discussed and thus enable the atlas to be used as an aid to identification. As a result of this classification, illustrations appeared from different steel types, from ferroalloys, refractories, slags and synthetic compounds, in no regular order based on origin. Of the steels, examples were taken from ingots, slabs, bars, sheets, wires, and finished products.

The author has felt that the inclusion atlas of parts I and II should be supplemented by a third part designed to give a more complete picture of the overall importance of inclusions in a steel product. To do this, it is necessary to consider the origins of inclusions in steel, their behaviour during working processes and their influence on the final product. The literature on this subject is large, but each investigation often yields only a small piece of information to fit the overall pattern. The present volume is based mainly on literature references, but some original research is also included. From this material the author has attempted to distil the most important results and to concentrate on those aspects which, in his opinion, are most important and fundamental to the future development of the composite product, steel with inclusions. Frequent reference has been made in the present work to figures, tables and text in the preceeding parts. In order to minimize the inconvenience of cross-references between the parts, the alphabetical listing of contents, figure and table numbers and references follow consecutively from part II. (Each part is, however, paginated separately.) Part III also contains, in section R, some information additional to the systematic descriptions of phases given in parts I and II. The author thinks, therefore, that the present part helps fulfill the primary aim of parts I and II. This aim was to give the ferrous metallurgists as useful a tool as possible for the identification of inclusions in steels. It is hoped also that it will enable metallographers to use to the full their existing facilities, especially the optical microscope.

Professor Roland Kiessling is Director of the Swedish Institute for Metal Research. The author is indebted to Dr Sten Bergh (of the Research Centre, Stora Kopparbergs Bergslags AB, Domnarvet), Dr Sten Gunnarson (of AB Volvo, Göteborg) and Mr F. B. Pickering (of the Swinden Laboratories, Rotherham) for interesting discussions and valuable criticism of part of the volume. Dr D. Dulieu (of BISRA, Sheffield) again assisted with the English text and Mr J.

Gustafsson assisted with several photomicrographs. Their help is gratefully acknowledged. Thanks are also due to Mrs G. Bergman and Mrs B. Nobel for valuable help with text and figures and to the author's wife for her great patience.

As was the case for parts I and II, part III will be printed also in Swedish in a series of communications in the Swedish journal Jernkontorets Annaler. *The printing of the English edition has been greatly facilitated by the loan of the blocks from Jernkontorets Annaler, and the help of the Editor of this journal is gratefully acknowledged.*

Stockholm, January 1968

0. The origin of non-metallic inclusions

'The most disturbing feature of non-metallic inclusions is the uncertainty as to where and in what form they may appear; therefore any evidence regarding their constitution and source and how to eliminate that source is valuable'.

Rait and Pinder 1946

O 1 Introduction

The present section deals with the origin of non-metallic inclusions in steel from a metallographer's point of view. The modern metallographic techniques of investigation give him greater possibilities than before to trace the origin of different inclusion types present in the solid steel. Comments on these techniques were given in part I, p.5 of this review and the systematic description of the different inclusion phases in parts I and II is based on such information.

The non-metallic inclusions naturally fall into two groups, those of *indigenous* and those of *exogenous* origin. The former group contains inclusions occurring as a result of reactions taking place in the molten or solidifying steel bath, whereas the latter contains those resulting from mechanical incorporation of slags, refractories or other materials with which the molten steel comes in contact. The indigenous inclusions, adopting Sims's definition,[67] are those that form by precipitation as a result of homogenous reactions in the steel. They are composed principally of oxides and sulphides and the reactions that form them may be induced either by additions to the steel or simply by changes in solubility during the cooling and freezing of the steel. Exogenous inclusions occur in a great variety but, for the most part, are readily distinguished from the indigenous inclusion. Characteristic features of exogenous inclusions include a generally larger size, sporadic occurrence, preferred location in ingot or casting, irregular shapes and complex structure. They are usually composed of oxides, a result of the compositions of potential exogenous materials such as slags and refractories.

Although the above division of inclusions provides a natural starting point for a discussion of inclusion origin, recent results from detailed analysis of inclusion composition with the electron probe have shown that this division is an oversimplification. It is a common phenomenon that indigenous precipitation occurs on exogenous nuclei during all the different stages of the steelmaking process. An inclusion nucleus, from the moment it enters the molten steel and independent of whether this occurs from homogenous nucleation or from an exogenous source, takes part in the future life of the steel. Many non-metallic inclusions therefore continuously change their composition in the steel bath, and even in the solid steel, until the moment when the diffusion rate is negligible. Several examples of this are given in the earlier parts of this review; compare for instance the inclusion in Fig.11 with that in Fig.13. The metallographer, who is faced with the problem of determining the origin of the resulting inclusion product in the steel matrix, is therefore faced with a much more complicated problem than this general classification of inclusions indicates. For example, a large multiphase inclusion with irregular shape and complex structure is not merely an exogenous piece of refractory which has entered the steel. The origin may be the refractory material, but the mean composition, and often therefore also the structure of this type of inclusion, usually differ considerably from the original refractory due to indigenous precipitation of oxygen as different oxides. Sulphides may also be nucleated on the oxide inclusion particle. Another example is given by the small glassy spherical inclusions in Figs.40, 125*a–c* and 134. These are only seldom indigenous SiO_2 particles. Their composition varies considerably within the system FeO–MnO–MgO–CaO–SiO_2–Al_2O_3–Cr_2O_3–TiO_2, which is an indication of different possible sources for the origin of these inclusions.

The composition and structure of the inclusions are not the only factors of importance in tracing their origin. The inclusion shape, size and position in the ingot are all highly relevant and are considered in section O 2. Discussion here has been based on recently published information, but methods of inclusion counting are not discussed. These have been reviewed in detail elsewhere, for example in the Iron and Steel Institute Special Report No.77, 'Clean Steel'.[169] This report dealt also with methods of inclusion examination, which were referred to briefly in part I (p.5).

The main emphasis in this section is on how the different inclusion types become located in the solidified ingot. The inclusions may be followed through the different environments of the bath, ladle, tapping process and solidification. This sequence has been introduced in part II, section M and Tables XXXVI–XXXVII, and the present section gives a more complete analysis of the relation between the inclusions and the steel during the manufacturing process.

The deoxidation mechanism is of fundamental importance for the formation of non-metallic oxide inclusions. Therefore thorough consideration has been given to the formation of oxide inclusions as a result of deoxidation (Section O 3). It is now often possible to trace the origin of different types of non-metallic inclusions. However their growth, coalescence, rise in the steel bath and detachment from the molten steel, that is, the general behaviour of inclusions in the molten steel, are still a matter of discussion. This is of special importance for the deoxidation reaction and some comments on recent theories are also given in section O 3. A more complete discussion is beyond the purpose of this review. In section O 4 the formation of non-metallic inclusions during the steelmaking process is discussed. In the following sections O 5–O 7, examples are given of the relation between the composition and microstructure of the inclusions and their origin. This relationship may be established with information from a combination of the techniques of electron probe analysis, microscopy and X-ray analysis. Frequent reference has been made to the figures and tables of parts I and II. The different inclusion sources considered have been summarized in Table XXXVIII,

TABLE XXXVIII **Possible inclusion sources discussed in the present review (and parts I and II) with reference to sections, tables and characteristic figures. Compare also Table XXXVII**

Possible inclusion source		Key elements	Section	Table	Figure
Furnace	Furnace slags	Ca	04,6	XIV	
	Furnace refractories	Ca	04,7	IX	146
	Ferroalloys	Cr, Al, Si	04,5		206
Tapping	Launder refractories	Mg, Ti, K	04,7	IX	
	Oxidation	FeO			
Ladle	Deoxidation		03,4	XIII	203
	Ladle slag	Ca, Mg	04,6	XIV	207
	Ladle refractories	Mg, Ti, K	04,7	IX	38
Teeming	Stopper and nozzle refractories	Mg, Ti, K	07	IX	
	Oxidation	FeO			
	Deoxidation		03,4	XIII	
Ingot mould	Refractories	Mg, Ti, K	07	IX	208, 209, 18
	Deoxidation		03,4	XIII	
Heat treatment and rolling	Surface oxidation	FeO			21, 22, 76–79
	Surface sulphurization	FeS	K2		155–158
	Inner oxidation	SiO_2			59
	Hot-shortness	FeS	Q4		232, 234
Welding	Welding slags	Ca, Ti	Q5	XXV	
	Electrode coatings	Ti, V	Q5	XXV	126, 130
	Steel inclusions		Q5		
	Hot tearing	S	Q5		

with reference to the appropriate sections. The further history of the inclusions from ingot to finished product will be followed in section P, while their influence on steel properties has been dealt with in section Q.

O 2 Inclusion size, quantity and distribution in the steel

'The cleanliness of steel, like beauty, is very much in the eye of the beholder'.
Melford 1962

The quantity and size of inclusions

Bergh[170] has found that approximately all oxygen in a silicon-killed steel can be recorded as visible oxide inclusions, and has made a detailed study of the number, size and distribution of these inclusions.

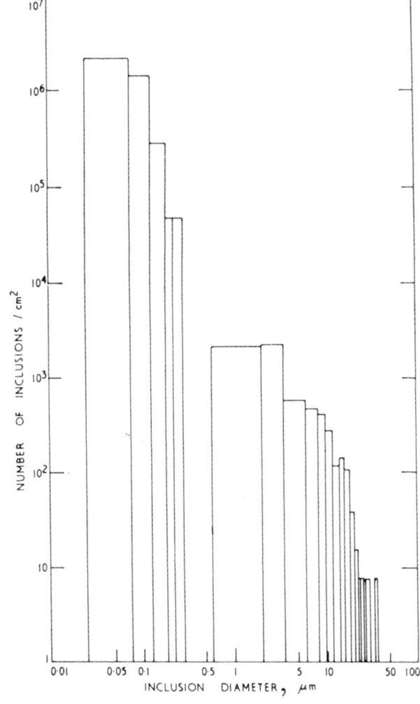

STEEL TYPE Basic electric steel, Si-killed. From centre of top section in 6-ton ingot, top poured, big end down
STEEL ANALYSIS % 0.16C, 0·35Si, 0·67Mn, 0·015P, 0·017S, 0·009N, ~100 ppm 0
INCLUSIONS The inclusion number was obtained from a surface of 10 cm², and the given diameter value is the middle value for the size fraction

COMMENT There is a great increase in the number of inclusions with decreasing inclusion size (logarithmic scale). The gap around 0·5 μm has no real significance, being due to limitations in the recording technique. Most of the oxide material, however, is concentrated in the fewer and larger inclusions, as shown by Fig.189. From ref.170.

188 Size distribution curve for the oxide inclusions in a Si-killed steel ingot

The result gives an interesting picture of the occurrence of indigenous oxide inclusions. The steel was a basic electric plain carbon steel with an oxygen content of 100 ppm, top poured (big end down) to 6 ton ingots. It was found that the number of inclusion particles rapidly increases with decreasing particle diameter. A histogram of the number of inclusion particles in different particle size fractions at the centre of the ingot top section is shown in Fig. 188. In this part of the ingot, about 98% of all inclusions are smaller than 0·2 μm. However, these particles only represent about 1–2% of the total oxygen in the steel. The oxygen content in different particle size fractions from different ingot positions is shown as a number of cumulative frequency curves in Fig.189. No inclusions

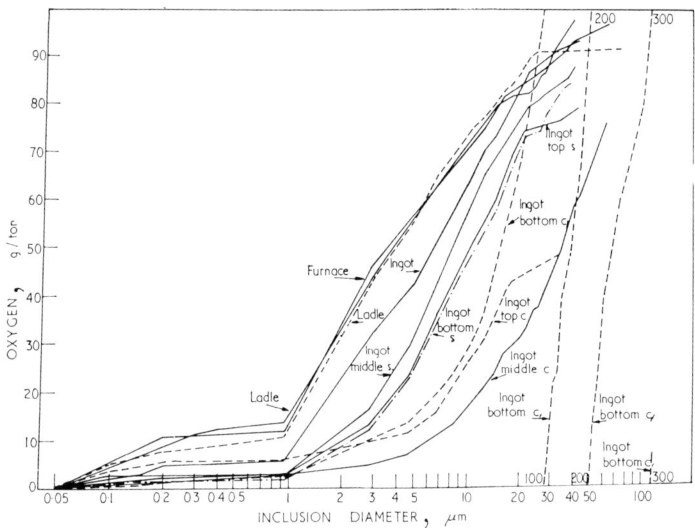

STEEL TYPE See Fig.188
STEEL ANALYSIS See Fig.188
INCLUSIONS The cumulative frequency curves for oxygen content in different particle size fractions have been given, expressed in grams of oxygen per ton steel. Different ingot positions are shown

COMMENT The main part of the oxygen is concentrated in the larger inclusion particles although their number is small (compare Fig.188). There is also a marked concentration of oxide inclusion material at the central root position of the ingot. From ref.170

189 Oxygen content in different particle size fractions of the inclusions in a Si-killed steel ingot

smaller than 300–400 Å were found, although an electron microscope replica technique with a particle resolution of about 100 Å was used. For the top section of the ingot, there was on average one inclusion for every steel surface element of 900 μm^2, that is, the mean distance between the inclusions was 30 μm, which corresponds to between

100 and 1 000 diameters of the most frequent inclusion size. In a 6 ton ingot there is, therefore, a total number of about $3 \cdot 10^{13}$ oxide inclusions! It is quite evident that it is impossible to produce steel that is entirely free from inclusions. With present knowledge it would seem that the figures given by Plöckinger[171] for ultimate inclusion contents of about 0·003% in vacuum-melted steels and of about 0·005–0·010% in steels subjected to orthodox deoxidation can be considered as being extremely small.

Several parameters have an influence on these figures. Most important for the steel properties are those which control the comparatively few larger inclusions. The deoxidation practice is one and some comments are given in section O 3. A study of the effect of vacuum degassing v. air melting on the size distribution of globular oxide inclusions in ball-bearing steel is summarized in Fig.190. The

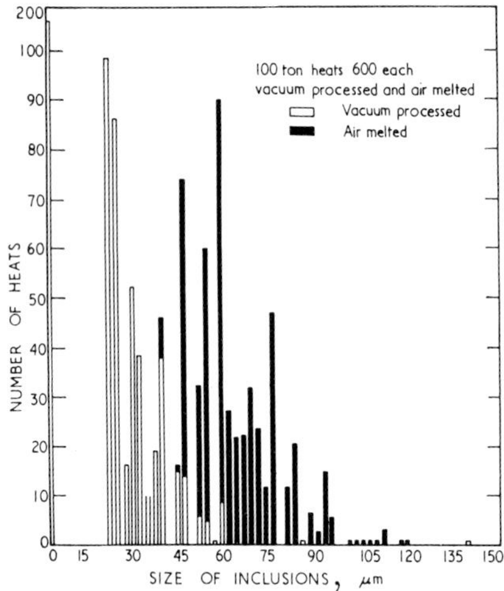

STEEL TYPE Bearing steel, air melted with CaSi deoxidation in the ladle as compared with ladle vacuum degassed (carbon deoxidation in vacuum)
STEEL ANALYSIS % 0·20C, 0·27Si, 0·81Mn, 0·009P, 0·013S, 0·55Ni, 0·51Cr, 0·20Mo for A.M., very similar for V.P.

INCLUSIONS The average diameter of globular oxide inclusions (D-type, compare Figs.112 and 117) in 600 heats of A.M. and V.P. steel were determined. The mean average diameter for A.M. steel was 60 μm against 20 μm for V.P. steel. Only inclusions larger than about 1μm were counted. From ref.124

190 Size distribution of globular type of oxide inclusions in air melted (A.M.) and ladle vacuum processed (V.P.) bearing steel

average oxide inclusion diameter was decreased from 60 μm to 20 μm and the frequency was considerably decreased due to the replacement of Ca–Si deoxidation in the ladle by carbon deoxidation in vacuum. The results are not comparable to those of Bergh's, since the inclusions smaller than about 1 μm were not counted. Nonetheless the figure illustrates how steel practice has a considerable influence on the size distribution of the larger inclusions, of fundamental importance for the effect of inclusions on several steel properties.

The most disastrous inclusions are usually the comparatively few exogenous inclusions originating from refractories and slags. Their number and size cannot be discussed systematically, as they depend on individual steelmaking practice.

For sulphide inclusions the information is more scanty, and the effect of different elements on their size and shape is less clearly understood than is the case for the oxide inclusions, as discussed in section Q 2. The size distribution curve of the mean diameters of MnS-inclusions in different positions in a top poured ingot of a resulphurized free-cutting steel is shown in Fig.191. This gives a

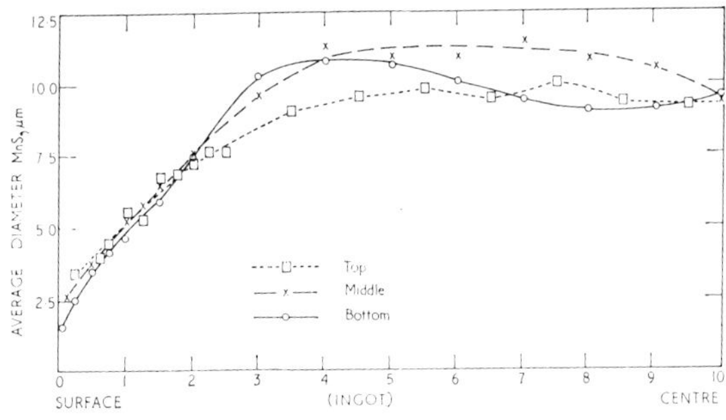

STEEL TYPE Commercial free-cutting steel ingot, 21in square, top poured, weighing approx. 3 ton
STEEL ANALYSIS Range of analysis for this steel type % 0·07–0·15C, 0·06Si max., 0·85–1·25Mn, 0·25–0·35S, 0·045P max.
INCLUSIONS In the region of the ingot surface (chilled zone and outer columnar zone) the inclusions are small and interdendritic. Across the columnar zone up to 4in from surface they coarsen rapidly, being of type II (Fig. 169). Towards the centre, solid solution inclusions of the (Mn,Fe)S type were observed (Section K 4). This should be compared with the volume fraction of sulphide inclusions from surface to centre, shown in Fig. 195. From ref.127

191 Mean diameter of MnS inclusions from surface to centre of free-cutting steel ingot

general idea of the sulphide size range. The mean inclusion diameter at the steel surface was about 2·5 μm, in the centre about 10 μm. If all the sulphur in this steel (0·30%S) was recorded as MnS inclusions with a mean diameter of 5 μm, there should be on the average about one such inclusion in every steel volume element of about 10000 μm^3. This gives an average distance between the sulphide inclusions of about 30 μm, the same as was found for the oxide inclusions which, however, were much smaller.

The spatial distribution of inclusions in steel

The spatial distribution of the oxide inclusions is, naturally, also of importance for their influence on steel properties. This distribution is related to the individual steelmaking practice and therefore varies, but some general investigations should be mentioned.

In top poured Si-killed plain carbon steel, the distribution of oxide inclusions is strongly dependent on the position in the ingot. The classic investigation of Dickenson[172] in 1926 showed that the main part of the oxygen was present as oxide inclusions in the lower central part of the ingot. The free steel dendrites which are heavier than the mother liquid sink to the bottom and catch the upward stream of inclusions. This results in a concentration of inclusions in the lower part of the ingot. The recent investigations by Bergh have again confirmed this, showing an oxygen content in the lower region of about 300 ppm against 100 ppm on the average. The oxygen content in different parts of the ingot, which is closely related to the inclusion content, is summarized in Fig.189. The top centre position had the lowest oxygen content (~50 ppm). This general inclusion distribution is also preserved after rolling.[173] Further information on the types of oxide inclusions present in different positions of a similar ingot is available from the work of Cox and Charles.[5,126]

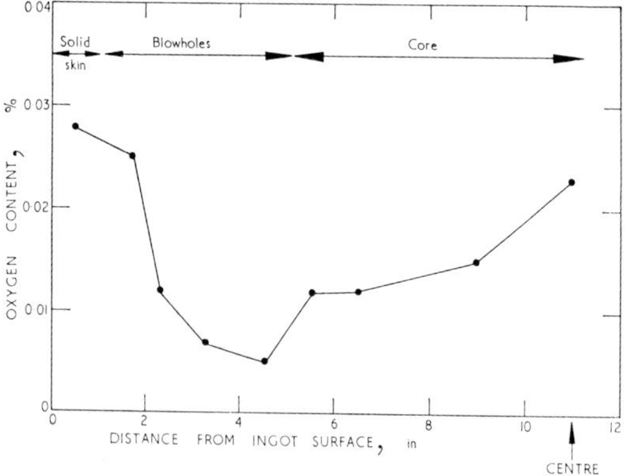

STEEL TYPE Rimming steel ingot, 22in square, weighing approx. 4 ton
STEEL ANALYSIS % 0·07–0·09C, 0·01–0·02Si, 0·3Mn, 250–300 ppm O.
INCLUSIONS The general distribution of the inclusions is readily revealed by the oxygen concentration across the ingot, here at mid-height. The distribution is similar at other positions in the ingot. The inclusions are predominantly silicates. From ref.174

192 Variation of oxygen content (and oxide inclusion content) across a rimming steel ingot

The general picture of oxygen distribution (and indigenous oxide inclusion distribution) given above is modified if the steelmaking practice is changed. For instance, Pickering[174] in his study of oxide inclusions in rimming steels has shown that the oxide inclusion concentration is high in the outer zone (chilled layers) of the ingot, decreases in the blowhole zone and again increases in the core (Fig. 192). This horizontal distribution is similar at different heights in the ingot. As shown by Brotzmann,[175] vacuum treatment changes the pattern of behaviour to a rather low and even distribution across the ingot in medium positions. However, in the case of vacuum deoxidation also there is a higher inclusion content in the axial position of the bottom part of the ingot, Fig.193. This tendency for maximum

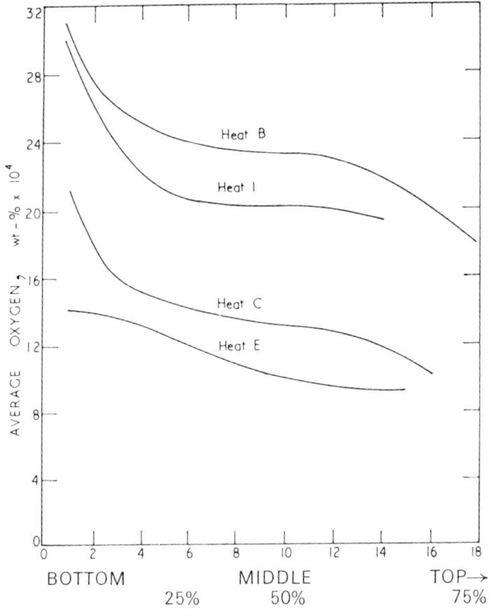

STEEL TYPE Basic electric steel, top poured 8 ton ingots, ladle-to-mould vacuum stream degassed with variations in melting and casting procedures (*see* ref.176)
STEEL ANALYSIS % 0·20C, 12–24 ppm O

INCLUSIONS The oxide inclusions distribution shows the same trend as the oxygen content, that is a marked increase in the central root position of the ingot

193 Variation of oxygen content (and oxide inclusion content) along the axial portion of carbon steel ingots, vacuum deoxidized in different ways

inclusion density to occur in the lower axial ingot position is independent of the deoxidation technique employed. This was shown by the work of Wick *et al.*[177] who studied the oxygen distribution in ingots deoxidized by different methods. As is evident from their results, shown in Fig.194, there is always a concentration of oxide inclusions in the central bottom part, but this tendency is decreased with increasing Al-content in the deoxidizing addition. Wick *et al.*

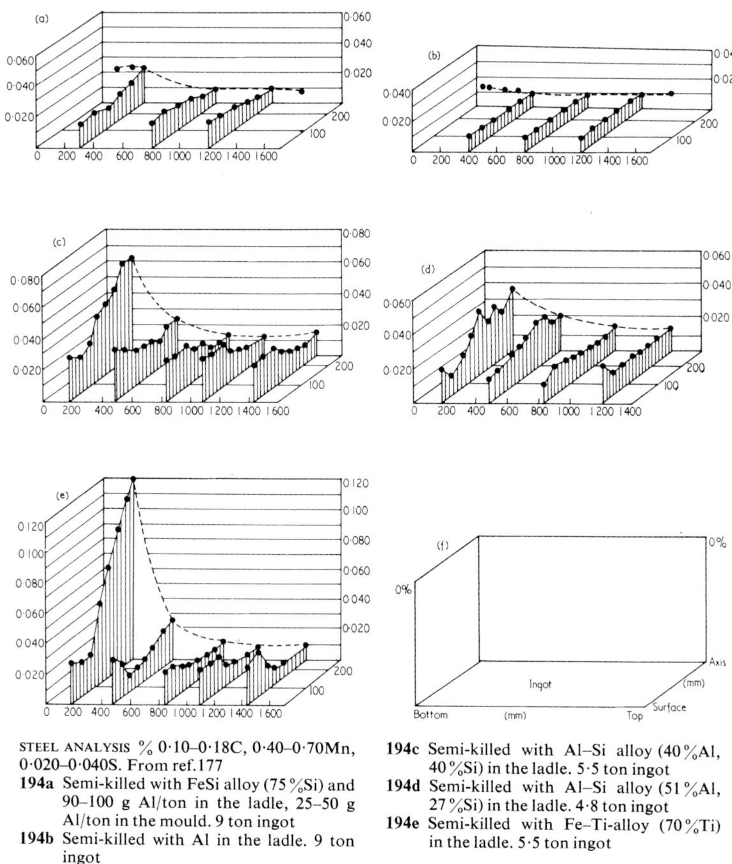

STEEL ANALYSIS % 0·10–0·18C, 0·40–0·70Mn, 0·020–0·040S. From ref.177

194a Semi-killed with FeSi alloy (75 %Si) and 90–100 g Al/ton in the ladle, 25–50 g Al/ton in the mould. 9 ton ingot

194b Semi-killed with Al in the ladle. 9 ton ingot

194c Semi-killed with Al–Si alloy (40 %Al, 40 %Si) in the ladle. 5·5 ton ingot

194d Semi-killed with Al–Si alloy (51 %Al, 27 %Si) in the ladle. 4·8 ton ingot

194e Semi-killed with Fe–Ti-alloy (70 %Ti) in the ladle. 5·5 ton ingot

194 Distribution of oxygen content (and oxide inclusion content) in OH-steel ingots deoxidized by different methods

gave as the reason for this that, with Al deoxidation, most of the inclusions are already separated from the steel in the ladle and that those remaining are mainly precipitated in the ingot itself.

The sulphides are usually localized with the highest concentration in the core of the ingot in rimmed steels, but in the outer regions of the ingots in killed steels, as discussed in section K 1. A typical distribution pattern of MnS-inclusions in a rimmed free-cutting steel ingot is shown in Fig.195. Further observations on MnS distribution are also given in Refs.5 and 126.

The author is well aware that these comments on inclusion size distribution and distribution of inclusions in the steel phase are

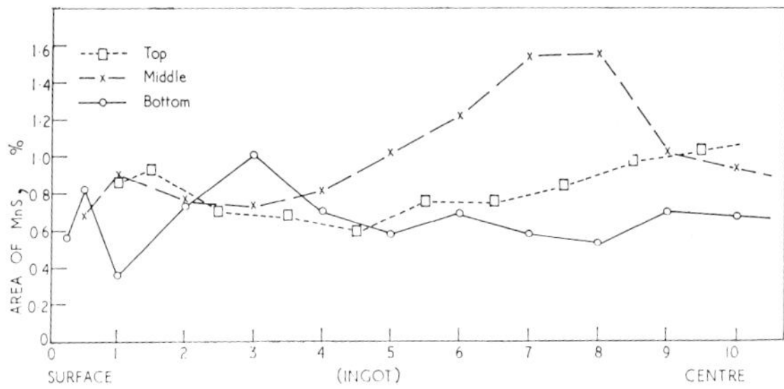

STEEL TYPE *See* Fig.191
STEEL ANALYSIS *See* Fig.191
INCLUSIONS The central zone of the middle position of the ingot has the highest amount of sulphides, Compare the size distribution of sulphide inclusions, Fig.191. From ref.127

195 Volume distribution of MnS inclusions from surface to centre of free-cutting steel ingot

incomplete, but they serve to give an idea of the general trend. When the influence on the properties of steels is discussed, it is important to realize that the inclusions are not evenly distributed within the steel and that they have a large size spectrum. Thus, as is realized by most metallurgists, quite different machinability or corrosion properties resulting from inclusions may be obtained from samples of the same steel charge, ingot or even billet, caused by the difference in inclusion concentration and type in different parts of the ingot.

O 3 The influence of deoxidizing elements upon inclusion formation

'In many instances deoxidation of the steel is not intended to achieve the smallest amount of inclusions but rather the formation of inclusions which have the least harmful effects on the technological properties of the steel.'

Plöckinger 1963

The behaviour of oxygen in pure iron

At the melting point of iron, the solubility of oxygen in the liquid phase is about 0·16%, and the solubility increases with temperature (Table XXXIX) The solubility in solid iron at the melting point is very much lower, in fact so low that no reliable determinations have been made. According to Sifferlen,[178] the oxygen solubility in solid iron tends to be zero with increasing purity and lattice perfection of the metal. The true oxygen solubility is probably below the working range of ordinary methods of analysis. If pure iron solidifies, a considerable part of the oxygen is precipitated as liquid FeO-inclusions, which will be visible as spherical inclusions in the

TABLE XXXIX **The solubility of oxygen in liquid iron at various temperatures. Two different equations have been used, namely**

$$\log |\%O| = -\frac{6320}{T} + 2{\cdot}734 \quad (1)$$

$$\log |\%O| = -\frac{5762}{T} + 2{\cdot}439 \quad (2)\ (\text{Ref.179})$$

Temperature (°C)	|%O|, eq. (1)	|%O|, eq. (2)
1550	0·185	0·190
1600	0·229	0·230
1650	0·281	0·277
1700	0·340	0·330
1750	0·409	0·388
1800	0·483	0·456

solidified iron. Examples of such inclusions of FeO (B 2, D 2) are shown in Fig.196.

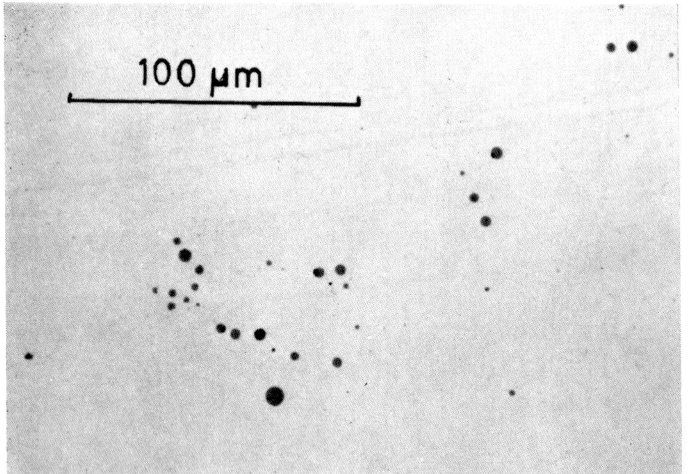

STEEL TYPE Electrolytically precipitated iron ('Orkla iron')
STEEL ANALYSIS ppm, 19–25C, 60–90S, 20N, 250–400O, 68Co, others < 10.
INCLUSIONS The iron was melted at just above the melting point and then left to solidify in argon. The main FeO nucleation and precipitation occurred during solidification, when the solubility of oxygen decreased considerably, but further increase in inclusion size also occurs during cooling of the solid iron

196 Inclusions of FeO, precipitated during solidification of pure iron

The behaviour of oxygen in steel

The oxygen content of liquid steel depends upon the nature of the elements present in addition to iron, the most important being carbon. Although the oxygen content of a steel melt tends to be lower than that of pure liquid iron, it remains high, being from 0·01–0·1% depending upon the composition. Most of the oxygen is in solid solution in the liquid steel, but a part may also be present as small oxide inclusions in the melt. The purpose of the deoxidation process is to decrease the amount of oxygen in the steel. This may be accomplished by adding to the steel elements with a higher affinity for oxygen than iron, precipitation deoxidation, and by the transfer of FeO to a suitable slag on the steel surface, diffusion deoxidation. Only the mechanism of precipitation deoxidation results in the formation of inclusions in the steel, however, and diffusion deoxidation through the steel/slag interface will not be considered further. A general picture of the oxygen affinity of different elements is given by the curves of the free energy of formation of their oxides as a function of temperature. Curves for the elements relevant to steelmaking have been given in Fig.80. In principle, all elements with a free energy of oxide formation lower than FeO may act as deoxidizers, but as was discussed in section K 1, the activity of the different elements when in solution in the liquid steel is different from the activity of the pure elements and this must be taken into consideration. In Fig.197 a comparison between the deoxidizing power of

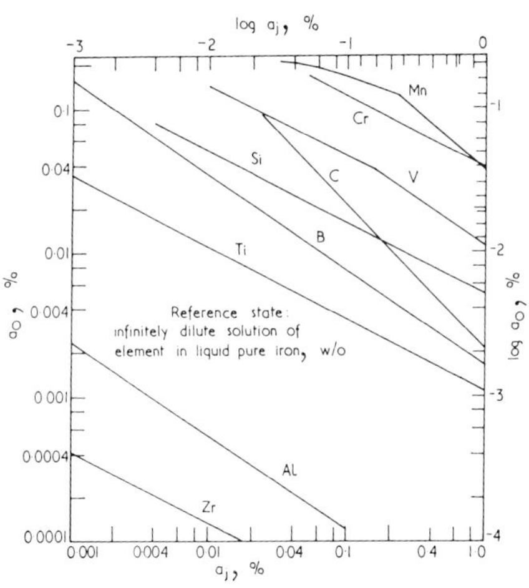

a_o = oxygen activity

a_j = activity of added element. From ref.179

197 Deoxidation power of different elements in steel at 1600°C

different elements at 1600°C is given from which the final oxygen activity in the steel resulting from different activities of the elements may be estimated.

Deoxidation may occur at constant temperature when the deoxidizing element is added to the melt. This results in the formation of *primary* deoxidation products. Deoxidation may also occur during cooling of both the liquid and solid steel at decreasing temperature, due to the action of the alloying elements in the liquid and solid steel and to the decreasing solubility of oxygen. This is called *secondary* deoxidation. The primary deoxidation products have a greater opportunity for escape from the steel than those which form by the secondary process on a falling temperature scale. Those inclusions found in the solidified steel which were formed by the deoxidation reactions are usually a result of both primary and secondary processes.

The main purpose of the present summary is not to discuss the complicated reactions involved in deoxidation with different elements. A recent literature survey with an extensive reference list covering the theories of deoxidation may be found elsewhere[179] and Turkdogan[180] has published a review article on this subject. The purpose of the present section is merely to discuss those inclusion types which might originate from the deoxidation process, their appearance and how they differ from inclusions formed from other sources. The various types of inclusions resulting from deoxidation with different elements are discussed separately. To provide a metallurgical background to observations of the inclusions, however, a summary of recent ideas on the mechanism of oxide particle formation during deoxidation is given first.

Formation of inclusions during deoxidation by alloying elements

Bergh's study[170] of the correspondence between oxygen content and the number of oxide inclusions in a silicon-killed steel enable some deductions to be made on the nucleation and growth of the oxide particles. It can be calculated that there were about $2{\cdot}5.10^7$ inclusions per cubic cm, larger than 0·5 μm at an oxygen content of 400 ppm in the steel. The smallest inclusions had a size down to about 0·04 μm. The mean distance between the inclusions was about 30–35 μm and the mean particle diameter about 5 μm. If such a particle had grown by diffusion, the time for this growth was about 0·5 seconds. It was also shown that there is a low probability for a further increase in size of the deoxidation inclusion products through diffusion of oxygen in the liquid steel and that the size increase due to diffusion is finished within some seconds. Turkdogan[180] has shown that the number of nuclei has a pronounced effect on the growth of oxide inclusions, on the extent of the deoxidation reaction, and on the rate of flotation of oxide inclusions out of the melt, and results of his

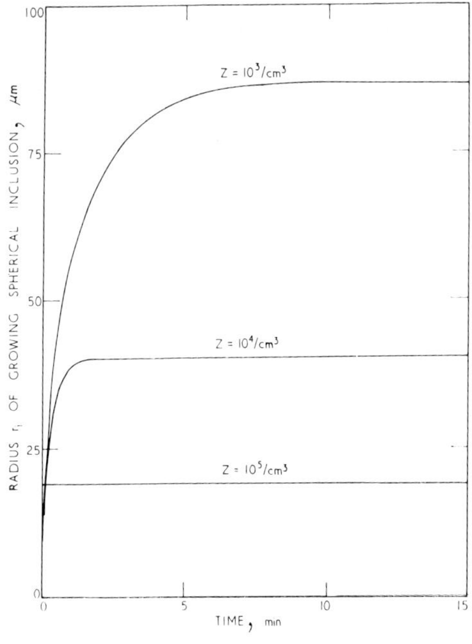

198a

The initial oxygen content of steel $[\%O]_0 = 0{\cdot}05\,\%$ and the final equilibrium oxygen $[\%O] = 0$. The figure illustrates the expected effect during deoxidation of the number of nuclei, in the liquid steel Z/cm³, on the rate of growth (*a*) and distance of rise with time (*b*) for various values of Z.

While for $Z = 10^5/cm^3$ growth is complete in about 10 s, about 6–7 min are required for a number of nuclei $Z = 10^3/cm^3$. As the inclusions grow and rise in the melt with increasing reaction time, the liquid steel is progressively depleted of inclusions from bottom to top. From ref. 180

198 The effect of number of nuclei in melt on the rate of growth of oxide inclusions as they rise in the melt and on the rate of ascent of growing inclusions (*See* next page for Fig. 198b)

calculations are given in Fig. 198a. These figures fit well with Bergh's observations and give further information. For instance, if the number of nuclei in the melt is $Z = 10^5/cm^3$, a typical figure, then the spherical oxide inclusions formed by deoxidation will grow such that the majority will have a final radius of 20 microns. 10^4 particles will reach a diameter of about 40 microns whilst 10^3 will grow to about 85 microns. The time required for the completion of this growth process increases successively, however, to 6–7 minutes.

The inclusions also rise in the melt, and in Fig. 198b the effect of the number of nuclei in the melt on the rate of ascent of growing inclusions is given. As the inclusions grow and rise in the melt with increasing reaction time, the liquid steel is progressively depleted of

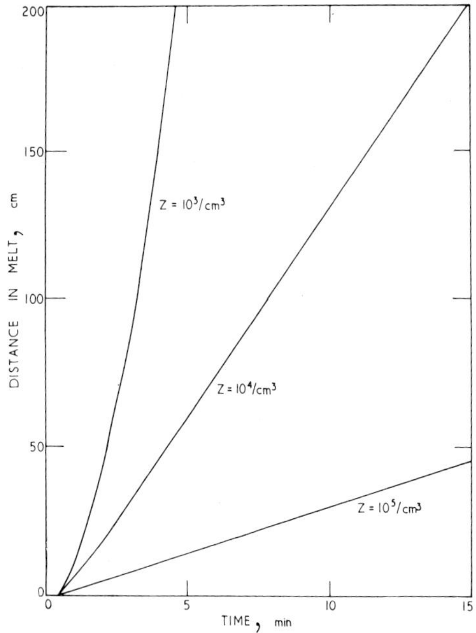

198b (*see* caption below Fig. 198a on previous page)

inclusions from bottom to top. Since there is insufficient supersaturation new nuclei cannot form in the inclusion-depleted part of the steel, and therefore the deoxidation reaction ceases before reaching the final equilibrium state. Turkdogan has given additional curves which summarize the net results of these effects. A significant point is his conclusion that, for any given deoxidation time, there exists a critical value Z_c of the number of inclusion nuclei. If the number of nuclei is less than this critical value, relatively large inclusions are formed which float out of the melt quickly; but a high residual oxygen content is left in solution. The steel may therefore be essentially free of inclusions at the time of pouring, but contain a high percentage of oxygen in solution leading to the formation of blowholes on solidification. For the case where the critical nucleus density is exceeded, the steel is effectively deoxidized but the inclusions are small and have a slow rate of ascent in the melt. Even after prolonged holding times the inclusion content of the melt is high and the solidified steel may become rich in inclusions.

It is evident from this short summary that the deoxidation of the steel from tap to solidification is a complex procedure, which is influenced by several factors. For a full discussion, the reader is referred to the monograph by Ward.[181] The resulting non-metallic inclusions may have a wide range of size, shape and composition. Some fundamental inclusion types may be distinguished, however, their occurrence depending on the nature of the deoxidation element. They will be indicated here with reference, where possible, to figures given in parts I and II.

Carbon

Carbon is an important deoxidation element for rimmed steel, but the deoxidation products are gaseous. It will therefore not be discussed in this review. It should be mentioned, however, that carbon in the liquid steel and oxygen from easily reduced oxide inclusions (mainly FeO) may react. Such gas precipitations are potential causes of *pinhole* porosity in steel castings.[182]

Manganese

This element is a weak deoxidizer, which is evident from Figs.80 and 197, and it is not, at least in a pure form, used as a deoxidizer. If Mn is added to an iron melt containing oxygen, inclusions of the type MnO–FeO are formed usually as solid solutions (A 6, B 2, D 2). This deoxidation process has been investigated by Sloman and Evans.[183] Typical examples of the appearance of the resulting inclusions are given in a further report.[184] The examples of inclusions of the solid solutions MnO–FeO, given in Figs.21, 22, 23 and 76 of the present review are not typical deoxidation products; they have been formed by oxidation of the steel. The deoxidation inclusions are smaller and evenly distributed in the steel, and their morphology depends on the MnO:FeO ratio of the solid solution, that is on the Mn-content of the resulting steel. When the melts from Ref.183 were studied it was found that if the steel had a low Mn-content, the resulting inclusions had a MnO content of less than about 30% and were spherical single phase inclusions, similar to those in Fig.196. With a higher Mn-content, the MnO-content of the inclusions increased but they were still mainly globular. As discussed in section B 2, they may sometimes be of a duplex character in spite of the complete solid solubility in the system MnO–FeO. For Mn-contents of the steel higher than about 0·7%, the inclusions were nearly pure MnO. These inclusions had solidified before the iron and their appearance therefore changed from being globular to having dendritic structure, whereas those inclusions richer in FeO solidified after the iron was solid. Uchiyama[185] has given a curve for the MnO-content of inclusions in iron deoxidized with different amounts of Mn (Fig.199) which is in good agreement with Pickering's results.[184]

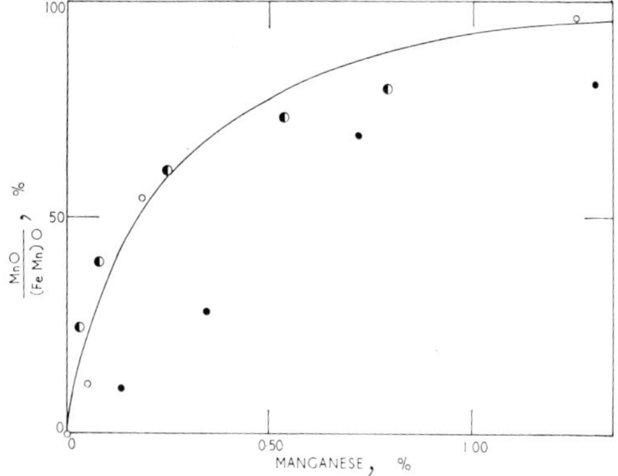

STEEL TYPE Electrolytic iron, 6 kg melt, deoxidized with Mn in high frequency furnace and cast to 5 kg ingot
STEEL ANALYSIS % 0·005–0·010C, 0·004–0·006Si, 0·003–0·005S, Mn (*see* figure)

INCLUSIONS The composition of inclusions >5 μm was determined by electron probe analysis, and the ratio MnO:(Mn,Fe)O was calculated. From ref.185

199 MnO-content of (Fe,Mn)O inclusions in iron deoxidized with different amounts of Mn

Silicon

As shown in Fig.80, silicon is a much more effective deoxidizer than manganese and it is often used in combination with that element. If pure iron is deoxidized with silicon, the deoxidation products are either iron silicates (B 3, 4), which are liquid at the melting point of iron (steel) or solid SiO_2 (A 3–5). According to Evans and Sloman,[186] pure SiO_2 is formed if the silicon content is higher than 0·08%. The SiO_2 inclusions are usually glassy and globular with diameters up to about 50 μm. They are similar to the types shown in Figs.40, 125 and 134, which were, however, silicates of different kinds. According to several authors, satellites of smaller inclusions are often found around the glassy SiO_2 inclusions, and a hypothesis has been advanced by Wahlster[187] that small droplets coalesce and grow to larger, glassy SiO_2 inclusions. Micrographs of such deoxidation products are given in his work, but it is debatable whether this is a correct assumption for their formation mechanism.

A more likely explanation has been given by Hultgren,[26] who has shown several microsections of such inclusions with satellites.[188] According to him there is a reduction of other iron and manganese oxides in the inclusions by silicon in the steel, giving a silica-rich surface layer of the inclusions, sometimes visible as a scale (Fig.11). The volume of the inclusion decreases as a result of the iron and manganese oxide reductions. The surrounding steel will be rich in oxygen but poor in silicon, whereas the inclusion surface is rich in silicon. The two elements will diffuse in opposite directions and small SiO_2 droplets are thus formed around the large inclusions.

As discussed in sections A 3–5 and O 7, it is essential to establish the SiO_2-modification present in the inclusions, if the origin has to be determined. SiO_2 may also be of exogenous origin, for example from refractories, and it is especially important to find whether or not low-quartz is present. If that modification is identified the SiO_2 cannot be a deoxidation product (see O 7, example II).

Silicon–manganese

Although deoxidation by silicon is more effective than by manganese, simultaneous deoxidation by these two elements gives much lower residual oxygen in solution. This was shown by Körber and Oelsen.[189] Normally in the deoxidation of steel both these elements are used together and the first stage of deoxidation is carried out by Si/Mn additions to the ladle during tapping. Curves for calculating the residual oxygen at different additions and temperatures have been given by Turkdogan.[180] Over the composition range applicable to the deoxidation of steel, the deoxidation product formed is essentially liquid manganese silicate or solid silica, with little or no iron oxide in solid solution. The non-metallic inclusions formed as a result of this deoxidation process therefore belong to the $MnO–SiO_2$ system, their composition depending on the ratio [%Si]/[%Mn] in the steel phase. This was discussed in section A 15 and Table VI gives a summary of the experimental results for a number of inclusions with reference to their microscopical appearance. Most of the inclusions discussed in that section are not a result of Si–Mn deoxidation only, but the inclusions shown in Figs. 11, 40 and 41 (left) are all typical examples of inclusions formed by this type of deoxidation. They are globular, either glassy or with SiO_2 (cristobalite, A 3) or rhodonite (A 9) crystallizing in a matrix of manganese silicate. The MnO-rich silicate tephroite (A 10) is, according to the author's experience, not a common inclusion phase and no example of this phase has been found as a primary deoxidation product from Si–Mn deoxidation only.

Turkdogan[180] has studied the equilibrium constant

$$K = \frac{[\%Si]}{[\%Mn]^2} \cdot \frac{(a_{MnO})^2}{(a_{SiO_2})}$$

He has concluded that there is a critical ratio $[\%Si]/[\%Mn]^2$ above which only solid silica forms as the deoxidation product ($a_{SiO_2} = 1$) and thus manganese does not participate in the deoxidation reaction. He has given curves for the critical silicon and manganese contents of steel in equilibrium with silica-saturated manganese silicate at various temperatures (Fig.200). From these curves it is possible to predict the deoxidation products and therefore the type of inclusions formed in the steel. If, for any particular temperature, the composition of steel lies above the curve in Fig.200, manganese

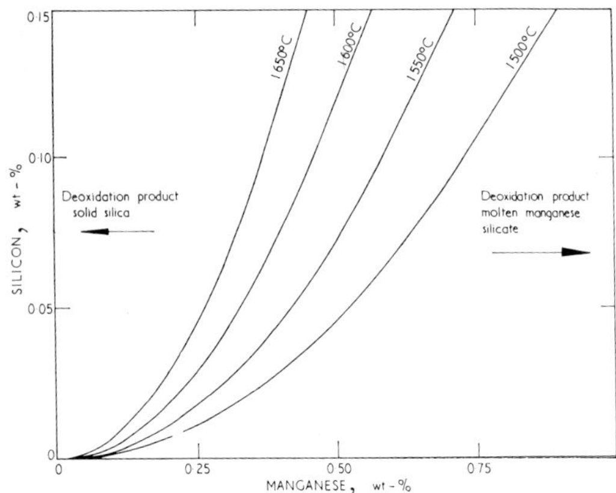

There is a critical ratio [%Si]/[%Mn] above which only solid silica forms as the deoxidation product. The figure gives the critical Si and Mn contents of steel in equilibrium with silica-saturated manganese silicate at various temperatures. If, for any particular temperature the composition of steel lies above the curve, Mn does not participate in the deoxidation reaction but solid SiO_2 is formed. In the region below the curve, the deoxidation product is molten Mn-silicate. From ref.180

200 Critical Si and Mn contents of steel in equilibrium with silica-saturated deoxidation product (manganese silicate) at different temperatures

does not participate in the deoxidation reaction and solid silica is formed. The inclusions to be expected should therefore be of the type shown in Fig.40, but their composition will consist only of SiO_2. If the composition of the steel lies below the curve, the primary deoxidation product belongs to the $MnO–SiO_2$ system and is molten. The composition is then determined by the ratio $[\%Si]/[\%Mn]^2$ in the steel. The resulting inclusions will be of the type and appearance shown in Figs.11, 40 and 41. It may be desired that the reaction product at all stages of deoxidation from 1650°C to 1500°C should be molten $MnO–SiO_2$ products. The manganese and silicon additions to the steel should then be so adjusted that, at the end of the first stage of deoxidation at tap (1650°C), the residual silicon and manganese concentrations are below those corresponding to the 1500°C isotherm. If so, subsequent secondary deoxidation during cooling will always yield molten $MnO–SiO_2$ products.

Aluminium

It is evident from Figs.80 and 197 that aluminium is a

strong deoxidizer, and it is commonly used for that purpose. The deoxidation 'constant' $K=[\%Al]^2\times[\%O]^3$ for aluminium and oxygen in solution in the steel phase has been determined by several authors with results varying from about 10^{-9} to about 10^{-14} at 1 600°C.[21,190] The discrepancy has been much discussed and recently Repetylo *et al.*[191] have found that the deoxidation of steel proceeds differently under purified argon from under an oxygen-containing gas. In the former case, kinetic experiments have shown that the initially high values of the deoxidation constant vary with time approaching the 'final' low value of $\sim 10^{-14}$ after 10 minutes. This behaviour is explained assuming that the alumina particles formed are self-eliminated (by upward movement, coalescence, and eventually physical reversion into the metal bath). It was shown that the coarser particles are eliminated in 7–10 minutes, but that a finer suspension (less than 70 μm) probably remains in the liquid steel even 20 minutes after the aluminium addition. Under an oxygen-containing atmosphere, however, an experimental value of about 10^{-7} was found for the deoxidation constant. The difference is certainly due to the alumina suspension continuously produced by reaction with absorbed oxygen in the latter atmosphere.

According to the phase diagram $FeO–Al_2O_3$ (Fig.201) three

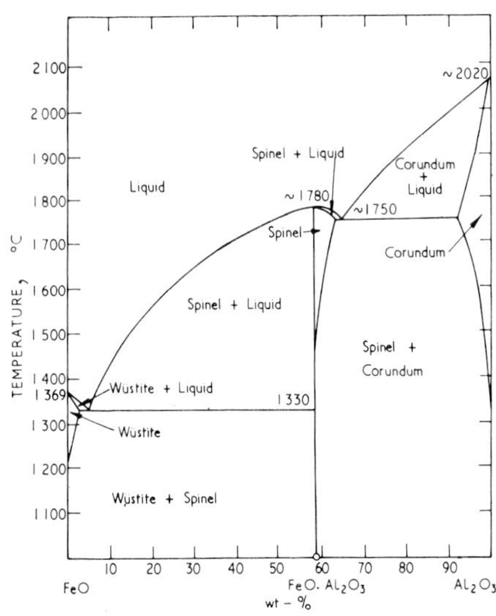

From ref.53

201 The equilibrium diagram for the pseudobinary system $FeO–Al_2O_3$

deoxidation products are possible as a result of the aluminium deoxidation of steel: liquid FeO–Al_2O_3 melts, solid hercynite (B 5, R 1) and solid alumina (A 2). In actual deoxidation practice, the formation of liquid FeO–Al_2O_3 melts is very unlikely because the oxygen content of the bath is never high enough. The two other compounds are, however, possible inclusion phases in steel originating from deoxidation. The most common of these two deoxidation products is Al_2O_3, usually in its α-modification, corundum (A 2). In Fig.202 the

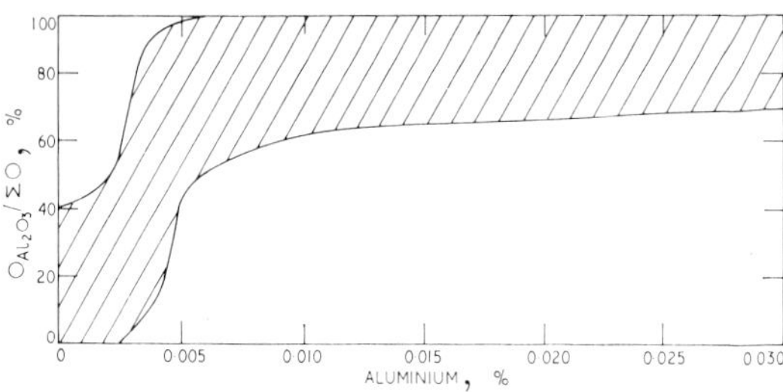

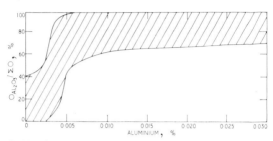

STEEL TYPE Killed steel for boilers
STEEL ANALYSIS Unknown
INCLUSIONS When the content of soluble Al in the steel increases above 0·002–0·004% the silicate inclusions gradually become richer in Al_2O_3, and at higher Al-contents α–Al_2O_3 inclusions appear. From ref.211

202 The relation between soluble Al in the steel and the appearance of Al_2O_3 in inclusions in killed steel

relation between soluble Al in the steel and the appearance of Al_2O_3 inclusions is shown for killed steels. Al_2O_3 inclusions in practice appear as clusters of Al_2O_3-particles, similar to those in Fig.6. They are rather small in size, mostly in the range of 1–5 μm, irregular in shape or partly molten, slightly greyish green in appearance, hard and sometimes glassy. They are usually trapped between the dendrites of the steel or extend from some part of the steel surface into the steel matrix. It has been shown by Torssell[192] that the particles in such clusters of Al_2O_3 are all in contact and form a three-dimensional unit. They are held together by the surface energy of the molten steel, an aggregate being a more stable unit than the cluster of separate particles due to the decrease in total surrounding steel surface. The Al_2O_3-particles therefore tend to stay together when they collide. Such larger units of Al_2O_3-particles are often attached to the steel surface in one corner of the aggregate due to the fact that as soon as one member of the aggregate reaches a steel surface, the surface energy decreases and the particle is kept there, anchoring the whole aggregate to that position. These three-dimensional units appear as clusters of Al_2O_3 in microsections of the steel.

The type of inclusions formed depends on the details of deoxidation practice. If aluminium is added on its own as the initial deoxidant, α–Al_2O_3 inclusions are formed; of these the majority are able to escape and therefore the result is a minimum quantity of primary α–Al_2O_3-clusters in the steel. Subsequent additions of less powerful deoxidant do not give rise to inclusions. If, as in conventional ladle deoxidation, the deoxidants are added more or less simultaneously, liquid manganese–aluminium–silicates are formed which are more stable in the steel and have a greater tendency to remain as inclusions. Such inclusions in cast steel are discussed in the example below and shown in Fig.203. In Fig.205a, the average change in Al_2O_3 content of the inclusions during the refining period is shown for different deoxidation procedures.

From current literature[179,193–195] it seems that the inclusions in the steel which originate from Al-deoxidation may be a result of three different deoxidation mechanisms (*a–c*), or from a reaction between high concentrations of Al-metal and oxygen (*d*). These all result in different possible inclusion types.

a. Deoxidation by homogenous nucleation: Very high supersaturations are necessary ($\sim 10^{14}$), nucleation is finished in less than 3 minutes and the size of the inclusions is determined by the supersaturation. The inclusions appear as clusters of finely dispersed α–Al_2O_3 particles, as in Fig.6, and movements in the liquid steel bath determine in which parts of the ingots these clusters will be situated. They rise in the steel bath and have a tendency to be anchored at a surface.

b. Deoxidation on small nuclei: Such nuclei are often present in the liquid steel bath, for example, from emulsified furnace slags (O 4), from reactions between the steel and the refractories (A15, O 4, O 7), or from surface oxidation of steel. Deoxidation can proceed on these nuclei without supersaturation. The resulting inclusions contain only negligible amounts of hercynite, but in addition to Al_2O_3 they may have variable amounts of other oxides from the nuclei, e.g. MnO, SiO_2, CaO. Inclusions which probably have this origin are shown in Figs.5, 136 and 137. However, part or all of the Al_2O_3 may also come from the chamotte refractory (A 15, O 4), especially for basic steel. To the author's knowledge there is no certain way to determine if the origin of this type is from deoxidation or refractory reactions, but small inclusions are more probably deoxidation products than larger ones. The Fe-content of the inclusions should be small if they are deoxidation products.

c. Deoxidation on large inclusions: Large inclusions can be formed by homogenous nucleation at the place of introduction of the aluminium into the melt. When these are carried into oxygen-rich, aluminium-poor parts of the melt, the resulting inclusions contain appreciable amounts of hercynite $FeO.Al_2O_3$ (B 5). This phase is not very common, and no good example was given in section B 5. In Fig.243, Section R of the present part, hercynite of this kind is shown, and examples are also to be found in the literature.[193,194]

d. Reaction between high concentrations of Al-metal and oxygen: Al_2O_3-inclusions of the type shown in Fig.6 are sometimes glassy and partly molten. They have therefore reached a temperature above the melting point of Al_2O_3 (2050°C). This is difficult to explain if they are primary deoxidation products, that is, reaction products between Al and O in the liquid steel. It has therefore been suggested[196] that they have been formed through a reaction between Al-metal at the temperature of the molten steel and oxygen at the steel surface. An alternative explanation would be that local overheating in the melt with high local temperature peaks results from the reaction between Al-metal and oxygen in the liquid state. Recent experimental evidence supports this explanation of the existence of partly molten Al_2O_3 inclusions.[278] The presence of such inclusions therefore indicates either an excess of Al-metal or an inefficient deoxidation technique.

Finally, some interesting observations by Adachi *et al.*[197] on the deoxidation mechanism of steel with metallic aluminium should be mentioned, results which are also supported by early observations of Sloman and Ewans.[198] Deoxidation products of laboratory iron melts with Al were studied after short times of deoxidation (seconds to minutes) and rapid solidification. In these small ingots, the presence of various low-temperature modifications of Al_2O_3, e.g. θ-, δ-, κ- and γ–Al_2O_3 were established. All these modifications are intermediate dehydration products between aluminium hydroxide and α–Al_2O_3, corundum.* To explain the occurrence of these phases it was therefore assumed that hydrogen, if present in the steel, may take a part in the deoxidation reaction and in such a case give aluminium hydroxide either as gibbsite (boehmite), or bayalite as a first deoxidation product. The hydroxide transforms via one or several of the possible transformation paths to α–Al_2O_3, corundum, as the final reaction product.

Generally speaking, α–Al_2O_3 clusters (Fig.6) and inclusions very rich in α–Al_2O_3 (Fig.5) are most probably of deoxidation origin. Inclusions with high amounts of Mn and Si, e.g. in Figs.3 and 4, are

---

* In section A 2 on corundum, these different modifications were not fully discussed. Therefore a supplement has been given in Section R (page III).

most probably a reaction product between steel and Al_2O_3-rich ceramics (O 4,7). The presence of the α–Al_2O_3 phase crystallizing in the inclusion matrix, is *not* proof of the Al_2O_3 phase being of deoxidation origin, whereas its crystallization directly in the steel matrix *is* an indication of deoxidation origin. It should be observed, however, that if the steel is plastically deformed, the Al_2O_3-phase in multiphase inclusions could be mechanically separated from its original inclusion material and appear isolated in the steel matrix as a result of working (Fig.5).

Example. The following study of inclusions in a cast steel heavily deoxidized with aluminium is an illustration of the complicated structure of Al_2O_3-rich macroinclusions, most probably resulting from deoxidation.

During machining of a cast 53 kg crankshaft, the carbide cutting tool broke down due to the presence of macroinclusions. Two types of these inclusions were observed. One type was rich in Al_2O_3 (Fig. 203) and will be discussed in the present section, whereas the other

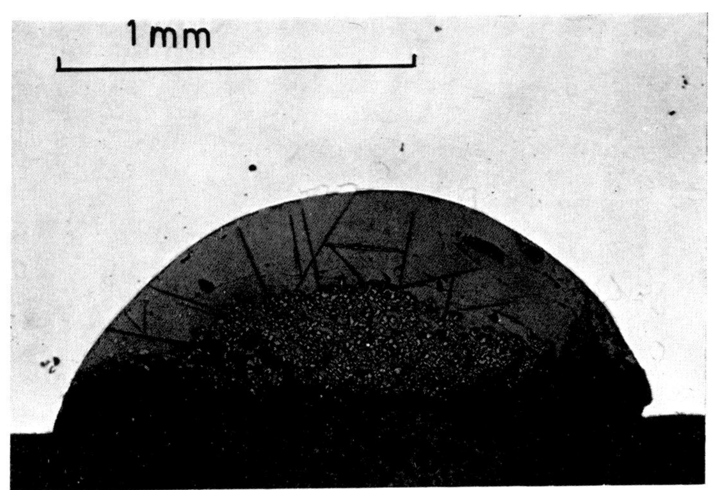

203a Large, spherical inclusion at a low magnification. The crankshaft was turned on a lathe, and the inclusion is situated in the worked steel surface. The inclusion has a central part with platelike corundum and an outer reaction zone with corundum laths

type (Fig.209) was rich in SiO_2 and is discussed in section O 7, example II. The cast steel had an analysis of (%) 0·13 C, 0·98 Si, 1·08 Mn, 0·04 P, 0·030 S, 0·11 Cr.

A microscopical investigation of the Al_2O_3-rich spherical inclusions shown in Fig.203a-d, combined with electron probe analysis of the phases and an X-ray diffraction study of extracted inclusions resulted in the following information.

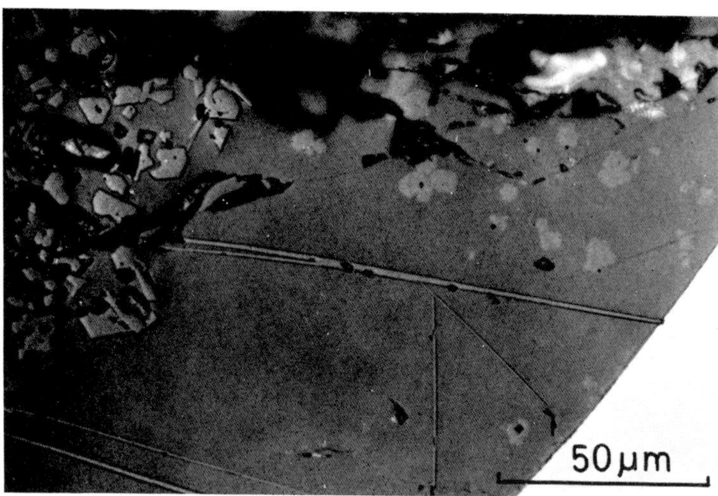

203b Part of the inclusion from Fig.203a at a higher magnification. Platelike corundum, precipitated from melt, together with laths of corundum and galaxite, precipitated from the solid matrix, are visible

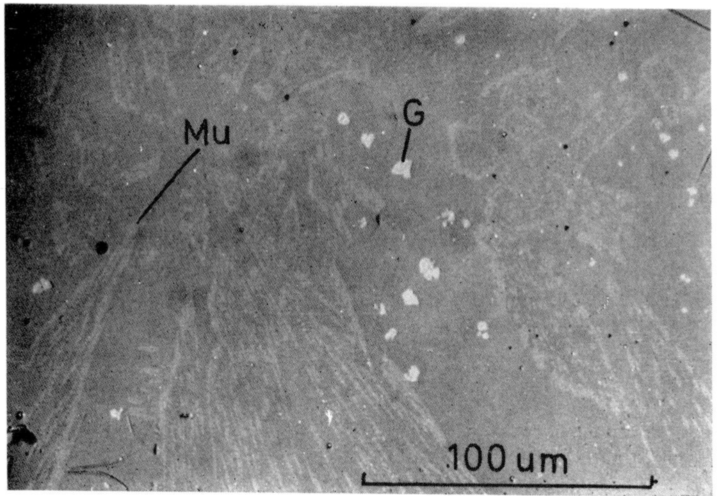

203c Part of the outer zone of the inclusion from Fig.203a. Mullite and galaxite have precipitated from the solid matrix

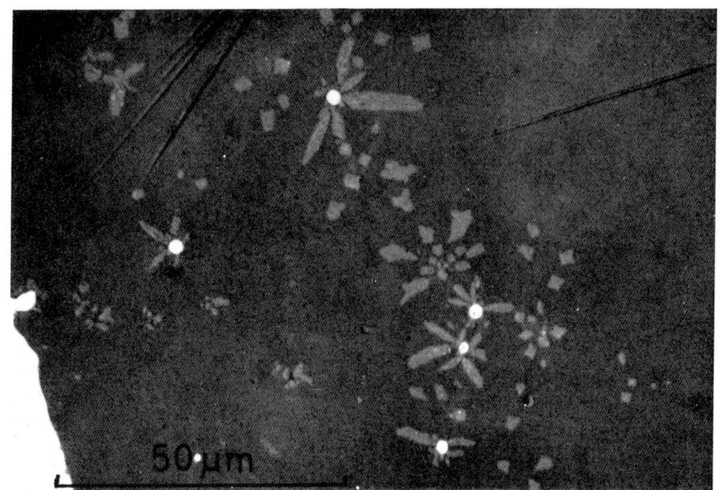

203d Part of the outer zone of the inclusion from Fig. 203a. Steel spheres have acted as nuclei for galaxite precipitation from the solid matrix in the outer reaction zone

STEEL TYPE Cast 53 kg crankshaft, heavily deoxidized with Al

STEEL ANALYSIS % 0·13C, 0·98Si, 1·08Mn, 0·040P, 0·030S, 0·11Cr

203 Inclusions in cast steel deoxidized with aluminium

The inclusions usually had a central part rich in corundum crystals with regular platelike sections in an Al_2O_3–SiO_2–MnO matrix. The central part was surrounded by a glassy matrix, which gradually changed composition becoming richer in MnO and SiO_2 when approaching the steel-inclusion contact surface. Corundum was also observed in this outer zone of the inclusions, but crystallized as long lathes. In addition, the phases mullite, galaxite and steel were observed in this outer matrix, and it was noted that steel spheres were often found in the centre of galaxite crystals.

This inclusion type probably originates from deoxidation. As a consequence of the aluminium addition, highly aluminous liquid silicates were formed when the steel was deoxidized. Small liquid steel droplets were dispersed in the liquid slag phase. The spherical slag droplets acted as large nuclei for MnO and SiO_2 precipitation during cooling of the steel. These precipitates increased in size with their outer part gradually becoming richer in MnO and SiO_2, but their central part remaining rich in Al_2O_3. The liquid droplets were included in the solid steel. When these inclusion droplets solidified, platelike Al_2O_3 was precipitated in the Al_2O_3-rich central part, which was surrounded by an outer glassy $MnO–SiO_2–Al_2O_3$ matrix. This matrix was, however, in a metastable condition and therefore further precipitation of corundum lathes, galaxite and mullite took place after solidification. This was often facilitated by the small solid steel spheres which acted as nuclei for the galaxite phase (203d). This type of precipitation is discussed in sections A 14 and P 2.

Chromium

This metal is a rather weak deoxidizer (Fig.80). Fundamental equilibrium constants have been given by Turkdogan[199] and studies on non-metallic inclusions from the Fe–Cr–O system in steels have been summarized by Adachi and Iwamoto.[200] Escolaite Cr_2O_3 (C 2), or chromite, $FeO.Cr_2O_3$ (C 4) are the possible deoxidation products, depending on chromium activity. Tetragonally deformed chromites have been observed (C 4, A 7) mainly as intermediate phases during annealing, the lattice deformation depending on both oxygen and chromium activity. The final product seems to be the cubic chromites. The existence of an inclusion phase Cr_3O_4, reported by Hilty *et al.*[190] has not been verified by later investigations, and is still debated. The inclusion type is dependent on the chromium content of the steel[200] in a manner shown in Table XL.

TABLE XL Phase analyses of isolated oxide inclusions from iron–chromium laboratory ingots

Cr (%)	Phases	Tetragonal def. of $FeO.Cr_2O_3$-phase (c/a)
3	$FeO.Cr_2O_3$	0·97
5	$FeO.Cr_2O_3$	0·95
8	$FeO.Cr_2O_3+Cr_2O_3$	0·95
13	$FeO.Cr_2O_3+Cr_2O_3$	0·90
18	$FeO.Cr_2O_3+Cr_2O_3$	0·90

The Cr_2O_3 inclusions for high Cr-contents are probably transformation products from chromite. The formation and dissolution of Cr-oxide inclusions in chromium has also been studied by electron microscopy.[201–202]

Titanium, zirconium

These metals are strong deoxidizers (Fig.80) and titanium is often present in calcium deoxidation alloys (Table XIII). The possible deoxidation products have been discussed in section H 2, with reference to examples in literature.

Calcium

The free energy of formation of CaO is the lowest among the common oxides (Fig.80). Calcium has therefore a high oxygen affinity and should be an effective deoxidant. On the other hand its solubility in steel is very low and, as discussed in Section G 1, the deoxidation process is complicated and not fully understood. Because of its low melting point Ca is usually added as different Ca–Si alloys, examples of which were given in Table XIII. The primary deoxidation inclusions are therefore usually different calcium silicates (G 4) but often also contain varying amounts of Al_2O_3, FeO and MnO in solid solution, depending on the composition of the deoxidation alloy and deoxidation practice. They are usually present in the ingots as small, glassy globules, which deform plastically if the steel is deformed. Typical examples of these inclusions with different compositions taken from ingots and plates are shown in Figs.125, 134, 135 and 142a.

Sometimes different combinations of Ca- and Al-deoxidation are used and this may result in spherical calcium aluminate inclusions (G 3). Such inclusions were shown in Figs.111 and 112. They do not deform readily during steel deformation.

It is not possible for Ca-inclusions which contain MgO to be primary deoxidation products even if they have a globular shape (e.g. Figs.110, 113, 117). They probably originate from a reaction of the calcium- and magnesium silicates in the ladle slag during Al deoxidation in the ladle, as discussed in Section G 6, point 4b. The presence of MgO may be easily established in polished sections through the characteristics of the spinel phase (F 3).

In Section O 4 various sources of the CaO-component in the inclusions are discussed, and in Fig.205b the average change in CaO-content of the inclusions during the refining period is shown for different deoxidation procedures.

Vacuum (carbon) deoxidation

Vacuum should also be included among the deoxidation procedures, as it has an influence on the oxygen level and therefore on the inclusions in the steel. Deoxidation is a result of the reaction of carbon with oxygen to form gaseous carbon monoxide, and the deoxidation effect of carbon in the steel is increased due to the decrease in carbon monoxide pressure resulting from vacuum treatment. Its effect is therefore more pronounced on steels with a high carbon and oxygen activity, that is on high-carbon as compared to low-carbon steels, and on silicon-killed as compared to aluminium or calcium-killed steels. The effect is, however, much dependent on slag and steel practice, methods of deoxidation and alloying; one therefore should be careful with general conclusions. The effect of vacuum casting on inclusion formation is, for instance, different from that of vacuum degassing in the ladle, where the molten steel is in contact with refractories during the vacuum treatment. Comprehensive summaries are available[203–205] and it is beyond the scope of this review to discuss all the aspects of vacuum treatment on steel properties. Expressed in general terms, the effects on non-metallic inclusions are:

1) The amount and size of indigenous oxide inclusions decreases due to the decreased oxygen level of the steel
2) The precipitation pattern of sulphide inclusions changes
3) There is an increased danger of large exogenous inclusions due to steel–refractory reactions or emulsified slags, since temperature and holding times are often increased as compared with conventional steelmaking, but in general an improvement is observed.
4) The influence on the inclusions is greater on high-carbon, low-alloy steel than on low-carbon high-alloy steel and on basic than on acid steel.

These schematic indications of the effect of vacuum treatment on non-metallic inclusions are further illustrated by the study of Church *et al.*[124] on the effect of ladle vacuum degassing a 0·5% carbon–chromium basic through-hardening bearing steel. They compared the same steel as air melted and CaSi deoxidized in the ladle (AM) and carbon-deoxidized in vacuum (VP). In both steels, three types of inclusions were found, namely globular oxide inclusions, sulphide inclusions and stringer type oxide inclusions. The first type was of multiphase character in the AM steel usually surrounded by a scale of (Mn,Ca) (SK 4) and consisting of galaxite (A 7), Ca-aluminates (G 3) and a Ca–Al–Si-oxide matrix (G 5). They were similar to the inclusions shown in Figs.87, 88, 110, 112 and 117. In the VP steel, these inclusions were smaller and fewer and they were less complex than in the AM-steel (Fig.190). They had no sulphide scale and they often only consisted of a nucleus of galaxite surrounded by a Ca–Al–Si oxide matrix. Their appearance was similar to the inclusion shown in Fig.113. This change in size and appearance for the globular oxide inclusions is a consequence of the lower amount of oxygen available for oxide formation after carbon deoxidation in vacuum as compared with CaSi deoxidation in the ladle. The galaxite acts as an exogenous nucleus for the indigenous Ca–Al–Si-oxide precipitation.

The sulphide inclusions in the AM steel were of two types, namely (Mn,Ca)Si as a rim around globular oxide inclusions and MnS of type I (K 3) with small amounts of Ca, Cr and Fe in solid solution. In the VP steel, only the later types was observed and in slightly greater amounts than in the AM steel. Church *et al.* relate the difference to deoxidation practice. In the AM steel part of the sulphur in the steel reacts with calcium on the surface of the globular inclusions, causing sulphides to nucleate there, whereas the sulphides nucleate freely in the VP steel. As sulphide scales on oxide inclusions are not recorded as sulphides by conventional inclusion counting methods, this is also an explanation of the increase in sulphide inclusions for the VP steel in spite of its decreased sulphur content.

The stringer type of inclusion was not influenced by vacuum degassing. They were multiphase inclusions with 70–90 % Al_2O_3, the balance being CaO, MgO and SiO_2. Often also sulphides were found randomly along the stringers. This type of inclusion is shown in Figs.90, 136 and 137. Similar inclusions have also been studied by Ockenhouse and Werner[176] in different steel types; they were found to be concentrated in the axial portion of the lower part of the ingots. This inclusion type has evidently an exogenous nucleus and is not primarily connected with vacuum treatment; the Al_2O_3 component results from the high-alumina refractories (O 4,7) and the calcium component from entrapped furnace slag. Modification in melting practice and refractories influence the Al_2O_3 component of the stringer-type inclusions while slag and deoxidation practice influence the CaO component. Therefore the inclusion type may be altered in appearance and morphology, as indicated in the two references given.

The effect of vacuum degassing on the inclusions in bearing steel has also been discussed by Uhrus[75] and Murray *et al.*[206]

O 4 The formation of non-metallic inclusions during the steelmaking process

The type, size and composition of the non-metallic inclusions in the *molten* steel changes at various stages during the steelmaking process. Small changes in several different operations during steelmaking may have a great effect on the inclusions resulting in the steel. Typical of the steelmaking parameters of importance are: the boiling time, refractory composition, the deoxidation practice, the composition of the deoxidizing alloys and details of the tapping and teeming operations. Unfortunately there are only a few investigations available dealing with these important factors which govern the occurrence of non-metallic inclusions in the solid steel.

Pickering[207] has made available to the author a detailed study on a high-carbon steel containing 0·65–0·75%C and 0·6–0·7%Mn. Because this work gives important information on the relation between different inclusion types and various stages of the steelmaking process, an extensive summary will be given in this section. The steel was made by a double slag basic electric arc process, using a 25 ton furnace. Different trials were made with the intention to determine at which stage of the steelmaking, tapping and teeming processes pick-up of the inclusions occurred. A particular aim was to find the origin of the calcium and aluminium in the silica inclusions. The results give an interesting summary of the changes in the inclusion composition during steelmaking.

The following effects were established:

a. At *melt out* the oxide inclusions were generally glassy manganese aluminium silicates resulting from the oxidation of the charge. They were of the type shown in Figs.40, 125 and 134, but at this early stage they were low in CaO and with about equal amounts of MnO, SiO_2 and Al_2O_3.

b. During *the boil*, calcium silicates were picked up from the slag due to the turbulence. The inclusions were still of the glassy silicate types shown in Figs.40, 125 and 134, but their CaO-content increased rapidly during the boil, as shown in Fig.204d. Their SiO_2-content showed little change throughout the boil (Fig.204b). The MnO-content of the inclusions remained reasonably high over the first part of the boil but then decreased to less than 10% at the end (Fig.204a). Their Al_2O_3-content decreased rapidly in the early stages of the boil and for most trials the final Al_2O_3-content of these glassy silicate inclusions at the end of the boil was less than 4% (Fig.204d).

c. During the *refining period* when deoxidation of the melt was carried out, the inclusion composition quickly reflected the ferroalloy additions made. In Fig.205a the average change in Al_2O_3-content of the inclusions during the refining period is shown for different deoxidation procedures. At the end of the boil, the inclusions are low in Al_2O_3. They then picked up Al_2O_3 during refining, but this depended on the deoxidation practice. It is evident that the Al-content of the ferroalloys (0·7–0·9% in FeSi and 1·2–1·7% in CaSi) is of great importance for the inclusion content of Al_2O_3. This is due to the very low value of the 'constant' $K=[\%Al]^2[\%O]^3$ for aluminium and oxygen in solution in the steel phase, according to which the alumina content of the inclusions may have a high value even for very low aluminium concentration in the steel phase. (*See* sections A 15, point 4 and 0 3, aluminium.) In most trials, the inclusions during the refining period were also of the glassy, globular types shown in Figs.40, 125 and 134, but with changes in composition

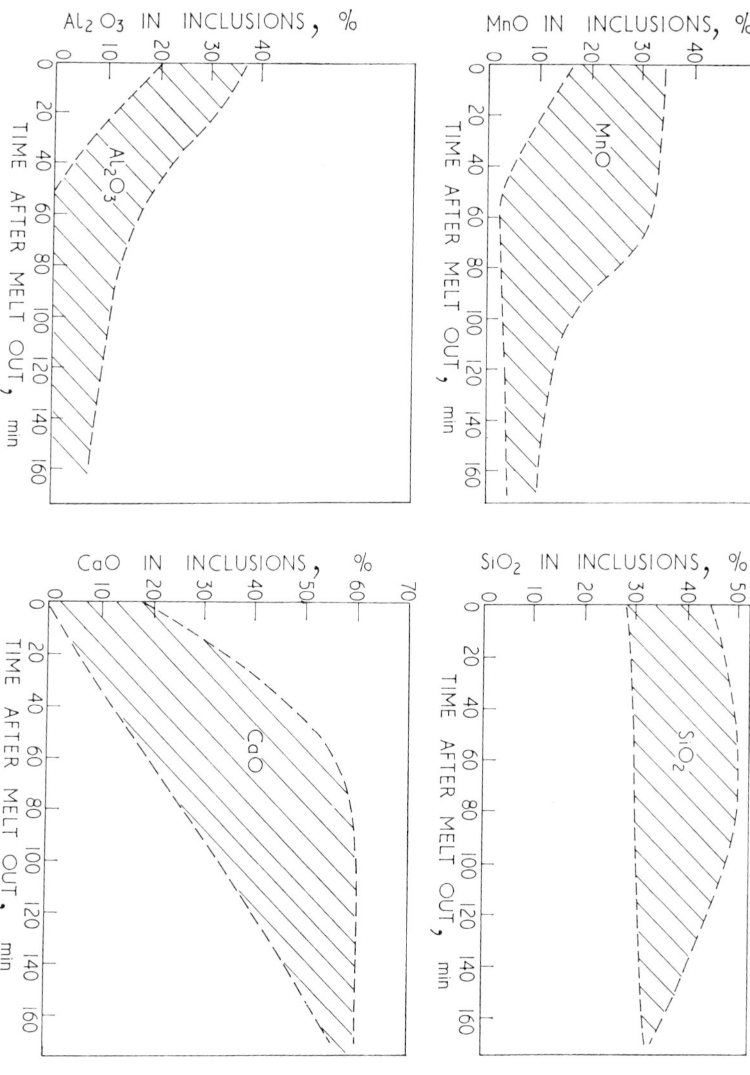

The detailed steel procedure was varied and the areas summarize results from 6 different trials. The variations during the refining period are given in Fig.205. From ref.207

204 Variation in the average content of different inclusion components during boil. Basic electric steel, double slag, steel analysis 0·65–0·70%C, 0·5–0·6%Mn, 25 ton furnace

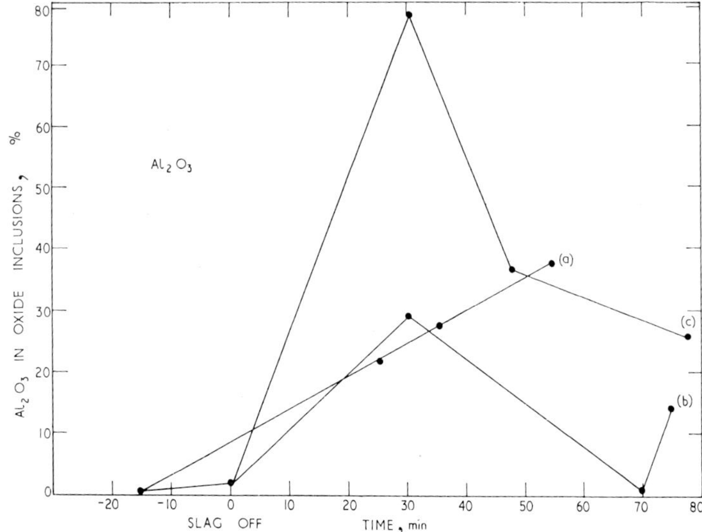

205a Variation in Al_2O_3 content

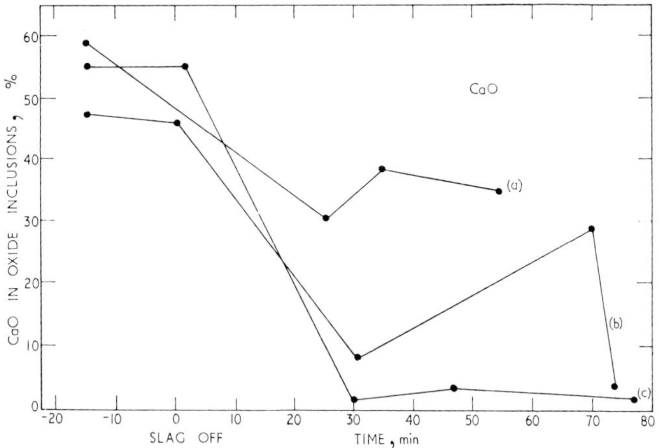

205b Variation in CaO content

Trial a: FeSi and CaSi added at slag off, CaSi also during the whole refining period

Trial b: FeSi added at slag off and 7 minutes after slag off

Trial c: FeSi and Al added at slag off

205 Variation in the average content of Al_2O_3 and CaO inclusion components during the refining period at different trials. The variation during the boil for the same steel is given in Fig.204. From ref 207

corresponding to the ferroalloy additions (Figs.205a-b). An exception was the trial where aluminium was added to the bath at the start of the refining. The average Al_2O_3-content of the inclusion in this trial increased to about 80% and the inclusions formed at this stage were either corundum, corundum and galaxite or highly aluminous silicates, similar to those shown in Figs.5 and 6.

In Fig.205b the average change in CaO-content of the inclusions during the refining period is shown for different deoxidation trials. At the end of the boil, the inclusions were high in CaO, mainly due to slag entrapment. During the refining period, the average CaO-content decreased by dilution effects, but if CaSi was used as a deoxidizer, the CaO-content of the inclusions was maintained at a high value throughout the refining. Secondary effects, mainly flotation of different inclusion types during the refining period as well as a pick-up of other slag inclusions in the steel also had an influence on the average inclusion composition. For instance, these were responsible for the second peak value after 65 minutes in one of the trials of Fig.205b.

d. During *tapping* two main phenomena influenced the inclusions; namely an increase in CaO-content of the glassy silicates due to CaO pick-up from the slag (slag and metal were poured together) and the appearance of silica rich $MnO–SiO_2–Al_2O_3$ inclusions, often containing cristobalite precipitation. Such inclusions are shown in Figs.11, 13 and 41a. These inclusions were found to be reaction products between the liquid steel and the launder refractories (the launder was rammed 'compo' with about 32%Al_2O_3, 63%SiO_2 and small amounts of CaO, MgO, TiO_2 and Fe_2O_3). (The reaction between metal and refractory was discussed in A 15, point 2 and further examples are given in Section O 5–8). The presence of TiO_2 in these inclusions was a valuable indication of their origin.

As reported by Ericsson,[208] considerable amounts of oxygen are also introduced into the steel during tapping. This oxygen does not influence the total amount of inclusions, however, as it rapidly disappears from the steel in the ladle.

e. In the *ladle* large calcareous silicates were entrapped during tapping, but a great part did float out. Therefore the average CaO-content of the inclusions again decreased during the teeming process, whereas the Al_2O_3-content of these inclusions increased if aluminium-bearing ferroalloys were used in the ladle. A further pick-up of Al_2O_3 by the silicates occurred in the ladle due to ladle refractory erosion, similar to the launder refractory erosion described above (*d*). In general, the inclusions in the ladle were much larger than those present at the end of the refining period.

f. In the *ingot* some flotation also can occur. Due to coalescence and growth by the further precipitation of deoxidation products, the ingot inclusions were always larger than those in the ladle stream. *g*. The *ingot scum* was not discussed in the investigation reported. It should not be forgotten, however, in a general metallographic discussion of inclusions like the present. It is a sink as well as a source for inclusions. In bottom casting, there is a close correspondence between the ingot inclusions and the ingot scum (see O 7, example *I*). In top casting, FeO-rich inclusions in the steel may originate from the oxide layer formed around the casting stream.

The investigation by Pickering *et al.* is of great value not only for the determination of inclusion origin but also for the overall influence of the inclusions on the steel properties. The physical properties of the silicate inclusions are dependent on their composition, and therefore their behaviour during steel deformation (Section P) and their influence on steel properties (Section Q) should be traced back to the steelmaking process.

O 5 Inclusions from ferroalloys

General. In principle there exists the possibility that inclusions in the ferroalloys are inherited as inclusions in the steel. The common ferroalloys are comparatively rich in non-metallic inclusions and several examples have been given in this monograph (Figs.8, 9, 12, 14, 31, 69, 70 and 138). Due to the high concentration of the metal alloyed with iron in these alloys, the inclusion phases are in general rich in the alloying element, for instance Cr in ferrochromium. The inclusions in ferroalloys have a characteristic and often complex structure (e.g. Figs.70 and 138). They therefore usually differ in structure and composition from those in the steel. In the author's experience inclusions in the solidified steel can rarely be directly related to inclusions in ferroalloys under normal steelmaking conditions. Thus it is unusual for inclusions to escape from ferroalloys, remain unchanged during the life of the melt and then appear in the solidified ingot. This is also supported by results published in the literature. Thus Schoberl and Straube,[209] from an extensive study of inclusions in ferrochromium and corresponding steels with different chromium contents, have concluded that oxide inclusions in the steel normally are not directly inherited from the ferrochromium. Therefore the use of ferrochromium alloys very low in inclusions is usually unnecessary. Only if the ferroalloys are added late in the steel process after deoxidation does there seem to be a possibility of direct inheritance, especially for ferro-vanadium high in oxygen.[210]

From the standpoint of inclusion formation, it is the aluminium and silicon contents of the ferroalloys which are more important than their inclusion content. Thus a silicon content higher than about 1 % in the ferrochromium decreases the total amount of oxide inclusions in the final steel. The resulting steel inclusions become larger and rounder (Fig.70) than if the ferrochromium is low in silicon, when they are sharp and irregular (Fig.31). Schoberl and Straube have also studied the composition and nature of the steel inclusions during different stages of the ferrochromium dissolution. If the ferrochromium is low in silicon, the first inclusion type in the steel during dissolution will be FeO (Fig.196). When the dissolved content of chromium in the steel melt is around 9·6 %, inclusions of FeO. Cr_2O_3 (C 4) appear in the steel, these increase in number with increasing chromium content of the steel. When the chromium content of the steel exceeds about 10 %, a third type of small inclusions appears which could not be identified. When MnSi is then added, the silicon content of the steel increases, resulting in a disappearance of the FeO and FeO.Cr_2O_3 inclusions. Another type of inclusion is now formed, belonging to the Cr_2O_3–SiO_2 system of the type shown in Fig.70. These are rounded and usually rather small. If the silicon content of the ferrochromium from the beginning was higher than about 1 %, the rounded Cr_2O_3–SiO_2-type inclusions appear much earlier in the history of the molten steel. They are the dominating inclusion type at silicon contents in the steel from 0·06–0·08 % and upwards.

Aluminium in the ferroalloys also has an influence on the steel inclusions, this was discussed in Section O 4c. As was mentioned, many ferroalloys have aluminium contents as high as 1–2 %. Therefore the resulting steel may have inclusions either of Al_2O_3 or belonging to the MnO–SiO_2–Al_2O_3-system, even if no aluminium has been added as deoxidant or entered from refractories.

Most evidence indicates that the inclusions in the ferroalloys usually float up to the molten steel surface and join the slag or else are dissolved. A direct inheritance of these inclusions seems to occur only accidentally; for example, if the ferroalloys have not been completely melted and dissolved, or if an unusually low temperature was used during melting.

Example. In a rolled sheet from an electric arc-melted high chromium steel, a large platelike inclusion was observed, elongated in the rolling direction. In Fig.206, a microsection parallel to the rolling direction and perpendicular to the surface of the steel is shown in different magnifications. At high magnification this section of the inclusion looked like a 'string of pearls', the 'pearls' being larger

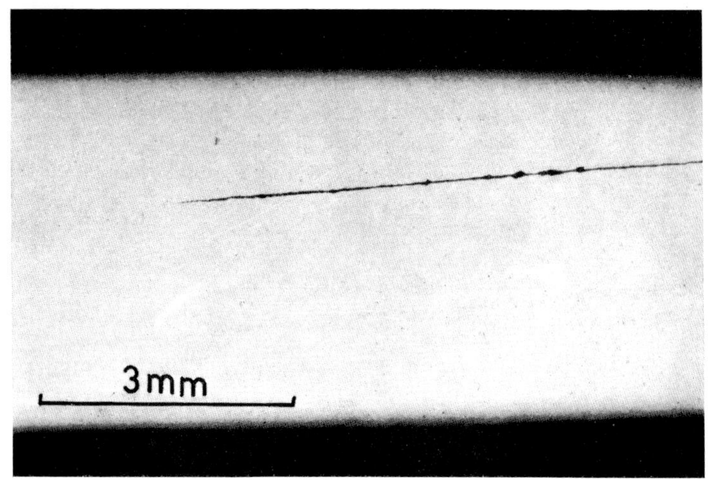

206a The cleavage in the steel at a low magnification

STEEL TYPE Basic arc-melted Cr-steel. From a broken roller
STEEL ANALYSIS % 1·96C, 0·31Si, 0·73Mn, 13·4Cr, 1·28W, 0·011P, 0·007S

INCLUSIONS The inclusions were aligned as a string of 'pearls' in this section of the cleavage in the steel roller

206 Ferrochromium inclusions, directly inherited by the steel as inclusions (*See* next page for Figs. 206b and 206c)

fragments joined along the string by a more deformable non-metallic phase. Due to the presence of the inclusion, a cleavage crack had been formed in the steel parallel to the surface of the sheet. The structure of the 'pearls' was very similar to the structure of spherical inclusions observed in the ferrochromium used for introducing the Cr into the steel (Fig.206b–c). They consisted of a nearly eutectic two-phase structure with the phases cristobalite (A 3) and Cr-galaxite (C 3). The composition as determined by electron probe analysis was very similar for both the steel inclusion and the ferrochromium inclusion. Thus the Cr-galaxite phase in both the steel and the ferrochromium had a small amount of TiO_2 in solid solution. The 'pearls' were joined by a deformable, glassy phase of Mn-silicate with a composition close to rhodonite (A 9).

It is the author's opinion that the probable history of the inclusion was as follows. A larger inclusion of the eutectic type, as shown in Fig.70a, was present in the ferrochromium. It did not dissolve or join the slag in the steel when the ferrochromium was dissolved, but acted as a nucleus for indigenous precipitation of MnO and SiO_2. In the steel ingot the eutectic ferrochromium inclusion was therefore surrounded by an outer zone, mainly consisting of Mn-silicate. When

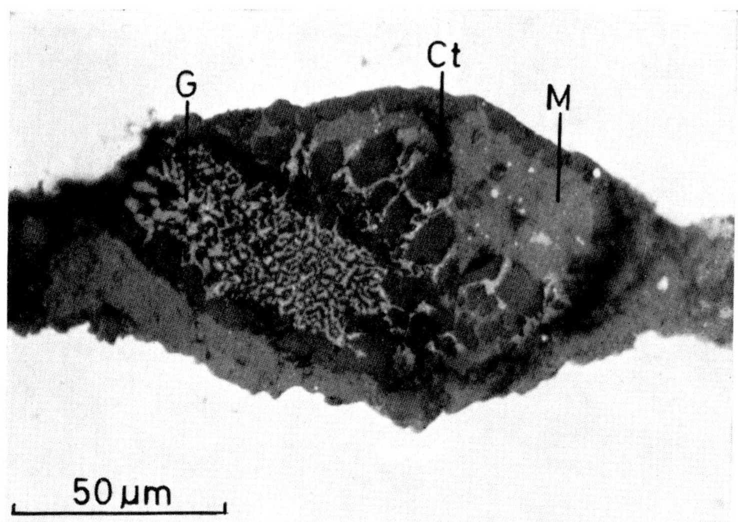

(Ct) Cristobalite with 100 % SiO_2 (A 3)
(G) Cr-galaxite (C 3) with 23MnO, $5Al_2O_3$, $45Cr_2O_3$, $3TiO_2$
(M) Matrix with 50MnO, $43SiO_2$, $4Al_2O_3$, $2Cr_2O_3$, acting as a 'binder' between the pearls; compare with rhodonite (A 9)

206b Inclusion 'pearl' from the string (*See* also main caption on previous page)

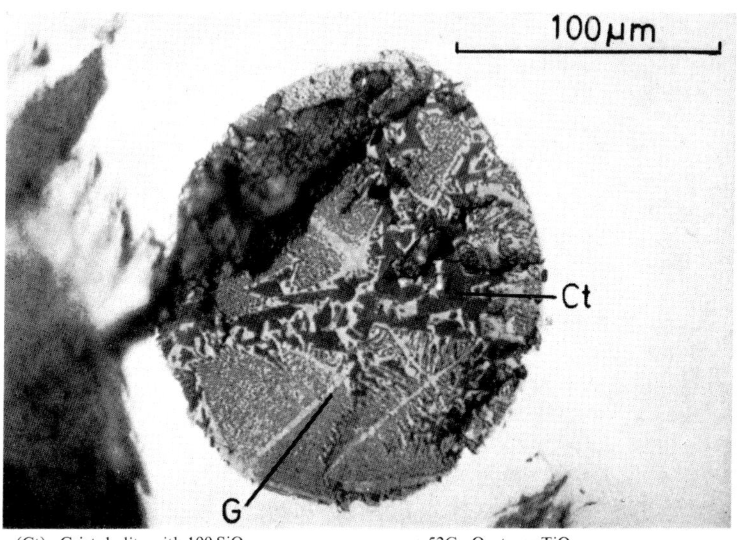

(Ct) Cristobalite with 100 SiO_2
(G) Cr-galaxite with 32MnO, trace Al_2O_3, $52Cr_2O_3$, trace TiO_2

206c Inclusion in ferrochromium suraffiné, used for alloying in the steel (*See* also main caption on previous page)

the steel ingot was rolled the brittle eutectic nucleus was crushed, whereas the spherical inclusion rim was deformed to a plate with fragments of the eutectic Cr-galaxite–cristobalite type, held together by a matrix of Mn-silicate.

O 6 Inclusions from furnace and ladle slag

As was discussed in Section O 4, furnace slag can be carried over into both the ladle and the ingot and thus be the origin of non-metallic inclusions. Also the ladle slag itself is an important exogenous source for ingot inclusions. It may be admixed with the molten steel and thus be carried over into the ingot, or it may adhere to the refractory lining of the ladle wall during teeming ('ladle glaze') and contaminate the following steel charge. Example of analyses for different slag types used in steelmaking are given in Table XIV.

In general, the furnace slag will tend to be calcium-rich, to contain only a little alumina and very small amounts of minor refractory additions such as TiO_2 and K_2O. The ladle slag, because it can be contaminated with ladle erosion products, tends to be richer in alumina and to contain more of TiO_2 and K_2O. This has an effect on the composition of the resulting ingot inclusions. Inclusions arising from the slags have compositions which are usually changed as compared with the slags. This is a result of indigenous precipitation, mainly of MnO and SiO_2, on the slag nuclei.

Example. Ultrasonic inspection of railway rails had indicated a defect zone in the rib of some rails, parallel to the rolling direction of the rail. A metallographic inspection of a section perpendicular to the rolling direction (Fig.207a) showed the existence of a zone with non-metallic inclusions in the central part of the rail. Further studies showed that the inclusions were glassy and deformed to plates parallel to the rolling direction (Fig.207b). They were often associated with small cracks and altogether constituted a discontinuity in the steel phase, extending along the central part of the rib and sometimes covering about $\frac{1}{3}$ of the longitudinal section.

In order to trace the origin of these inclusions, their composition was determined by electron probe analysis and compared with analyses of possible sources for non-metallic inclusions, considering also the steelmaking practice used for the rail steel. This steel was silicon killed, vacuum degassed in the ladle and then rolled to rails. The steel analysis was 0·62 %C, 0·47 %Si, 1·78 %Mn, 0·034 %P, 0·01 %S, 0·007 %N.

An electron probe analysis of the inclusions gave as a result that they consisted mainly of a glassy matrix (A). A small number of

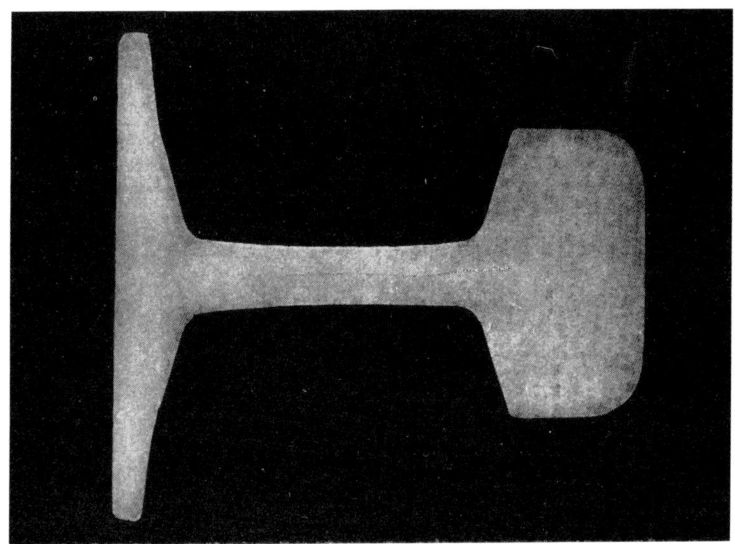

207a Transverse section through rail, photomacrograph with defective zone visible in the central part of the rib

STEEL TYPE Silicon-killed basic electric, vacuum degassed in the ladle and rolled to rails

STEEL ANALYSIS % 0·62C, 0·47Si, 1·78Mn, 0·034P, 0·018S, 0·007N

207 Inclusions in railway rail, originating from ladle slag (ladle glaze) (*See* next page for Figs. 207b and 207c)

regular, grey-white crystals (B) were also observed precipitated in the matrix. The composition was the following:

Matrix (A)	13 % MnO, 23 % CaO, 41 % SiO_2, 20 % Al_2O_3, 7 % MgO
Precipitate (B)	12 % MnO, trace of CaO, 62 % Al_2O_3, 25 % MgO
Mean analysis	13 % MnO, 23 % CaO, 41 % SiO_2, 20 % Al_2O_3, 7 % MgO.

Phase B is thus a double oxide of spinel type (A 7, F 3), with the typical appearance of these phases (*see*, for example Figs. 25–28 and 86). It has no influence on the mean composition of the glassy inclusions and merely indicates that they have started to crystallize due to the energy introduced during deformation of the inclusions in the steel.

The presence of MgO and CaO in the inclusions indicates an exogenous source, and the high content of these oxides is a further indication that the exogenous nuclei have a considerable influence on the composition of the inclusions; that is, they are introduced rather late in the steelmaking process. The most likely sources are slags or

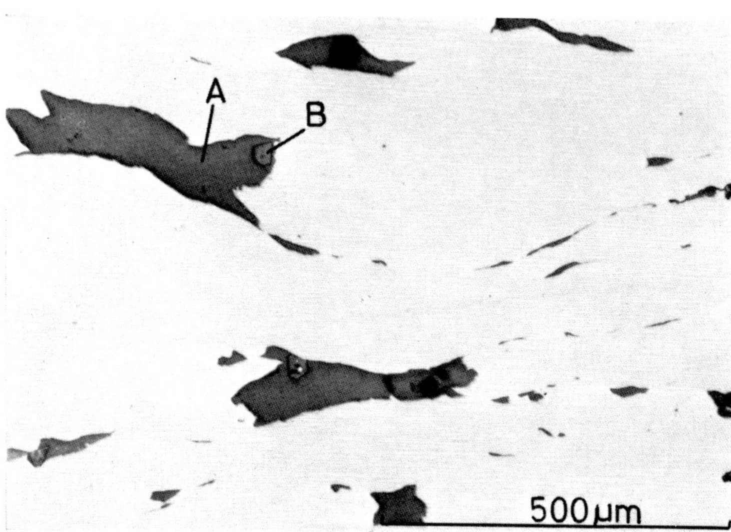

INCLUSIONS (A) Matrix with 13MnO, 23CaO, 7MgO, $41SiO_2$, $20Al_2O_3$
(B) Galaxite with 12MnO, 25MgO, $62Al_2O_3$

Mean analysis (calc) 13MnO, 23CaO 7MgO, $41SiO_2$, $20Al_2O_3$

207b Inclusions in the defective zone (Inclusion I, Table XLI)

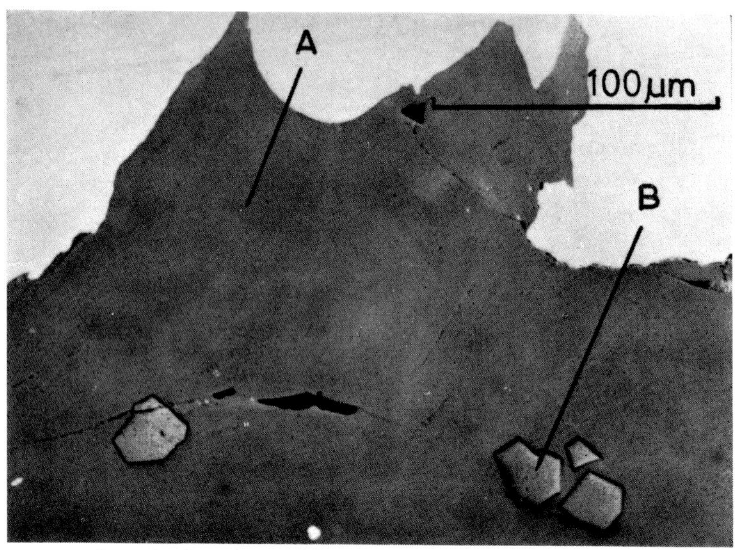

COMMENT Comparing the analyses of different possible exogenous inclusions sources, Table XLI, with the analysis of the inclusions (I) it was concluded that the inclusions originated from ladle slag

207c Same inclusion type of higher magnification

refractories (Table XXXVII), as no deoxidation alloys containing Ca were used.

Spectrochemical analyses of the different possible exogenous sources are summarized in Table XLI. The mean analyses of the inclusion

TABLE XLI Spectrochemical analyses of the composition of different possible exogenous inclusion sources as compared with inclusions *in situ*

Exogenous source		MnO	FeO	CaO	MgO	Al_2O_3	SiO_2	TiO_2	Fig.
Furnace slag		10	7	45	10	13·5	15	—	
Ladle slag:									
Before degassing		11	2·5	30	8	9·5	34	—	
After degassing		10	+	27	6	14·5	37	—	
Ladle refractories:									
Top lining		—	+	+	+	75	22	+	
Main lining		—	2	+	+	33	62	+	
Ingot mould refractories		—	2	—	20	38	40	—	
Ingot scum		22	6	+	12·5	12·5	40	—	
Inclusion I	Mean analysis	13	—	23	7	20	41	—	207
Inclusion II	Mean analysis	8	+	+	34	10	46	+	208a
Inclusion III	Mean analysis	31	—	4	14	8	43	+	208b

discussed above and two other types to be described in Section O 7 are also given.

There is a close correspondence between the composition of the ladle slag after degassing and the inclusion composition. The higher Al_2O_3 and SiO_2 contents of the inclusion also indicate a contribution from the top lining of the ladle, but as no FeO was observed, this part is comparatively small. The relative increase in Al_2O_3 is much higher than in SiO_2, therefore the top lining seemed to have been more attacked than the general ladle refractories.

It was concluded that the inclusions originated from the ladle slag. When the ladle was poured during teeming, slag was left and adhered to the ladle wall ('ladle glaze'), especially around the upper steel surface. This slag layer was not removed before the next pour, and therefore the succeeding ladle charge was contaminated by ladle slag from the earlier tapping.

O 7 Inclusions from refractories

> 'During production, molten steel and slag must come in contact with a large number of refractory materials. – As a result of their erosion and corrosion by the molten slag and metal, these refractories offer a very real and ever present source of non-metallic inclusions.'
>
> *Rait and Pinder 1948*

Refractories are an important exogenous source of origin for different non-metallic inclusions. During the whole steelmaking process molten metal and refractories are in contact and surface reactions between the two phases, as well as erosion effects, are possible. It is important to realize that inclusions originating from refractories are not merely pieces of refractory, which have been accidentally eroded from the refractory bricks by the molten steel and then appear embedded in the steel matrix. Erosion products act as nuclei for indigenous precipitation from the moment they are included in the molten metal and therefore continuously change their composition as compared with the original refractory. Inclusions formed by surface reactions have a structure which often has no direct relation to the refractory structure and their composition may also be quite different. Components of refractories such as MgO, TiO_2 and K_2O serve as valuable inactive tracers, however, in tracing inclusion origin.

Erosion products

Refractories and their relation to non-metallic inclusions have been discussed in many sections of the present review. In Table IX typical analyses for different refractories used in steelmaking were given. Microsections of burnt magnesite are shown in Figs.83, 84 and 128, of dolomite and burnt dolomite in Figs.85, 103, 104, and 127, and of olivine bricks in Figs.97 and 98. Mullite–sillimanite inclusions from the bottom smearing of the ladle are shown in Fig.38 and inclusions with quartz and other phases from refractories on the discharge side are illustrated in Figs.18 and 185. General references to refractories have also been given. [31, 54, 55, 57] The microsections discussed in all these figures show the structure of refractories, or of inclusions which are erosion products and have a structure and composition closely related to the refractories. The determination of the origin of these inclusions is usually not difficult for the metallographer, provided that the steelmaking process is known in detail. Difficulties arise if the erosion products have been included in the molten metal in such a finely dispersed state, or so early in the steelmaking process, that their composition has changed considerably due to indigenous precipitation.

Reaction products from refractories

Most of the inclusions of refractory origin are not of this simple type but are products of surface reactions between refractory materials and the molten metal or molten slags. In Table X a summary has been given of the different phases which have been identified in refractories after service for some time in steelmaking furnaces. Most of these phases were not present in the original refractories; they were formed as reaction products during steelmaking. A microsection of the reaction zone, steel–slag–burnt dolomite, is shown in Fig.141. In Figs.143 and 146 the reaction zones between ladle slags and ladle refractories and of alumina and burnt dolomite have been illustrated. It is evident that such reactions all result in oxide reaction products, which sometimes appear as non-metallic inclusions in the steel and are closely related to other non-metallic inclusions. In section O 4, points *d* and *e*, some comments on Al_2O_3 pick-up in the inclusions from launder and ladle refractories are given. The Al_2O_3 pick-up from chamotte (Table IX) is probably the most important among the metal–refractory reactions in common steelmaking practice, and it is discussed in Section A 15, point 2. Of special importance are the nozzle and stopper refractories. It has been reported [208] that as much as 30% of the macroinclusions originate from these two details if they are made of chamotte. The resulting inclusions are manganese oxide–silica–alumina particles, for instance of the types shown in Figs.5 and 13. The determination of their origin is a difficult problem for the metallographer as the same oxides also are usually the main constituents of indigenous inclusions. Minor additions present in the refractories like TiO_2, ZrO_2 and K_2O are valuable trace elements in establishing a relation between the inclusions and refractories. The inclusions should therefore be thoroughly checked for the presence of these elements. Some indications are given in Section M.

The following examples where inclusions have most probably originated from refractories, are given in order to illustrate a systematic deduction of their origin, based on analyses of different exogenous materials.

Example I

Ultrasonic inspection of railway rails had indicated a defective zone in the rib of some rails, which had been rolled from the bottom part of the ingot. The defective zone was parallel to the rolling direction. A microscopical study of the transverse sections of such defective rails revealed large coherent multiphase non-metallic inclusions in the central part of the rails (Fig.208a). It also showed a second type of smaller, elongated multiphase inclusions. These were in the vicinity of the larger coherent inclusions and orientated parallel to them. The inclusion zone was often distributed over the main part of the rail section with a special concentration at its central parts. This constituted a serious source for internal cracks in the rails as well as a general weakness zone.

The steel was silicon killed, vacuum degassed in the ladle and top-poured into steel moulds, big end down. The ingots were then rolled

INCLUSIONS:
(F) Forsterite with 8MnO, 3FeO, 55MgO, $37SiO_2$
(G) Spinel with 7MnO, 25MgO, $70Al_2O_3$
(M) Matrix with 7MnO, 3CaO, 6MgO, $57SiO_2$, $22Al_2O_3$
(Me) Steel
Mean analysis (calc) 8MnO, 1FeO 1CaO, 34MgO, $46SiO_2$, $10A_2lO_3$

COMMENT From a systematic investigation of the analyses of different possible exogenous sources as compared with the inclusions (II) in Table XLI it was concluded that these inclusions originated from the ceramic covering of the bottom plate of the steel moulds due to the stirring action of the molten steel

208a Large coherent inclusions in central part of rail (inclusion II, Table XLI)

to rails. It was a plain carbon steel with 0·59%C, 0·39%Si and 1·53%Mn.

The microscopical study of the inclusion types was completed by electron probe analyses of the inclusion phases. The following result was obtained for the coherent stringer type inclusions (inclusion II Table XLI). The two main phases were found to be crystals of forsterite (F 5) in a glassy silicate matrix. In addition small amounts of spinel (F 3) as well as small, rounded droplets of steel were observed. The composition of the phases was found to be: (F) Forsterite with 8%MnO, 3%FeO, 55%MgO, 37%SiO_2; (M) Matrix with 7% MnO, 3%CaO, 6%MgO, 22%Al_2O_3, 57%SiO_2; (G) Spinel with 7%MnO, 25%MgO, 70%Al_2O_3; (Me) Steel. Mean analysis: 8% MnO, 1%FeO, 1%CaO, 34%MgO, 10%Al_2O_3, 46%SiO_2. The smaller, separate, elongated, multiphase inclusions (inclusion III, Table XLI) consisted of enstatite (F 4) in a glassy silicate matrix. MnS (K 3) was often observed in the silicate matrix near the tips of the elongated inclusion section. The composition of the phases was

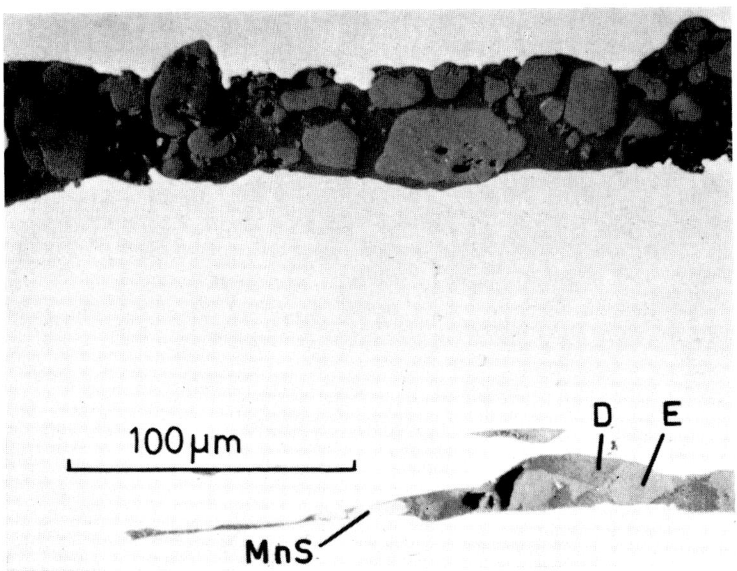

INCLUSIONS:
(E) Enstatite with $34MnO$, $26MgO$, $39SiO_2$
(D) Matrix with $27MnO$, $7CaO$, $4MgO$, $45SiO_2$, $14Al_2O_3$
(Sd) MnS
Mean analysis (calc) $31MnO$, $4CaO$, $14MgO$, $43SiO_2$, $8Al_2O_3$

COMMENT The higher MnO- and lower MgO-contents, as compared with the inclusions in Fig.208a, indicate a larger amount of indigenous precipitation. Thus the fragments of the bottom covering had been present in the molten steel for a longer time, and had also acted as nuclei for MnS precipitation.

208b Separate, elongated, smaller inclusions (III, Table XLI) in central part of rail

STEEL TYPE *See* Fig.207. The steel was top poured big end down, into steel moulds with protective ceramic covering on the bottom plate of the mould
STEEL ANALYSIS % 0·59C, 0·39Si, 1·53Mn
LOCATION IN RAIL *See* Fig.207a

208 Inclusions in railway rail, originating from ingot mould refractories

as follows: (E) Enstatite with 34 %MnO, 26 %MgO, 39 %SiO_2; (D) Matrix with 27 %MnO, 7 %CaO, 4 %MgO, 14 %Al_2O_3, 45 %SiO_2; (Sd) MnS.

Mean analysis: 31 %MnO, 4 %CaO, 14 %MgO, 8 %Al_2O_3, 43 %SiO_2.

The general microscopical appearance of the large coherent inclusions in Fig.208, as well as their high MgO-content, are proof of an exogenous origin. They must occur at a late stage of the steel-making process but before solidification and the subsequent rolling of the steel. This is evident since the inclusions have a moderate MnO-content and also as molten steel droplets were present in their interior.

Spectrochemical analyses of furnace and ladle slags as well as ladle refractories from similar steel charges were available (Table XLI). The low CaO-content (traces only) of the inclusions denoted as II is a strong indication that the slag contribution to their composition is negligible, and that their MgO-component must originate from some other source. A trace of TiO_2 indicates a contribution from the ladle refractories, but this cannot be a main source as these refractories are free of MgO. A further search for MgO-sources near the ingot end of the steelmaking process revealed that the bottom plate of the steel moulds was protected by a MgO-rich ceramic mass, applied before each teeming and with a composition as shown in Table XLI. It was concluded that the large inclusion type of Fig.208 mainly resulted from the stirring action of the molten steel phase on the ceramic mass in the bottom of the mould. Rather large fragments of this covering were dispersed in the steel as molten drops. Before solidification of the steel some precipitation of MnO and SiO_2 occurred on these drops. A small contribution from ladle refractories is also conceivable (trace of TiO_2). During rolling, the inclusions were deformed to plates or strings parallel to the rolling direction.

The main elements in the smaller inclusions in Fig.208 are the same as for the larger inclusions, but the MnO content is higher and the MgO-content lower. This indicates that the exogenous source is the same but that the influence of indigenous precipitation on the mean composition is greater. These particles were therefore probably dispersed in the steel bath as small particles and not as bigger fragments during the teeming. They also acted as nuclei for MnS precipitation.

Finally, it is of interest to note that the ingot scum (Table XLI) has about the same composition as the smaller inclusions and therefore was probably formed by particles from the bottom covering. These floated up through the steel bath in the ingot mould and changed their composition due to indigenous precipitation.

Example II

In the cast steel crankshaft, discussed in section O 3, there were two types of large inclusion. One is shown in Fig.203 and is discussed in that section, the other type is shown in Fig.209. The main structural components of this inclusion type are large quartz grains (A 5). An outer matrix with MnO, SiO_2 and some Al_2O_3 surrounded the quartz grains. The SiO_2-modification cristobalite (A 3), as well as small amounts of galaxite and mullite, were also identified in this outer zone.

STEEL TYPE *See* Fig.203
STEEL ANALYSIS *See* Fig.203
INCLUSIONS, %
(K) Quartz with $100SiO_2$ (A 5)
(T) Tridymite with $100SiO_2$ (A 4)
(Ct) Cristobalite with $100SiO_2$ (A 3)
(G) Galaxite, $MnO.Al_2O_3$ (A 7)
(M) Matrix with $MnO–SiO_2–Al_2O_2$
The MnO-content near the steel surface was about 20%, gradually decreasing to about 5% near the quartz surface

COMMENT The quartz has been partly dissolved in the $MnO–SiO_2–Al_2O_3$ matrix at the temperature of the molten steel. During cooling, cristobalite, which is the high-temperature modification of SiO_2, was precipitated. It does not transform during cooling of the steel but remains in a metastable condition at room temperature

209 Inclusion from the quartz lining of the casting mould

This inclusion type has an exogenous origin, as is evident from the presence of the quartz modification of SiO_2. As discussed in section A 5, quartz, if introduced into molten steel, transforms only very slowly to the high-temperature modifications tridymite or cristobalite. Conversely, if cristobalite or tridymite are formed indigenously in the steel as a result of Si-deoxidation, they do not transform to quartz at cooling rates met in steelmaking, but remain in a metastable condition. This is a consequence of the slow transformation rate. This is also the case if cristobalite is precipitated from a SiO_2-rich matrix in an inclusion. Several such inclusions are shown, for instance in Figs. 11, 12 and 13.

The origin of the inclusion type in Fig.209 seems to be the quartz sand from the casting mould. When quartz grains are dispersed in the molten steel, they are heated but do not melt and transform only slowly, in the present example to tridymite. They also act as nuclei for indigenous precipitation of MnO, SiO_2 and Al_2O_3. This outer matrix is liquid and attacks the quartz, dissolving part of the SiO_2.

During cooling, SiO_2 again is precipitated from the matrix but now in its high-temperature modification cristobalite, which is easily recognized by its 'rosette' morphology. This modification does not transform but remains in a metastable condition. Also small amounts of other phases from the MnO–SiO_2–Al_2O_3 system are precipitated from the outer matrix, e.g. galaxite, mullite, and spessartite.

P. The behaviour of non-metallic inclusions in wrought steel

'There is evidence to show that the internal structure of the inclusion affects its plasticity and therefore the way in which the inclusions are disseminated throughout the metal during hot working . . . The multi-phase nature of inclusions is important in the way in which they behave during hot working, and also in determining their effects on mechanical properties.'

Pickering 1962

P 1 Introduction

The inclusions in solid steel constitute only a very small part of the steel ingot and they are usually finely dispersed. It therefore follows that their thermal history has been imposed upon them by the cooling cycle of the steel. The nucleation of phases within the individual inclusions is thus often incomplete and their structure does not always correspond to equilibrium conditions. Furthermore, many of the inclusion phases belong to systems with complex crystal structures, with several modifications and slow transformation rates. During heat treatment of the steel further structural changes may be produced in the ingot inclusions; which may then transform to stable phases or crystallize from a glassy condition. These changes are accelerated by the energy introduced into the inclusions by the working of the steel phase. Some examples were discussed in section A 14 and a more complete discussion follows in section P 2.

A steel ingot is usually not the final steel product. It has to be worked and heat-treated in different ways. The parameters for rolling, forging, drawing, machining and other shaping processes for steel are all based on the behaviour and properties of the steel phase. The inclusions, which are dispersed in this phase, have physical properties which differ from those of the parent steel and it is important to know how far different inclusion phases adapt themselves to the various working parameters of their steel matrix. Both the steel phase and the inclusions are usually multiphase structures. However, with the present state of knowledge, the steel matrix may be regarded as a homogeneous phase since the dispersion of the structural components is usually so much finer in comparison with the coarser inclusion structures. The nature of the inclusions is of importance as the different inclusion phases have a wide variety of physical properties. For instance, the melting point of several of the iron and manganese silicates is in the range of 1 100–1 300°C whereas corundum melts at 2050°C and escolaite at 2265°C. MnS at room temperature has a microhardness of about 170 kp/mm², the silicates in the range of 700–1000 kp/mm² and corundum 3000–4500 kp/mm². Considering that these inhomogeneities in the steel are irregularly distributed with a size varying between one tenth of a micron and several hundreds of microns and that, furthermore, the physical properties of the inclusion phases change considerably with temperature, the complexity of the problem is evident. It is the author's opinion that the most important physical property of the different inclusion phases is their plasticity as compared with the plasticity of steel at different temperatures. If this property was known for all the different inclusion phases, the steels could be much better 'tailored' for different working operations such as rolling, forging, deep-drawing and turning. Further, the number of 'point defects' in the finished steel due to unyielding inclusions could be considerably reduced. Even if a full knowledge is still a pipe dream, a more systematic research effort in this field is in progress and recent findings will be discussed in section P 3, the deformability of inclusions, as well as in section Q 2, the influence of inclusions on machinability.

P 2 Behaviour of inclusions during heat treatment of the steel

Transformation to stable modifications

In Fig.17 two similar inclusions, both rich in silica, were shown. The inclusion in Fig.17a was from a steel ingot and the two principal phases present were cristobalite and rhodonite. The inclusion shown in Fig.17b was from hot-rolled sheet and contained tridymite, the modification of silica stable in the temperature range 870–1470°C. The transition between these two inclusion types is an example of an inclusion transformation during heat treatment of the steel. Thus cristobalite, as discussed in section A 3, is only stable above the temperature of steel solidification, but its rate of transformation to low temperature modifications is slow. The cooling rate in the solidifying ingot was too fast for the cristobalite–tridymite transformation to occur and therefore the former was retained in the solidified ingot in a metastable condition. The combination of reheating and the energy of deformation during hot-rolling caused the transformation to take place. However, once again the cooling conditions after rolling were too fast for the further phase change, tridymite–quartz, the modification stable at room temperature, to take place.

The silica modifications are valuable tracers for the origin of inclusions. Quartz (Figs.18, 185, 186 and 209) is of special interest, as its presence usually indicates an exogenous origin of the inclusion (see O 7, example II). The temperature cycle of the steel is in general such that the necessary conditions for a transformation of silica to quartz nevero ccur, whereas a partial transformation of quartz to tridymite or cristobalite is possible. It has been reported in literature that quartz can also be formed directly as a deoxidation product of Si, but the findings have been debated. It has been suggested that this direct formation of quartz or its formation as a transformation product of cristobalite in the steel, is due to extraordinary circumstances.[212]

Crystallization from glassy phases

Silicates in general have a slow crystallization rate and silicate phases are often glassy when present in steel ingot inclusions. In general heat treatment and working of the steel introduce more energy in the inclusions; the nucleation rate is increased and this often results in a crystallization of glassy phases. Several such examples have already been given. For instance, the glassy spherical single phase type of inclusions shown in Figs.40, 125a–c, and 134 often already show different stages of crystallization in the ingot. They are all rich in SiO_2, 35–50 %, and in addition contain different amounts of MnO, CaO and Al_2O_3. However, their crystallization rate is not always influenced by ordinary treatments of the steel and they sometimes remain glassy even after deformation of the steel, as shown in Fig.135.

The ternary silica phases spessartite (A 11) and anorthite (A 12, G 5) as well as mullite (A 8) are more sensitive to thermal and deformation-induced crystallization. It is common for one or several of the phases mentioned to precipitate from a silica-rich inclusion matrix during heat-treatment and working of the steel. Several examples of this have been given. In Fig.53, spessartite and mullite are precipitated from the silica-rich (45 % SiO_2) matrix of an inclusion in a rolled billet after heat treatment of the steel for 2 h at 1 100°C. In Fig.54 spessartite and Mn-anorthite crystallize in the glassy matrix of an ingot inclusion after a similar heat treatment. The calcia-rich inclusion in Fig.86 is mainly glassy, but after heat treatment of the steel grossularite (G 5) and anorthite are abundant within the whole matrix. A similar structure was produced by heat treatment of a glassy inclusion free from calcia (Fig.142b), where the precipitated phases are now spessartite (isostructural with grossularite) and Mn-anorthite (isostructural with anorthite).

Precipitation from supersaturated solid solutions

The temperature cycle which is forced upon the inclusions by the environment of the steel may also result in formation of supersaturated solid solutions among the inclusion phases. The possibility that such solutions are formed is greatest among the sulphides. As was discussed in section K 4, MnS has a wide range of substitutional solid solubility for other metals, especially Fe and Cr. In steelmaking practice MnS inclusions in the steel may have an Fe- or Cr-content higher than the equilibrium value at temperatures below the solidifying temperature of the steel. If the steel is heat treated, FeS or other sulphides may precipitate within the MnS phase. This type of precipitation has been observed by Chao *et al.*[109] and by Matsubara.[114] FeS often precipitates in a Widmanstätten-pattern and this precipitation morphology was shown in Fig.154 in a synthetic Fe–Mn–S solid solution.

Precipitation is also possible in oxide inclusions. Figs.14 and 70a show inclusions rich in Cr_2O_3 and SiO_2. The inclusion in Fig.14 consists of primary Cr_2O_3, crystallizing in a matrix with Cr_2O_3 in solution in SiO_2. The corresponding inclusion in Fig.70a has a fine precipitate of Cr_2O_3. As was discussed in section A 2 and C 2, there is a wide range of solid solubility between Al_2O_3–Cr_2O_3–Fe_2O_3. The solid solutions are unstable at lower temperatures and precipitation of different solid solutions may occur as a result of heat treatment of the steel phase. The principal phenomenon was illustrated for a synthetic Al_2O_3–Cr_2O_3 slag in Fig.71, and may also include precipitation from double oxides, for instance $FeO.Cr_2O_3$ (B 3). As reported by Koch[48] the lattices of inclusions of the cubic iron chromite are sometimes observed to deform tetragonally, depending on the heat treatment of the steel. This inclusion behaviour has been further investigated by Adachi *et al.*[200] who found a variation in the c/a ratio of the chromite depending on the cooling velocity of the steel. They concluded that the reason for the parameter variation was a variation of the Cr and O activity in the steel, influencing the solubility of these elements in the chromite inclusions.

P 3 Deformability of steel inclusions

As discussed in P 1 the plasticity of the inclusions, as compared with the plasticity of the steel phase, has a strong influence on the behaviour of the steel. If the steel phase and the inclusions are not working together during all the steelworking operations, the inclusions will be a potential source of future defects in the finished steel product. In contrast, inclusions may also enhance steel properties such as machinability, by their ability to participate in the plastic flow of the steel phase. This is one of the reasons why a knowledge of the metallography of the inclusions is so important. If much more was known about the behaviour of the different inclusion types during different steel operations, the composite product steel–inclusions could be more closely tailored to the different operations and purposes such that considerable economic and technical advantage could be achieved.

A basic approach to the problem of the deformation of inclusions *in situ* resulting from deformation of the steel phase is difficult and complex. Not only is a knowledge of the physical properties of the different inclusion phases necessary; the influence of temperature and pressure on these properties must be known. Also, an analysis of the complicated stress–strain pattern at the steel/inclusions interface is required. Even this approach is an oversimplification, as the steel in itself is composed of several phases with different properties. Nevertheless, through a more detailed knowledge of the metallography of inclusions and the basic properties of their phases, as well as through a more fundamental approach to the deformation mechanism of inclusions, a better understanding of the complex steel/inclusions interaction during deformation should soon be possible.

A knowledge of the *absolute* plasticity of the inclusion phases is still lacking; the amount of information available is too small. For technical purposes an interesting approach to the influence of inclusions on the workability of steel is to regard the *relative* plasticity of the steel and the inclusions at different temperatures and working conditions. This is now possible, mainly through the works of Pickering,[213] Scheil and Schnell,[214] and Malkiewicz and Rudnik.[215,216] In the present section the author will first summarize their work. Then information on the deformability of different inclusion phases has been considered, and where possible illustrated by microsections of inclusions in the present review or from the literature. Finally some general remarks are given. The section Q 2 on the influence of inclusions on machinability also deals with deformation of inclusions and reference is often made to that section.

Plastic deformation of inclusions relative to the steel matrix (the concept of 'index of deformability')

A simple method of comparing the deformability of inclusions with the deformability of steel was used by Scheil and Schnell.[214] These authors studied compressed steel samples with oxide and sulphide inclusions before and after different amounts of steel deformation. The deformation ratio for the compressed steel samples was compared with the deformation ratio for globular inclusions which deformed to ellipsoids. By this method, the deformability of oxide and sulphide inclusion types could be compared with the deformability of the steel (at this type of deformation) at different temperatures. Curves of the type shown in Fig.210 were obtained. Their results will

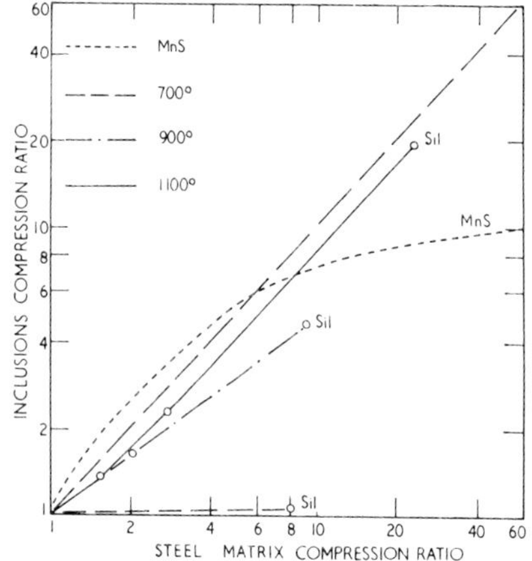

The oxide inclusions (Fe–Mn–silicates with unknown composition) are undeformed at lower compression temperatures, but their deformation behaviour becomes more similar to that of the steel with increasing temperature, independently of the compression ratio for the steel phase. The sulphide inclusions (MnS) deform in a manner similar to the steel matrix at all temperatures for lower compression ratios, but for compression ratios higher than about 6 the sulphide inclusions deform less than the steel matrix. This difference increases with increasing compression ratio for the steel and should be compared with the results from rolling. Fig. 212. From ref.214

210 Compression ratios for oxide and sulphide inclusions v. compression ratio for the steel matrix at different temperatures

be discussed in detail below. Pickering,[213] in his studies of the effect of hot working on various types of inclusions, measured the deformation of the inclusions in rolled bars on microsections parallel to the rolling direction, where the deformed inclusions were visible as elongated ellipses. It can be shown that when a spherical inclusion is deformed by hot deformation into an ellipsoid within the mass of the steel, the length:width ratio of the inclusion is constant for any section parallel to the major axis of the ellipsoid, that is, the longitudinal direction. By measuring the ratio of major to minor axes (λ=b:a) for such longitudinally sectioned inclusions, a ratio can be obtained which is compared with a similar ratio for the steel as a whole. For the steel itself it can be shown that if a reduction in cross-sectional area of the original ingot is from F_0 to F_1, then the length: width ratio is given as b:a, where b:a$=F_0^{3/2}:F_1^{3/2}$. A comparison was then made between the length:width ratio for the inclusions and for the steel, giving a measure of the relative inclusion–steel plasticity. Pickering's results on the effect of temperature and rolling reduction on the plasticity of FeO and silicates will be discussed below. These ideas have been further developed by Malkiewicz and Rudnik,[215] who have defined an index of deformability (ν) for the inclusions

$$\nu=\frac{\epsilon_i}{\epsilon_s}=\frac{2}{3}\cdot\frac{\ln\lambda}{\ln h}=\frac{2}{3}\cdot\frac{\log\lambda}{\log h}$$

where $\epsilon_i=\ln\ \lambda=\ln\frac{b}{a}$ is a measure of the true elongation of the inclusions. $\epsilon_s=\frac{3}{2}\ln\ h=\frac{3}{2}\ln\frac{F_0}{F_1}$ is an expression for the true elongation of the steel where F_0 is the initial cross-section of the ingot and F_1 the cross-section of the billet, bar etc. after working of the steel. The value of $\nu=\epsilon_i:\epsilon_s$ of the deformability index can change from zero for inclusions not changing at all during working of the steel, to unity for inclusions which elongate equally to the steel. Even values greater than unity are possible, for inclusions which elongate to a greater degree than the steel.

It is useful to study the deformability index as a function of temperature for different inclusion types. The author has collected available information. Curves of the type shown in Fig.211 are obtained. This figure is of fundamental importance for a discussion of the composite product steel with inclusions: and will be referred to in several sections.

Curves showing the deformability index of inclusions v. ϵ_s, the degree of steel deformation, may also be plotted. These give useful information about the behaviour of inclusions when the steel is deformed in different ways and typical curves are shown in Fig.212.

The relation between inclusion deformation and steel deformation for a selected inclusion type with changing steel composition is also advantageously summarized in this way. Curves showing the variation of index of deformability for the inclusion type with steel compo sition are shown in Fig.213.

If during steel deformation the non-metallic inclusions lengthen less than the steel, stresses are built up which may lead to cracking or other discontinuities between the inclusions and the steel phase. A systematic study of this important phenomenon, closely related to the plastic properties of steel and inclusions, is difficult, but also here the studies of the index of deformability give interesting information. Rudnik[216] has studied discontinuities in hot-rolled steel caused by non-metallic inclusions for inclusion phases with different deformability indices. For a deformability index close to 1, the inclusions lengthen in the same way as the steel. The binding forces at the inclusion/steel interface are never broken, and the inclusions appear

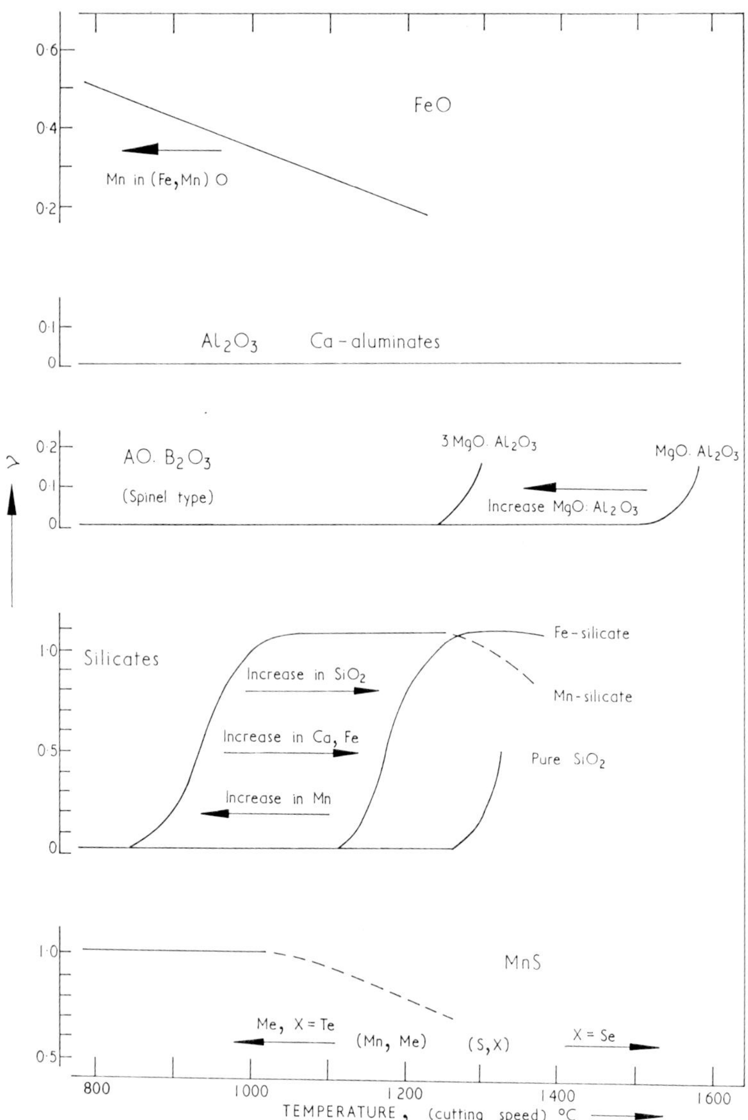

The plastic behaviour of different inclusion types can be measured relative to the deformation of the steel matrix and expressed as the index of deformability, *v* (*see* P 3). In this figure curves for the major inclusion types show the dependence of deformability index on temperature. In all except for the undeformable aluminates, the trend of plasticity with compositional variation is also indicated. These curves should be regarded only as a qualitative indication of the differences in behaviour between the inclusion types. (Information used in compiling the figure was obtained from various sources and thus numerical values for *v* may not be precisely comparable.) Since the temperature dependence of inclusion plasticity is of particular importance in machining operations it should be noted that an increase in cutting speed is equivalent to an increase in temperature

211 Influence of temperature on the plastic deformation of different inclusion types as compared with steel

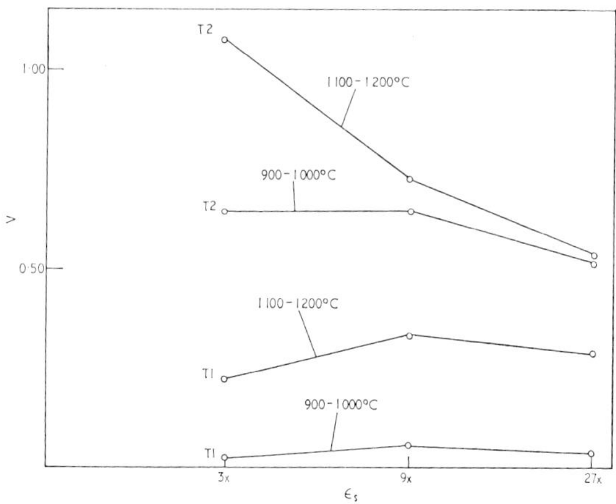

ν = index of deformability, ϵ_s = deformation ratio for the steel matrix

STEEL TYPE Basic arc-melted steel oxidizing slag only, 250 kg ingot. Middle section rolled to square bars under different conditions (*see* fig.) T1 deoxidized with FeSi followed by FeMn. T2 deoxidized withMnSi

STEEL ANALYSIS T1: % 0·08C, 0·34Mn, 0·27Si
T2: % 0·10C, 0·48Mn, 0·36Si

INCLUSIONS The plastic deformation of silicate inclusions, as compared with the steel matrix, was studied after different rolling reductions. The glassy silicates examined were of the same type as those shown in Fig.40.

The analysis was not determined *in situ*, but it was concluded that the inclusions in T1 were rich in FeO, those in T2 in MnO

COMMENT The variation of the index of deformability with steel deformation at different temperature shows that Mn-rich silicates deform more readily than Fe-rich silicates and also that the deformability of these oxide inclusion types decreases with an increase in the degree of deformation of the steel on rolling. The results are therefore not in complete agreement with those reported for the deformation behaviour under compression (Fig.210). From ref.215

212 Influence of steel deformation during rolling on the plastic deformation of iron- and manganese-silicate inclusions at different temperatures

in microsections as ellipses without the occurrence of discontinuities in the steel caused by the inclusions (Fig.214a). If the index of deformability decreases, the inclusions do not elongate uniformly during rolling of the steel. Stresses are built up at the inclusion/steel interface which may lead to cracking. In principle, cracking occurs between the inclusion and the steel perpendicular to the flow of the steel. However, the direction of the cracks is determined by the simultaneous flow of the steel, leading to the formation of a conical gap at the steel/inclusion interface, as shown in Figs.214b and c. The base of the cone is resting on this interface and the apex pointing in the direction of steel flow. The size of the conical gap depends on the inclusion plasticity, on rolling temperature and rate of deformation. Such a gap at a spherical inclusion in rolled steel is visible in Fig.111 (rolling direction upwards in the figure). Due to the compressive stresses in the steel phase perpendicular to the direction of flow, the

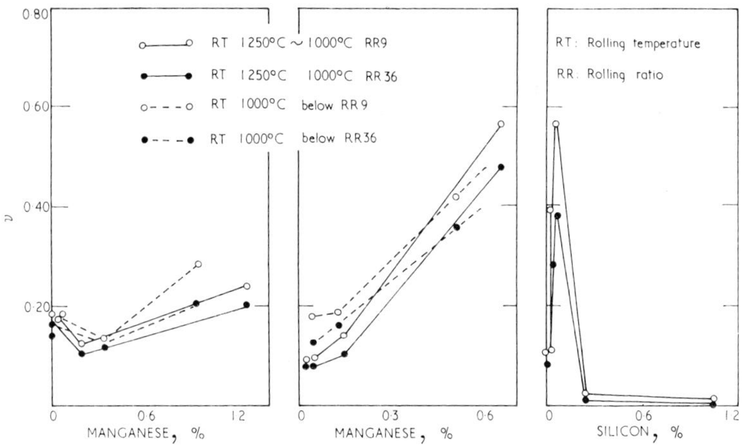

STEEL TYPE Laboratory 6 kg ingots of electrolytic iron, molten in a high frequency induction furnace in magnesia crucibles and deoxidized with manganese (left), silicon-manganese (middle) and silicon (right)
STEEL ANALYSIS % (left) 0·006–0·010C, 0·004–0·006Si, nil P, 0·003–0·005S. 0 < Mn < 1·25
(Middle) 0·002–0·018C, 0·002–0·005P, 0·005–0·006S, 0·019 < Mn < 0·66, 0·003 < Si < 0·087
(Right) 0·002–0·007C, nil Mn, 0·001–0·006P, 0·002–0·006S, 0·002 < Si < 1·175
From ref.185

213 Relation between the index of deformability (ν) for silicate inclusions and the Mn- and Si-content of the steel

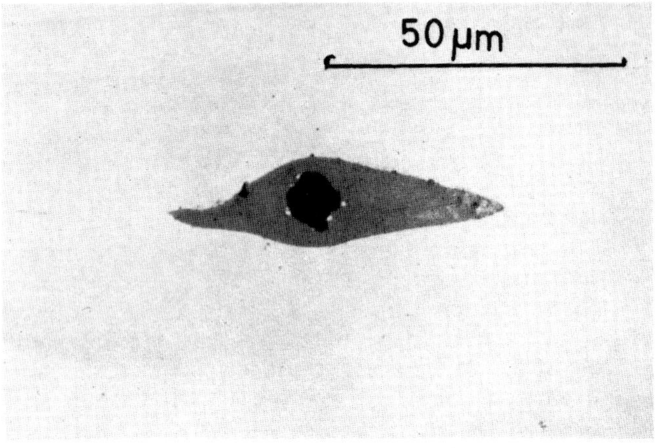

214a Highly deformable inclusion ($\nu = 1$) in rolled steel

The eliptically deformed inclusion, in the same steel sample as the inclusion in Fig.214c, is MnS (K3) with a small nucleus of calcium aluminate. No discontinuities occur between the inclusion and the steel phase

214 Crack formation at the metal/inclusion interface (*See* next page for Figs.214b, 214c and 214d)

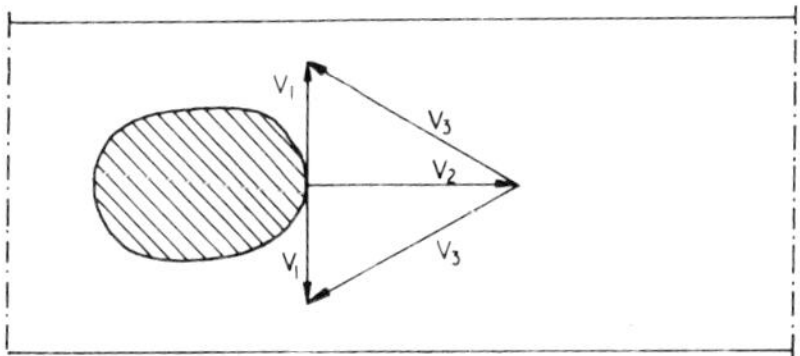

From ref.216

214b Schematic indication of crack formation at the metal/inclusion interface, causing a conical gap

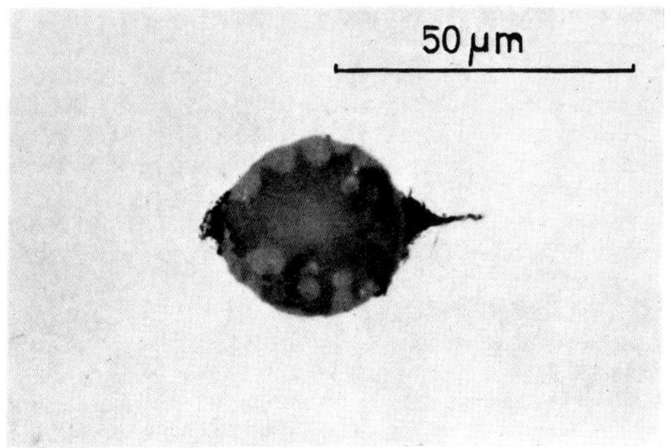

The spherical inclusion is a calcium aluminate (G3), which was not deformed at the rolling temperature of the steel. The rolling direction is indicated by the cavity. White areas in the inclusion are MnS precipitates at the inclusion surface. *See* also Figs. 111 and 117

214c Non-deformable inclusion ($\nu=0$) in rolled steel with conical gap in the steel phase

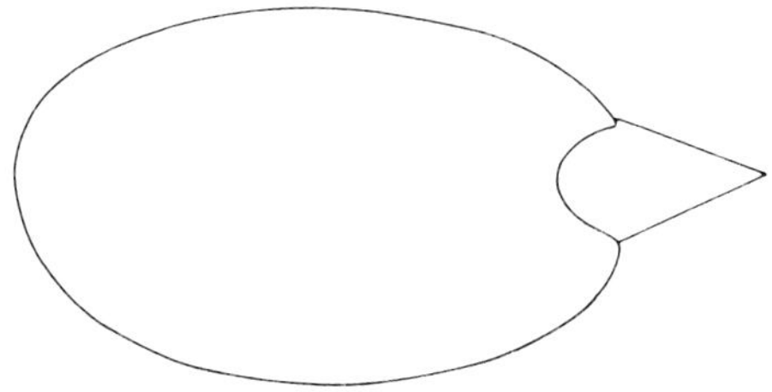

From ref.216

214d Schematic indication of fishtail inclusion with cracks

conical gap may be filled with steel leaving only the cracks shown in Fig.117, but if the steel is cold worked, the conical voids may be left unfilled. An intermediate stage in (relative) inclusion plasticity is represented by those inclusions which have after rolling a fishtail shape. The outer layers of these inclusions have lengthened in the direction of the steel flow before the bonds at the inclusion/steel interface have been broken. This results in a fishtail inclusion shape and simultaneously a gap is formed between steel and inclusion as indicated in Fig.214d. Rudnik[216] has shown several microsections of inclusions in hot-rolled steel initiating these types of microcracks in the surrounding steel matrix. He concluded from experimental evidence that inclusions with an index of deformability ν=0·5–1 deformed normally with a low frequency of microcracks in the inclusion/steel interface. Those with ν=0·03–0·3 often gave fishtails with conical gaps, more frequently at the lower range of ν. For underformed inclusions (ν=0) conical gaps and hot tears were frequently observed. A necessary condition for an inclusion type to be harmless, or even to be of advantage during hot-working operations of the steel, is that it participates in the plastic flow of steel during the specific hot-working operation. It should therefore have an index of deformability high enough to prevent crack formation. The work of Rudnik indicates that this index should preferably be ν=0·5–1. The inclusion phase should not be molten at the hot-working temperature as it then deforms to thin films which flow away (hot-tearing, Q 5). The behaviour of inclusions during machining of steel is thoroughly discussed in section Q 2. There are several similarities between inclusion–steel behaviour during machining and hot-working, and as most detailed work has been connected with machining, the reader is referred to that section.

Summarizing. The observations on inclusion deformability are important and indicate that inclusion plasticity should not be studied as an absolute value, but in relation to the plasticity of the surrounding steel matrix in the same conditions.

The importance of the index of deformability lies in the fact that it gives an indication of the behaviour of the inclusion *in situ*, that is, in its actual steel environment. It is therefore possible to connect direct metallographic studies of individual inclusions with their mechanical behaviour in the steel.

Influence of reduction

The relative plasticity values for different inclusion phases are in general higher for small than for large reductions. Apart from the difficulty of comparing correctly the plasticity of the small inclusion volume and the large steel volume at high reductions, this is also due to the fact that the deformed inclusions offer less resistance to the flow of metal than the spherical ones. From the results on compression samples[214] it also seems that the decrease in plasticity of the inclusions at high reductions is greater for sulphides than for oxide inclusions. For sulphides, there is a difference depending on sulphide type.[217]

Influence of particle size

This also has an effect, presumably due to the forces at the metal/inclusion interface. Small particles tend to deform less readily than do larger ones, and they generally only start to deform at low temperatures and large reductions when the forces acting upon them are large. Uchiyam and Sumita[185] have given curves for the correspondence between inclusion deformation and inclusion size (Fig.215) for

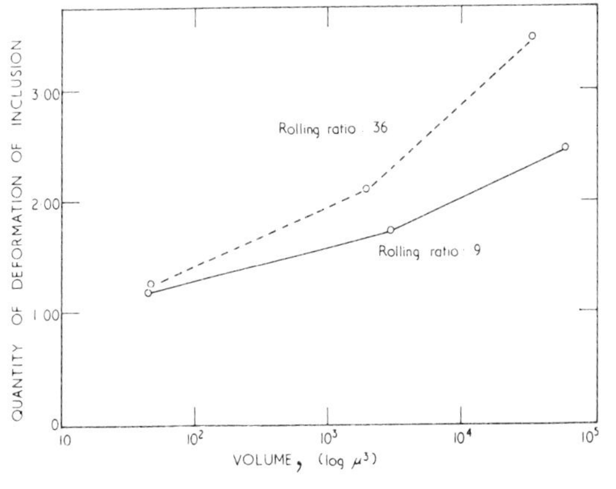

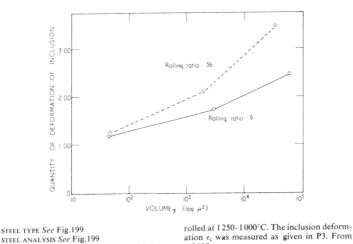

STEEL TYPE *See* Fig.199
STEEL ANALYSIS *See* Fig.199
INCLUSIONS Inclusions of the (Fe,Mn)O-type, rolled at 1 250–1 000°C. The inclusion deformation ϵ_i was measured as given in P3. From ref.185

215 Influence of inclusion size on oxide inclusion deformation

spherical inclusions of the (Fe,Mn)O type in iron deoxidized with manganese.

Deformability of different inclusion phases

The work of Scheil and Schnell[214] showed that there was a fundamental difference in the behaviour of oxide and sulphide inclusions.

The oxide inclusions were brittle at low temperature, but at temperatures from 850–1 250°C, depending on SiO_2 content, their deformability in compression samples was comparable to that of steel (Fig. 210). The sulphides deformed plastically independent of temperature from −80°C–+1 260°C, and their deformability in compression samples was at all temperatures comparable to that of steel. A further evaluation of these results is not possible as the precise identity of the inclusion phases was not known, but from various references a clearer picture of the relative plasticity of different inclusion phases compared to steel can now be drawn. This was summarized in Fig.211.

Inclusion phases of the FeO type (A6, B2, D2)

From the experimental work of Pickering[213] and Uchiyama and Sumita[185] on iron deoxidized with different amounts of Mn, it seems that spherical inclusions of the FeO type (Fig.196) become decreasingly plastic, as compared with iron, as the temperature increases. The deformability index decreases from about 0·50 at 800°C to about 0·20 at 1 200°C. Furthermore, with an increase in the Mn-content of the (Fe,Mn)O solid solutions their plasticity decreases. For steels with a higher Mn-content this is not quite evident if the index of deformability is studied, because the plasticity of the steel itself decreases more rapidly than the plasticity of the (Fe,Mn)O inclusions (Fig.213). Finally, the plasticity of FeO is higher at low reductions than at high reductions, that is, the inclusions are more severely deformed in the early stages of hot rolling.

The relative plastic deformation of (Fe,Mn)O as compared with other inclusion phases is indicated in Fig.211.

Microsections of FeO and (Fe,Mn)O inclusions in different states of deformation are shown elsewhere.[218]

If the steel is deformed under such conditions that the (Fe,Mn)O inclusions are not within the range of plastic deformation, they may fracture without deformation due to the shearing action of the increased differential forces at the steel/inclusion interface. The smaller parts of the fractured inclusions are then sometimes plastically deformed. Such examples have been shown in literature.[219, 220]

The influence of FeO inclusions on the fatigue properties of steel is discussed in section Q 4.

Inclusion phases of the corundum type, Me_2O_3 (A2, C2, D4)

All these phases are hard at all possible temperatures of steel deformation, and the deformability index is $\nu=0$ at all temperatures of interest (Fig.211). Several examples have already been given to illustrate the fact that corundum and isostructural inclusion phases are not deformed as single or multiphase inclusions, even if the steel itself is heavily deformed, e.g. Figs.3, 5, 7 and 175. At heavy steel deformations the unyielding corundum phase is sometimes mechanically separated from a more deformable matrix, as in Fig.5. Corundum inclusions sometimes have a rounded shape, as in Figs.6 and 8; this is not due to plastic deformation but to partial melting, occurring during deoxidation from a reaction between high concentrations of Al-metal and oxygen. This phenomenon is discussed in section O 3, aluminium.

Inclusion phases of the spinel type, $AO.B_2O_3$ *(A7, F3)*
These phases are all hard and unyielding at the normal working temperatures for steel. They therefore behave similarly to corundum during deformation and are often found separated from their original inclusion matrix. Many examples of different inclusion types with double-oxides of the spinel type after steel deformation have been given in Figs.15, 16, 27, 29, 30, 32, 55, 72, 73, 89, 91, 114, 139 and 168. A characteristic feature of the spinel-type phase is that it is never plastically deformed and often appears separated from its orginal more plastic inclusion matrix. After heavy deformations brittle fracture of the double oxide phase is sometimes observed, as in Figs. 55 and 114.

Even if the index of deformability for the common spinel type inclusions is $\nu=0$ for the hot-working temperatures now in use, recent results from fundamental studies on the micromechanical properties of double oxides indicate an increase in their deformability indices at higher temperatures. It is therefore possible that these oxide phases may deform plastically below the melting point of steel and thus participate in plastic deformation during certain hot-working operations or influence machinability at very high cutting rates.

Thus Lewis[221] has studied the defect structure and mechanical properties of spinel single crystals, $MgO.Al_2O_3$ (F 3). This oxide phase deforms through gliding and slip in the temperature range of 1 300–1 500°C. In the [110] directions, partial dislocations appear, separated by cation stacking faults with a width of 500–1 000Å. Newey[222] has shown that the mode of deformation depends on the $MgO:Al_2O_3$ ratio. These spinels deform plastically above a characteristic temperature and the yield stress was also determined, as shown in Table XLII. Newey has also reported that the 1:1 and 1:3 spinels deform through a greater variety of slip systems than do the 1:2 spinels.

TABLE XLII **Lowest temperature for observed plasticity ($\nu > 0$) of the spinel phase with different ratios MgO:Al_2O_3**

Atomic ratio MgO:Al_2O_3	Temperature °C	Yield stress kg/mm^2	Slip system
1:1	1552	65	{111} <110>
2:1	1277	65	{110} <110>
3:1	1252	40	{111} <110>

According to Newey's results, therefore, a deviation from stoichiometry lowers the temperature for plastic deformation to nearly within the range of hot-working temperatures for steel. As reported earlier (section A 15, point 5) double oxides of the spinel type found in slag inclusions usually have a composition deviating from stoichiometry. This should therefore influence their mechanical properties and make plastic yielding possible in certain hot-working operations, even if the double oxides in general are hard and brittle at steel working temperatures (Figs.55 and 114).

Inclusion phases of calcium aluminates (G3)
The calcium aluminates observed in steel inclusions, that is CA, CA_2 and CA_6 have a high melting point and are comparatively hard, the hardness increasing with Al_2O_3 content (Table XVI). All the observations made on deformed steel indicate that the calcium aluminate phases never deform, that is, they have an index of deformability $\nu=0$. Examples of these inclusion phases in deformed steel have been shown in Figs.89, 91, 114 and 214c. The calcium aluminates often appear together with double-oxides of the spinel type, and in heavily deformed steel the aluminate inclusions may be crushed and appear in stringers as in Fig.89.

Pure SiO_2 (A3–5)
Pure silica is probably brittle with $\nu=0$ for the whole temperature range of interest, but some difference seems to exist between the various modifications. Available information indicates that tridymite (A 4) and quartz (A 5) are brittle at all working temperatures of steel ($\nu=0$) and an example of tridymite with cracks in rolled steel is shown in Fig.17b. For cristobalite (A 3) a slight plasticity at hot-working temperatures has been indicated, but most information, for instance Fig.216, points to a low index of deformability for this phase also. The transformations between the different SiO_2 -modifications, induced by temperature and pressure from the surrounding steel matrix during hot working, are of greater importance than the deforma-

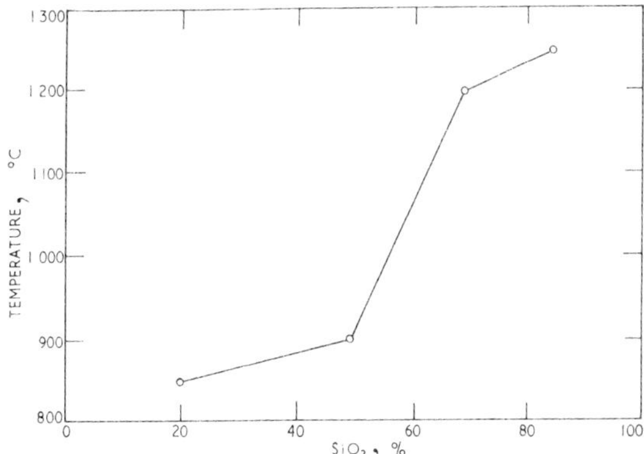

The lowest temperature at which inclusions deformed plastically was estimated from microscopical studies of silicate inclusions in different steel samples deformed by compression. The SiO_2-content of the inclusions is the mean composition of inclusion isolates from the different steel samples. From ref.214

216 Oxide inclusions with SiO_2; The lower temperature of plasticity v. SiO_2 content

tion of pure SiO_2-inclusions. This topic has been discussed separately above (P 2). Many glassy inclusions studied before electron probe analysis was available have been reported as 'silica' but are most probably different glassy silicates. Therefore deformability information on 'silica' must be evaluated critically.

Silicates

Glassy single-phase inclusions with varying amounts of silica are frequently found in steel, and multiphase inclusions usually have a matrix containing varying amounts of SiO_2. A characteristic feature of the silicate phases in inclusions is that they have a low index of deformability at lower temperatures, but that this index increases rapidly above a characteristic temperature between about 800°C and 1 300°C, depending on their composition. The index may even approach unity. Silicates are thus brittle at cold-working temperatures of steel but deform plastically within the temperature range of hot-working. The composition controls the plasticity of the silicate phase mainly through its influence on the melting point of the constituent silicates, but the early precipitation of hard components like Al_2O_3 or spinel-type phases probably also has some effect. The following results on the influence of composition on silicate inclusion plasticity have been collected from various references.[185, 213, 214, 223]

The temperature when ν starts to increase from zero for Fe–Mn silicates with different SiO_2-contents, that is, the onset of plasticity, is given in Fig.216. It is evident that an increase in SiO_2-content considerably increases this temperature. At temperatures below the critical limit, the silicates are brittle during deformation, but above this temperature they are plastically deformed. Also, the ratio FeO: MnO in silicates with a similar SiO_2-content is of importance, the plasticity increasing with increasing MnO. Some numerical values of ν at different temperatures and steel deformation have been collected in Table XLIII. There is some indication that at temperatures above

TABLE XLIII Deformability index (ν) for Fe- and (Fe,Mn)-silicate inclusions at different steel rolling temperatures and steel deformation

Type of inclusions	Rolling temp. °C	Steel def. h	Deformability index ν
Fe-silicates	1 150	×3	0·23
		×9	0·33
		×27	0·28
(Fe,Mn)-silicates	1 150	×3	1·07
		×9	0·71
		×27	0·52
Fe-silicates	950	×3	0·03
		×9	0·05
		×27	0·04
(Fe,Mn)-silicates	950	×3	0·64
		×9	0·64
		×27	0·51

1 250°C the plasticity of Mn-silicates again decreases.

Most silicate inclusion phases have several oxide components, for example MnO, FeO, CaO, Al_2O_3 and SiO_2, and the plastic properties of the silicate phase at any temperature are dependent on its composition. The deoxidation practice has been intensively developed recently and it has a great influence on the composition of silicate inclusions (O 3). The use of Ca-alloys has had as a consequence the occurrence of many silicate inclusions belonging to the CaO–SiO_2–Al_2O_3 system (G 4–6). These silicates have a comparatively high melting point but are still plastically deformed over a comparatively wide temperature range. This is of interest for their machinability (Q 2) as well as for different hot-working operations. Additions to the steel such as boron seem to lower the temperature where $\nu > 0$ for these silicate inclusions without lowering their melting point,[224] and are therefore of some interest for hot-working operations. The general trend of deformation *in situ* for different silicates v. temperature is shown in Fig.211.

Microsections of silicate inclusions *in situ* in steel deformed at different temperatures and to different deformation ratios have been shown in literature and in the present monograph. The author has tried to select some examples which illustrate the deformation behaviour of silicate inclusions under different conditions and the relative deformation of silicate inclusions v. other inclusion types. A comparison between the behaviour of a pure glassy SiO_2 inclusion and a glassy silicate inclusion in the same sample of rolled steel is shown in the literature.[225] The former is undeformed, the latter has been plastically deformed to an ellipsoid.

The effect of rolling temperature on the deformation behaviour of Fe-silicate inclusions shows fracture and displaced silicate chippings below 800°C, slight deformation at 900–950°C, plastic deformation and fishtail effects at 1 000–1 050°C and plastic deformation at 1 100–1 150°C.[226]

In the present work the deformation behaviour of glassy monophase silicate inclusions during hot-working of the steel is indicated by comparison of Figs.134 and 135. Ingot and deformed multiphase inclusions with a silicate matrix are shown in Figs.28–29.

At normal rolling temperatures the silicates have a higher index of deformability than most other phases, for example, galaxite (Figs.27, 29, 72, 139) and corundum (Figs.3 and 5). The high index of deformability for silicates, comparable to that of steel if the rolling temperature is correct (from the inclusion point of view) is also evident from the close contact between steel and matrix, for instance in Figs.29, 36, and 142a.

The relative deformabilities of silicates and MnS are dependent upon temperature and perhaps also pressure. At high temperatures the silicate phase is more plastic than the MnS phase; an example is shown in Fig.173. In contrast, there are also inclusions where MnS has deformed more than the silicate phase.[227] The two phases often behave in a similar manner, however, as shown in Fig.172.

MnS (*K 3*)

In general, MnS must be rated as a highly deformable inclusion phase with an index of deformability near to unity which is independent of temperature over a wide range from room temperature and upwards. This is evident for instance from the experiments on inclusion deformation in compression samples[214] or from the hot-hardness curves of MnS and steel (Fig.217). Under favourable loading conditions MnS will deform more than the surrounding steel. It has also been shown that MnS and steel have sufficiently comparable deformation characteristics so that slip which has started in the steel

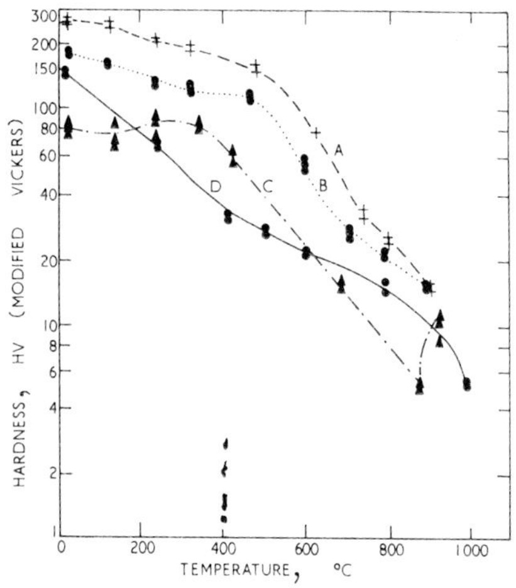

A 1080 steel % 0·75–0·88C, 0·60–0·90Mn
B 1040 steel % 0·37–0·44C, 0·60–0·90Mn
C Iron of high purity
D Single crystal of MnS

(This may be compared with the deformation of MnS inclusions at different temperatures, Fig.218)
From ref.109

217 Hot hardness values of MnS as compared with iron and steel

may be continued through the sulphide inclusion and back into the steel, if the crystal orientations are favourable.[228] The plastic deformation of MnS is of great importance for the machinability of steel; this is discussed further in section Q 2. In the present section the more fundamental aspects of MnS-deformation will be summarized based mainly on the work by v.Vlack *et al.*[228–231]

The deformation pattern of MnS has been thoroughly analysed. The primary glide system is <110> {110}; secondary slip occurs on the {111} plane whereas no evidence has been found for <100> deformation common among other NaCl-type structures. The {100} cleavage is predominant with crushing, but the {110} planes become the primary fracture planes when MnS is loaded by surface indentation. The deformation mechanism and the possible fracture are therefore dependent on the orientation of the MnS-crystals. Some comments on fracture will be given in section Q 3.

The relative deformation of the MnS inclusions and surrounding steel are closely dependent upon the relative hardness of the two phases. The hot-hardness curves for MnS show that the phase is as hard as pure ferrite below 1000°C but softer than the microstructures encountered in carbon-containing steels. Further, the relative hardness between MnS and different steel types often changes over the temperature range from 20°C to 1000°C. Thus MnS is harder than pure iron below 200°C, softer between 200° and 600°, harder to about 900°C and again softer above this temperature.

The hardness of MnS is also influenced by solid solubility (*see* sections K 4 and L 1) and therefore the relative deformation of MnS in steel is complicated. Fundamental studies on single crystal MnS inclusions in sintered powder compacts of low-carbon steel have shown that the deformation pattern changes with temperature and crystal orientation (Fig.218a–d). The MnS inclusions will deform more than the steel when the [100] direction is aligned with the compression axis but less than the steel when the [111] direction is parallel to this axis. The influence of temperature is related to the relative hardness of the two phases in the different temperature ranges mentioned above. The fundamental knowledge of the deformation mechanism for MnS is of great importance and has a bearing on the development of steels for different purposes, such as enhanced machinability (Q 2), as well as on the knowledge of the role of inclusions in crack formation in steel (Q 3).

The plasticity of MnS and of steel are thus very similar over a wide temperature range. Small changes in the phases such as solid solubility of metals in MnS, a change in the orientation of MnS inclusions or variation in steel type, composition or treatment, have a great influence on the relative deformation of the two phases and therefore on the behaviour of steel with MnS-inclusions during working at different temperatures. Also the sulphide type (I, II, or III, section K 3) is of importance for the relative MnS–steel deformation. According to Dahl *et al.*,[217] the deformation of the sulphide phase in steel during rolling in the temperature range of 850–1 300°C increases in the order of type I to II to III. As the morphology of the MnS phase depends on the oxygen content of the steel, this gives a further possibility to influence the properties of this important inclusion phase. Information about the hot hardness of MnS and different steel types in the temperature range of 1000°–1 300°C is unfortunately not yet available but is urgently needed, as this temperature range is of special importance for many hot-working operations. The influence of sulphur on these operations is not at present fully understood. A full knowledge of the behaviour of the composite product MnS-steel might give new perspectives on the machining, hot-working and crack susceptibility of steel.

Microsections of single phase MnS inclusions in steel deformed in different ways are shown in Figs.162, 171, 172 and 225. Multiphase deformed inclusions of this kind are shown in Figs.173 and 175, and 214a illustrating the difference in deformation behaviour of MnS and oxide phases.

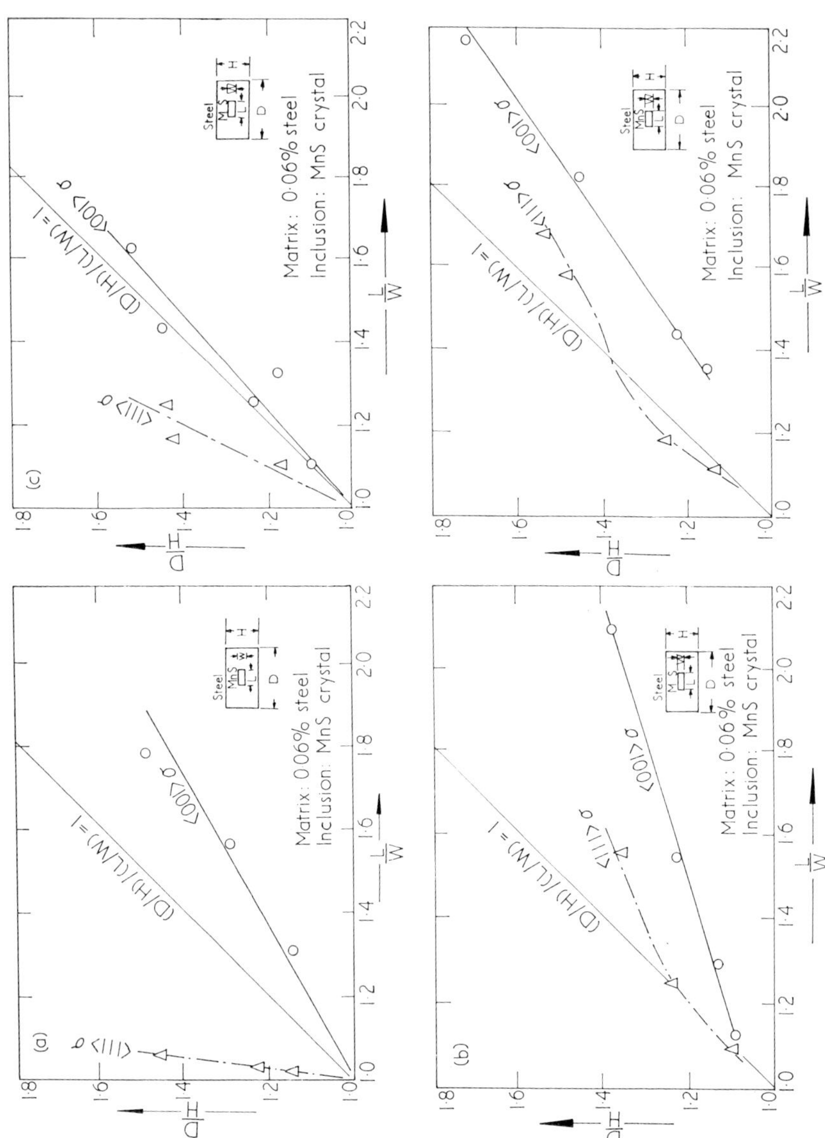

Small MnS inclusions with known crystallographic orientations were placed inside powder compacts of low-carbon steel. The deformations for the inclusions and the steel were compared after axial compression. ⟨001⟩ indicates an axial stress applied in the [001] direction of MnS, ⟨111⟩ a stress parallel to ⟨111⟩

218 Deformation of MnS inclusions v. deformation of steel at different temperatures and with different orientations

218a Deformation temperature 20°C. The MnS is harder than the steel at this temperature

218b Deformation temperature 260°C. The MnS is softer than the steel at this temperature

218c Deformation temperature 690°C. The MnS is harder than the steel at this temperature

218d Deformation temperature 950°C. The MnS is softer than the steel at this temperature

COMMENT MnS always deforms more than the steel when a [001] direction is aligned with the compression axis. The inclusions may deform more or less than the metal when a [111] direction is aligned with the compression axis, depending on temperature. (Compare with the hot hardness values, Fig.217). From ref.231

The principal deformation behaviour of MnS at different temperatures, as compared with other inclusion types, is given in Fig.211, and is discussed also in section Q 2 on machinability, in Q 3 on crack formation, in Q 4 on fatigue properties and in Q 6 on impact properties.

Q. The influence of non-metallic inclusions on the properties of steel

'An inclusion that may be damaging in one context may not be so bad in another.'

Russell 1962

Q 1 Introduction

The ultimate purpose of the continuous work on improvement of the cleanness of steel is naturally to obtain steel with improved properties. As far as *exogenous* inclusions are concerned the question is simple: they should be avoided. For *indigenous* inclusions the problem is much more complicated. These inclusions are structural components of the steel mainly being formed by the oxygen and sulphur present. Their quantity, size, shape, distribution and composition can be modified but the presence of these inclusions can never be entirely avoided. Furthermore, inclusion phases are not always detrimental to the steel properties, they may even improve certain properties. In the present section the author has tried to summarize available information concerning the influence of different inclusion types on steel properties. Unfortunately, much of the information in the literature on this subject is difficult to evaluate as the inclusion types have not been known. The author has mainly dealt with indigenous inclusions. The effects of exogenous macroinclusions have not been dealt with in this section as it is obvious that a large piece of foreign material which has been included in the steel in an uncontrolled way is detrimental to its properties.

Q 2 Influence of inclusions on machinability of steel

'I think that the study of inclusions over the next few years is going to become important from an entirely different angle, namely that of machinability.'

D. A. Oliver 1946

Machinability parameters

'Machinability', according to the Metals Handbook,[232] is the relative ease of machining a metal. The term 'machinability of steel' is difficult to define for a metallographer. It is a complex operation, depending on several factors such as the deformation behaviour of the different structural components of the steel, the properties of the different structural components of the tool, the exact nature of the machining operation, the tool geometry, speed, temperature, lubricant etc. This is also a field where the metallographer and the mechanical engineer tend to use a different language. The reader who is interested in an overall picture of machinability is referred to the papers from the joint conference on machinability in London.[233] In the present section machinability will be regarded only in relation to steel inclusions, and even with this severe limitation, only some general comments are possible with the present state of knowledge. A machining operation in general involves cutting of the steel with a tool under conditions which may vary over a wide range, depending on the exact nature of the operation. The factors of most general interest are the life of the tool, the quality of the finish of the generated steel surface and the magnitude of the cutting forces. The inclusions have an influence on all these factors, but their effect must be discussed in relation to other variables such as steel composition and structure and the extent of segregation, since these sometimes have a greater influence on machinability.

When discussing the influence of non-metallic inclusions on the machinability of steel, it is necessary to consider the metal cutting process in some detail. An excellent summary of the metallurgical features of metal cutting has been given by Trent[234] who also has contributed much to the understanding of this process, and his approach to the problem has been used in the present section.

In a metal cutting operation a tool in the basic form of a wedge is forced asymmetrically into the work material (Fig.219). A thin layer

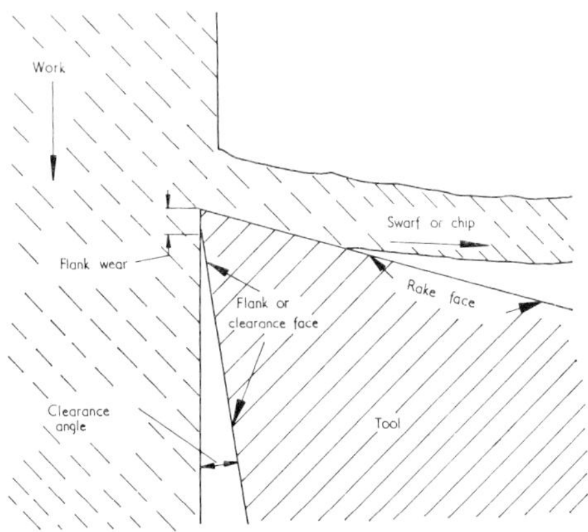

From ref.234

219 Basic features of the metal cutting process

is removed from the more rigid work material, bears in the form of fragments or a continuous ribbon on the rake face of the tool and passes over it. The more rigid body of the work material bears against and passes over the flank of the tool. Wear of the tool takes place on the flank (flank wear) and often also on the rake face, usually at a certain distance from the tool tip, where a crater forms on the rake face of the tool (crater wear). In steel cutting a fragment of the work material also often adheres at the edge of the tool throughout the cutting process, causing a built-up edge. The conventional description of the metal cutting process is indicated by Fig.220a. The work

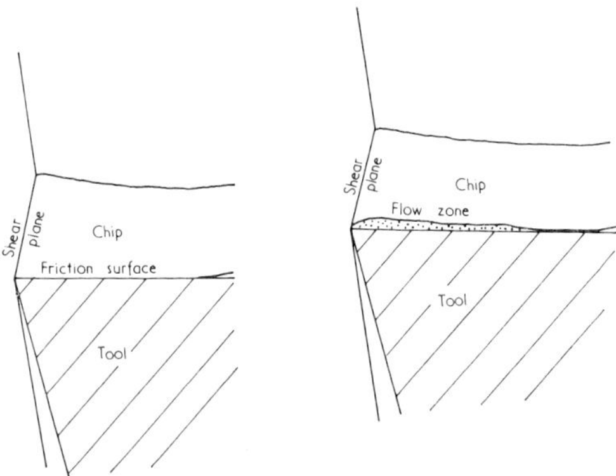

220 Left: Conventional terms which describe the metal cutting process, according to the friction 'surface' model

Right: Terms proposed by Trent for describing metal cutting process. The existence of a flow zone with extremely intense conditions of strain and temperature are essential. In this flow zone, the work material may be stationary or moving slowly at the tool surface but with the same speed as the body of the chip at a distance of only a few microns from the tool surface. From ref. 234

material is sheared along a shear plane to form the swarf, which glides continuously over the rake face of the tool. According to earlier theories, there exists a friction zone between the work material in the swarf and the tool surface. The new concept of Trent is, however, that no friction zone exists but that a condition of seizure arises between tool and work material in the swarf. Therefore a zone of intense shear exists near but not at the tool surface (Fig.220b). This zone is called the flow zone, and the conditions in this zone play an even more important role in the machining process than the conditions on the shear plane. This flow zone may be regarded as a built-up edge of material continuous with the work material and the flow pattern of the metal near the cutting edge at different cutting speeds is indicated in Fig.221. The flow zone has a thickness of 10–20μm,

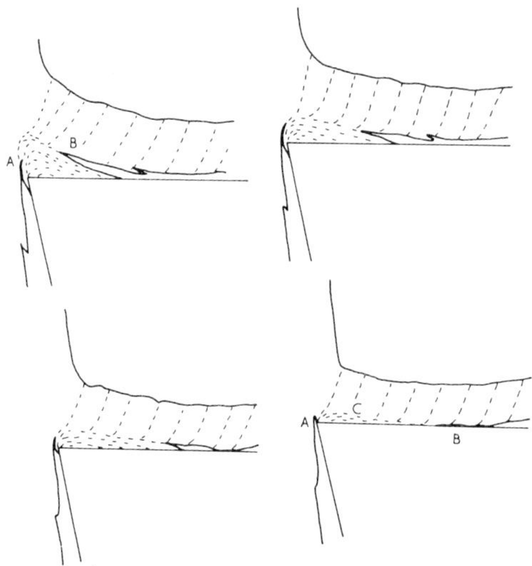

Flow zone changes from A–B to A–C–B. From ref.234

221 Built-up edge and flow pattern near cutting edge. The change with increasing cutting speed is indicated from above left to below right

the pressures are of the order of 100 kg/mm^2, and temperatures of 800°C and higher have been measured. Therefore the material in the flow zone under these high isostatic pressures and high temperatures can withstand an extraordinary amount of strain and behaves more like a viscous fluid than a normal plastic metal. The conditions peculiar to metal cutting are, according to Trent, that this pseudo-fluid layer can be maintained through the continuous generation of heat concentrated in a small interfacial area. Values of the shear stress in the shear plane and flow zone have been calculated for an alloy steel at various speeds (tool temperatures). The results are summarized in Fig.222, showing that the shearing stress in the flow zone was about $\frac{1}{2}$ to $\frac{1}{3}$ that in the shear plane and decreased as the cutting speed was raised.

When discussing the influence of different types of non-metallic inclusions on the machinability of steel, it is therefore essential to discuss their influence on the shear process of the steel both in the shear plane and in the flow zone at different temperatures. It is these two shearing processes that determine the behaviour of the steel during cutting.

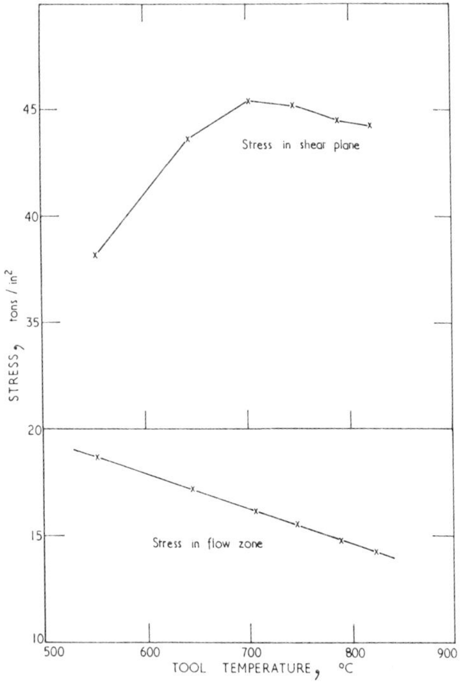

From ref.234

222 Shear stress in shear plane and flow zone of tools used for cutting an alloy steel at various speeds (Compare Fig.220)

An additional factor is of importance in the machining of steel, namely the tool life, and here also the non-metallic inclusions exert an important effect. Wear of the tool occurs at the flank and at the rake face. Measurements have shown that the temperature at the flank of the tool is several hundred degrees lower than at the rake face. Metallographic evidence shows that more than one wear process is involved in flank wear; the effect of non-metallic inclusions is probably mainly due to their abrasive effect when the edge of the tool generates the new metal surface of the work piece. The general physical properties of inclusions *at room temperature*, as well as their size and distribution, are therefore of importance.

The rake wear by cratering only occurs at relatively high cutting speeds (high temperatures) and results from a wear process based on diffusion. Different types of non-metallic inclusions may act as diffusion barriers and the behaviour of non-metallic inclusions at high temperature is of importance. Thin layers of different phases have been experimentally recorded on the rake face of the tools, having their origin in the steel inclusions.

Much work is at present concentrated on the effect of different types of non-metallic inclusions on the rake wear. The correspondence between methods of deoxidation of the steel in order to produce certain inclusion types and rake wear of the tool is intensively studied in order to improve tool life. Important contributions have been made by Opitz *et al.* and some of their results will be discussed below.[235-237]

The influence of non-metallic inclusions on the machinability of steel is complex. Inclusion phases, depending on their physical properties, may simultaneously both increase and decrease different parameters of importance for the machinability of steel. Before discussing the effect of different inclusion phases, the present author has therefore summarized the conditions, often conflicting, which an inclusion type should meet in order to improve the machinability parameters of steel.

1. The inclusions should act as stress raisers in the shear plane of the swarf so initiating crack formation and embrittling the chips. The chip–tool contact length then decreases which is of advantage. They should not, however, be such strong stress-raisers that the work piece cracks.
2. The inclusions should participate in the flow of metal in the flow zone, increasing the shear of the metal, but should not cut through the plastic flow of metal and thus damage the tool surface.
3. The inclusions should form a diffusion barrier on the rake face of the tool at the temperature of the tool/chip interface. This temperature depends on several variables, especially the cutting speed.
4. The inclusions should give a smooth work piece surface and not act as abrasives on the flank face of the tool.

Different inclusion phases influence these four conditions differently, to be elaborated on below. The overall influence of the inclusions on machinability is determined by such variables as

5. Inclusion quantity
6. Inclusion size
7. Inclusion shape
8. Inclusion distribution
9. Interinclusion distances.

Sulphide inclusions – MnS (K 3)

It is well known that an increased sulphur content of the steel is of advantage for its machinability; especially if the tool life is regarded as a criterion.[238] This is shown in Fig.223. Also in milling the same

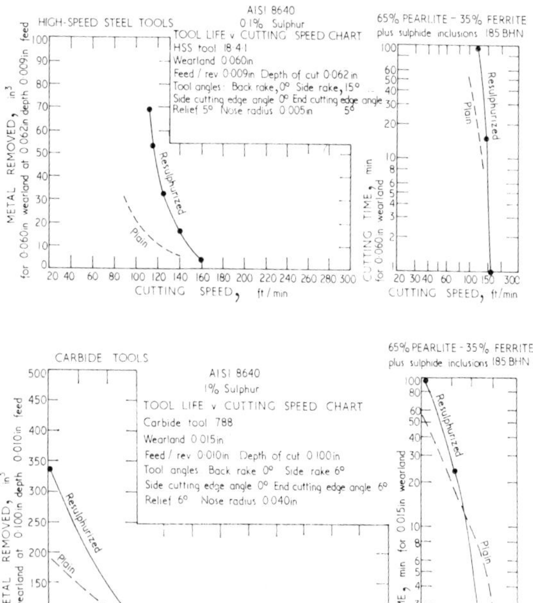

STEEL TYPE AISI 8640 with 65% pearlite and 35% ferrite. The resulphurized quality had 0·1%S
223a High speed steel tools
223b Carbide tools

COMMENT A greater tool life is possible for the resulphurized steel quality, particularly with high-speed steel tools. With carbide tools, the greatest tool-life gains occur in the region of lower cutting speeds. From ref.238

223 Influence or resulphurizing on tool life for different cutting speeds.

effect has been noted[239] (Fig.224). The free-machining or free-cutting steels have been developed in order to make machining of the steel easy by promoting the formation of small chips. They are resulphurized carbon or alloy steels with sulphur contents up to about 0·3%, and the main part of the sulphur is combined with manganese, forming MnS inclusions. Therefore sulphide inclusions of the MnS-type are considered to be of advantage for the machinability of the steel. As discussed in Section P 3, MnS has an index of deformability not too far from $\nu=1$ over a wide temperature range (Fig.211). When steel ingots with MnS-inclusions are rolled to bars, these inclusions are plastically deformed to ellipsoids with their longest axes in the

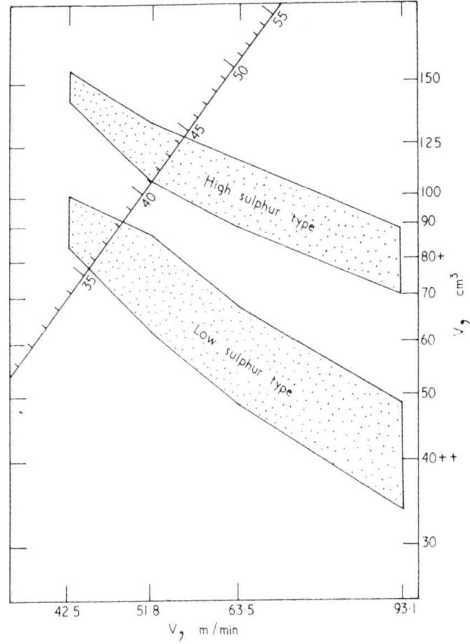

Vcm³ = amount removed for Vm/min feed. From ref.239

224 Influence of sulphur content on the machinability (milling) of case hardening steel

rolling direction of the bars (Figs.162a, 171a and b). If, for instance, such a bar is machined on a lathe, the ellipsoidal MnS-inclusions are deformed perpendicular to the long ellipsoid axes. The long ellipsoids are flattened along the shear planes of the steel chips, and their circular transverse sections (Fig.171b) are deformed to plate-like sections, following the shear plane of the swarf. This is shown in Fig.225. Depending on the original inclusion size, the thickness of these sections may vary. If the original MnS inclusions are small (type II), the ellipsoids may be flattened to nearly two-dimensional surfaces, whereas MnS-inclusions of type I usually give thicker sections.[217] A change in the plasticity of the MnS inclusions by for instance solid solutions of the (Mn,Me)S-type (K 4) or Mn(S,Se)- or Mn(S,Te)-type (L 1, L 2) also has an influence on the deformation pattern of the inclusions. Oxygen is also of importance, and duplex MnS-inclusions with oxides are not as plastically deformable as monophase sulphide inclusions.

The influence of MnS inclusions on machinability will now be discussed using the basic criteria 1–4 above. MnS-inclusions act as almost perfect stress raisers in the shear plane of the swarf (point 1).

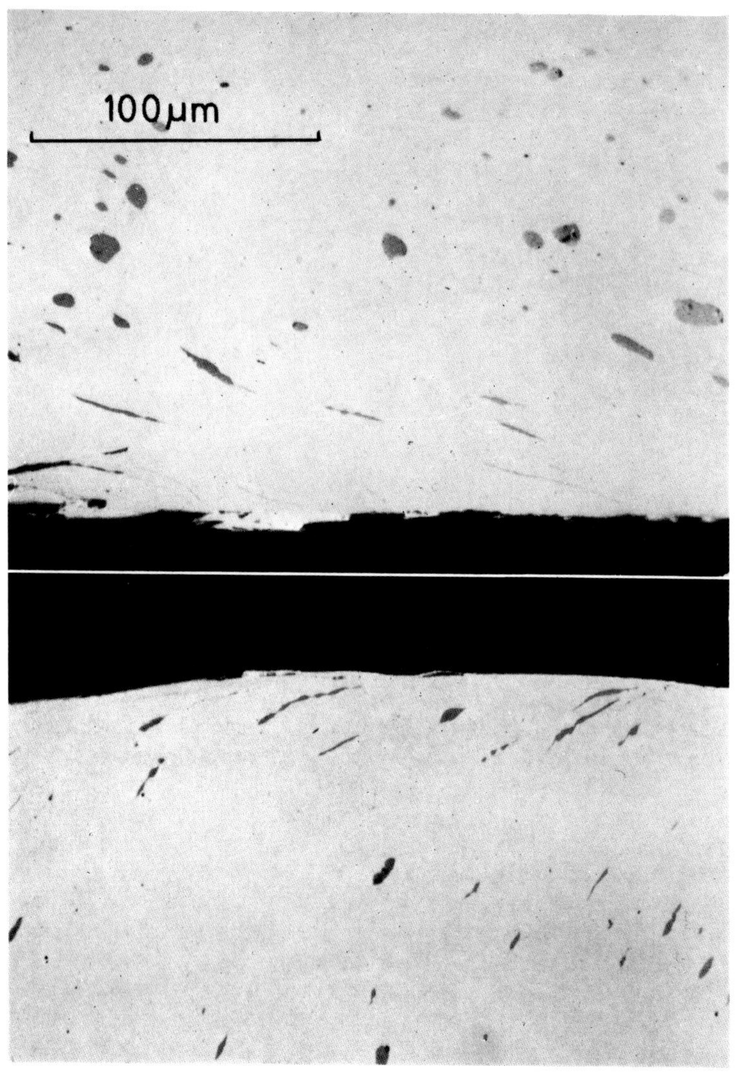

The figure shows the surface zone of the bulk material (above) and chip (below) at a high magnification. The free-cutting steel (0·1 %S) was turned on a lathe with the edge of the tool coming from the left in the figure. The MnS inclusions, which in the bulk material are spherical, participate in the flow of the metal, forming elongated plates of sulphide. (Dark material is bakelite)

225 MnS-inclusion, deformation occurring during machining of a resulphurized free-cutting steel.

This effect has been studied by several authors, for example Trent,[234] Shaw[240] and Rubenstein,[241] The last two have concluded that this is the main effect of sulphur, which influences the machinability of steel by virtue of its ability to embrittle. The stress raising effect depends on the shape of the inclusions, as well as on their size. Inclusions with elongated sections give higher values of the ratio between the maximum stress and the mean stress than do inclusions with circular sections and should therefore be advantageous, as long as they do not become too thin. Very small inclusions, or inclusions which do not deform along the shear planes of the steel, are not good stress raisers, whereas an increased hardness is advantageous. Several observations on the size and shape effect of MnS on machinability could be explained as due to their stress-raising properties. For instance, Boulger[196] has compared two nearly identical free cutting steels with a sulphur content of 0·20% but one with large, rather globular MnS-inclusions and the other with elongated stringers and plates. The former gave higher machinability ratings. Bearing in mind that the steel inclusions are cut perpendicular to the stringers during machining the explanation for the difference in behaviour should be that the globular MnS inclusions during machining deform along shear planes of the swarf to elongated, stress raising plates whereas the small stringers by further deformation become so thin that their stress-raising effect becomes negligible. This discussion is until now only of a qualitative nature, and no quantitative calculations have been reported, to the author's knowledge.

Similar observations on the favourable size effect of MnS on machinability have also been made by Gladman and Pickering,[242] whose relation between the number of MnS-inclusions and machinability in a free-cutting steel is given in Fig.226. The results show that the machinability is much improved by an increasing size (decreasing number) of MnS inclusions. Further observations on the size effect are also reported by Blank and Johnson,[127] who found that the machinability near the surface of a bar with extremely small MnS inclusions was inferior to that at the centre where they were larger.

Additions to sulphur-rich, free cutting steels also influence the morphology and properties of MnS. Kiessling *et al.*[243] have studied the influence of Se and found that the MnS-inclusions become larger and globular through Mn(S,Se) formation; giving an explanation for the improved machinability of stainless steels with Se addition. Bellot and Herzog[244] have found a similar effect on certain steels from small Te additions. The MnS-inclusions become harder through the Mn(S,Te) formation and the machinability is increased.*

* There is a small solid solubility of Te in MnS.[245]

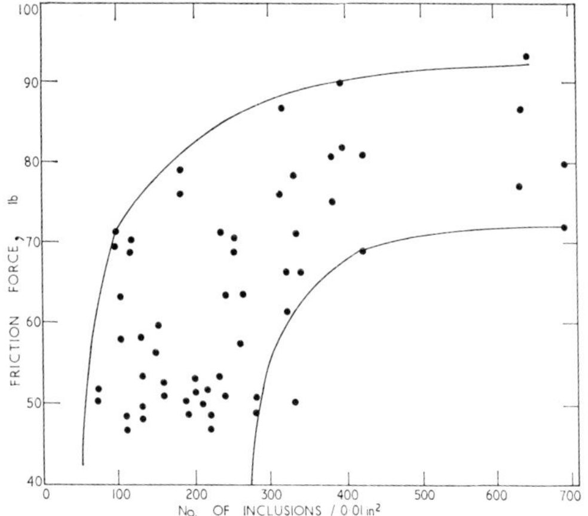

There is a constant volume fraction of MnS in the steel. The larger the sulphide particle size (i.e. the smaller the number of sulphides) the better the machinability. From ref.242

226 Relationship between machinability and number of sulphide inclusions in free cutting steel

This last observation is of special interest as it verifies the importance of a stress-raising effect of MnS for machinability since the MnS has become harder. The 'lubricating' effect of MnS, as suggested by Trent[234] but opposed by Shaw,[240] should therefore not exist.

A second criterion for the effect of inclusions on machinability is their behaviour in the flow zone of the steel (point 2). In this zone there is plastic flow of the metal and temperatures reach 800°C and higher. The stress-raising effect of inclusions is no longer the important effect since the shear stress is low and no chips should be formed. The most important feature of the inclusions in the flow zone is their ability to deform plastically with the steel over a *wide range of temperature*. MnS has the unique property among inclusion phases of a high index of deformability over a wide temperature range (Fig.211). The studies of hardness v. temperature for MnS by Chao *et al.*[109] shown in Fig.217, show a decrease in hardness from 0–900°C comparable to that of steel; that is, the index of deformability is approximately constant. Therefore MnS-inclusions should participate in the steel flow over the whole flow zone with its very high temperature gradient.

The possibility that MnS-inclusions from the steel will adhere to the rake face of the tool and deform plastically to a thin, protective diffusion barrier (point 3) has been experimentally verified at low and medium cutting speeds by Trent,[234] Opitz and Köning[235] and Wicher.[246] At high cutting speeds giving high temperatures the experimental evidence has shown that the MnS-layer disappears. It is known that the advantage in machinability of resulphurized steels over plain steels decreases with higher cutting speeds and finally disappears (Fig.223), and the two observations are therefore evidently related. A hypothesis by the present author is that at a certain critical temperature ($>900°C$) the plasticity of MnS in relation to steel changes, its index of deformability decreases and its enhanced effect on the machinability of steel disappears (compare the discussion on deformability of inclusions, section P 3 and Fig.211). The advantage of MnS-inclusions on the machinability of steel is therefore limited to low and medium cutting speeds.

Regarding the flank wear of the tool (point 4) MnS should be of advantage as compared with other inclusion phases due to its high index of deformability. As the temperature at the flank face is much lower than at the rake face, the influence of MnS on flank wear will not be noticeable even at high cutting speeds.

Summarizing the effect of MnS-inclusions on the machinability of steel, they are in general of advantage with regard to all machinability properties with the important limitation, however, that they only prevent cratering of the tool at low and medium cutting speeds. The shape and size of MnS-inclusions (points 5–9) are of great importance. Factors which have an influence on these properties, such as Se-, Te- and other additions, as well as variations in the deoxidation mechanism, have an influence which depends on steel type and tool composition. This is an important field for future research.

Oxide inclusions

These inclusions have usually been regarded as being detrimental to the machinability of steel. Recent progress in the methods of deoxidising steel, as well as the trend towards higher cutting speeds (higher temperatures in the flow zone) and new types of hard metal qualities for tool materials have, however, indicated that a difference should be made between different oxide phases.

Machinability is impaired by those oxide phases which never deform plastically at any temperature equal to or lower than the highest possible temperature in the flow zone of the steel; that is those which have an index of deformability near zero at all such temperatures. Important inclusion phases in this category are corundum (A 2), escolaite (C 2), spinels (A 7, F 3) and the calcium-aluminates (G 3). As discussed in section P 3 and shown in several figures (Figs.5, 55 and 114), these phases never deform at any temperature below the melting point of steel perhaps with exception of the spinels. They therefore do not fulfil any of the four basic points essential for inclusion phases which increase machinability.

Of *some advantage* are those oxide inclusion phases which deform plastically. As mentioned in section P 3, inclusions within certain composition ranges in the $MnO–SiO_2–Al_2O_3$ and $CaO–SiO_2–Al_2O_3$ systems have an index of deformability which is low at lower and medium temperature but which increases and even approaches $\nu=1$ in the temperature range of 800–1 200°C, temperatures which are achieved in the flow zone of the steel at high cutting speeds (Fig.211). Opitz and König[235] and Wicher[246] have shown that the crater wear and the thickness of the protective layer at high cutting speeds depend on the deoxidation practice, and their results are summarized in Fig.227.

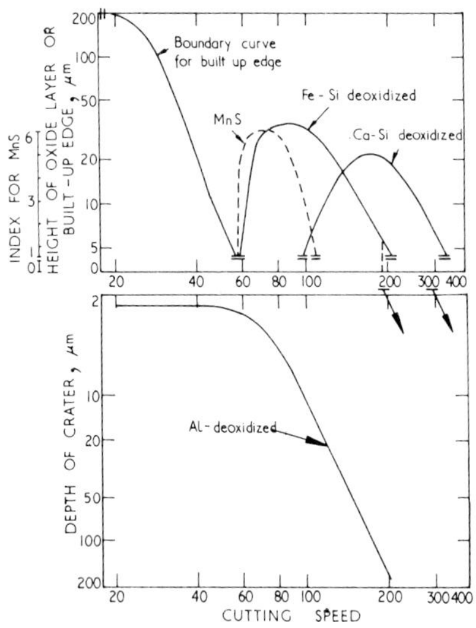

From ref.235

227 Effect of alloy deoxidant on cutting speed limits

Summarizing the effect of oxide inclusions on machinability of steel; only those which deform plastically are of advantage. These prevent crater wear (point 3, above) and are even more advantageous than MnS at higher cutting speeds in this respect. Their influence on the three other criteria for good machinability is more difficult to judge with the present state of knowledge. Their properties as stress raisers (point 1) are probably not very good, as the temperature in the shear plane of the swarf is below the temperature of plasticity. Their ability to participate in the flow zone of the steel at high cutting speeds (point 2) should be better than MnS in those parts of the flow zone where the temperature is high but worse in the other parts. Their ability to prevent flank wear (point 4) should only be better than MnS at very high cutting speeds. The recent results concerning the effect of deoxidation of the steel on its machinability indicate the increasing importance of control over the composition of the oxide inclusions in the steel. This will apply especially under more severe and extreme cutting conditions.

Finally, the general curves of the variation of the deformability index for different inclusion types with temperature, given in section P 3 (Fig.211), should be also of interest for machinability. They give an indication of the relative effect of different inclusion types on the machinability at different cutting speeds since the temperature in the flow zone increases with increasing cutting speed. The curves in Fig.211 also indicate that a third type of oxide inclusion exists, namely FeO and MnO, which deform plastically at lower temperatures but lose their plasticity at higher temperatures. These therefore might be of interest for medium cutting speeds between the range for MnS and for silicate oxide inclusions.

Q 3 Influence of inclusions on the fatigue properties of steel

> 'I feel that there has been a tendency to be rather emotional about the size of inclusions and to neglect other aspects. – Failure might be governed more by factors which influence the probability of an actual inclusion being at the point of maximum stress than by aspects of the inclusion itself.'
>
> *Sewell 1962*

The influence of non-metallic inclusions on the fatigue properties of steel has been studied mainly for high strength and ball-bearing steels. There has been much discussion as to how detrimental the inclusions are for the fatigue properties, and the concept of 'clean steel' has been much debated. In the past, all inclusions have been regarded as harmful and one of the chief causes of fatigue failure in steel. The present picture is much more complicated. Even if at present it is not possible to give a full theoretical treatment of the subject there is still much experimental information available on the influence of different inclusion types on fatigue. This section is a summary of what can be concluded, based mainly on the papers by Atkinson,[247] Duckworth and Ineson,[248] Johnson and Sewell,[249] Kawada *et al.*,[250] Murray and Johnson[206] and Uhrus.[75]

The number and dispersion of inhomogeneities in the steel play an important part in the initiation of fatigue cracks. The non-metallic inclusions act as stress raisers; cavities in the steel also have the same effect. Tensile strength is incompletely maintained by the inclusions or the cavities and the stress is concentrated in the surrounding steel in accordance with the shape of the defect: thus shear deformation and crack nucleation are induced in the steel. The appearance of a fatigue failure induced by an inclusion is shown in Fig.33. Excellent photomicrographs of fatigue cracks which have started at oxide inclusions are also shown by Uhrus.[75]

The difficulty when discussing the influence of inclusions on fatigue is that the fatigue life of a material is a statistical concept, one which has to be defined in a manner consistent with the wide dispersion generally observed in the fatigue life of individual test pieces. The usual methods of inclusion counting are also not sensitive enough to account for differences in the shape and type of inclusions. These have a considerable effect on fatigue life and therefore many of the results reported are conflicting. A development of inclusion counting methods has therefore been necessary, as summarized in ref.169, with special emphasis on the estimation of these inclusion parameters.

The results have shown that it is not only the number of inclusions which matters. Also their physical properties, especially plasticity, as well as their size, shape and position in the steel are of importance when discussing the influence of inclusions on the fatigue properties.

Influence of physical properties (*inclusion type*)

It seems reasonable to assume that inclusions which have a low index of deformability may induce fatigue cracks in the steel in two ways. One depends on their inability to transduce stresses existing in the steel matrix which means that critical peak stresses may be reached around these inclusions during the working conditions for the steel; that is, they have a direct nucleating effect on crack formation during the conditions of fatigue. This inclusion type may also have an indirect effect. If the inclusions have a low index of deformability during hot or cold working of the steel they may cause the introduction of microcracks at the steel/inclusion interface during manufacture, as discussed in sections P 3 and Q and shown in Figs.111, 113, 117 and 237. These microcracks, which thus are already present in the steel from the beginning of service, may then be the origin of later fatigue failure. In analysing the cause of a failure it is difficult to decide whether a fatigue failure was induced through crack nucleation at the inclusion during service, or crack propagation from a pre-existing microcrack. This difficulty constitutes an important reason why it is not yet possible to obtain a clear picture of the influence of inclusions on fatigue properties. From all experimental evidence it seems clear that inclusions which are brittle as compared with the steel and retain their shape during working processes ($\nu=0$) are much more detrimental to the fatigue properties than inclusions which deform plastically ($\nu>0$). Inclusions which have an index of deformability $\nu=1$ are the least harmful. Fig.211, which summarizes the index of deformability for different inclusion types, is therefore a convenient starting point for a discussion of their influence on the fatigue failure of the steel matrix.

Sulphide inclusions have a high index of deformability at all temperatures. No cracks are developed at the inclusion/steel interface during working of the steel and the inclusions participate in different stages of steel deformation, changing their shape according to the same pattern as the surrounding steel matrix. The bonding forces at the interface between the inclusions and the steel are never broken and there is no tendency for cavity formation, as visible from for instance Figs.168, 171, 172 and 175. This is in agreement with the experimental observations in refs.67, 75 and 206, that sulphide inclusions in ball bearing steel are not harmful for the fatigue properties. It is also reported that the presence of sulphide phase in the silicate inclusions decreases their detrimental effect on fatigue.

Oxide inclusions are in general harmful to the fatigue properties, but differences exist between different oxide inclusion phases and their size and position are also of importance. Those most detrimental to the fatigue properties are the spherical, non-deformable Ca-aluminate inclusions[247] and the Al_2O_3 inclusions,[206] that is, oxide inclusions with an index of deformability $\nu=0$. Both these inclusion types retain their shape during steel deformation. Cavities may be developed around the spherical Ca-aluminate inclusion as in Fig. 214c. The Al_2O_3-inclusions are often angular and similar cavities have not been reported but it has been observed that such sharp-edged inclusions are the nuclei of cracks. The siliceous types of inclusions seem to have an intermediate position between the non-deformable oxide inclusions and the plastically deformable sulphide inclusions, in accordance with their intermediate index of deformability. It is evident from Fig.211 that such inclusions change their index of deformability in the temperature range of 800–1000°C, depending on composition, from $\nu=0$ to $\nu \sim 0{\cdot}5$–1, and therefore may be plastically deformable at hot working temperatures. For instance, Murray and Johnson[206] have arranged the inclusions detrimental to fatigue in the decreasing order of Al_2O_3, SiO_2, (TiN) and further indicated that the presence of sulphide in silicate inclusions is of advantage and makes these inclusions less harmful to the fatigue properties. The effect of oxide inclusions on the flaking of rig tested bearings is shown in Fig.228 and the effect of Al_2O_3 inclu-

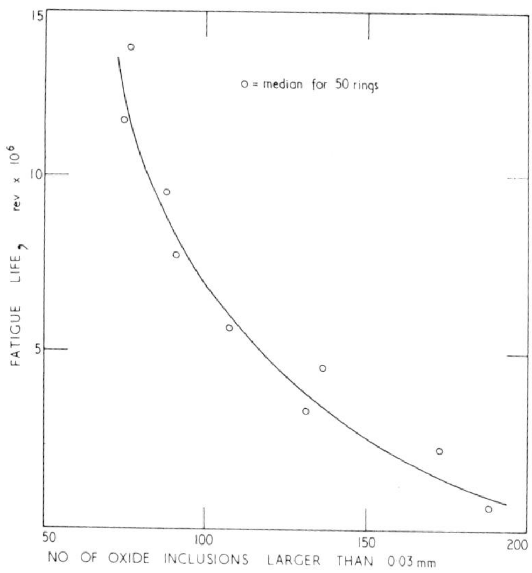

STEEL TYPE Through hardening steel for ball-bearings

STEEL ANALYSIS % 1C, 0·5Mn, 1·5Cr

228 The effect of number of large oxide inclusions on the flaking of rig-tested bearings (1309 outer rings). From ref. 75

sions on life is shown in Fig.229. For FeO inclusions the effect is unclear but it seems that these inclusions do not exert much influence on the growth of fatigue cracks.[251] It is the present author's opinion that this behaviour, which differs from that of other oxide inclusion phases, is related to the plastic properties of this inclusion phase. FeO has a higher index of deformability at lower temperatures than at higher, contrary to other oxide inclusion phases (Fig.211). At the comparatively low temperature of fatigue failure it therefore should behave in a manner more similar to the plastic MnS, with unbroken bonds between inclusion phase and steel matrix, than to other oxide inclusion phases which have a low index of deformability at the lower temperatures.

Large, exogenous oxide inclusions

These inclusions of slag or refractory origin are always detrimental

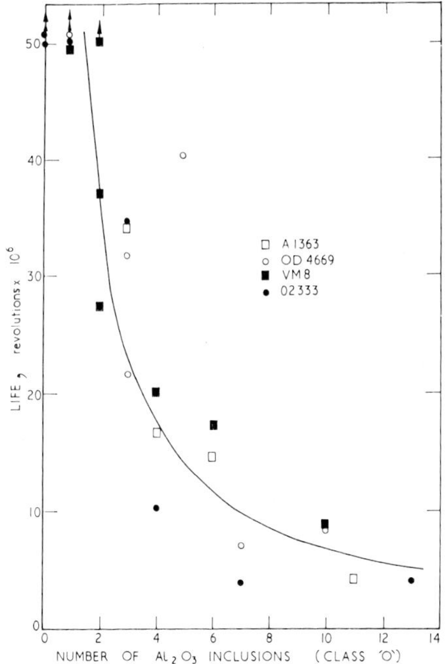

STEEL TYPE Standard basic electric arc steel, two air-melted and two vacuum melted casts

STEEL ANALYSIS % 0·95–1·05C, 0·3–0·6Mn, 1·3–1·6Cr

229 Effect of number of Al_2O_3 inclusions on life of rig-tested bearing steel. From ref.206

to the fatigue properties because of their size and irregular shape. An example of such an inclusion is shown in Fig.18. According to Uhrus,[75] this type of inclusion in through-hardening ball bearing steels is dependent on the steel process and the 'total length' of such inclusions is greater in basic electric steel than in acid OH steel. With vacuum remelting, the number of both micro- and macro-inclusions is further decreased and this is also reflected in the fatigue life, which is considerably increased, as shown in Fig.230. Vacuum-arc remelting seems to be of particular value for basic electric steel since the early failures are substantially delayed, no doubt because of the improved cleanness of the steel after the removal of large oxide inclusions.

Influence of size, shape and position of inclusions

A fundamental study on the influence of the size of Al_2O_3-inclusions

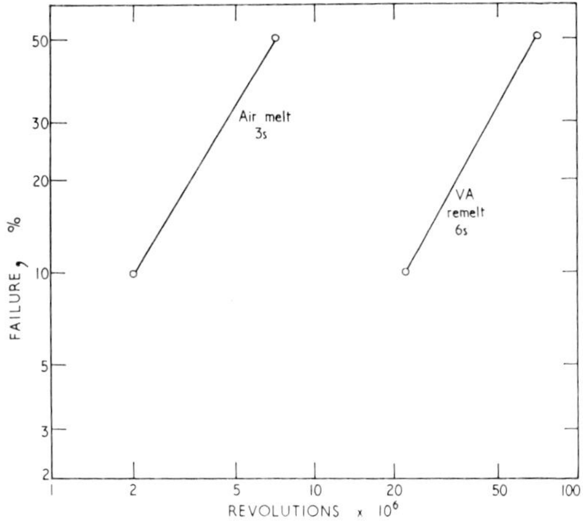

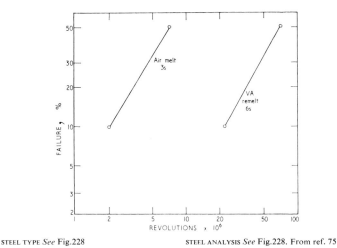

STEEL TYPE *See* Fig.228 STEEL ANALYSIS *See* Fig.228. From ref. 75

230 The effect of vacuum melting on fatigue life of basic electric ball-bearing steel (1309 outer rings, 30 bearings in each series)

on fatigue failure was carried out by Duckworth and Ineson[248] using artificially introduced Al_2O_3 inclusions. They found a correlation between inclusion size and fatigue life provided that the inclusions are not situated in the steel surface and that they are above a critical size. If so, the strength-reducing effect (K_f) of an Al_2O_3 inclusion is proportional to the cube root of the inclusion diameter (i) or

$$i = k \cdot K_f^3$$

The strength reduction factor K_f is the ratio of the fatigue limits in a steel without and with inclusions of a diameter i. The critical inclusion size was found to depend on the distance below the steel surface and increased from 10 μm just below the surface to 30 μm about 100 μm below the surface. If the inclusion size was below this critical value, no effect was observed on fatigue life. Kawada *et al.*,[250] in a study of the influence of inclusion size on the fatigue strength, also found a similar value for the critical inclusion size (8 μm) and an increase in this value with the inclusion distance from the surface. From a study of a series of fatigue failures in different bearing steels they obtained the values summarized in Fig.231 and were able to indicate the relation between critical inclusion size and depth below the steel surface for inclusions which could start a fatigue crack.

Regarding the shape of inclusions, it seems as if angular Al_2O_3 inclusions are more damaging in fatigue than rounded ones.[75] But

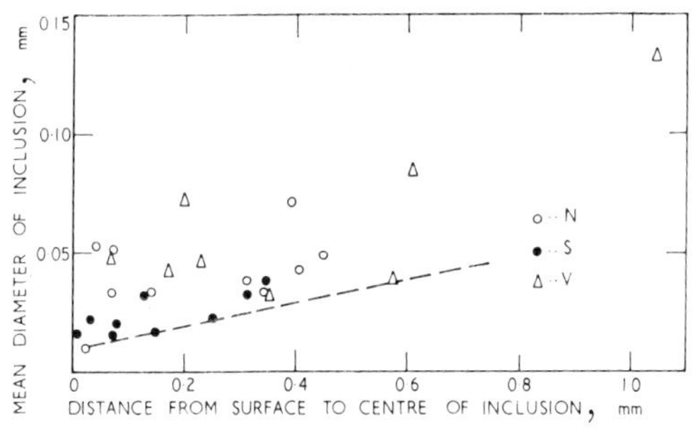

Steel Analysis

	%	C	Si	Mn	P	S	Cr	Ni	Cu
Steel	N	1·04	0·27	0·40	0·013	0·011	1·43	0·08	0·15
Steel	S	1·04	0·31	0·35	0·020	0·020	1·65	..	..
Steel	V	1·08	0·29	0·35	0·012	0·006	1·39	0·17	0·14

From ref. 250

231 Fatigue fracture of bearing steel due to inclusions in rotary bending. Distance from steel surface v. diameter of inclusions which have initiated fatigue fracture

it is also reported that spherical Ca-aluminate inclusions are very dangerous and initiate cavities whereas this has not been observed for angular Al_2O_3 or galaxite inclusions: compare for instance Figs.5 and 55 with Fig.214c. It therefore seems difficult at present to give an analysis of this parameter.

For cast steel, the influence of metallurgical defect size on the fatigue properties has been studied by de Kazinczy.[252] He found that micro-shrinkage cavities were the most frequent defects, followed by different types of inclusions (spherical oxides, (Ca,Mn)S phase and $CaO.6Al_2O_3$ slag layers), but no difference was found between the different types of defects with regard to their effect on fatigue properties. The fatigue behaviour of mild steel butt welds with slag inclusions has also been discussed.[253]

The 'white spot'

In fatigue fractures initiated by inclusions, an area with a roughly circular shape and of lighter colour than the remainder of the fracture is often visible, the so-called 'white spot'. Such an area is shown in Fig.33 with an inclusion remaining at its centre. The fracture appearance of fatigue failures without inclusions is different and therefore the white spot seems to be associated with the inclusion. Duckworth and Ineson have studied the size and frequency of such white spots, but no explanation of their nature was given.

The white spot in the fractured surface is a circular area of silvery brightness with a nearly plane surface and with the inclusion (or the cavity left by the inclusion) at its centre. Its brightness is, in fact, very similar to that from flakes in steel and according to Gunnarson[254] the white spot is caused by the hydrogen content of the inclusions. It is known from Troiano's work[255] that the solubility of hydrogen in steel increases with tensile stress. The inclusions are known to hold an amount of hydrogen in solid solution (residual hydrogen). They therefore serve as small internal reservoirs of hydrogen inside the steel matrix, and each time the surrounding steel is in tension, a small amount of hydrogen is released from the inclusion into the steel. This hypothesis explains the similarity between the white spot and hydrogen flakes. It also gives an explanation for the circular shape of this section of the fractured surface with an inclusion in the centre, from which hydrogen has diffused out.

Summarizing the present knowledge of the effect of inclusions on fatigue, it seems that it is the crack initiation characteristics of inclusions which are most important. Probably two main criteria must be fulfilled for an inclusion to be a potential source for a fatigue failure; namely a critical size, depending on the depth below the steel surface, and a low index of deformability at the actual temperature of fatigue. Dangerous inclusions are therefore single phase Al_2O_3, spinels, and Ca-aluminates with a size above about 10 μm, and the least harmful are MnS-inclusions. In multiphase inclusions it seems that the presence of a phase with a high index of deformability, such as MnS, modifies the dangerous properties of one with a low deformability index. For instance a sulphide rim around a Ca-aluminate particle as in Fig.112, may be of advantage.

The 'white spot' which is often visible in fatigue fractures initiated by inclusions is probably caused by the hydrogen content of the inclusions.

Q 4 Hot-shortness (red-shortness)

> Den allmännaste och säkraste orsaken till rödbräcka är vitriol eller svavelsyran, som uti mineralriket sällan tryter: som är den mest eldhärdiga och som till metalliskt järn har den starkaste attraktionen. – Detta är av allmän erfarenhet mera kunnigt, än att något bevis därtill tycks vara nödigt. . . .*
>
> *Rinman 1782*

* Free translation: 'Hot-shortness is caused by sulphuric acid which is abundant in nature, fire proof and strongly bound to iron. – This is so familiar from experience that no further proof is necessary.'

Hot-shortness (red-shortness) in steel is its tendency to brittle fracture in the hot forming temperature range. This phenomenon has long been attributed to sulphur, but oxygen has also been considered at least to enhance the effect of sulphur, whereas it is well known that manganese counteracts hot-shortness. The failure of the steel is related to its sulphide inclusions, and in its original and pronounced form, hot-shortness is caused by the precipitation of FeS in the austenite grain boundaries of the steel as described in section K 2. In the present section hot-shortness will be considered more fundamentally, and a practical example will also be given. The discussion is mainly based on the paper by Josefsson *et al.*[256] *See* also Q 5, hot tearing.

General

Hot-shortness is a fracture phenomenon, and therefore the fracture characteristics of metals have to be considered. The principal mechanisms for high temperature deformation of metals are

a) dislocation movement on slip planes within the grains

b) grain boundary sliding or 'viscous flow' at the grain boundaries.

When grain boundary sliding occurs shear stresses at grain boundaries are relaxed and, as the flow proceeds, stress concentrations are created at certain 'locked' points at the boundaries, e.g. at inclusions and triple points. The stress concentrations may initiate cracks that spread along the grain boundaries as shown in Fig.232. This type of

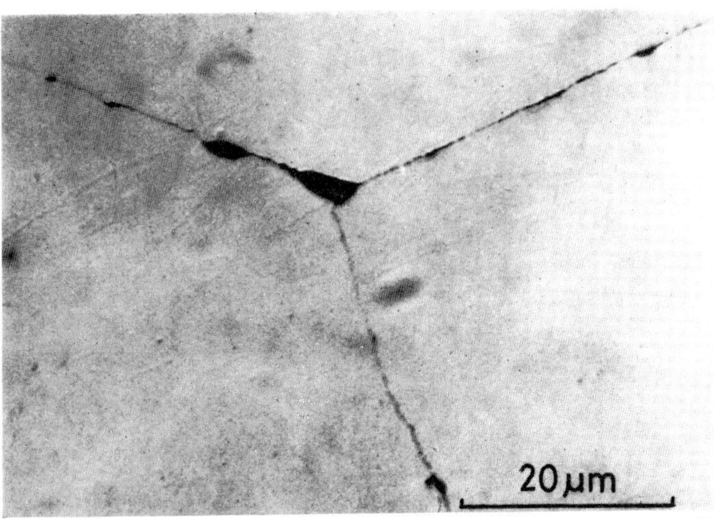

STEEL TYPE Laboratory ingot. Vacuum melted in 15 kg induction furnace. Heated to 1250°C for 15 min then transferred to a furnace at 850°C and held for 4 min

STEEL ANALYSIS % 0·59C, nil Mn, 0·035S, 0·010O

INCLUSIONS FeS in austenite grain boundaries

232 Hot-shortness. Cracks along austenite grain boundaries with FeS. From ref. 256

intercrystalline failure is likely to be favoured by an increase in the resistance to the dislocation movements within the grains as well as by a decrease in the shear strength of the grain boundaries. An increase in resistance to dislocation movements is caused by such phenomena as very fine precipitates or solid solution strengthening in the metal matrix. In contrast a decrease in shear strength of the grain boundaries may occur at high temperatures if phases or eutectics with a lower melting point than the metal matrix are present in these regions.

Sulphur

The maximum solid solubility of sulphur in pure iron is about 0·050% for the γ-phase and decreases with decreasing temperature. The solubility in α-iron is much lower. If gamma iron with sulphur in solid solution is cooled, sulphur may be precipitated in two ways; namely precipitation of FeS at the γ-grain boundaries and decomposition of the supersaturated solid solution to give intragranular precipitation of FeS on transformation to α-iron. The relative extent of precipitation in the two forms depends upon the cooling rate and both may occur simultaneously. Grain boundary FeS is shown in Fig.232 and the amount increases with increasing sulphur content. At high sulphur contents in pure iron, more or less continuous sulphide networks may be observed in microsections, outlining the γ-iron grain boundaries. As discussed in section K 2, FeS has a low melting point (1 190°C) and also forms low-melting eutectics in the temperature range of 900–1 200°C. In pure iron sulphur therefore induces hot-shortness in two ways. The formation of low melting-point phases at the grain boundaries decreases their shear strength, while solid solution of sulphur or, at low temperatures, sulphide precipitation, strengthen the austenite. Thus the tendency to crack formation in the grain boundaries is increased.

Carbon extends the γ-range of iron to lower temperatures and also lowers the sulphur solubility in γ-iron to about 70–80% of the value for pure iron. It therefore indirectly counteracts to some extent the effect of sulphur.

Manganese

If iron has manganese in solution, which is the case in almost all commercial steels, the solid solubility limit for sulphur in the γ-phase is considerably lowered. This is shown in Fig.233. The lower amount of sulphur in solid solution will then result in a lower resistance to dislocation movements within the metal matrix. Secondly, as discussed in section K 2, the sulphur will be bound to manganese instead

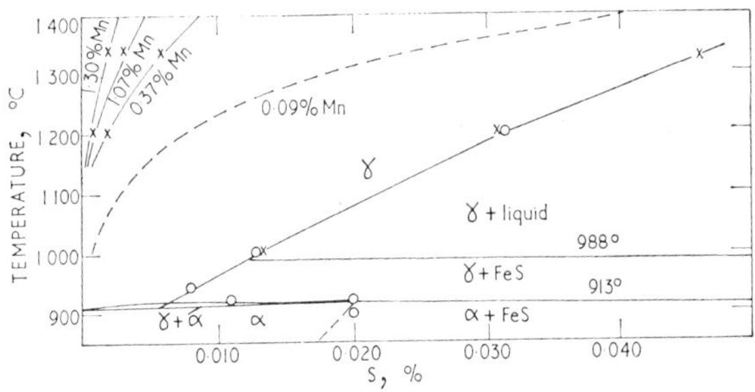

The curve for 0·09%Mn is computed from data in literature, assuming equilibrium with pure MnS. From ref.256

233 Part of Fe-S phase diagram including solubility curves for sulphur in austenite with 1·30, 1·07, 0·37 and 0·09%Mn

of iron and precipitated as MnS if the manganese activity is high enough. MnS has a higher melting point than FeS (1 610°C) and the possible MnS eutectics also have higher melting points than those with FeS. The shear strength of the grain boundaries is therefore not decreased as much as with FeS, and hot-shortness may be avoided if the manganese concentration is high enough. Manganese therefore counteracts hot-shortness in two ways by neutralizing both the two embrittling mechanisms attributed to sulphur. According to Anderson *et al.*[257] the concentration of manganese necessary to prevent hot-shortness is represented by the empirical formula

$$(\%\mathrm{Mn}) = 1{\cdot}25(\%\mathrm{S}) + 0{\cdot}03$$

This formula is only valid if the steel is low in oxygen.

Oxygen has a low solubility in γ-iron, and it therefore cannot be expected to increase the resistance to dislocation movements or to form oxide precipitates in the grain boundaries. It also to some extent lowers the sulphur solubility in γ-iron. It has been shown experimentally that it does not cause hot-shortness in manganese free iron at low sulphur contents. In the presence of manganese it counteracts to some extent the beneficial effect of that element since it lowers the effective manganese content in the austenite by forming (Fe,Mn)O inclusions, and thus promotes hot-shortness. Large oxide inclusions caused by inhomogenous absorbed oxygen will be very effective in developing local hot-shortness.

Heat-treatment

The thermal history of the steel has a marked influence on hot-shortness, and the effect of sulphur may be more or less eliminated by suitable heat-treatment.

Example
After hot-rolling of a rimmed carbon steel, surface cracks were observed at the edges of the rolled billets (Fig.234a). The steel had

STEEL TYPE Rimmed carbon steel. Hot-rolled billet
STEEL ANALYSIS % 0·04C, nil Si, 0·14Mn, 0·012P, 0·017S, 0·003N, 0·064O
COMMENT The oxygen as well as the sulphur activities are high in this low-Mn steel. Part of the Mn in the steel is bonded to oxygen in (Fe,Mn)O inclusions. The Mn activity of the steel is decreased, resulting in local hot-shortness

234a Hot-shortness. Surface cracks on a rolled billet

an analysis of (%) 0·04 C, 0·14 Mn, 0·012 P, 0·017 S, 0·003 N, 0·064 O.

A microscopical investigation showed the presence of several multiphase non-metallic inclusions of the type shown in Fig.234b, mainly in the interior of the billets. At the surface, smaller inclusions were present and oxidation along grain boundaries was observed at the surface cracks (Fig.234c). The composition of the different phases in the multiphase inclusions was determined by electron probe analysis, showing the presence of (Fe,Mn)O, (Fe,Mn)S, (Mn,Fe)S and $(Fe,Mn)_3PO_4$. The oxides at the grain boundaries near the surface were of the (Fe,Mn)O type with a high Mn-content. (Fe:Mn was about 1·5:1.)

The ratio Mn:S is low in this steel, but still sufficient according to the formula given above for steels low in oxygen, but the steel in

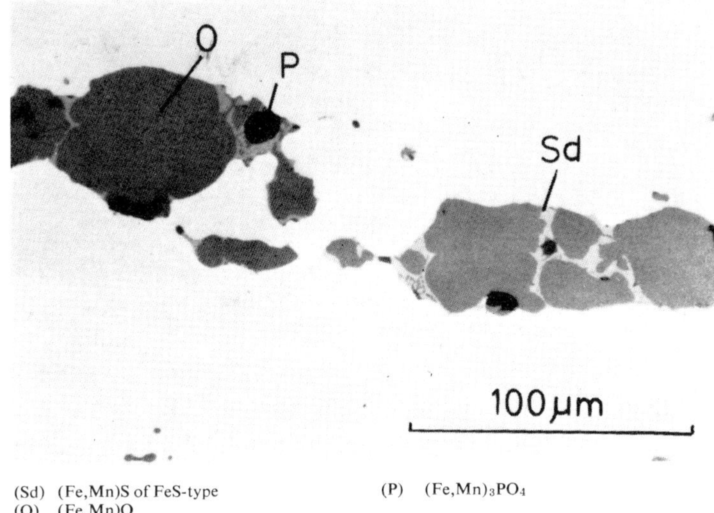

(Sd) (Fe,Mn)S of FeS-type
(O) (Fe,Mn)O
(P) $(Fe,Mn)_3PO_4$

234b Multiphase inclusions in the interior of the billet of Fig.234a

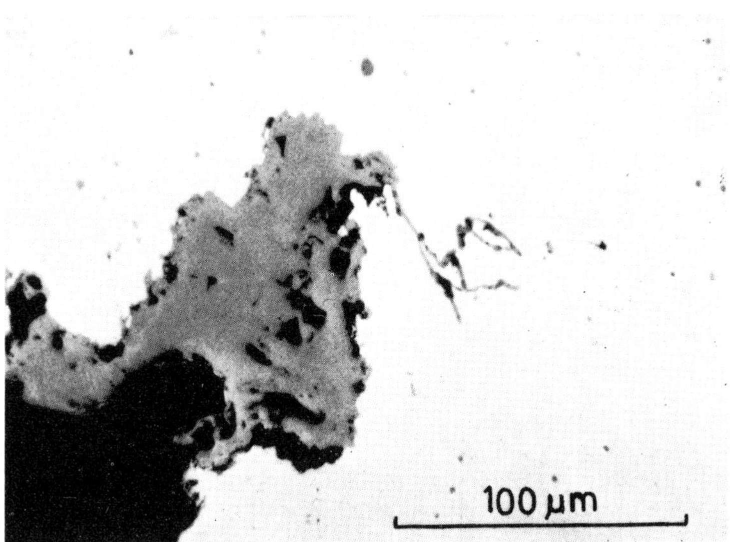

The oxide is of the (Fe,Mn)O type with Fe:Mn about 1·5:1

234c Grain boundary oxide at surface crack. Transverse section of the billet of Fig.234a

question was not deoxidized and had a rather high oxygen content. Part of the manganese in the steel was therefore bonded to oxygen in (Fe,Mn)O inclusions, lowering the manganese activity. The cracking was a result of hot-shortness. The phosphate phase was probably formed by a re-reduction from the slag phase and phosphate inclusions might have acted as nuclei for the other inclusion phases. (The inclusion shown in Fig.159 is of the same type and was therefore not formed from an oxide scale, as was indicated in that figure.)

Hot-shortness in the presence of manganese has also been observed in steels which have been annealed in oil-heated furnaces and exposed to the oxidation products from the oil which often have a comparatively high sulphur content. Oxide scales from such steels are shown in Figs.155–158.

Q 5 Influence of inclusions on various steel properties

In section Q 2–Q 4 a thorough discussion has been given on the influence of inclusions on machinability, fatigue and hot-shortness. The effects in relation to other steel properties have not received as much attention and only a few references are available. In the present section, the author has collected some of these literature references but only given very brief comments.

Surface finish

The lustre of a polished steel surface is sensitive to surface irregularities. The presence of different inclusion types in the steel surface therefore has an influence on the degree of finish attainable. A systematic study of the influence of deoxidation on the finish of high-chromium alloyed steel has been reported by Schöberl *et al.*[258] They tried seven different deoxidizing methods but did not study in detail the inclusions formed. According to their results, all the deoxidation methods, Fe–Cr–Si, Fe–Si, Ca–Si, Ca–Al, Al, Al in the ladle and Al+Ca–Mn–Si in the ingot mould, gave as a result globular oxide inclusions with diameters up to about 10 μm, often surrounded by sulphide scales (*see* Fig.112). The composition was not studied in detail, but the inclusions had different amounts of SiO_2, Al_2O_3, CaO, MnO and MgO, depending on the deoxidation practice. In spite of the high Cr-content of the steel, chromites (C 4) were not very common as inclusion components. The authors concluded that the nature of the inclusions is of little importance, but that their number is the important factor. This number is closely connected with the oxygen content of the steel sheet, and therefore those deoxidation methods which gave the lowest oxygen content (e.g. Al and Ca–Mn–Si) should be used in order to give the best surface finish.

Weldability

Non-metallic inclusions may be introduced into weld metal from several sources. They may originate from the parent metal, from the electrode material, from the non-metallic coating of the electrode or by entrapment of the electrode slag. Such inclusions are shown in Figs.126a–b and 130a–d, and some references to important trace elements are given in Table XXXVII. The inclusions present in the parent steel are also sometimes affected by the temperature cycle of the heat-affected zone. The effect of such non-metallic inclusions on the cracking susceptibility of the steel in the heat-affected zone and in the fusion weld has been discussed by Tremlett.[259] Possible effects are laminations in the unmelted heat-affected zone, burning and hot tearing from sulphides (below) as well as crack formation from larger oxide inclusions. However, because the non-metallic matter absorbed in the weld pool affects the weld chemistry, the situation is complex and no broad generalization can be made at present as to the practical significance of different inclusion types on weldability.

The influence of non-metallic inclusions on the fatigue behaviour of mild steel butt welds has also been studied.[253]

Hot tearing

When steels are welded, the temperatures in the heat-affected zone increase up to the melting range at the fusion boundary. Small microcracks sometimes appear at the prior austenite grain boundaries in the heat-affected zone which are similar to those observed in burned steels. Such microcracks are referred to as hot tears. The mechanism for the formation of such hot tears has been studied by Boniszewski *et al.*[260,261] They have shown that MnS inclusions present in the heat-affected zone can dissolve in austenite grain boundaries. The liquid MnS films at the prior austenite grain boundaries can lead to hot tearing when the heat-affected zone shrinks on cooling from the peak temperatures. Also phosphides and arsenides may be of importance. Compare hot-shortness, Section Q 5.

Q 6 Basic studies on the influence of inclusions on mechanical properties

In sections Q 2–Q 5 the influence of inclusions on such steel properties which are of direct technical importance was discussed. Those fields were selected for which it was possible to give a more or less complete picture of the properties of the composite product steel with inclusions. Fundamental studies on the different mechanical properties of iron and steel have as an important aspect the influence of defects and a determination of the critical defect size for failure. Non-metallic inclusions in the steel matrix should therefore also be discussed from this point of view and in the present section the author has summarized available information on this topic.

Tensile properties

Tensile properties of sintered compacts of iron with Al_2O_3–particles from 1–7 volume per cent and in the size range of 15–35μm have been compared with those of sintered pure iron.[262,263] Although the amount of non-metallic inclusions is higher than in steels, the results are of interest as they indicate the trend of the effect of inclusions. The effect of these synthetically introduced Al_2O_5 particles on the tensile properties of pure iron are summarized as follows (Fig.235):

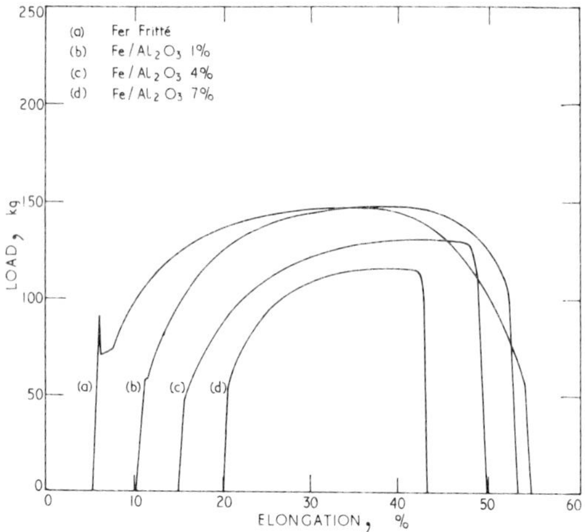

The Al_2O_3-inclusions were in the size range of 15–35μm and prepared synthetically. Compared with the tensile curve for sintered iron (a), an increase in Al_2O_3 content from 1 to 7 vol.% (b–d) results in the disappearance of the upper yield point, a decrease in the elastic limit and ultimate tensile strength and a general decrease in ductility. From ref.262.

235 Curves showing the tensile properties of sintered compacts of iron with different amounts of Al_2O_3

a) The upper yield point is suppressed
b) The elastic limit and the ultimate tensile strength decrease with an increasing number of Al_2O_3 inclusions
c) The ductility also decreases.

The effect of the particle size was found to be less important than that of the volume fraction of the non-metallic particles.

The particle shape also has an influence which can be concluded from studies of the tensile properties in transverse and longitudinal directions of rolled steel plates.[217] For steels with inclusions deformed to stringers or plates during rolling, the difference between transverse and longitudinal tensile specimens was much larger than for steels with spherical inclusions. This difference could be reduced if the stringer inclusions were broken up to smaller units.

Impact resistance
There are indications that the impact resistance decreases with an increasing amount of inclusions but that the transition temperature remains constant.[217, 263] The anisotropy of the notch toughness in rolled steels is closely related to the anisotropy in shape of the inclusions. It is more pronounced for sulphides and oxides which are deformed to stringers or plates than for non-deformed spherical inclusions.

Crack formation
It is a well known fact that large exogenous inclusions in the steel are likely to lead to the formation of cracks. It is usually a matter of routine control for the steelworks to prevent such large inclusions from occurring in the finished steel product. For instance, it is obvious that inclusions of the type shown in Fig.38 constitute an area of weakness in the steel which may lead to crack formation and fracture independent of the nature and physical properties of the inclusion phases. The problem is, therefore, to find the origin of these macro-inclusions and avoid them, whereas the crack formation and fracture in the steel phase is a trivial consequence of the large amount of foreign material present. This type of fracture will not be considered further in the present review. In recent years, however, increasing attention has been paid to the influence of the microinclusions on crack formation and crack propagation in the steel. These inclusions are normal constituents of the steel, which are always present but whose nature, shape, size and distribution may be influenced in different ways. Fundamental knowledge of the mechanisms by which cracks are nucleated in the steel by different types of non-metallic inclusions is therefore important for the development of high-quality steel.

There are different ways in which non-metallic inclusions can nucleate cleavage cracks in the steel. They depend on the relative plasticity between the inclusion phase and the steel phase.

Hard spherical particles, $v=0$ (for instance calcium aluminates, section G 3, Figs.111–113, 117). One way, in which such inclusions can nucleate cleavage cracks has been experimentally and theoretically considered by Hull.[264] Single crystals of Fe–3·25%Si were deformed in uniaxial tension at a strain rate of $10^{-4}s^{-1}$. At 20°K these crystals were completely brittle, cleavage being initiated at the intersection deformation twins. At 293°K the crystals were completely ductile, but at 77°K cleavage fracture was initiated after about 20% plastic strain by the formation of a longitudinal cleavage crack parallel to (001). The source of the crack was a spherical inclusion of about 2·5 μm in diameter. During plastic deformation of the steel, an elongated ellipsoid cavity is formed around the inclusion (Fig. 236). When a neck is formed in the tensile specimen, non-longitudinal

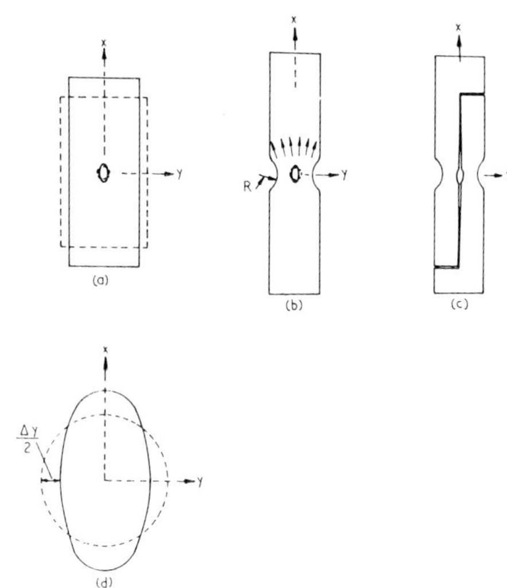

STEEL TYPE Single crystal of iron–3·25%Si alloy. The plane of the diagram is parallel to (110)

a. Uniform extension $E_x = -E_y$. A circle shown in section changes to an ellipse (*See* d.)
b. Formation of a neck and development of non-longitudinal normal stresses
c. Ellipsoidal cavity at the inclusion propagates as an (001) cleavage crack
d. Representation of wedge displacement produced by a hard unyielding particle in a plastic cavity. Compare Fig.237. From ref.264

236 Schematic representation of the sequence of events leading to fracture initiation at an inclusion

normal stresses successively increase and cause the ellipsoidal cavity around the inclusion to propagate as a cleavage crack along the (001) cleavage plane. If the shape of the spherical inclusion in the cavity is not affected by deformation of the iron, it acts as a wedge which tends to force the cleavage planes apart. For the inclusion studied in the present example Hull has estimated the presence of the inclusion being equivalent to introducing about 4×10^4 [001] edge dislocations between the cleavage planes. The process of cleavage crack formation by non-deformed spherical inclusions according to these experimental observations is, therefore, essentially one of slip in which the inclusion is effective in concentrating the plastic deformation. The unusual feature of the experimental observations is that the formation and propagation of cleavage cracks occurs along the axis of tension. Although these observations were all made on single crystals of iron–silicon deformed in tension, a similar analysis of stress distribution in polycrystalline steel deformed in different hot- and cold-working operations should soon be possible. This would throw new light on the formation and propagation of cleavage cracks due to undeformed inclusions. For instance, the cavity around the spherical calcium aluminate inclusion in a hot-rolled steel in Fig.237 is of a similar nat-

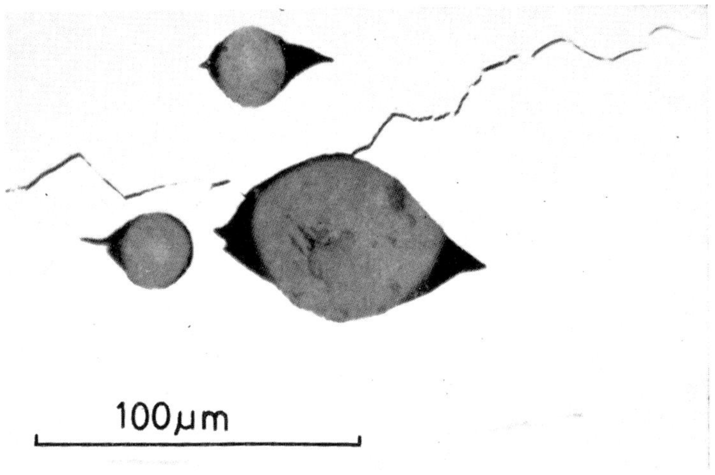

STEEL TYPE Unknown
STEEL ANALYSIS Unknown
INCLUSIONS Spherical calcium aluminate inclusions with 38CaO, $54Al_2O_3$, $8SiO_2$, trace MgO (CA-type, section G3)
MnS inclusions, elongated in the rolling direction of the steel are also present

COMMENT Compare Fig.236. Rolling direction of steel parallel to the dark, elongated cavities. The crack was associated with several inclusion cavities but it was not possible to determine, if it had been initiated as suggested in Fig.236

237 Wedge displacement in steel matrix, produced by a spherical calcium aluminate inclusion ($\upsilon = 0$) due to tensions during rolling

ure to the cavities discussed above. The early stages of cavity formation at spherical inclusions, forming a conical gap in the steel phase at the inclusion interface, is also discussed in section P 3, Fig.214.

Inclusions with a deformability index $0 < \nu < 0{\cdot}5$ (for instance silicates at lower temperatures). As was discussed in section P 3, these types of inclusions are deformed to some extent during steel deformation. Cavity formation is therefore not as pronounced as for the undeformed inclusions. The possibility that these particles act as wedges is also much smaller; they do not introduce the same high stresses in possible cleavage planes. Those inclusions with a deformability index around 0·2 or less are often deformed to incomplete ellipsoids, which may be a source for cavity formation in the steel as discussed in section P 3 (Fig.214). Further analysis of the role of these cavities in crack formation is at present not possible.

Inclusions with a deformability index $\sim 0{\cdot}5 < v < 1$ (for instance silicates at higher temperatures and MnS). These inclusion types usually deform plastically with the steel. As shown by several examples in section P 3 and Q 2 they should not be expected to cause crack formation, although they act as stress raisers in the slip planes of the steel.

Inclusions with a deformability index $v > 1$. (for instance MnS at certain temperatures ranges, molten silicates.) The general attitude is to regard such highly plastic inclusions as being of advantage for steel deformation. They do not act as crack inducers in the parent steel phase, if such phenomena as grain boundary embrittlement due to molten phases are excluded from the discussion. However, tension cracks may arise from axial compression in an iron matrix with highly deformable inclusions. This has been shown for sintered iron compacts with glass as an inclusion phase.[230] The lateral displacement of the metal by the highly deformable inclusion phase may introduce tensile forces in the metal phase above and below the inclusion which may lead to tension cracks in the metal (Fig.238).

As discussed in Sections P 3 and Q 2, much fundamental work was also carried out by v.Vlack *et al.* on the fracturing and deformation of the important MnS-phase which relates to the role of this phase in metal fracture. The MnS deforms plastically at hot-working temperatures, but during cold working considerable fracturing can and does occur which is influenced in extent by the orientation of the MnS-inclusions. The fracture patterns for MnS have been analysed.[229] The authors conclude that although fracturing may be initiated within the MnS inclusions by compressive loads, it is premature to conclude that these fractures will continue into the surrounding steel. For example, in Fig.239 inclusions with fractured MnS are shown, where no visible continuation of the fractures is visible in the steel phase. The fractured inclusions, however, serve an important role in initiating failure of the steel when subsequent stressing is reoriented from the initial compressive pattern. Fracture

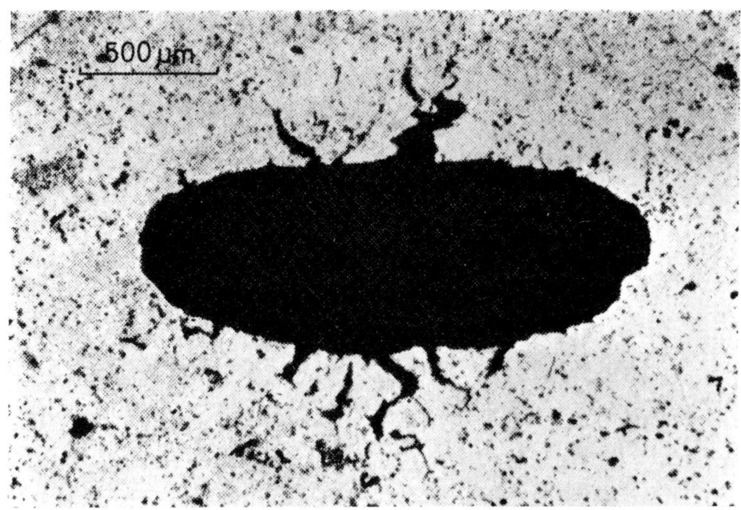

The total specimen received 20 per cent axial deformation at 940°C, but the soft glass underwent 50 per cent axial deformation. The lateral displacement of the metal by the highly deformable glass introduced tensile forces above and below the glass. From ref.230

238 Tension cracks in iron, arising from axial compression (vertical) around a highly deformable inclusion phase (glass, $\upsilon > 1$)

surfaces in sintered iron–Al_2O_3 compacts have been studied by means of microfractography.[262] The fracture was initiated by void formation at inclusions either at the boundary surface between the iron matrix and the inclusions or as a brittle fracture in the inclusions.

Fracture toughness

The ability of a material to arrest an existing crack and prevent the onset of rapid crack propagation at stress levels below the yield stress is called fracture toughness. [265, 279–281] This new concept will no doubt have a great influence on the future research on non-metallic inclusions, as the inclusions constitute an ever-present number of existing defects in the steel. When enough information is available on those material constants which determine fracture toughness, it should be possible to calculate a critical size limit for different inclusion shapes and positions in a steel. Inclusions larger than this limit constitute a dangerous flaw in the steel, independent of their nature. At sizes below this limit, however, the beneficial properties of certain inclusion phases could be used for improving the properties of the composite product steel with inclusions.

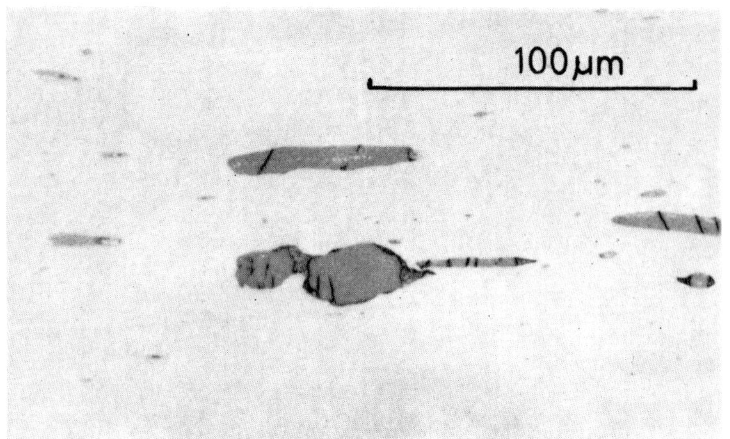

STEEL TYPE Free cutting high-sulphur steel. From billet
STEEL ANALYSIS % 0·09C, 0·60Mn, 0·06P, 0·30S
INCLUSIONS MnS with fractures. Small amounts of silicate are also visible, mainly at the tips of some MnS inclusions
COMMENT MnS inclusions usually deform plastically at the hot-working temperatures of steel production, but relative deformation of the inclusions and the metal is closely dependent upon MnS orientation and relative hardness of the two phases. (Figs.217 and 218). It was not possible to establish the reason for the MnS fracturing in this sample, as the exact rolling temperatures of the steel were not known, but the figure shows that fractures in the MnS inclusions do not necessarily continue into the steel phase

239 Fractures in MnS inclusions in rolled steel, which have not continued into the steel phase

Precipitation phenonema and grain growth

The discussions in earlier sections have dealt mainly with the influence on steel properties of those inclusions which are clearly visible in the microscope and which can be analysed with the electron probe. Therefore the discussions have been limited principally to inclusions with a diameter larger than about 1 μm. It is conceivable that the direct influence of these large inclusions on steel properties is the most important and that the main research effort at present must be concentrated on their effect. However, as mentioned in section Q 2, Fig.188, about 98% of all oxide inclusions are smaller than 0·2 μm. Although these small inclusions only represent about 1–2% of the total oxygen in the steel, the presence of such a great number of particles from other phases must have a fundamental influence on many phenomena in the steel matrix. These submicroscopic particles may act as nuclei for precipitation of other phases such as carbides, hydrides, nitrides and borides. Many steel properties which are dependent on the formation, size and distribution of these important compounds of non-metallic elements may therefore indirectly be related to the submicroscopic non-metallic oxide inclusions. No systematic research has been reported but some observations are of interest. As a result of electron microscope studies on the embrittlement in an Fe–Cr alloy it has been observed that preferential precipitation of Cr_2N occurs not only at grain boundaries and dislocations but also at submicroscopic inclusions[266] (Fig.240). Another

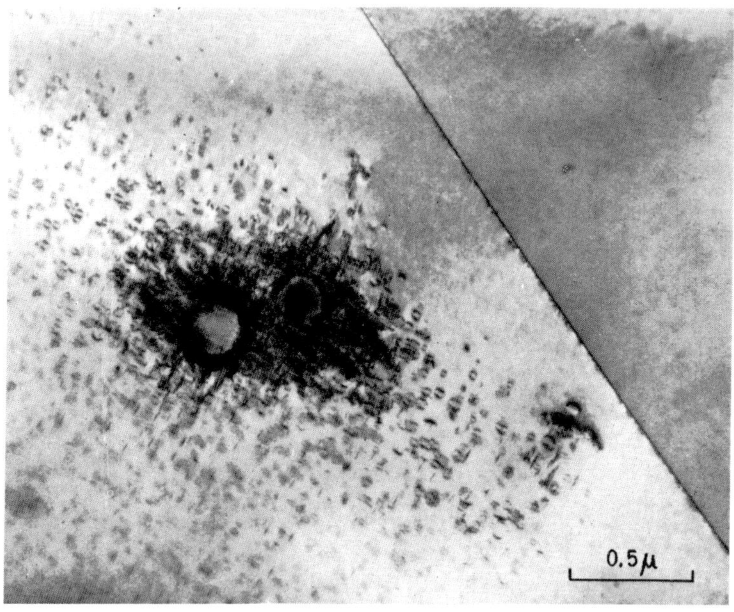

STEEL TYPE Fe–Cr alloy, induction melted and remelted in a high-vacuum consumable arc furnace. From hot forged bar
STEEL ANALYSIS % 0·006C, 0·05Mn, 0·004P, 0·010S, 29·8Cr, 0·03Ni, 0·10Mn, 0·01Co, 0·01Cu, 0·004Al, 0·025N, 0·045O
COMMENT The alloy was annealed for ½ h at 1050°C, quenched and aged for 1h at 550°C Transmission electron micrograph. From ref. 266

240 Preferential precipitation of Cr_2N at inclusions

example has been given by Langer, [267] who studied the quench aging process in iron. Transmission microscopy of thin foils from a 0·02C, 0·0186N, 0·37Mn-iron alloy showed that ϵ carbide aggregates sometimes nucleated at non-metallic inclusions in the matrix. Aggregates of this hexagonal iron carbide were observed after long aging times at temperatures where only this carbide was present, and their perfect symmetry was a result of nucleation at non-metallic inclusions. Such aggregates are shown in Fig.241. Both these observations are important for future work on the smallest size ranges of non-metallic inclusions.

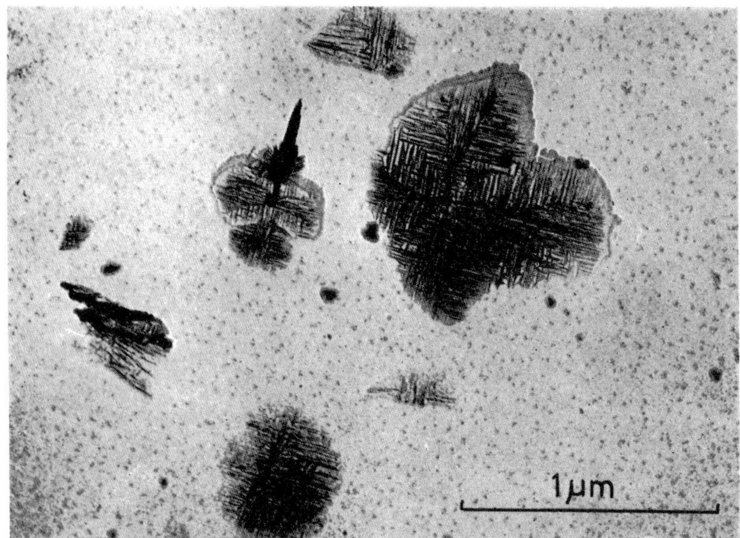

STEEL TYPE Converter steel, from rolled bar
STEEL ANALYSIS % 0·02C, 0·37Mn, 0·0186N
COMMENT The steel was normalized at 950°C, cooled in furnace to 680°C, held there for 1h and quenched. Aged at 206°C for 215 min. Transmission electron micrograph. From ref.267

241 Preferential growth of ϵ-carbide at inclusions

Grain growth and grain boundary movement in the steel are also phenomena which may be influenced by the presence of non-metallic phases. This has been investigated mainly for carbides and nitrides, but oxides and sulphides should also have an influence. Some effects have been observed by comparing air-melted and vacuum-treated steel.[124,268] A summary of the influence of inclusions on austenitic grain growth is given by Eckstein and Lüdemann.[269]

R. Additions to Parts I and II

A 2 Corundum

Alpha phase (α), *corundum*, is the stable modification of Al_2O_3. It was discussed in section A 2 and is that modification which appears in non-metallic inclusions, either formed directly or as the end product of transformations from hydrates or metastable modifications during the early stages of deoxidation as discussed in Section O 3.

Several metastable modifications of Al_2O_3 have been reported, but their true nature and their relations are still unclear. Summaries of the nature of these phases have been given.[197, 270, 271] They have formed the basis of the note given below on aspects of importance for the information of inclusions. d-spacings and crystallographic parameters from most of the modifications are given elsewhere.[197]

Beta phase (β) is not a pure Al_2O_3 modification, it is a compound.
Gamma phase (γ) has a spinel type lattice and forms solid solutions with these important inclusion phases. It is also an intermediate dehydration product from the hydrates bayerite and boehmite and may occur in the deoxidation process of steel with aluminium. The phase relationships are set out below.
Delta phase (δ) has been suggested as a possible intermediate dehydration product from bayerite, but its existence is not definitely established.
Eta phase (η) is a cubic phase, probably without importance for inclusion formation.
Theta phase (θ) is a monoclinic phase, isostructural with β-Ga_2O_3. It is a dehydration product from bayerite and may occur in the deoxidation process of steel with aluminium.
Kappa phase (κ) is a dehydration product of gibbsite and may occur in the deoxidation process of steel with aluminum.
Xi phase (ξ) is not a pure Al_2O_3 modification.

Chi phase (x) is a dehydration product of gibbsite, which may occur in the deoxidation process of steel with aluminium.

According to Adachi[197], if hydrogen is present in the molten steel, different hydrates of alumina are formed at the moment of the addition of metallic aluminium during deoxidation. These then transform to α–Al_2O_3, corundum, via different dehydration paths, as follows:

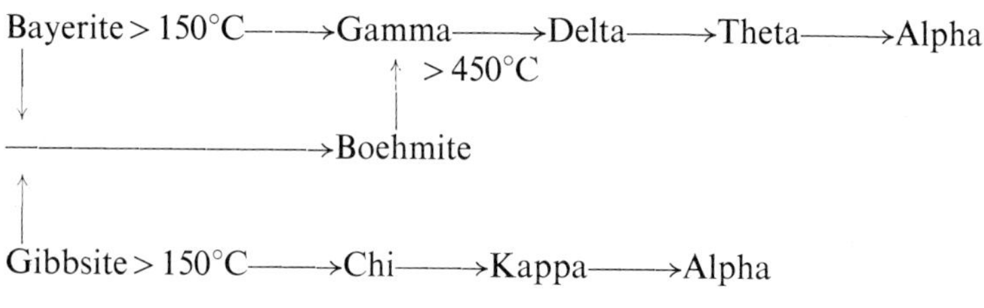

A3–5: SiO_2

The transformation temperatures were omitted. They are:

quartz $\xrightleftharpoons{573^\circ C}$ high-quartz $\xrightleftharpoons{870^\circ C}$ tridymite $\xrightleftharpoons{1470^\circ C}$ cristobalite $\xrightleftharpoons{1723^\circ C}$ melt

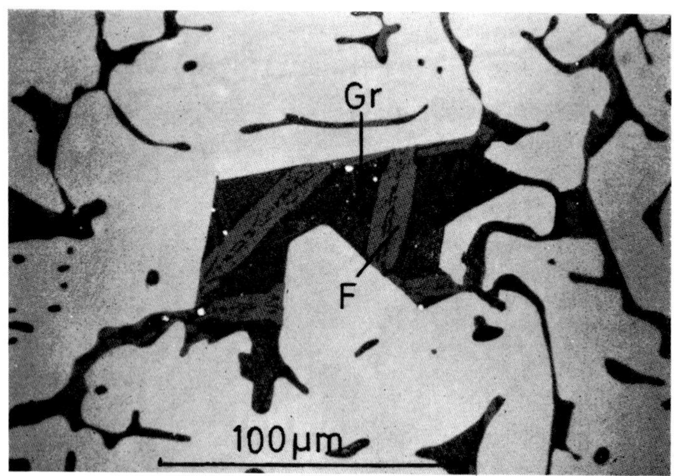

MATERIAL Magnetite electrode with approx. 90% Fe O_4 and 10% siliceous binder
MATERIAL ANALYSIS Silicate % $75SiO_2$, 6·7Pb, 8Ca, 4·8Al, 2·5 Na_2O, 0·5K_2O, 0·5Zn, Cu+, Ba+, Mg+
BINDER (Gr) Grunerite with 24FeO, 12CaO, 5PbO, $52SiO_2$, $3Al_2O_3$
(F) Fayalite with 63FeO, 3CaO, $30SiO_2$
(M) Matrix with 19FeO, 13PbO, $58SiO_2$, $5Al_2O_3$
(Me) Metal with 7Fe, 9Pb, 85Cu
COMMENT The $FeSiO_3$-phase has been stabilized by CaO. *See* B4

242 Grunerite ($FeSiO_3$, B4) and fayalite (Fe_2SiO_4, B3)

A 6: MnO
Solid solutions of the type $(Fe,Mn)_{1-y}O$ have been systematically studied with regard to the vacancy concentration.[272]

A 8: Mullite
A comprehensive study on mullite has been published.[273]

B 4: Grunerite, '$FeSiO_3$'
It has been reported that pure $FeSiO_3$ can be synthesized at pressures of 18–45 kb at temperatures of 1150–1400°C. Two modifications, one monoclinic and the other orthorhombic, have been reported.[274] A microsection of grunerite with CaO in solid solution is shown in Fig.242.

B 5: Hercynite
An inclusion with hercynite is shown in Fig.243.

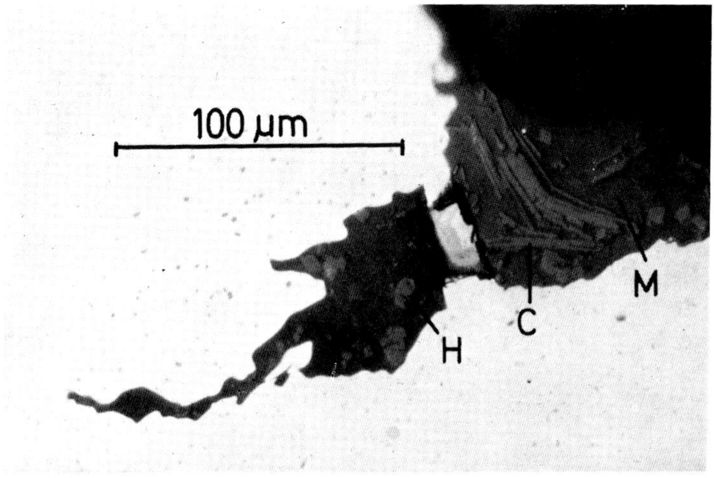

STEEL TYPE Unknown
STEEL ANALYSIS Unknown
INCLUSION:
(H) Hercynite with % 5MnO, 43FeO, $48Al_2O_3$
(C) Corundum
(M) Matrix with 11MnO, 5FeO, $48SiO_2$, $23Al_2O_3$
COMMENT The hercynite phase is similar to other double oxides of the spinel type, for instance galaxite (A 7)

243 Oxide inclusions in steel: Hercynite (B 5) in oxide scale

G 3. The calcium aluminates
Since part II was published, several inclusions with CA_6 have been

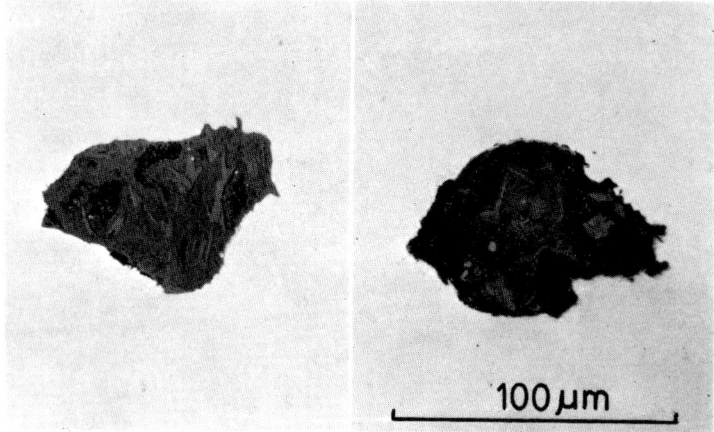

STEEL TYPE Carbon steel. deoxidized with SiCa in the ladle and Al in the mould. From rolled billet
STEEL ANALYSIS % 0·30C, 0·22Si, 1·03Mn, 0·027P, 0·018S, 0·010N
INCLUSIONS After etching in 2%NaOH (*see* ref.9), the three different aluminates could be separated.
Dark phase is CA with 32CaO and $59Al_2O_3$
Grey areas are CA_2 with 20CaO and $73Al_2O_3$
Light grey precipitates are CA_6 with 8CaO and $91Al_2O_3$

244 Oxide inclusions in steel: the calcium aluminates CA, CA_2 and CA_6 (G 3)

found in steel and the phase has also been observed in cast steel.[252] Etching with NaOH or HF according to the methods given by Trojer[9] is a useful method of distinguishing between the different calcium aluminates. In Fig.244, two inclusions with the phases CA, CA_2 and CA_6 are shown, which have been etched with 2% NaOH.

L 2. Inclusions with tellurium

Te is only slightly soluble in MnS, whereas Se shows a complete solid solubility (L 1). In free cutting steels with both Te and S, multiphase inclusions with MnS and MnTe are often formed.[244,275] Such inclusions are shown in Fig.245. The MnS phase often has Fe and Cr in solid solution, as well as a small amount of Te. MnTe often has small amounts of S and Cr in solid solution. MnTe has a low melting point as compared with MnS and MnSe (about 1 150°C[276]). In steels deformed above this temperature, this telluride is therefore molten and inclusions of the types shown in Fig.245 are often highly deformed. Salter *et al.*[277] have observed inclusions with the phases MnTe, FeTe and PbTe in different types of free cutting steels.

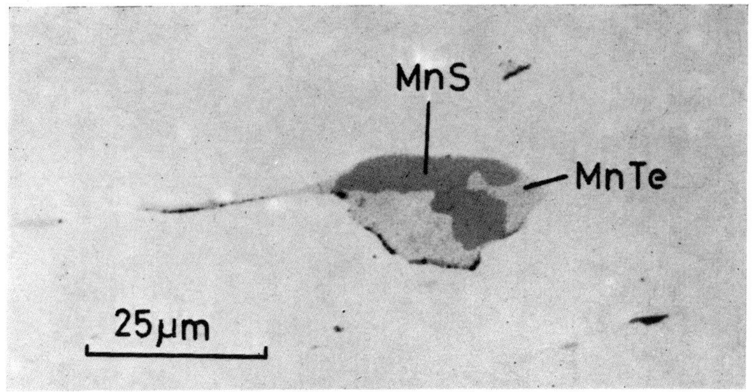

245a

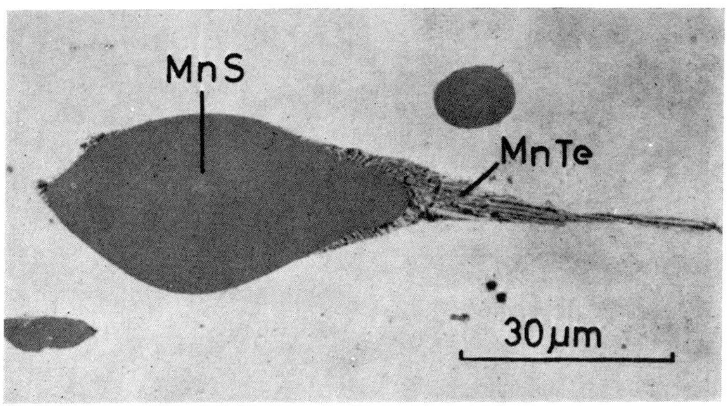

245b

STEEL TYPE (*a*) High-chromium steel with tellurium, ferritic (*b*) Mild steel, resulphurized with tellurium. From billets

STEEL ANALYSIS % (*a*) 0·125C, 0·73Mn, 0·415Si, 0·016P, 0·016S, 0·13Ni, 16·7Cr, 0·10Te
(*b*) 0·10C, 1·14Mn, 0·025Si, 0·074P, 0·342S, 0·040Te

INCLUSIONS (*a*) MnTe with 36Mn, 2·5S, 62Te, MnS with 50Mn, 34S, 1·2Te, 3·5Fe (*b*) MnTe with 29Mn, 3S, 60Te, 5Cr MnS with 53Mn, 36S, 2·5Te, 5·5Cr, Fe+ In the eutectic structure with MnTe there also is a silicate phase with approx. composition 23MnO, 3FeO, 60SiO_2, 9Cr_2O_3, 4Te (?)

245 Telluride inclusions in steel

REFERENCES

167. R. Kiessling and N. Lange: 'Non-metallic inclusions in steel', Part I, *ISI Spec. Rep. 90*, 1964, London.
168. R. Kiessling and N. Lange: 'Non-metallic inclusions in steel' Part II, *ISI Publ. 100*, 1966, London.
169. 'Clean Steel', *ISI Spec. Rep. 77*, 1963, London.
170. S. Bergh: *Jernkont Ann.*, 1962, **146**, 748–762.
171. E. Plöckinger: in *ISI Spec. Rep. 77*, 1963, London, 51–56.
172. H. S. Dickson: *JISI*, 1926, **113**, 177–211.
173. G. Persson: *Jernkont. Ann.*, in press.
174. F. B. Pickering: *Jernkont. Ann.*, 1964, **148**, 845–872.
175. K. Brotzmann: in *ISI Spec. Rep. 92*, 1965, London, 148–153.
176. G. H. Ockenhouse and J. E. Werner: *J. Metals*, 1966, **18**, 52–58.
177. K. Wick, E. Büchel and E. Hofmann: *Arch. Eisenhüttenw.*, 1967, **38**, 701–710.
178. R. Sifferlen: *Compt. Rend.*, 1955, **240**, 2526–2528 and 1957, **244**, 1192–1193.
179. M. Bergqvist, A. Bratli, G. Grünbaum, O. Nyquist, H. Pettersson, L. Tiberg and K. Torssell: *Medd. Jernkont. Tekn. Råd*, 1966, **21**, 649–725.
180. E. T. Turkdogan: *J. Metals*, 1967, **19**, 38–44.
181. R. G. Ward: 'An introduction to the physical chemistry of iron and steelmaking', 1962, Edw. Arnold Ltd, London.
182. S. Bergh and Å. Josefsson: *JISI*, 1964, **202**, 356–366.
183. H. A. Sloman and E. L. Evans: *JISI*, 1951, **169**, 145–152.
184. R. E. Lismer and F. B. Pickering: *JISI*, 1952, **170**, 48–50.
185. I. Uchiyama and M. Sumita: *Trans. Nat. Res. Inst. Metals* (*Japan*), 1965, **7**, 233–241.
186. E. L. Evans and H. A. Sloman: *JISI*, 1952, **172**, 296–300.
187. M. Wahlster: *Freiberger Forschungsb.*, 1962, **68**, 47–70.
188. A. Hultgren: *Jernkont. Ann.*, 1945, **129**, Figs. 10–14, 638.
189. F. Körber and W. Oelsen: *Mitt. Kais-Wilh. Inst. Eisenforchung*, 1933, **15**, 271–309.
190. D. C. Hilty, W. D. Forgeng and R. L. Folkman: *Trans. Met. Soc. A.I.Min. Met. Pet. Eng.*, 1955, **203**, 253.
191. O. Repetylo, M. Olette and P. Kozakevitch: *J. Metals*, 1967, **19**, 45–49.
192. K. Torssell: Private communication.
193. F. H. Woehlbier and G. W. P. Rengstorff: *J. Metals*, 1967, **19**, 50–53.
194. A. McLean and R. G. Ward: *J. Metals*, 1965, **17**, 526–528.
195. I. Uchiyama and T. Saito: *Trans. Nat. Res. Inst. Metals* (*Japan*), 1966, **8**, 42–43.
196. F. W. Boulger: *ASTE*, 1958, **58**, paper No. 80.
197. A. Adachi, N. Iwamoto and M. Ueda: *Trans. ISIJ*, 1966, **6**, 24–30.
198. H. A. Sloman and E. L. Evans: *JISI*, 1950, **165**, 81–90.
199. E. T. Turkdogan: *JISI*, 1954, **178**, 278–283.
200. A. Adachi and N. Iwamoto: *Techn. Rep. Osaka Univ.*, 1963, **13**, 417–422, 1964, **14**, 635–642, 1965, **16**, 281–287.
201. R. E. Hook and A. M. Adair: *Trans. AIME*, 1964, **230**, 1278–1283.
202. P. W. Clark and F. C. Martin: *JISI*, 1965, **203**, 2–17.
203. 'Vacuum degassing', *ISI Spec. Rep. 92*, 1965, London.
204. J. C. C. Leach: SR2/66, *Engl. Steel Corp. Ltd*, Group Res. Lab., 1966, Sheffield.
205. S. Eketorp: *Jernkont. Ann.*, 1966, **150**, 585–648.

206. J. D. Murray and R. F. Johnson: in *ISI Spec. Rep. 77*, 1963, London, 110–118.
207. E. L. Morgan, J. R. Blank, W. J. H. Salter and F. B. Pickering: *JISI*, in press.
208. C. Ericsson: *Jernkont. Forskning, Serie C*, nr 311, 23.10. 1967.
209. A. Schöberl and H. Straube: *Radex-Rundschau*, 1966, 34–50.
210. R. Cummins and T. K. Jones: R. 22/64, *Engl. Steel Corp. Ltd, Group Res. Lab.*, 1964, Sheffield.
211. Y. Shimokawa: *Tetsu-to-Hagané Overseas*, 1963, **3**, 349–351.
212. I. Uchiyama and T. Saito: *Proc. Mem. Lect. Meeting 10th Anniv. Found., Nat. Res. Inst. Metals*, 1966, Tokyo, 122–124.
213. F. B. Pickering: *JISI*, 1958, **189**, 148–159.
214. E. Scheil and R. Schnell: *Stahl u. Eisen*, 1952, **72**, 683–687.
215. T. Malkiewicz and S. Rudnik: *JISI*, 1963, **201**, 33–38.
216. S. Rudnik: *JISI*, 1966, **204**, 374–376.
217. W. Dahl, H. Hengstenberg and C. Düren: *Stahl u. Eisen*, 1966, **86**, 796–817.
218. F. B. Pickering: *JISI*, 1958, **189**, Fig.7, 153.
219. F. B. Pickering: *ibid.*, Fig.1, 149.
220. I. Uchiyama and M. Sumita: *Trans. Nat. Res. Inst. Metals* (*Japan*), 1965, **7**, Fig.2, 236.
221. M. H. Lewis: Int. Symp. 'Anisotropy in single-crystal refractory compounds', Dayton, Ohio, 13–15.6.67.
222. C. W. A. Newey: *ibid.*
223. A. Schöberl, W. Holzgruber and H. Raisky: *Radex-Rundschau*, 1964, 251–265.
224. K. A. Ridal and J. D. H. Hughes: *JISI*, 1967, **205**, 183–185.
225. A. M. Portevin and R. Castro: *JISI*, 1935, **132**, Fig.35, plate XLIII, 240.
226. F. B. Pickering: *JISI*, 1958, **189**, Figs.8–11, 154–157.
227. E. Scheil and R. Schnell: *Stahl u. Eisen*, 1952, **72**, Fig.14, 686.
228. H. C. Chao, L. Thomassen and L. H. van Vlack: *Trans ASM*, 1964, **57**, 386–398.
229. H. C. Chao and L. H. van Vlack: *Trans ASM*, 1965, **58**, 335–340.
230. H. C. Chao and L. H. van Vlack: *Nat. Res. St.*, 1965, **5**, 611–613.
231. H. C. Chao and L. H. v. Vlack: *Trans AIME*, 1965, **233**, 1227–1231.
232. 'Metals Handbook', 8th ed., ASM 1961, Novelty, Ohio, 23.
233. 'Machinability', *ISI Spec. Rep.94*, 1967, London.
234. E. M. Trent: in *ISI Spec. Rep.94*, 1967, London, 11–18, 77–87, 179–184.
235. H. Opitz and W. König: in *ISI Spec. Rep.94*, 1967, London, 35–41.
236. A. Wicher and R. Pape: *Stahl u. Eisen*, 1967, **87**, 1169–1178.
237. A. Wicher and R. Pape: *Stahl u. Eisen*, 1967, **87**, 1262–1269.
238. 'Control of Materials, Tools, Machines' (U.S. Air Force, Curtiss Wright Corp. Wood-Ridge, N.J.), 1951, **2**, 112.
239. G. Niklasson: *Proc. 3rd Int. MTDR Conf.*, Birmingham, Sept. 1962, 1963, Pergamon Press, 65.
240. M. C. Shaw, N. H. Cook and P. A. Smith. Free Machining Steel: *Trans. ASME, B*, J. *Engin. for Industry*, 1961, **83**, 163–179.
241. C. Rubenstein: in *ISI Spec. Rep.94*, 1967, London, 49–53.
242. T. Gladman and F. B. Pickering: *Steel and Coal*, Dec. 21, 1962, 1178–1187.
243. R. Kiessling, B. Hässler and C. Westman: *JISI*, 1967, **205**, 531–534.
244. J. Bellot, M. Hugo and E. Herzog: 2 ème Colloque de Portorož, 5–8 June, 1967, 9.

245. R. Kiessling: To be published.
246. A. Wicher: *Radex-Rundschau*, 1965, 432–438.
247. M. Atkinson: *JISI*, 1960, **195,** 64–75.
248. W. E. Duckworth and E. Ineson: in *ISI Spec. Rep.77*, 1963, London, 87–103.
249. R. F. Johnson and J. F. Sewell, *JISI*, 1960, **196,** 414–444.
250. Y. Kawada, H. Nakazawa and S. Kodama: *Mem. Fac. Techn. Tokyo Metropol. Univ.*, 1965, **15,** 1163–1176.
251. M. Sumita, I. Uchiyama and T. Araki: *Trans. Nat. Res. Inst. Met.* (*Japan*), 1966, **8,** 220–221.
252. F. de Kazinczy: *Jernkont. Ann.*, 1966, **150,** 493–506.
253. R. P. Newman and T. R. Gurney: *Brit. Welding J.*, 1964, **11,** (No. 11), 341–352.
254. S. Gunnarson: To be published.
255. J. G. Morlett, H. H. Johnson and A. R. Troiano: *JISI*, 1958, **189,** 42.
256. Å. Josefsson, Y. Koeneman and G. Lagerberg: *JISI*, 1959, **191,** 240–250.
257. C. T. Anderson, R. W. Kimball and F. R. Cattoir: *Trans AIME*, 1954, **200,** 835–837.
258. A. Schöberl *et al.*: *Radex-Rundschau*, 1966, 170–183.
259. H. F. Tremlett: in *ISI Spec. Rep.77*, 1963, London, 119–122.
260. T. Boniszewski: *BWRA Bulletin*, 1965, 144–152.
261. D. J. Widgery and T. Boniszewski: *JISI*, 1966, **204,** 53–54.
262. L. Roesch, G. Henry, M. Endier and J. Plateau: *Mem. Sci. Rev. Mét.*, 1966, **63,** 927–940.
263. L. Roesch, G. Henry and J. Plateau: *Mem. Sci. Rev. Mét.*, 1966, **63,** 941–950.
264. D. Hull: *Proc. Roy. Soc. A*, 1965, **285,** 148–150.
265. P. Kenny and J. D. Campbell: *Prog. Mat. Sci.*, 1967, **13,** 135–181.
266. R. Lagneborg: *ASM Trans. Quart.*, 1967, **60,** 67–78.
267. E. W. Langer: Dr. diss., Dept. of Metallurgy, Techn. Univ. of Denmark, 1967, Copenhagen.
268. A. Rose, A. Wicheler and H. Ketteler: *Arch. Eisenhüttenw.*, 1963, **34,** 617–624.
269. H.-J. Eckstein and K.-F. Lüdemann: *Freiberger Forschungsh.*, 1966, **B125,** 90–101.
270. J. A. Kohn, G. Katz and J. D. Broder: *Am. Miner.*, 1957, **42,** 398–407.
271. G. Fink: *Naturw.*, 1965, **52,** 32.
272. J. Voeltzel and J. Manenc: *Mem. Sci. Rev. Mét.*, 1967, **64,** 191–194.
273. Ber. Deutsch. Keram. Ges., 1963, **40,** 279–354.
274. L. H. Lindsley, B. T. C. Davis and J. D. McGregor: *Science*, 1964, **144,** Nr. 3614, 73–74.
275. R. Kiessling: To be published.
276. L. H. van Vlack: Private communication.
277. W. J. M. Salter and F. B. Pickering: *JISI*, 1967, **205,** 973–974.
278. F. B. Pickering: private communication.
279. M. H. May and G. F. Walker: BISRA/ISI Conference on 'Low-alloy Steels', April 1968, Paper 16. To be published.
280. C. L. M. Cottrell and P. F. Langstone: *ibid.*, Paper 17. To be published.
281. M. H. May, C. E. Nicholson and A. H. Priest: *ibid.*, Paper 18. To be published.

Part IV

Supplement to Parts I–III including literature survey 1968–1976

Experimental methods

This inclusion atlas, as it first appeared some 13 years ago, was based on experimental observations of non-metallic inclusions in steels, observed and studied *in situ*. The revolutionary method at that time was electron-probe microanalysis, and this technique for composition determination on the microscale was combined with optical microscopy and X-ray diffraction. The combined techniques made it possible to identify the different phases present in inclusions, confirming their identity by comparison with phases in experimentally prepared synthetic slags. This approach allowed such important phenomena as the extent of solid solutions and homogeneity ranges in different inclusion phases to be studied.

Since the first edition, additional experimental methods have come into widespread use for inclusion studies. These methods are not of the same significance for our present knowledge of inclusions as was electron-probe microanalysis in the early 1960s. However, they give valuable additional information, particularly in the area relating inclusions to the properties of the parent steel. It is appropriate, therefore, to describe the major additional techniques briefly.

The Scanning Electron Microscope (*SEM*) has become a standard tool in most research laboratories. Its main advantage for the study of inclusions is its high resolution combined with a relatively large depth of focus, even at high magnifications. Unlike the transmission electron microscope, which offers the same combination of properties in principle, the SEM is not limited to an electron transparent specimen. Much of the versatility of the instrument derives from its ability to relate features such as inclusions to the surrounding matrix, in specimens which may be fracture surfaces, conventionally prepared plane sections, or specimens deeply etched to reveal complex inclusion particle morphologies. These applications of the instrument are illustrated in Fig.246, which shows a fatigue failure in a low-alloy Cr steel, starting at a spherical calcium aluminate inclusion.[282] This Fig. may be compared with the optical micrographs shown in Fig.33.

Figure 247b–d illustrates the value of deep etching techniques in revealing inclusion particles. A particularly important example of the use of the SEM in this context was the work of Rege *et al.*[283] In an

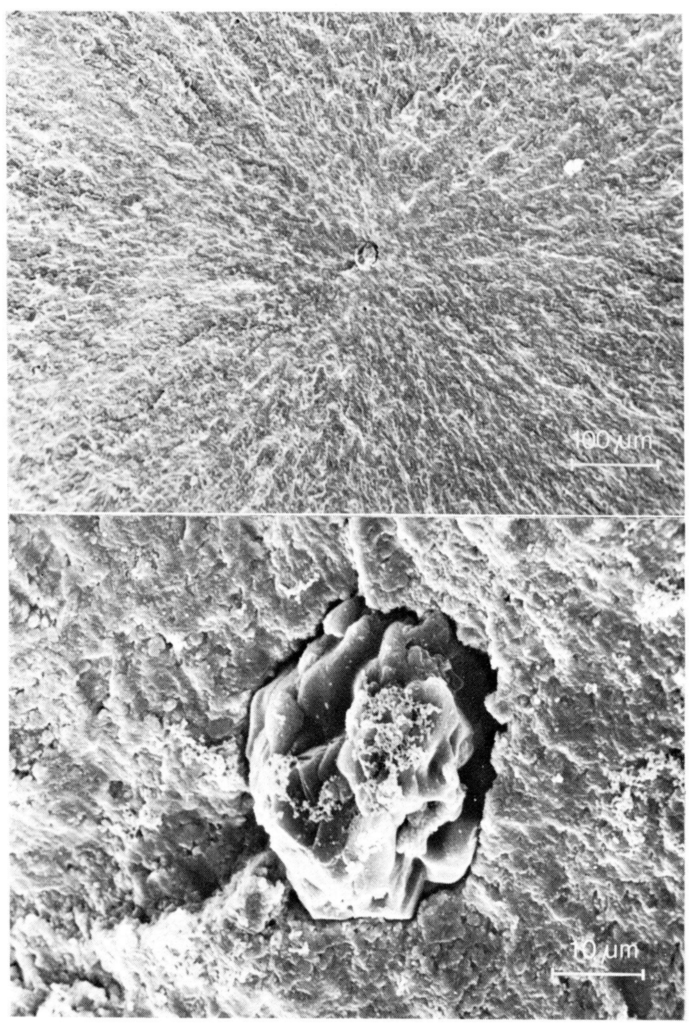

STEEL low-alloy chromium steel
INCLUSION calcium aluminate
COMMENT photograph illustrates depth of focus at high magnification given by SEM and topographic details in surrounding steel matrix. Compare I:40, Fig.33

246 Scanning electron micrographs of fatigue failure initiated at an inclusion[282]

optical micrograph of a prepared surface Al_2O_3 inclusions appeared as a cluster of small particles. Etching and examination in the SEM demonstrated that the particles were connected and formed a single Al_2O_3 dendrite.

The SEM may also be used to obtain three-dimensional images for stereometric analysis and is particularly valuable for the examination of particle morphology, features of the particle/matrix interface, and the formation of voids or cracks at particles as a result of deformation. Of the specialized SEM techniques, semi-quantitative *in situ* chemical analysis is perhaps the most useful in inclusion studies. Although not yet refined to the standards of quantitative analysis permitted by the wavelength-dispersive X-ray detectors used in conventional microprobe analysis, non-dispersive X-ray energy analysers are of great value when combined with the SEM. In particular, their high count rates permit the qualitative analysis of small particles in rough surfaces such as fractures.

Scanning electron microscopy has been used as a complementary method in studies of many inclusion phenomena. These include the location and effect of inclusions on fracture surfaces (Fig.246), the precipitation patterns and soldification mechanisms for sulphides (Fig.247), identification of inclusions at very high magnification,[283, 284] and identification of electrolytically isolated inclusions.[285] Other studies embrace the relation between microcracks and inclusions,[286] the formation of voids around inclusions (*see* page 47), and corrosion around inclusions.[287]

A new development of the SEM technique has recently been reported.[288] The object has been to perform *tensile tests in situ in the scanning electron microscope* by simultaneous video recording of the microscope image and a load–elongation curve. The nucleation and growth of cavities around surface inclusions has thus been monitored quantitatively (Fig.260).[289] The technique is of particular interest for inclusion studies, as it makes possible dynamic observation of inclusion behaviour at very high magnifications. Further studies are now in progress in order to observe the effects of temperature on inclusion deformation.

The simultaneous recording of the SEM image and the load–elongation curve in order to relate the various mechanical and microstructural events occurring in the same experiment is an important development. There are, however, limitations to the application of the method for the study of void nucleation at sub-surface inclusion particles.

The scanning principle can also be used with thin-film samples in transmission; this is referred to as STEM, which offers further possibilities for inclusion studies. For a general review of the development of electron microscopy and other advanced tools several conference proceedings are available, for instance Ref. 290.

Automatic inclusion counting methods have been developed to a high degree of refinement for the electronic processing of images.[291] The most important feature of these methods is that the relations between different physical properties of the steel and the size, shape, and distribution of inclusions can be evaluated, based on a larger population of inclusions than can be studied using manual techniques. Correlations can thus be established on a firmer statistical basis.

As an example of the importance of the time saved in routine inclusion assessment, manual microscopical assessment of inclusions by the method of comparing fields on a sample with the JK scale takes about 15 min for 100 fields each of area 0·5 mm^2, giving a total surface area examined of about 50 mm^2. A typical automatic technique requires only 50 s to calculate and print out oxide and sulphide inclusion area fraction data from 500 fields each of 0·32 mm^2, giving a total area examined of 160 mm^2. Manual counting gives a typical rate of about 10 inclusions counted per minute, compared with about 3500 inclusions in the same time using an automatic image analyser. Examples of the application of automatic techniques given in the references include: the assessment of inclusions in relation to fatigue,[286] surface cracks in cold-rolled tubing,[292] and the shape control of sulphides in free-machining steels.[293] Results regarding the distribution of inclusions based on automatic inclusion counting are given on page 28.

It must be stressed, however, that most automatic inclusion assessment methods rely on conventional optical imaging. They are limited, therefore, to the resolution and contrast conditions given by suitable combinations of objective, microscope, and metallographic preparation. There are some aspects of inclusion morphology which are of considerable importance and cannot be dealt with adequately using optical methods. For the study of very clean steels, increasing use will be made in the future of SEM imaging techniques in combination with automatic image analysis.

A. The MnO-SiO_2-Al_2O_3 system

A 2 MnO–SiO_2–Al_2O_3: Corundum

In parts I:15 and III: 22 it is stated that Al_2O_3 inclusions in practice often appear as clusters of Al_2O_3 particles similar to those in Fig.6. It was mentioned that the clusters are often three-dimensional units of small Al_2O_3 inclusions, assumed to be held together by the surface energy of the molten steel. Experimental evidence indicates that such clusters in fact could be three-dimensional Al_2O_3 dendrites, or clusters of such dendrites.[192, 283]

A 7 MnO–SiO_2–Al_2O_3: Galaxite

Galaxite ($MnO \cdot Al_2O_3$) is a double oxide of the spinel type. The spinels have the general formula $AO \cdot B_2O_3$ and are common oxide inclusion constituents in a wide variety of different steels, as discussed in I:29. A summary of the chemical and physical properties of spinel compounds has been published,[294] with a discussion of the formation of spinel-type inclusions during deoxidation and casting. Isolated spinel-type inclusions are usually less than 25 μm in size and generally have little effect on steel performance. It is their agglomeration into macroinclusions, held together within a lower melting point silicate, which gives rise to damage. In order to avoid dangerous inclusions it is therefore more effective to prevent the formation of silicate than the formation of spinels. During teeming, the incidence of spinel inclusions can be decreased by adjusting conditions in order to minimize splashing and reoxidation, for instance by redesign of the refractory nozzle.

A 13 MnO–SiO_2–Al_2O_3: Mn-cordierite

Inclusions corresponding to the composition of Mn-cordierite have been reported in ingots of 0·2%C steel, deoxidized in the ladle with FeSi+FeMn. The majority of the inclusions were found to be homogeneous glassy silicates with a composition corresponding to Mn-cordierite.[295] If the deoxidation practice was changed, corundum and spessartite were the main inclusion components.

J. Oxide inclusions: Tabular summary of phases

König and Ernst have given an extensive summary of possible oxide inclusion phases and the origin of oxide inclusions in steels.[296] This was based on petrographical and chemical analysis of inclusion isolates and of different ceramics and minerals used in steelmaking. The paper gives comprehensive optical and morphological data, as well as *d*-values for a great number of oxide phases of interest in steelmaking. The tables given form a useful complement to Tables XXVI–XXIX (II:90–93).

K. Sulphide inclusions

K 1 General

Recently, important contributions to the understanding of sulphide inclusions and their formation have appeared. The author has tried to cover the developments since parts II and III first appeared in the 1960s, by referring to some general reviews and also by describing in more detail important work under the same special headings used in parts II and III.

Thermodynamic data for the sulphides, as given in II:98, should be completed with the corresponding values which have been published for the rare-earth sulphides.[297, 298]

The field of sulphide inclusions in steel is reviewed in the proceedings of an international symposium held in 1974 at Port Chester, N.Y.[299] The monograph covers the subject of sulphide inclusions up to 1974 by means of a series of invited papers and includes many references to recent work on sulphides. It deals with the formation, identification, and properties of sulphide inclusions, and also with their influence on steel properties such as formability, weldability, mechanical properties, machinability, and corrosion.

Important new information relevant to the formation of sulphide inclusions has been published for the Fe–Mn–S, the FeS–MnS, and the CaS–MnS systems, as well as for the lesser known chromium sulphides. Also, there have been many studies of the effects of sulphide inclusions on various steel properties. More details are given below under respective headings.

K 3 Sulphide inclusions: The manganese–sulphur system

The *melting point of pure MnS* has been determined at 1655±5°C and the monotectic temperature in the Mn–MnS system at 1570±5°C.[300] The Mn–MnS phase diagram has been recalculated using these new temperature data by applying a regular solution model. Lower values

247a

reported earlier are attributed to the presence of oxygen. (In II:115 the value 1610°C is given for the melting point.)

The *morphology and formation of the MnS phase* has been studied by means of directional solidification in the work of Fredriksson and Hillert.[301] They studied a series of synthetic Fe–Mn–S alloys with low oxygen contents and various amounts of carbon and aluminium. The sulphides were classified according to their mechanism of formation in the melt, and four different sulphide morphologies were identified.

Type I (Sims type I) are drop-like with a random distribution in

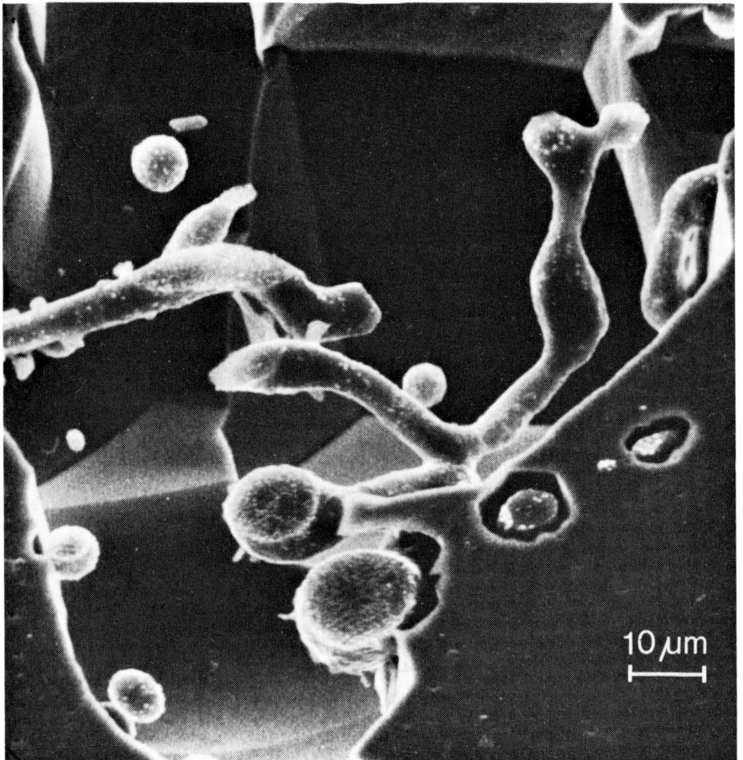

247b

relation to the steel structure, often with other metals in solid solution. The authors suggest that these sulphides are formed by a degenerate monotectic reaction where MnS forms as a liquid phase.

Type II (Sims type II) are rod-like and distributed as chain-like formations or thin precipitates in the primary ingot grain boundaries of the steel. It is suggested that the inclusions are formed by a cooperative monotectic reaction where MnS forms as a liquid phase together with the solid Fe-rich phase. There is a conflict, however, with the recent findings of Bigelow and Flemings[302] who, from new data on the Fe–Mn–S diagram, concluded that type II sulphides result from eutectic and not monotectic solidification.

Type III (Sims type III) are equiaxed and bounded by facets. This sulphide type is always monophase, randomly distributed in the steel, and sometimes occurs with an internal dendritic growth pattern.

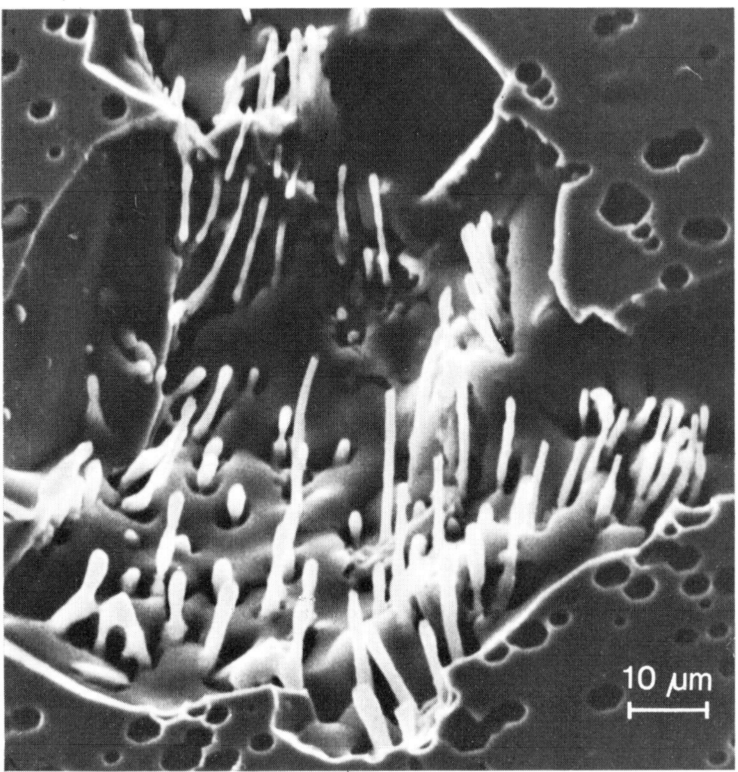

247c

Fredriksson and Hillert suggest that MnS forms as a crystalline phase by a degenerate eutectic reaction.

Type IV crystallize in a ribbon-shaped pattern with plate-like sulphides. It is suggested that MnS forms as a crystalline phase together with the solid Fe-rich phase by a cooperative eutectic reaction. This type has not been described earlier.

The authors only observed the monotectic reactions (types I and II) in ternary Fe–Mn–S alloys whereas the generally accepted phase diagram predicts a eutectic reaction. The phase diagram may therefore be incorrect. The effect of carbon and aluminium is to favour the eutectic reaction, i.e. types III and IV (compare II:125).

Examples of types I–III are shown in Figs.168–170. A characteristic structure of type IV is shown in Fig.247*a*. SEM photographs of sulphides of types I, II, and IV have been included in order to show

247d

STEEL TYPE 150 kg test ingots of resulphurized 18Cr2MoTi steel with different Mn and Ti contents
INCLUSIONS *a* titanium sulphides type IV, micrograph
b (Mn,Cr)S type I, SEM
c titanium sulphides type II, SEM
d titanium sulphides type IV, SEM
COMMENT The Mn:Ti ratio was varied and the general sulphide composition for the different ratios can be deduced from the different areas in Fig.250:
a 0·20 %Mn, 0·57 %Ti
b 0·19 %Mn, 0·06 %Ti
c 0·20 %Mn, 0·35 %Ti
d 0·20 %Mn, 0·57 %Ti
The sulphides in *c* were probably τ–Ti_2S which is one sulphide component in that area of the diagram in Fig.250

247 Sulphide inclusions in steel: sulphide of type IV and SEM micrographs of sulphides of types I, II, and IV

the characteristic morphology of the sulphide types at high magnification.

Other morphologies of MnS are known. Figure 248 shows Widmanstätten MnS precipitates found in a continuously cast 0·13 %C–0·039 %S–0·51 %Mn steel. The MnS platelets are visible in the as-cast

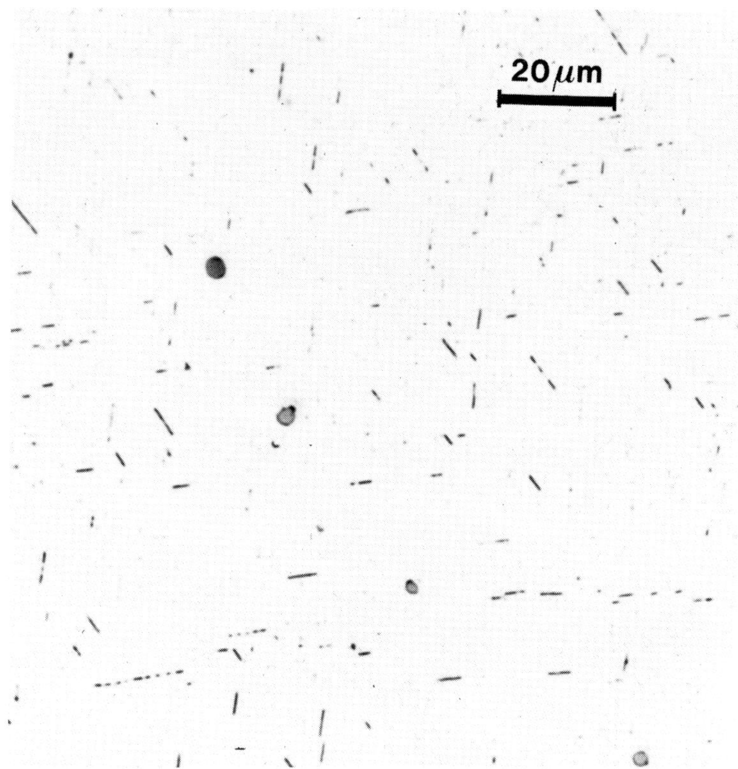

COMMENT Samples from this steel were heated to 1400°C, cooled at 20°C/min from 1400°C to 1000°C, and then brine-quenched. Etched in 2% Nital. (Micrograph supplied by Dr. D. Melford, Tube Investments Ltd, Hinxton Hall, England)

248 Widmanstätten MnS platelets formed in continuously cast 0·13%C–0·039%S–0·51%Mn steel

structure but to develop the precipitate further, samples were heat-treated (*see* caption). The Widmanstätten platelets are only about 1 μm thick.

Part II:115, dealing with the lesser known β, β', γ modifications of MnS, requires replacement in the light of a critical literature review. In addition to the well-known cubic α-MnS of NaCl type with a=5·226 Å (ASTM 6–0518) there exist two other modifications, both of which were called β by Mehmed and Haraldsen.[121] One is cubic of zincblende type with a=5·59 Å (3–1065) and one hexagonal of wurzite type with the lattice parameters a=3·988 Å, c=6·433 Å (2–1268). The hexagonal β phase is sometimes also called γ (3–1062) or β'. Both α and the two β phases were originally identified as precipitates in aqueous solutions of different manganese (II) salts.[121]

α-MnS is the common inclusion type in steel. Recently the hexagonal β phase was observed in a resulphurized ferritic 18Cr–2Mo steel.[303]

K 4 Solid solutions of the (Mn, Me)S type

From the work reported in part II on the FeS, MnS, and (Mn, Fe)S inclusions, certain observations were not readily explained. Thus, FeS as an inclusion phase always has the structure of troilite (II:106), the more stochiometric modification of FeS, and never of pyrrhotite which has a simpler structure and a higher sulphur content. Also, at that time it had not been explained why the maximum iron content of (Mn, *Me*)S found in inclusions is usually much lower than the solubility limit for FeS in MnS. An explanation has now been given by Mann and Van Vlack,[304] who have re-examined *the FeS–MnS system both with and without excess iron.* When excess iron is present, which is true for sulphide inclusions in steel, they observed that the pseudobinary system contains a peritectic rather than the previously assumed eutectic invariant. The maximum solubility limits at 997 ± 3°C in the two solid phases are: 7·5 wt-%MnS in FeS and 73·5 wt-%FeS in MnS (compare II:128). The two solid sulphide phases are nearly stochiometric in the presence of excess iron. The peritectic liquid contains 66 wt-%Fe, 34 wt-%S, and 0·4 wt-%Mn.

These authors were able to be more specific than earlier investigators about the Fe–FeS–MnS–Mn–region of the Fe–Mn–S ternary diagram, finding that the end of the miscibility gap in this system does not cross the invariant line between primary solid metal and (Mn, Fe)S(s) phases. They found that there is no eutectic invariant point in this part of the ternary system in contrast to the FeS–MnS system without excess iron. Resulting from their studies of the FeS–MnS system with excess iron, they conclude the following for iron- and maganese-sulphide inclusions.

FeS has a troilite NiAs(B8) type of crystal structure in the presence of metallic iron, that is, as FeS inclusions. The FeS–MnS phase diagram without excess iron, however, contains FeS with the pyrrhotite structure.

The solid solubility limits and invariant temperature in the FeS–MnS system with excess iron differs considerably from the system without excess iron. This explains why the iron content of sulphide inclusions in most steels is expected to be low, although the solid solubility limit of FeS in MnS is as high as about 65–70 wt-% (II:128).

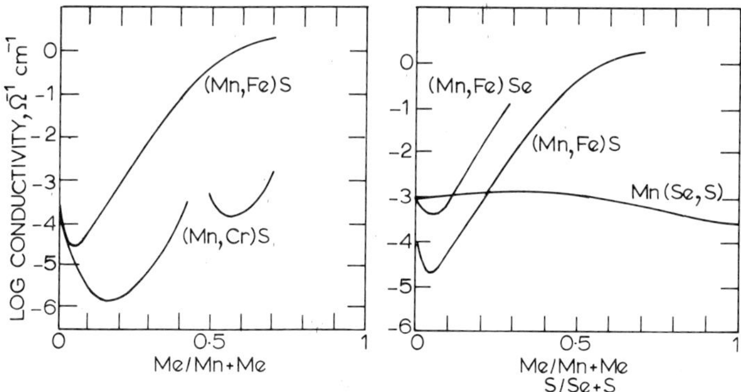

249 Electrical conductivity (in $\Omega^{-1}cm^{-1}$) for (Mn,Fe)S, (Mn,Cr)S, (Mn,Fe)Se, and Mn (S,Se) at different Fe, Cr, and Se contents of sulphides. The conductivity was determined for sintered powder compacts of the solid solutions[306]

The ternary phase relationships with excess iron, as in steel, explain why lower Mn/S ratios are possible in resulphurized steel than in low-sulphur steels while still avoiding hot-shortness (III:94). The general manganese requirements for a sulphur-containing steel are expressed as: $\%Mn \geq 0{\cdot}4\% + 2(\%S)$.

However, manganese in excess of the amount calculated from the equation may be required if surface scaling is encountered in re-heating operations and a liquid containing small oxide particles is introduced into the subscale region.

The *solubility of the second long-period transition metals in MnS* has been investigated.[305] Of the second long-period elements, yttrium is highly soluble in α-MnS, whereas the solid solubility of the transition metals zirconium to palladium is nil. The behaviour of the second long-period elements thus differs completely from that of the first long-period elements (Fig.176). For yttrium, the solid solubility limit is about 35% of the total metal atoms.

The *MnS–CaS system* has been studied (*see* page 18).

Conductivity measurements on solid solutions of the (Mn,Fe)S type have shown that there is a substantial variation with composition giving a minimum in conductivity when 5–10 at-% of the metal atoms are Fe in (Mn,Fe)S (Fig.249).[306] This is important for the corrosion behaviour of steel with sulphide inclusions (*see* page 52).

K 5 Sulphides of Group III (the lanthanides)

Much new information is available on the sulphides and oxysul-

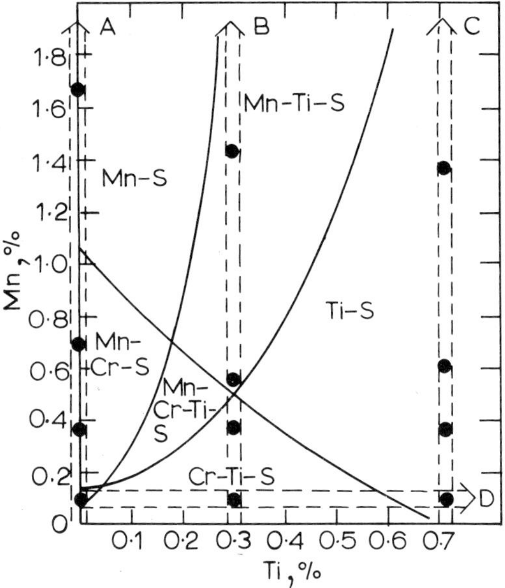

COMMENT The results were based on electron-probe microanalysis of 32 150 kg test ingots with compositions of about 0·012%C, 0·7%Si, 0·2%S, 18%Cr, 0·07%Ni, 2·3%Mo and varying levels of Mn and Ti. The low-Mn and low-Ti corner was not fully explored and it is not possible to decide whether an area with pure Cr sulphides exists. The sample series *A*, *B*, *C*, and *D* refer to the breakdown potential measurements reported in Table LI

250 Influence of Mn and Ti content of 18Cr–2Mo resulphurized steel on sulphide composition[299]

phides of the rare-earth elements, but it deals mainly with the roles of these elements in desulphurization and deoxidation and their influence on steel properties (sulphide shape control). Recent summaries are reported on page 29, to which the reader is referred. Microphotographs of RE-sulphide and oxysulphide inclusions have been published (Table XLVIII) and are given in Figs.253 and 254.

K 6–8 Sulphide inclusions with transition metals of Groups IV (Ti, Zr, Hf), V (V, Nb, Ta), and VI (Cr, Mo, W)

General

As mentioned in II:137–140, sulphides of these metals are not very common as inclusions in steels, although the sulphide systems and the intermediate phases have been studied with X-ray methods and are well-known. The main reason is that the Mn content of modern steels is balanced so that the sulphides usually are of the MnS type. For example, as mentioned in II:140, chromium sulphides are seldom observed in steels even if the chromium content is high. If, however, the manganese content is low and the sulphur content is high, these sulphides can be precipitated in the solidifying steel. Also, in welded structures of Ti-stabilized steels, Ti sulphides may appear.

The necessary conditions for the precipitation of different Mn, Ti, and Cr sulphides in ferritic, high-sulphur steels of the 18Cr–2Mo type have been outlined (Ref. 299, p. 104–122). The levels of Mn and Ti were varied systematically and the different sulphides identified by means of electron-probe analyses *in situ* and X-ray diffraction analyses of isolates from a series of such steels. The results are summarized in Fig.250. The composition and structure of the sulphides were dependent on the Mn and Ti content of these steels and different areas were distinguished (for Cr and Mo contents constant at 18 % and 2 % respectively).

The main sulphides observed in the different areas of Fig.250 were:

low-Ti high-Mn: MnS phase, type I
low-Ti medium-Mn: (Mn,Cr)S phase, type I
low-Ti low-Mn: not investigated
medium-Ti high-Mn: MnS and Ti sulphides, type II
medium-Ti medium-Mn: (Mn,Cr)S, Ti, and Cr sulphides, type II
medium-Ti low-Mn: Ti and Cr sulphides, type II
high-Ti: Ti sulphides, type IV.

The low-Ti low-Mn area is not fully explored and it has not been possible to decide if an area with pure Cr sulphides exists. The breakdown potential of a number of alloys with different sulphides according to the arrows *A–D* has been measured (Table LI) and is discussed on page 54.

The hexagonal chromium sulphide $Cr_{1-x}S$ and the monoclinic CrS

The identification of the chromium sulphides presents a problem. It was found that they usually have a hexagonal structure and could be described by the formula $Cr_{1-x}S$.[303] According to previous studies of the Cr–S system and as reported in the ASTM index, several chromium sulphides exist with various formulae, but most chromium sulphides are hexagonal and of the type $Cr_{1-x}S$ (ASTM 10–344) with the parameters $a=3{\cdot}464$ Å and $c=5{\cdot}763$ Å. Also, phases with different structures exist but these can be described by hexagonal pseudocells. All hexagonal chromium sulphides described in the literature (Cr_7S_8, Cr_2S_5, $CrS_{1.17}$, Cr_2S_3, and Cr_5S_6) seem to be closely related. The hexagonal chromium sulphide inclusions, which form the majority of the inclusions observed, should therefore be given the formula $Cr_{1-x}S$. In addition some inclusions are monoclinic and of the probable composition CrS (ASTM 11–343).

TABLE XLIV Hexagonal Cr $_{1-x}$S type chromium sulphide inclusions with other metals in solid solution identified in resulphurized 18:2 ferritic steels (from Ref. 303)

Metal analysis, %						Inclusion analysis, EPA		
Mn	Cr	Mo	Ti	Zr	Nb		*a*, Å	*c*, Å
0·19	17·9	2·3	0·6	...	...	$Cr_{0\cdot7}Mn_{0\cdot2}S$	3·474	5·854
0·06	17·9	2·3	0·8	...	...	$Cr_{0\cdot8}Ti_{0\cdot1}S$	3·464	5·838
0·06	18·2	2·4	...	0·05	...	$Cr_{0\cdot7}Zr_{0\cdot2}S$	3·507	5·886
0·06	18·0	2·3	...	...	0·36	$Cr_{0\cdot8}Nb_{0\cdot01}S$	3·470	5·824
0·05	17·8	2·3	...	...	0·12	$Cr_{0\cdot9}Nb_{0\cdot02}S$	3·475	5·843
0·06	18·2	2·3	...	...	0·62	$Cr_{0\cdot8}Nb_{0\cdot06}S$	3·475	5·865
0·06	18·2	2·3	...	...	0·96	$Cr_{0\cdot8}Nb_{0\cdot07}S$	3·478	5·869
ASTM 10–344						...	3·464	5·763
ASTM 11–7						...	3·431	5·594

Solid solutions of the $(Cr,Me)_{1-x}S$ *type*

In resulphurized 18Cr–2Mo steels with low Mn contents, a series of chromium sulphide inclusions of the hexagonal $Cr_{1-x}S$ type with other metals in solid solution were observed.[303] The steels had minor additions of Ti, Nb, and Zr. Examples of such inclusions are shown in Table XLIV.

There thus exists a second solid solution series of sulphide inclusions in addition to (Mn,*Me*)S of the MnS type (II:128). As discussed in II:128 it is essential to recognize such solutions because the properties of the sulphides change with composition and this may influence the corrosion resistance and machinability of the parent steel.

The solid solubility limits for the $(Cr,Me)_{1-x}S$ solid solutions have been determined using synthetically prepared sulphides with *Me* one of the metals Ce, Ti, Zr, V, Nb, Mn, and Fe[307]. The results are summarized in Table XLV.

K 11 Sulphides of Group II (Mg, Zn, Cd, Hg, Ca, Ba, Sr)

The CaS–MnS system has been investigated.[308] The system shows complete solid solubility above 1200°C, but at lower temperature a miscibility gap develops (Fig.251). At 1000°C there is an equilibrium between a Mn-rich sulphide with 13·5 wt-%Ca and a Ca-rich sulphide with about 42·7 wt-%Ca. The two sulphide phases are isostructural with an α-MnS lattice but with different *d* values, which are given in Ref. 308. The two phases look alike, but have different corrosion behaviour and the high-Ca sulphides have been identified

TABLE XLV **Solid solubility limits and corresponding lattice parameters of synthetically made hexagonal sulphides of the type $(Cr,Me)_{1-x}S$, where Me = Ti, V, Mn, Fe, Zr, Nb or Ce. The solubility limit is based on electron microprobe analyses. All the sulphides seem to have metal vacancies in analogy with the (Mn,Me)S sulphides (II: 128). Therefore the variation in lattice parameters is small and not consistent with the Me/Me + Cr ratio (from Ref. 307)**

Me	Sulphide formula at solubility limit	*Me*/(*Me*+Cr) in at-% at solubility limit		Lattice parameters at solubility limit *a*, Å	*c*, Å
Ti	Between $Cr_{0\cdot24}Ti_{0\cdot60}S$	between	71	$\leqslant 3\cdot4$	$\geqslant 5\cdot98$
	and $Cr_{0\cdot10}Ti_{0\cdot75}S$	and	88		
V	Between $Cr_{0\cdot42}V_{0\cdot43}$	between	51	$\leqslant 3\cdot41$	$\geqslant 5\cdot75$
	and $Cr_{0\cdot10}V_{0\cdot69}$	and	87		
Mn	$Cr_{0\cdot49}Mn_{0\cdot30}S$		38	3·45	5·76
Fe	$Cr_{0\cdot60}Fe_{0\cdot15}S$		20	?	?
Zr	$Cr_{0\cdot76}Zr_{0\cdot05}S$		6	3·45	5·78
Nb	$Cr_{0\cdot31}Nb_{0\cdot45}S$		59	3·47	5·86
Ce	$Cr_{0\cdot88}Ce_{0}S$		0	3·47	5·77

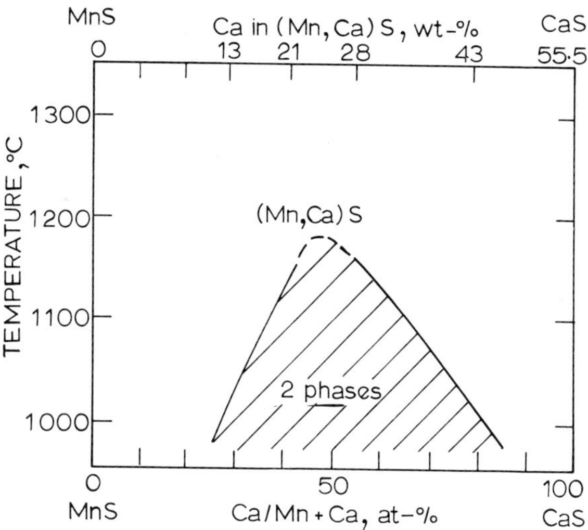

COMMENT Measurements were made on synthetically prepared sulphides[308]

251 The MnS–CaS system. Approximate extension of miscibility gap

as being active in the initial corrosion attack around slag inclusions in steel.[287]

(Mn,Ca)S inclusions in 1 %C–Cr steels have been observed.[309] They were almost always associated with Ca-bearing oxides of the calcium aluminate type, usually as a peripheral phase to the oxide (Fig.111). The mutual solubility of MnS, CaS, and FeS is indicated.

The MnS–BaS, MnS–SrS, BaS–SrS, and CaS–SrS systems are being studied.[310] BaS and SrS are both cubic with the *B*1 structure and *d* values 6·38 Å and 6·04 Å respectively. They are thus isostructural with α-MnS. Preliminary results show that there is very little, if any, solid solubility of Ba and Sr in α-MnS and also of Mn in BaS and SrS. There is a wide solid solubility between CaS and SrS but a miscibility gap exists between 40 and 50 at-%SrS at 1100°C. The solid solubility of Ba in SrS and Sr in BaS is small. The systems mentioned are of practical importance since Ba- or Sr-bearing CaSi alloys are used in deoxidation practice. It has been found that these alloys seem to have a favourable effect on the final deoxidation of Al-bearing low-alloy steels.[311] They allow shape control of the sulphides and the alumina inclusions, which change from stringers to uniformly distributed globular calcium aluminate inclusions with sulphide.

L. Inclusions with selenium and tellurium

L 1 Inclusions with selenium

Selenium-containing inclusions in 18Cr–8Ni steels[312] and in carbon steels[313] have been studied. In the 18:8 steel selenium was present as MnSe in multiphase inclusions, which also contained Mn–Cr silicates and oxides. In carbon steels with low Mn content, Fe selenides were present in addition to MnSe. Iron–manganese selenides have also been studied under conditions where excess iron was present, a situation similar to that in steel.[314]

Mann and Van Vlack have reported that in the presence of excess metal the FeSe–MnSe pseudobinary has a peritectic at 976$\pm$3°C with the following phases: Fe(<0·3 at-%Mn) (s), (Fe,Mn)Se(2·5 at-% MnSe) (s), (Mn,Fe)Se(48·3 at-%FeSe) (s), and a selenide liquid (5·2Mn, 45·4Fe, 49·4 at-%Se). A minimum was observed in the liquidus surface at about 920°C and FeSe–9 wt%-MnSe. A qualitative liquidus surface was proposed for the Fe–FeSe–MnSe–Mn section.

Conductivity measurements of the (Mn,Fe)Se and Mn(S,Se) solid solutions have been carried out. These show a minimum in conductivity for about 5%Fe in total metal atom sites in (Mn,Fe)Se but a relatively constant conductivity for the Mn(S,Se) solid solutions over the whole solubility range (Fig.249).[306]

L 2 Inclusions with tellurium

The phase diagram MnS–MnTe has been studied.[276] It shows that in addition to the hexagonal MnTe phase there also exists at high temperature a MnTe modification with the NaCl-type structure, isostructural with α-MnS and α-MnSe. This phase is only stable above 1040°C. MnS is slightly soluble in the hexagonal, low-temperature modification of MnTe and a little more so in the high-temperature form (up to about 9 mol. %). Examples of Se- and Te-containing inclusions in carbon steels have been published.[313] The authors observed that in addition to MnS and MnTe, iron selenides and tellurides may be present in steels with moderate Mn contents, in particular after rapid cooling. The influence of Te additions in free-machining leaded steels on MnS morphology has been studied.[293]

Phase relationships in the iron telluride–manganese telluride system in the presence of excess metallic iron are reported.[315] This portion of the Fe–Mn–Te system has a eutectic at 830±6°C. The equilibrium phases at this temperature are Fe(<0·1 at-%Mn) (s), $(Fe,Mn)_{1\cdot2}Te$(<0·1 mole % MnTe) (s), (Mn,Fe)Te(93·5 mole % MnTe) (s), and an iron telluride-rich liquid (~0·5Mn, 55Fe, and 44·5 at-%Te). A monotectic is at 895±3°C where the equilibrium phases are Fe (<0·1 at %Mn) (s), (Mn,Fe)Te(91 mole %MnTe) (s), an iron telluride-rich liquid (5·9Mn, 47·4Fe, and 46·7 at-%Te), and a manganese telluride-rich liquid (46·2Mn, 9·4Fe, and 44·4 at-%Te).

Additional information is also presented on the Mn–Te system and the iron-rich part of the Fe–Te system. A qualitative liquidus surface is proposed for the Fe–$Fe_{1\cdot2}Te$–MnTe–Mn part of the Fe–Mn–Te ternary.

Pit corrosion measurements have been carried out on stainless steels with Se and Te additions.[369]

O. The origin of non-metallic inclusions

O 2 Inclusion size, quantity, and distribution in steel

Quantity and size of inclusions
Improvements in steelmaking techniques since parts I–III were written have resulted in lower oxygen and sulphur contents than were possible earlier. Oxygen contents of 10 ppm are not unrealistic and sulphur contents of less than 10 ppm have been reported. For instance, laboratory studies have been made in order to reduce the sulphur content in structural steels to below 0·001% with REM additions, at the same time keeping the oxide inclusions to a minimum.[316] REM additions were placed in the centres of solid steel blocks, which were remelted in a vacuum induction furnace under argon. The low-melting centre traversed the solid block and the remaining inclusions were transported to the surface and the crucible wall. The resulting remelted steel was 'very low in inclusions' and its mechanical properties were found to be isotropic. The usual decrease in transverse properties due to inclusions was absent.

The mechanical properties of different steels with extra-low sulphur (10 ppm or less) have been investigated.[317] This study used production ingots of a 0·45%C steel, a CrMo and a CrNiMo steel, a 5%Cr die tool steel, a 5%Cr super high-strength steel, VAR, and a 12%Cr steel. All these steels were vacuum deoxidized and REM-treated. The results show that sulphur contents of 0·003% or less will improve the impact value for all these steel types, especially in the transverse direction. All visible inclusions remaining in the low-sulphur steels were small, hard, and brittle, whereas in corresponding steels with a higher sulphide content (160 and 320 ppm respectively) the inclusions were the larger stringer-type plastic sulphides and silicates.

It will certainly be possible in the future to decrease further the amount of oxygen and sulphur in the steel, thereby decreasing the effects of the inclusions on mechanical properties. The total number of oxide and sulphide inclusions in 1 t of a steel with as little as

TABLE XLVI Influence of inclusion size on inclusion numbers in a hypothetical steel with 1 ppm oxygen and 1 ppm sulphur as spherical Al_2O_3 and MnS inclusions

Inclusion diameter, μm	No. of inclusions/t	Volume of steel per inclusion	Mean distance between inclusions
1 000	$2{\cdot}3\times10^{3}$	55 cm³	3·8 cm
100	$2{\cdot}3\times10^{6}$	55 mm³	3·8 mm
10	$2{\cdot}3\times10^{9}$	55×10^{6} μm³	380 μm
1	$2{\cdot}3\times10^{12}$	55×10^{3} μm³	38 μm
10^{-1}	$2{\cdot}3\times10^{15}$	55 μm³	3·8 μm
10^{-2}	$2{\cdot}3\times10^{18}$	55×10^{-3} μm³	0·38 μm
10 Å	$2{\cdot}3\times10^{21}$	55×10^{-6} μm³	380 Å

1 ppm of each of these elements will still be about 10^{12} if the inclusions are distributed as 1 μm inclusions (Table XLVI). If their size increases to 10 μm or 100 μm their number decreases to about 10^{9} or 10^{6} respectively, but their local effect on steel properties increases drastically. The main effect of the low oxygen and sulphur content therefore seems to be a decrease not in number but in size of the inclusions. A steel 'very low in inclusions' should therefore mean a steel containing very few inclusions above about 10–20 μm. The basic concept of the present atlas, that the inclusions are a natural structural component of the steel, therefore holds even for those steels very low in oxygen and sulphur. However, the large inclusions are the most harmful (*see* pages 5, 43, and 44).

Much of the information given in parts I–III was based on inclusions in OH steels and basic electric steels. Some reviews have appeared since then dealing with the inclusion levels in steels made by other steelmaking methods. A few comments on the general inclusion level for steels made with the latest steelmaking methods should, therefore, be of interest.

The AOD process for stainless steel combines the stirring action of the gaseous innoculants with a purging action of the final argon stir. If the deoxidation practice is carefully controlled the AOD process results in significant improvements in the cleanness of stainless steels. It is reported[318] that in converter melts the level of oxides of all lengths has been reduced by approximately 50% when compared with arc melts. In the case of extra-low carbon types the decrease in inclusion levels for all sizes is of the order of 70%.

The ESR technique generally results in low inclusion levels. The mechanism seems to be that oxide inclusions go into solution in the molten electrode tip during ac electroslag remelting.[319] The mechanism of inclusion removal suggested is that oxide inclusions dissolve in the liquid metal at the electrode tip, which has a high temperature, and that oxygen and deoxidants and/or the deoxidation products are transferred to the slag by slag–metal reactions taking place at the tip interface, the droplet interface, and the metal/pool interface. Inclusions then reprecipitate during the solidification of the ingot but usually as a more even distribution of smaller inclusions than before. For example, in type 316 stainless steel it is reported[320] that both the alumina clusters and larger, globular oxides are reduced in size by a factor of two. Remelting, particularly with ac power, reduces sulphide as well as silicate inclusions. The relationship between electrode material and slag chemistry is critical for a low inclusion content and the lowest levels are achieved when the furnace atmosphere is free of oxygen.[321]

A comparison of the cleanness of steels from ESR and VAR processes has been carried out.[322] It was found that VAR steels exhibit a greater freedom from oxide inclusions while ESR steels are better as regards freedom from sulphides. However, ESR steels exhibit properties superior to those of conventionally melted steels. This is illustrated in Fig.252 showing the oxide and sulphide inclusion contents in a series of 316 steels arc-melted in air, ESR, and VAR respectively. The minimum obtainable oxygen in VAR steel is said to be limited by melt–refractory interactions and not by the oxygen content of the vacuum.[323] The cleanness of steel produced by VAR and ESR depends much on the detailed steel melting practice and the two methods should not be judged by this single investigation only. Both methods give improved cleanness.

Several studies have been made concerning the nature and distribution of inclusions in continuously cast steel, as compared with ingot-cast steel. Generally, the cleanness of continuously cast steel depends on the cleanness of the steel in the furnace and ladle and a thorough deoxidation practice is essential. The cleanness of the steel will not be improved through continuous casting and the inclusion content in the solid steel depends much on the operation of the casting machine. There is a different distribution of the inclusions in different types of casting machines. The large inclusions consist of deoxidation, and tundish or nozzle erosion products, reoxidation products of the molten steel, and scums entrapped in the mould.[324] The inclusion distribution depends, however, on several factors: principally nozzle and tundish construction, casting speed, and bath level in the tundish. Inclusion distribution diagrams for different variations of these parameters have been determined.[325] Nippon Kokan report[326] that continuously cast material contains fewer and

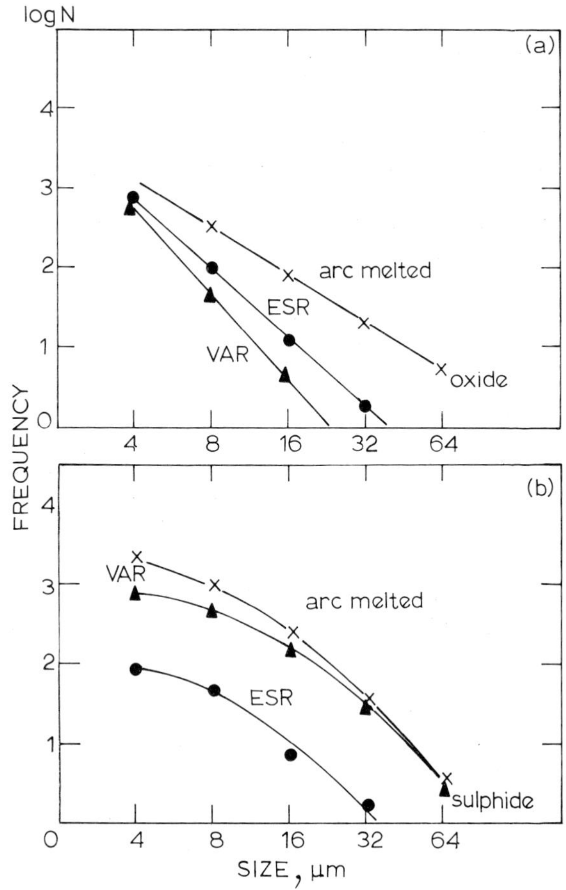

a number of oxide particles per cm² *b* number of sulphide particles per cm²

252 Comparison of inclusion content and size distribution in type 316 steel remelted in different ways

shorter sulphides than ingot-cast material, but that there is a tendency for entrapment of inclusions on the solidifying wall of the inner radius of the curved slab (bow-type machine). This has also been reported by other authors.[327, 328] Different methods to improve the cleanness are: use of ladle-to-tundish stream protection, a special tundish and tundish nozzle, control of tundish temperature, and the use of casting powders.

A comparison of uphill- and direct-teemed Mn-containing, Al-killed and RH-treated carbon steel has shown that uphill- is better than direct-teeming for achieving cleaner sub-surfaces and better internal quality in the 10–30 t ingots investigated.[329] About 70% of the large inclusions in such steels are a result of refractory erosion and about 30% result from reoxidation. The showering of equiaxed crystals from the ingot top during the early stage of solidification prevents the floating up of inclusions and contributes to the formation of the inclusion concentration in the central bottom part of the ingots (Fig.194). A method of preventing this concentration is to apply quick-igniting, fast-burning, highly exothermic fluxes to the top of the melt immediately after filling.

Inclusion distribution

Several investigations on inclusion distribution in different types of steel have appeared since part III was published. Automatic inclusion assessment techniques were used in several of these investigations; therefore the results are based on large inclusion populations. Usually the deoxidation practice has been varied and the resulting changes in oxygen content, inclusion distribution, and inclusion nature followed.

The ingot oxygen distribution in ladle deoxidized 0·2%C steel was studied by Franklin and Evans.[295] They determined the level of oxygen, the inclusion size distribution, and the formation of bottom-cone segregation of oxides after deoxidation with various silicon–manganese alloys and with ferro-aluminium before silicon–manganese. The results show that after silicon–manganese deoxidation, the bottom cone region is higher in oxygen than the ingot mean, whereas the upper central chill region has a significantly lower content. An initial addition of ferro-aluminium produced a lower level of oxygen and no bottom-cone segregation of oxygen. The number of microinclusions decreased but the number of large inclusions remained at the same level.

The difference in inclusion distribution in 200 kg ingots of a low-alloy carbon steel, deoxidized with aluminium and Ca–Si or with aluminium, Ca–Si, and REM (Fe–Ce), has been reported.[330] Inclusion counts for many of these small ingots showed that the addition of REM decreased the number and size of visible inclusions. The distribution of the inclusions was also changed from the common segregation distribution to one more uniform over the different ingot sections. As already mentioned,[329] the application of exothermic fluxes to the top of the mould immediately after filling can also diminish the bottom-cone segregation of inclusions.

In rare-earth treated, fully killed ingots of 0·1%C–1·35%Mn–

0·025%S steels there is a marked bottom-cone inclusion segregation,[331] explained by the sedimentation of high melting point and high-density REM inclusions. The cone-shaped segregate cannot be entirely avoided but will be confined to a small region at the base of the ingot if excessive RE additions are avoided.

The size and distribution of oxide and sulphide inclusions in non-stabilized stainless steel for the sequence ingot–slab–plate has been investigated, using automatic inclusion counting methods.[332] The major concentration of large inclusions was found in the central bottom part of the direct-teemed ingots. Microinclusions were more abundant in the immediate vicinity of the surface than deep inside the ingot. A more uniform distribution of the large inclusions was noted in uphill-teemed ingots which also contained fewer such inclusions.

In the slabs, the total number of larger inclusions was lower than in the ingots. These inclusions were also concentrated in the centre. In the cross-sections they are located midway between the two rolled surfaces. In the plates, the assessed inclusion content is again lower, but the number of microinclusions increased due to continued breakdown of the oxide inclusions during rolling. Finally, the investigations showed that ESR-treated steel was considerably cleaner than oxygen-blown basic electric steel.

Attempts to develop *theoretical models* for inclusion distributions in ingots have been made and comparison between theory and experiments seems to be fairly good. The following two references are examples of these modelling exercises.

The distribution of oxide inclusions in a 7 t carbon steel ingot, expressed as a function of oxygen content, distance from the centre, and height above the bottom, has been calculated using stepwise regression analysis of microscopical inclusion counts.[173] The agreement between experimental results and calculated values was fairly successful for the oxygen distribution, with the usual concentration of oxide inclusions in the bottom central part of the ingot. The method was less successful for prediction of the volume fraction of oxide inclusions.

A theoretical treatment of the distribution of endogenous oxide inclusions along the axis of an Al-killed steel ingot has been made, based on hydrodynamic transport in the molten core of the solidifying ingot and the absorption capacity of the slag in the meniscus.[333] The solution described the distribution of oxide inclusions from top to bottom for various aluminium deoxidation methods and different casting practices. The agreement between experiments and theoretical results appeared to be satisfactory.

TABLE XLVII Typical rare-earth alloy compositions, wt %

	Mischmetal	Rare-earth silicide	
iron	3·0 max.	iron	30–35
residual	1·0 max.	silicon	30–35
total REM	96·0 min.	REM	30–35
REM content of mischmetal		*REM content of silicide*	
cerium	45·0–51·0	cerium	48–50
lanthanum	23·0–26·0	lanthanum	32–34
neodynium	15·0–19·0	neodynium	13–14
praseodynium	4·0–6·0		
other REM	1·0–2·0		

O 3 Influence of deoxidizing elements upon inclusion formation

Aluminium

In III: 22 it was mentioned that the clusters of Al_2O_3 which are frequently observed in aluminium deoxidized steel, are in fact three-dimensional units which when sectioned look like clusters. It has been shown that the Al_2O_3 particles form during deoxidation by dendritic growth and SEM pictures of such dendrites have been published.[283] There are circulating currents in the ingot during teeming. Before solidification, therefore, particles will be carried around with a high probability of Al_2O_3 dendrites becoming joined to larger units, according to the mechanism given in III: 22 and described by Torsell and Olette.[192]

Rare-earth metals

When the first edition of this atlas of inclusions was written, the rare-earth metals (REM) were regarded as potential deoxidizers and desulphurizers for liquid steel because of their high affinity for oxygen and sulphur. Many of their oxide, sulphide, and oxysulphide phases had been identified and studied by X-ray crystallography (II:133–37) but little was known of their use in steelmaking, the nature of REM inclusions, and the effect of these inclusions on the properties of steel. Sincc then rare-earth metals, as 'mischmetal' and other compounds (Table XLVII) have been used widely in increasing amounts for deoxidation, desulphurization, and shape control (or modification) of inclusions.

The thermodynamics and phase equilibria of those rare-earth metals which are used in steel practice have been summarized.[297, 298] The standard free-energy data for their oxides, oxysulphides, and sulphides are also given in these references, and the diagrams complete those for other metal oxides and sulphides given in part II (Figs.80 and 147). A review of their overall effects through the different steps in steelmaking, hot rolling, mechanical working, and welding has been given by Luyckx.[334]

When rare-earth metals are added to molten steel, simultaneous deoxidation and desulphurization occur. There are several ways in which they may be introduced into the molten steel.[298] They may be added to the ladle during the latter part of the tapping operation, generally after substantial aluminium additions for predeoxidation or vacuum deoxidation. They may be plunged through the slag layer after the tap or inserted through the open meniscus created at the slag/metal interface by gas purging. If the steel is subjected to a vacuum degassing treatment, additions may be made directly to the degassing vessel. Rare-earth metals may also be added to the mould during or immediately after casting. Their main effects, as far as inclusions are concerned, are: significant desulphurization, combined with modification (*see* page 41) of the remaining sulphides and deoxidation with formation of complex oxide and oxysulphide inclusions. There is a complex relationship between the oxygen and sulphur levels and the rare-earth metals content, and the calculation of the correct REM addition is difficult. In practice, the amount of recovery of the rare-earth metals and the degree of sulphide shape control obtained are affected by both the REM/sulphur ratio and the initial oxygen content of the steel at the time of the addition. The extent of reaction between rare earths in the steel and the ladle slag or refractories and reoxidation of the steel during casting will affect final recovery and inclusion modification. It is also known that the recovery of REM in steels deoxidized previously with aluminium, at least to a level sufficient to provide grain size control, is better than when the steel is modestly deoxidized with silicon. Nomograms for determining the rare-earth requirement in order to obtain a particular residual sulphur level have been constructed.[298]

Thermodynamic data indicate that the inclusions formed as a result of REM deoxidation are first oxides, followed by the precipitation of RE oxysulphides and these in turn by the precipitation of RE sulphides.[297] Cerium from the mischmetal is usually concentrated in the sulphide inclusions, whereas lanthanum is enriched in the oxysulphide inclusions. Physical data for RE sulphides and oxysulphides are given in Table XXXIV (II:135). Micrographs and electron-probe analyses of RE inclusion types in RE-treated steels have been published by Wilson.[335] He investigated different deoxidation practices where RE metals are used and found the following inclusion types.

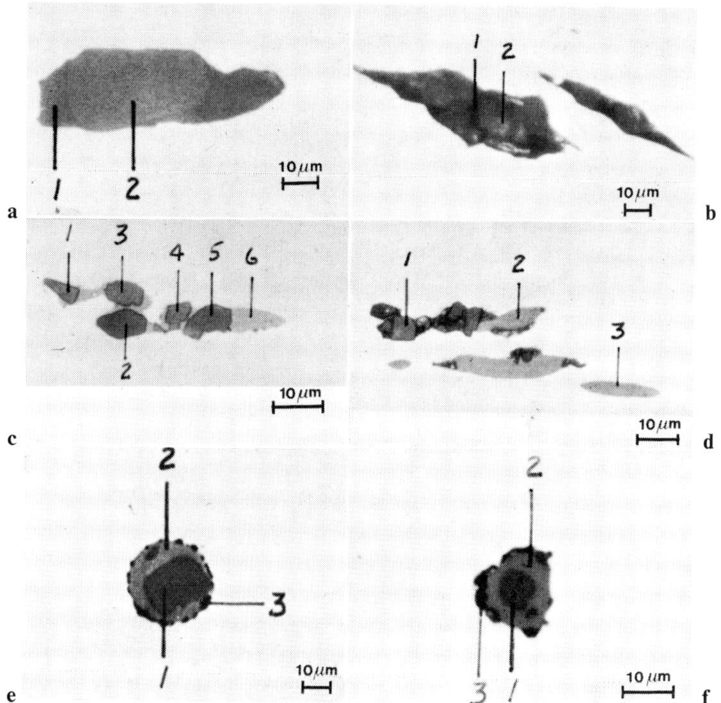

a STEEL TYPE construction steel (AISI 1038), deoxidized with Si and REM; 0·35–0·6 kg REM/t added in ladle; residual REM in steel ≤0·02%
INCLUSIONS, % (1) $37SiO_2$, 21MnO, $17Al_2O_3$, $19La_2O_3$, $11(RE)_2O_3$, small amounts of FeO, CaO, TiO_2 and S; (2) $25SiO_2$, $37La_2O_3$, $28(RE)_2O_3$, small amounts of MnO, FeO, Al_2O_3 and S

b STEEL TYPE see 253*a*
INCLUSIONS, % (1) $42SiO_2$, 25MnO, $8Al_2O_3$, 6CaO, $12(RE)_2O_3$, small amounts of FeO, TiO_2 and S; (2) $20SiO_2$, 7MnO, $46(RE)_2O_3$, small amounts of FeO, Al_2O_3, CaO and S

c STEEL TYPE low-alloy construction steel (AISI 8620), deoxidized with Al and REM (0·008–0·002% REM) in ladle
INCLUSIONS, % (1) $63(RE)_2O_3$, $27Al_2O_3$, small amounts of FeO and CaO; (2) $14(RE)_2O_3$, $79Al_2O_3$, small amounts of FeO, CaO and MgO; (3) $64(RE)_2O_3$, $27Al_2O_3$, small amounts of FeO and CaO; (4) $64(RE)_2O_3$, $28Al_2O_3$, small amounts of FeO and CaO; (5) $15(RE)_2O_3$, $77Al_2O_3$, small amounts of FeO, CaO and MgO; (6) 61Mn, 33S, 3Fe

d STEEL TYPE see 253*c*
INCLUSIONS, % (1) $59(RE)_2O_3$, $32Al_2O_3$, small amounts of FeO and CaO; (2) 60Mn, 35S, 3Fe; (3) 59Mn, 34S, 5Fe

e STEEL TYPE construction steel, deoxidized with Al and REM (0·02 REM, RE:S = 1·18) in ladle
INCLUSIONS, % (1) 75RE, 14S, oxysulphide; (2) 67RE, 33S; (3) Al_2O_3

f STEEL TYPE construction steel, deoxidized with Al and REM (0·03 REM, RE:S = 3·5) in ladle
INCLUSIONS, % (1) 64RE, 16S, oxysulphide; (2) 82RE, 20S; (3) Al_2O_3
COMMENT The higher RE/S ratio as compared with 253*e* did not change the type of sulphide-oxysulphide inclusions but their size was reduced. Also the vol. % of inclusions decreased with the higher RE/S ratio and no MnS phase was formed

253 Inclusion types given by REM additions to construction steels[335]; *see* also note on page 35*

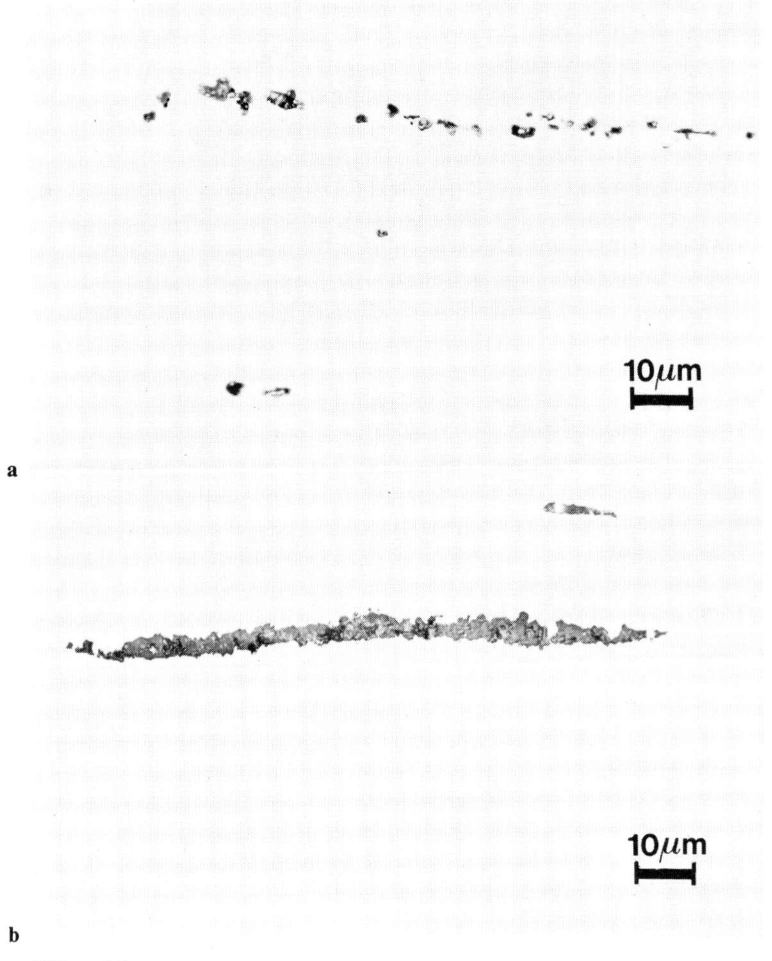

254a and b

In continuously cast carbon constructional steels deoxidized with silicon and manganese, the inclusions were mainly highly deformable Mn silicates. If REM were added as deoxidizers in the ladle, the oxide inclusions changed to silicates with a high content of RE_2O_3 components (Fig.253*a* and *b*). The silicates were less deformable and smaller, and long silicate plates did not appear. They were broken

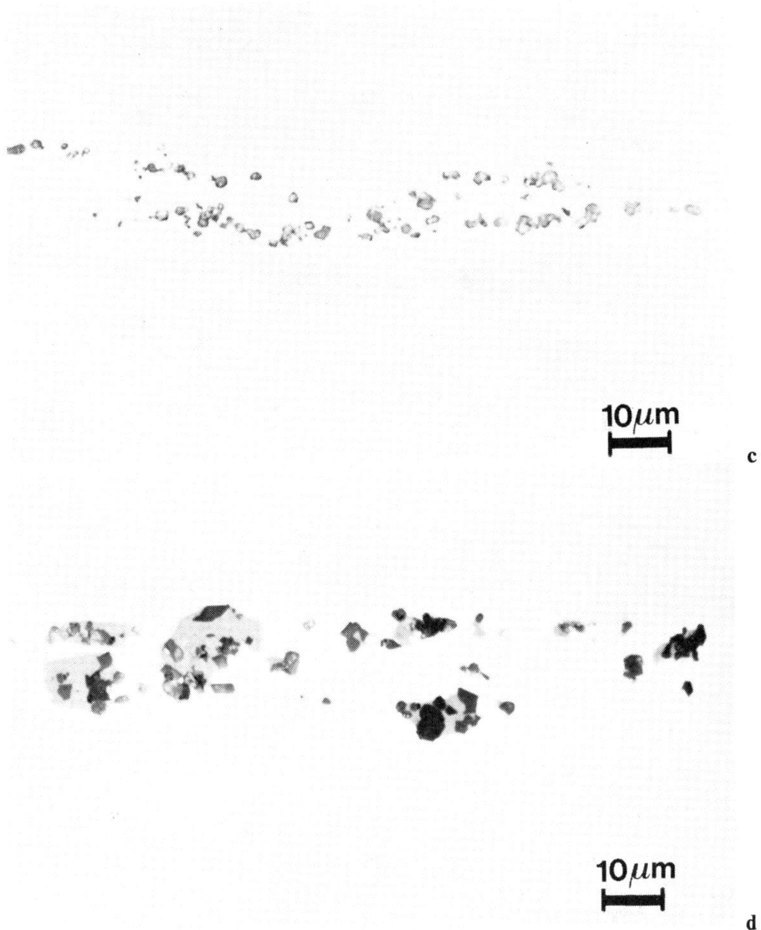

254c and d

down in the melt to smaller, more globular units, giving improvements in the ductility and formability of the steel.

In low-alloy, (Cr, Ni, Mo), fine-grained steels deoxidized with aluminium, the inclusions after hot rolling were deformed silicates, strings of Al_2O_3 particles and Ca aluminates, and highly deformed plates or ellipsoids of MnS. If the steels were deoxidized also with

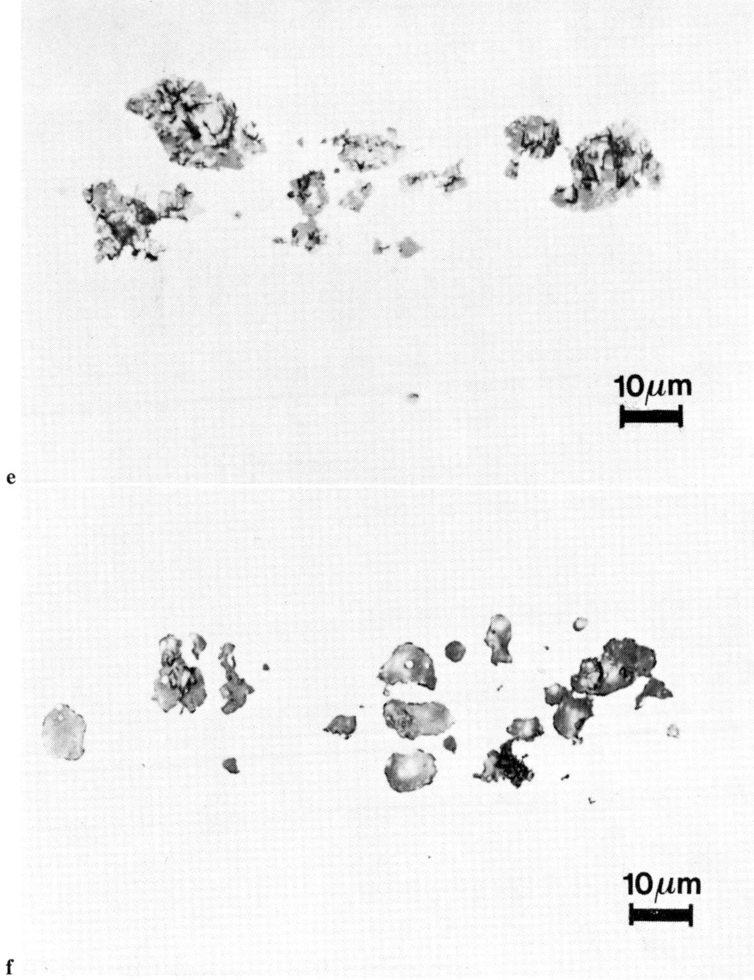

254e and f

a STEEL TYPE, %, 0·79C, 0·28Si, 0·39Mn, 0·003S, 0·008Ce, 0·010Al_{tot}, 0·006Al_{sol}, 0·0078O; deoxidized with Al + REM in feeder
INCLUSIONS stringer of small $(RE)Al_{11}O_{18}$ inclusions

b STEEL TYPE, %, 0·80C, 0·19Si, 0·31Mn, 0·006S, 0·006Ce, 0·004Al_{tot}, 0·001Al_{sol}, 0·0068O; deoxidized with Al in furnace and Al + REM in feeder
INCLUSIONS $(RE)(Al,Si)_{11}O_{18}$ or $(RE)(Al,Si)O_3$ as elongated inclusions

c STEEL TYPE, %, 0·83C, 0·30Si, 0·49Mn, 0·004S, 0·009Ce, 0·009Al_{tot}, 0·008Al_{sol}, 0·0056O; deoxidized with Al + REM in feeder
INCLUSIONS stringer of small $(RE)AlO_3$ inclusions

d STEEL TYPE, %, 0·81C, 0·19Si, 0·30Mn, 0·011S, 0·010Ce, 0·018Al_{tot}, 0·015Al_{sol}, 0·00480; deoxidized with Al in furnace and Al + REM in feeder
INCLUSIONS MnS (gray) in duplex with $(RE)Al_{11}O_{18}$ and $(RE)AlO_3$

e STEEL TYPE, %, 0·78C, 0·30Si, 0·36Mn, 0·008S, 0·006Ce, 0·019Al_{tot}, 0·014Al_{sol}, 0·0054O; deoxidized with Al in furnace and Al + REM in feeder
INCLUSIONS stringer of medium-sized $(RE)AlO_2S$

f STEEL TYPE, %, 0·83C, 0·19Si, 0·51Mn, 0·0125S, 0·0085Ce, 0·0015Al_{tot}, 0·005Al_{sol}, 0·0046O; deoxidized with REM in ladle
INCLUSIONS stringer of medium sized globular $(RE)SiO_2S$ inclusions

254 Inclusion types given by REM additions to high-carbon steels[336] (*see* also note on page 35).* 6 t melts of 0·7–1·3 %C, produced in the Uddacon converter; from slabs; deoxidation with Al and REM

smaller additions of REM (0·008–0·02 %) the silicate and aluminate inclusions were replaced by RE–Al oxide inclusions. These were mainly of two types, $(RE)AlO_3$ and $(RE)Al_{11}O_{18}$. They either appeared isolated in the steel or as multiphase inclusions with MnS. It was also noted that the sulphides were more globular after REM addition. Examples are shown in Fig.253*c* and *d*.

If the amount of REM was increased above 0·02 % but the REM/S ratio was below 3·0, new types of inclusions could be identified. They consisted of the RE_2O_2S phase, often surrounded by the RE_xS_y phase (Fig.253*e*). The MnS phase was still present but the sulphides were smaller and much less deformed than in REM-free steel. If the REM/S ratio was larger than 3·0 the MnS phase disappeared. The oxide–sulphide inclusions were small and multiphase, having oxysulphide nuclei surrounded by the RE_xS_y phase (Fig.253*f*). Also Al_2O_3 often appeared at the surface of the inclusions. The general tendency was a decreased volume fraction of inclusions with an increased initial REM/sulphur ratio.

REM inclusion types and morphologies in 6 t high-carbon steel ingots deoxidized with various amounts of aluminium and mischmetal have been studied by Malm.[336] The aim of the investigation was to develop general relations between steel compositions and inclusion types and morphologies.

In Fig.254, characteristic RE inclusion morphologies from Malm's study are shown. The morphology of the RE oxides is similar to that of Al_2O_3, i.e. stringers with small individual particles. The RE oxysulphides are generally larger than the oxides and Si-alloyed RE inclusions are more plastic than Al-alloyed RE inclusions. In comparison with earlier investigations, it was observed that Al can replace RE in RE_2O_2S and that Si can substitute Al in oxide as well as oxysulphide RE inclusions. It was also found that the (RE) $Al_{11}O_{18}$ inclusions have a wide RE/Al homogeneity. The degree of sulphide shape control increases with increasing amount of RE oxysulphides in the steels. It is suggested that this is caused by a decrease in the number of elongated MnS particles, due to the sulphur consumption on precipitation of the RE oxysulphides. No rare-earth elements have been detected in the MnS phase. A semi-theoretical method for

*Dr W. Wilson of Molybdenum Corp. of America has supplied the author with the micrographs in Fig.253, which were originally published in Ref. 335. The microprobe analyses of the inclusions were made by Mr A. Miller of the Babcock & Wilcox Res. Lab., Alliance, Ohio, USA. The micrographs in Fig.254 were published in Ref. 336 and the originals supplied by Mr S. Malm of Uddeholm AB, Hagfors, Sweden.

TABLE XLVIII REM phases observed in steel inclusions, resulting from REM additions

Inclusion phase (probable ideal composition)	Remarks	Ref. no. and micrograph
(RE)$Al_{11}O_{18}$	wide range RE/Al	Figs.253*c*, 254*a* Ref. 335, Fig.11, p. 31 (in colour)
(RE) $(Al,Si)_{11}O_{18}$ or (RE) $(Al,Si)O_3$	Al in steel ≤0·004%	Figs.253*a* and *b*; 254*b*
(RE)Al O_3		Figs.253*c* and *d*; 254*c* Ref. 335, Fig.12, p. 31 (in colour)
$(RE)_2O_2S$		Figs.253*e* and *f*; 254*e* Ref. 335, Fig.10, p. 31 (in colour)
(RE) Al O_2S		Fig.254*d* and *e*
(RE)Si O_2S	Al in steel ≤0·004%	Fig.254*f*
$(RE)_xS_y$		Fig.253*e* and *f* Ref. 335, Fig.14 and 15, p. 31 (in colour)
(RE)S		Ref. 335, Fig.15, p. 31 (in colour)

estimating inclusion types has been developed, from which the predominant inclusion phases can be predicted from average chemical composition for the steels. Also RE inclusion diagrams have been constructed, from which predictions of inclusion type can be made directly from average Ce, Al, and S contents if the 'phase boundaries' have been calculated for appropriate oxygen contents.

In Table XLVIII, reference is given to different micrographs of REM inclusions in steel.

P. The behaviour of non-metallic inclusions in wrought steel

P 3 Deformability of steel inclusions

The deformation behaviour of inclusions during hot working of steels is of fundamental importance for steel properties. The information available at the time of the first edition was mainly of a qualitative nature. It was possible, however, to summarize the general behaviour of the different inclusion phases by using the index of deformability v (III: 57), which expresses the deformation of the inclusion relative to the steel matrix. The value of v changes from zero for inclusions which do not change at all during working of the steel, to unity for those inclusions whose plasticity equals that of the steel. Values of v greater than unity are possible for inclusions which elongate to a greater degree than the steel. Figure 211 can still be regarded as a general guideline to the deformation behaviour of different inclusion types at different temperatures but is based on qualitative information.

Several later papers have dealt with the plasticity of inclusions during hot working. Waudby[337] studied the plasticity of silicate inclusions. As shown in Fig.211, the inclusions are brittle at lower temperatures but above a certain critical temperature their plasticity increases markedly. This temperature is sensitive to the composition of the silicates and Waudby has studied the main features controlling this temperature dependence. It was found that there was a correlation between the minimum temperature of observed plasticity for the silicates and their melting temperature. The average melting temperatures for different silicate inclusions, as given by Waudby, are summarized in Table XLIX. There is, however, no general effect of composition on the temperature at which silicates become plastic, which was assumed in Fig.211. An increase in a particular oxide constituent of the silicate can lead to either an increase or a decrease in the melting and softening temperature range, depending on the

TABLE XLIX Average melting temperature for different silicate inclusions (from Ref. 337)

Composition of silicate inclusion, wt-%	Minimum plasticity temperature, °C	Solidus temperature, °C	Liquidus temperature, °C	Average melting temperature, °C
70 FeO, 30 SiO_2	~940	1 180	1 210	1 195
71 MnO, 29 SiO_2	~1 020	1 250	1 350	1 300
59 FeO, 3 MnO, 3 Al_2O_3, 32 SiO_2	~970	1 180	1 300	1 240
48 FeO, 4 MnO, 5 Al_2O_3, 34 SiO_2, 8 MgO	~1 070	1 220	1 350	1 285
20 SiO_2, 80 (FeMn)O	~850	1 240	1 330	1 285
50 SiO_2, 50 (FeMn)O	~900	1 240	1 660	1 450
75 SiO_2, 25 (FeMn)O	~1 200	1 240	1 700	1 470
85 SiO_2, 15 (FeMn)O	~1 250	1 240	1 700	1 470
45 SiO_2, 15 Al_2O_3, 40 FeO	~825	1 080	1 150	1 115
5 FeO, 50 MnO, 45 SiO_2	~1 000	1 250	1 400	1 325

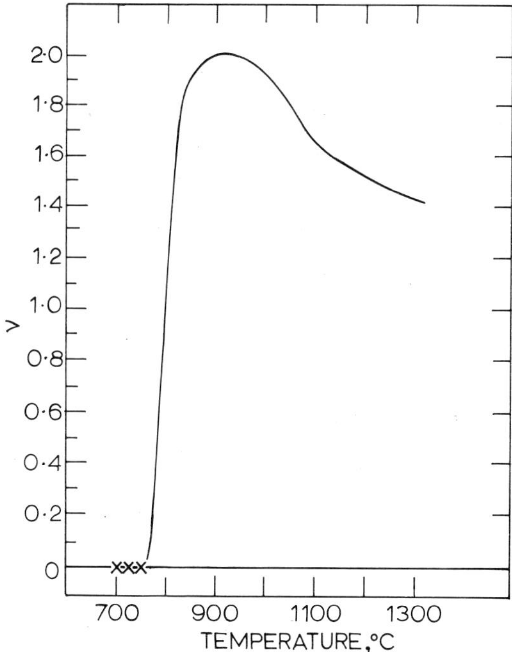

COMMENT The deformation index (III:57) changes within a narrow temperature range and also exceeds 1, i.e. the silicate inclusions are more deformable than the surrounding steel phase[338]

255 Influence of temperature on plastic deformation of silicate inclusions as compared with steel, the deformation index ν.

initial composition of the inclusion or the presence of other phases in the silicate inclusions.

Ekerot[338] also studied the behaviour of silicate inclusions in steel during hot working. Small ingots, with homogeneous silicate inclusions having compositions in the system SiO_2–Al_2O_3–MnO–CaO, were deformed at different temperatures with 50% plain-strain compression and the deformation index ν was measured. The existence was confirmed of a critical temperature range (called the transition temperature), where the inclusions changed from brittle to highly plastic behaviour and it was observed also that this temperature range was narrow. The ν v. temperature curves obtained were therefore usually steep, as shown in Fig.255. Ekerot pointed to the importance of the silicate viscosity and described the transition temperature as that at which the viscosity of the inclusion reaches a critical value of about $10^{7.5}$ poise. He also observed that ν often exceeded 1 and found a maximum value of about 2; such inclusions would cause severe anisotropy in a steel. A typical, experimentally determined ν v. temperature curve is given in Fig.255. There is often a downward trend of the curves, which cannot be explained from the expected changes in the flow stress of the steel or in terms of viscosity for the inclusions. Ekerot suggested that the decrease in ν value is a result of cooperative changes in stress–strain relationships.

A summary of the basic factors controlling the deformation behaviour of non-metallic inclusions during hot working of steels has been published by Baker *et al.*[339] They measured the microhardness of the inclusions *in situ* together with that of the surrounding matrix and found that hardness gives a reliable correlation with inclusion deformation at comparatively low strain or under conditions of negligible strain hardening. This gives a simple experimental method to determine ν according to the relation:

$$\nu \simeq 2 - h_i/h_m$$

where h_i is the inclusion hardness and h_m the matrix hardness. An inclusion will thus undergo little or no deformation if it is more than twice as hard as the matrix and will extend little more than twice as much as the corresponding matrix strain if it is fluid. The practical implication is that in order to achieve maximum toughness and minimum anisotropy in hot-rolled steels, the hardness of the inclusions should preferably be double that of the steel during hot working temperature so as to minimize their deformation.

These authors also suggest that a convenient single parameter for assessing the effect of non-metallic inclusions on steel toughness should be the total projected length of inclusions per unit area P. This parameter for a random array of elliptical inclusions is given by:

$$P = \frac{2f\lambda}{3\pi a}$$

where λ is the ellipsoidal aspect ratio and a, b, and c the semi-axes of the elliptical inclusions present at a total volume fraction f. The total inclusion projection should be measured on a section normal to the c-axis and in the direction of a. The advantage of this method is that provided the sulphides can be resolved P is amenable to automatic inclusion-counting techniques. Baker *et al.'s* work was based on the rather specialized case of a high-sulphur steel tested in the short-transverse direction. However, there has been additionally considerable work which has established relationships between inclusion projected length and short-transverse ductility, especially in heavy structural plates.

The deformation of sulphides at different temperatures is generally

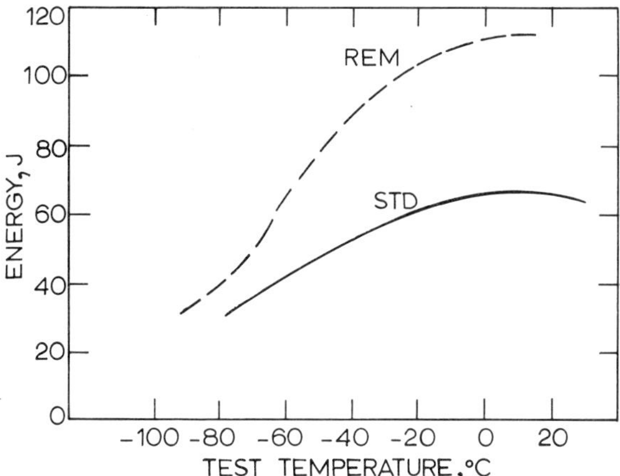

256 Improvement in transverse notch toughness (Charpy V) of high-strength low-alloy steel as a result of inclusion shape control with REM (from Ref. 342)

of the type described in Fig.211, but the detailed variation with temperature of the relative plasticity ν is more complicated due to changes in the surrounding steel matrix (α–γ transformation, *see* also Figs.217 and 218). They also emphasize the importance of sulphide shape control (*see* below) by additions of strong sulphide formers like Ti, Zr, Ca, and rare-earth elements, and conclude that toughness anisotropy arising from inclusion shape effects can now be practically eliminated.

P 4 Sulphide shape control

Directionality of mechancial properties is typical of hot-rolled steels. In aluminium-killed steels this results mainly from elongated manganese sulphide inclusions. Directionality can be reduced if the sulphides retain a globular shape and this can be accomplished by promoting the formation of sulphides which are less deformable, smaller, and have higher melting points than that of manganese sulphide. Thermodynamic considerations show that additions of Ti, Zr, Ca, Mg, Nb, V, and REM are suitable for this purpose. A marked improvement in toughness and formability of hot-rolled HSLA steel in the transverse direction has been reported as a result of successful sulphide shape control.[340] Patents exist for the production of semi-killed steels, which are free from elongated sulphide (and silicate) inclusions with the remaining sulphides of a globular type and which are claimed to have enhanced bendability properties.[341] In resulphurized ferritic 18Cr–2Mo steels it has been reported that the free-machining properties are enhanced if the sulphides change from large elongated (Mn,Cr)S inclusions of type I to small, less deformable Ti and Cr inclusions of types II and IV (Ref. 299, p. 104). The shape control is maintained through a balanced Mn:Ti ratio (Fig.250).

Improvement in transverse notch toughness has been achieved with sulphide control for carbon, high-strength low-alloy and alloy electric furnace steels.[342] REM additions were used and also magnesium as magnesium–nickel alloys for nickel-bearing steels. The increase in transverse notch toughness is due to a change in sulphide shape from the normal stringer-type sulphide normally found in hot-rolled steel to globular sulphide (Fig.256).

Q. The influence of non-metallic inclusions on the properties of steel

Q 2 Influence of inclusions on machinability of steel

Extensive coverage of this subject is given in the proceedings of a conference held in Tokyo in September 1977.[368]

Q 3 Fatigue properties

The *effect of inclusions on fatigue crack initiation and propagation* has been studied by Shih and Araki on high-strength carbon steels.[343] The conclusions are based on studies of fatigue tests of such steels mainly on smooth testpieces by alternate bending with a unidirectional stress amplitude. The results are mainly in agreement with the conclusions in III: 87, but are more detailed and based on a wider range of experimental evidence.

The authors concluded that the oxide inclusions have a deleterious effect on the fatigue fracture resistance of high-strength steels. Crack initiation depended on their size and shape, but was independent of their composition. There existed a critical minimum inclusion size for crack initiation. Spherical oxide inclusions larger than 20 μm and embedded in or close to the steel surface were potential nucleation sites for the main crack (compare III: 92 and Fig.231). The critical minimum size decreased if the oxide particles (Al_2O_3) were angular or of the stringer type aligned perpendicular to the stress axis. Failure initiated at microcracks, resulting from fracture in the inclusion itself, by separation of an internal boundary in multiphase inclusions, or by separation of the matrix/inclusion interface such as in Fig.214*c*. Sulphur probably has a beneficial effect on the fatigue properties (compare III: 89).

Smaller inclusions were found to be unimportant for crack nucleation but may have contributed to fatigue crack propagation. It was found that the propagation rate decreased with increasing cleanness of the steel but was independent of the inclusion composition. The tempering temperature of the steel also had an effect, the inclusions being more dangerous at lower tempering temperatures than at higher. The authors have related the crack propagation rate to the stress intensity factor and suggest an 'alternating dimple-rupture and cleavage-rupture mechanism' for fatigue crack propagation, observing that the dimples are always associated with inclusions or carbides smaller than 0·5 μm in diameter.

Q 6 Basic studies on influence of inclusions on mechanical properties

In 1968, when the first edition of this atlas was printed, there were few publications on the more fundamental aspects of the influence of inclusions on the mechanical properties of steel (III:101–110). This is an area which has since attracted much interest and several publications have appeared describing both theoretical and experimental studies.

A comprehensive review of the subject is clearly beyond the scope of the present work, which has as its main aim the description and characterization of inclusion phases. However, the author has chosen to summarize certain recent results relating inclusions to mechanical properties, because they are relevant to some of the main themes developed in this atlas. A more extensive coverage of the subject may be found in the papers given in the proceedings of recent conferences devoted to the subject of inclusions and their effects on steel properties.[299, 344, 345]

The concept of fracture mechanics gives rise to the specification of a critical defect size within a material. In III:107 it was mentioned that if it were possible to calculate this defect size for different inclusion shapes and positions within the steel, this would have a tremendous impact on the definition of the term 'clean' steel. This term would be given some quantitative significance, since the upper bounds of inclusion type and size constituting a flaw could be defined for a specified stress level applied to a steel component. Inclusions below these bounds could be neglected from the quality and design standpoint and might even be used to improve certain properties of the composite product steel with inclusions. That such a critical limit exists for fatigue is evident and recent findings are discussed on page 43. Also, the attempts to make 'ultraclean' steel, mentioned on page 23, clearly demonstrate that it is not the number but the size of the inclusions which is critical.

Several authors have tried to calculate a critical size limit for the

TABLE L **Critical defect size at $\sigma = \sigma_Y$ for different steels. Reference to K_{IC}-values is given in the table. Diameter of inner spherical defect (*A*) and depth of ellipsoidal surface crack (*B*) are also given** [346]

Steel type	σ_Y, kp/mm^2	K_{IC}, kp × mm$^{-3/2}$	*A*, mm	*B*, mm
AISI 4340	177	195	1·90	0·37
AISI 4340 (675°F/4^h)	147	172	2·13	0·41
AISI 4340 (475°F/4^h)	157	171	1·85	0·36
D6A (650°/4^h)	170	217	2·58	0·49
D6A (950°/4^h)	153	337	7·65	1·45
Si-modif. Ni–Cr (1 600°F/oil + 600°F/1 + 1^h)	170	142	1·11	0·21
Si-modif. Ni–Cr (+750°F/1 + 1^h)	137	174	2·59	0·48
Low-alloy Ni–Cr–Mo (400°F/1^h)	167	130	0·96	0·18
Low–alloy Ni–Cr–Mo (800°F/1^h)	149	196	2·73	0·52
Mar 18Ni (at 250°C)	176	306	4·8	0·91
Mo-alloyed (at −150°C)	62	102	4·3	0·81
Carbon steel (at −25 to −40°C)	27	190–273	5–160	

inclusions by comparing them with cracks or other flaws in the steel. Kiessling and Nordberg[346] estimated the critical defect size for stresses equal to the yield stress of different steels. It was found that the critical size was in the range of 0·2–1·0 mm for defects intersecting the surface and 1·0–5·0 mm for internal defects (Table L). They also concluded that a spherical or cylindrical cavity can be regarded as equivalent to a sharp crack if cracks are present in the steel matrix originating in the cavity surface and with a size of 0·15–0·20 *a* (*a*= cavity radius). The critical values are obviously 10–100 times too large compared with inclusions; experimental results point to critical inclusion sizes in the order of 50–500 μm depending on stress state. Despite this discrepancy, the general concept is important since it links quantitatively the control of inclusion size and shape in processing and their measurement in inspection to service requirements.

The role of inclusions as crack initiators has been discussed by several authors. Brooksbank and Andrews[347] have shown that internal stresses can be generated around inclusions due to differences in the thermal expansion coefficients of the inclusion and the matrix. They categorize the inclusions in terms of the stresses generated at the matrix/inclusion interface on cooling. These stresses may be either positive in the steel, or a void may form at the interface, depending

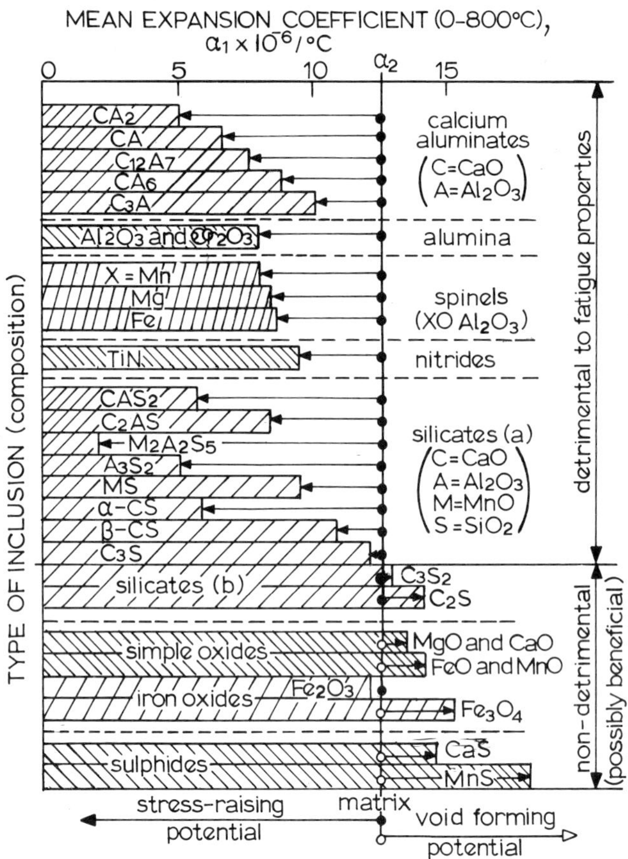

257 Stress-raising properties of inclusions in 1%C–Cr-bearing steel (from Ref. 347)

upon whether the coefficient of thermal expansion of the inclusion phase is smaller or greater than that of the matrix (Fig.257). Inclusions like sulphides give rise to voids at the inclusion/metal interface, (*see* below). Most oxides cause dilatational stresses. The properties of the matrix around such inclusions may be greatly altered and localized yielding can occur. The volume of steel undergoing yielding is dependent upon the number of inclusions, their shape, and the magnitude of the dilatation. This is shown in Fig.258, where stress fields around planar inclusions at different oxygen levels have been calculated on a correct scale.[348] Pickering[349] has pointed out that these considerations are of importance for the performance of rolling and

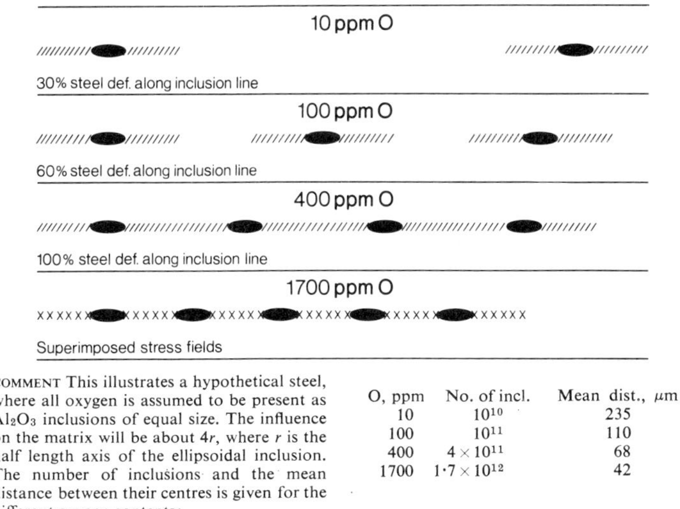

COMMENT This illustrates a hypothetical steel, where all oxygen is assumed to be present as Al_2O_3 inclusions of equal size. The influence on the matrix will be about $4r$, where r is the half length axis of the ellipsoidal inclusion. The number of inclusions and the mean distance between their centres is given for the different oxygen contents:

O, ppm	No. of incl.	Mean dist., μm
10	10^{10}	235
100	10^{11}	110
400	4×10^{11}	68
1700	$1{\cdot}7\times10^{12}$	42

258 Stress fields around stringers of ellipsoidal inclusions

other hot forming operations and also make it necessary to employ methods for assessing inclusions not only for their size and distribution but also in terms of their shape. The ease of formation of voids at the inclusion/matrix interface is of importance for the ductile failure of steels with sulphide inclusions (*see* page 48).

Several theories and experimental investigations deal with the role of inclusions in ductile fracture. A paper by Gladman summarizing the behaviour of sulphide inclusions is given in Ref. 299, p. 273–285 and several papers are also to be found in Ref. 344. The process of ductile failure involves void nucleation, growth, and coalescence. SEM techniques have been invaluable in the study of these processes. For example, Gladman followed void nucleation and growth by interrupting tensile tests at various strains, mostly after the onset of necking. The specimens were then cooled to −196°C and fractured through the neck. The cleavage surface revealed the presence of inclusions and/or voids in the cleavage facets. The void and inclusion fraction could then be determined by point counting.

As indicated above (*Experimental methods*), an experimental technique has been devised to follow void development around inclusions in the steel surface during deformation by carrying out tensile tests within a SEM.[289] Nucleation and growth of cavities was monitored quantitatively by simultaneous video recording of the microscope image and a load–elongation curve. Miniature flat tensile specimens

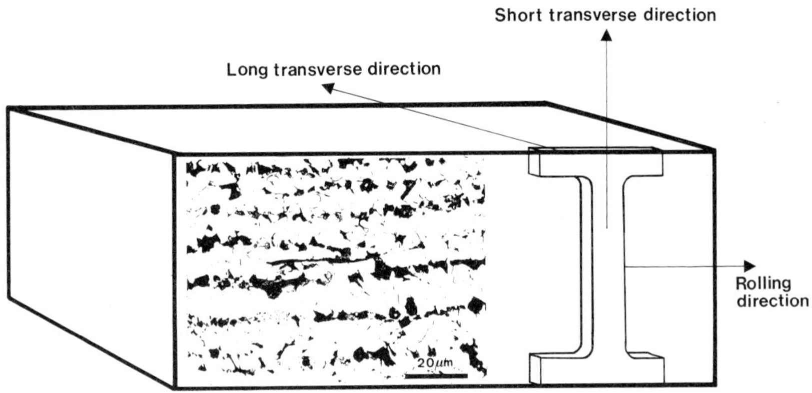

259 Orientation of miniature tensile specimens for *in situ* dynamic SEM inclusion studies (Fig.260)

were used, orientated as shown in Fig.259. The microalloyed carbon steels contained MnS inclusions with various shapes, depending on steel treatment. The progress of deformation in a single area with one or more inclusions was monitored and the load–elongation curve recorded simultaneously. A sequence of micrographs from such tensile tests with elongated or spherical sulphides is shown in Fig.260.

Gladman concluded that the effects of sulphides on ductile fracture are consistent with the traditional views of void nucleation, growth, and coalescence. The sequence of events leading to ductile failure in a carbon steel with sulphide inclusions should be the following. Voids form at the sulphide/ferrite interface at small strains due to the weak nature of the interfacial bonding. The growth of the voids is generally in the direction of the principal strain and very sensitive to the particle shape. The elongation of voids results in a decrease in the lateral spacing of voids and no evidence of lateral void growth was observed until the onset of void coalescence.

There is a critical volume fraction of voids at which coalescence will occur. Lateral growth of voids is generally associated with the later stage of void coalescence, which occurs over a limited strain range and results in ductile rupture. For a steel with sulphide inclusions this means that the ductility is sensitive to the sulphide shape and to the volume fraction of sulphide inclusions. Gladman found that disc-shaped sulphides with their planes normal to the tensile axis are far more detrimental than sulphides elongated in the test direction. The inclusion size, however, has little significant effect on ductility when tests are carried out at constant volume fraction. The criteria for the onset of void coalescence and ductile rupture are

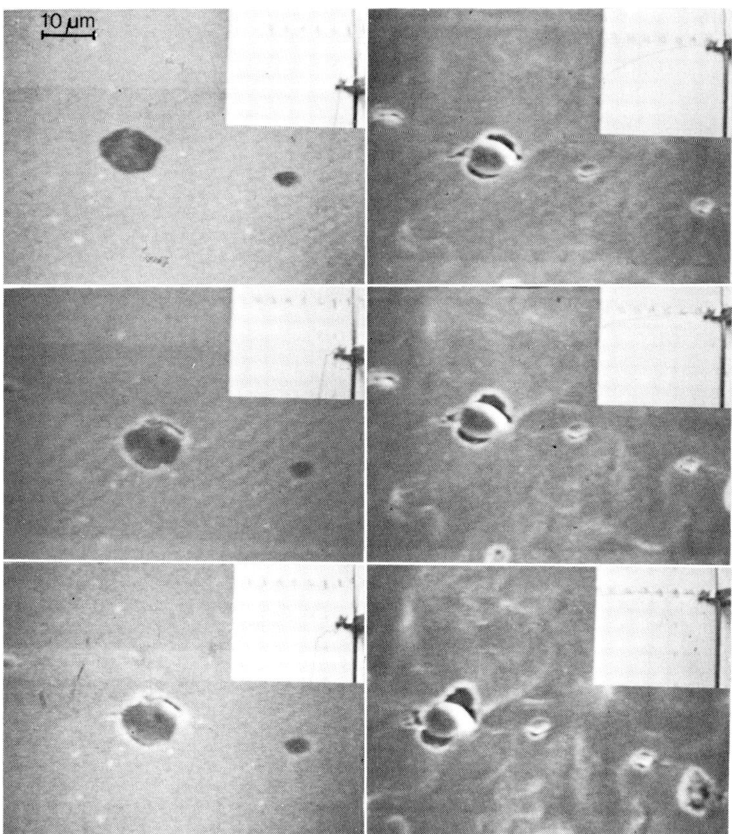

260a

significantly influenced by the inclusion spacing. The importance of other void nucleants was also considered and oxide particles appear to behave in the same manner as sulphides for a given particle shape and fraction. For steels with high carbon contents, the increase in flow stress over lower-carbon materials may be sufficient to cause void nucleation at other second-phase particles, such as carbides, in the vicinity of sulphide particles. This would result in a decrease in ductility with increasing carbon content at a constant inclusion volume fraction. The effects of stress-dependent void nucleation processes on ductile failure require further study. The subject is important, since it demonstrates that levels of steel cleanness can only be

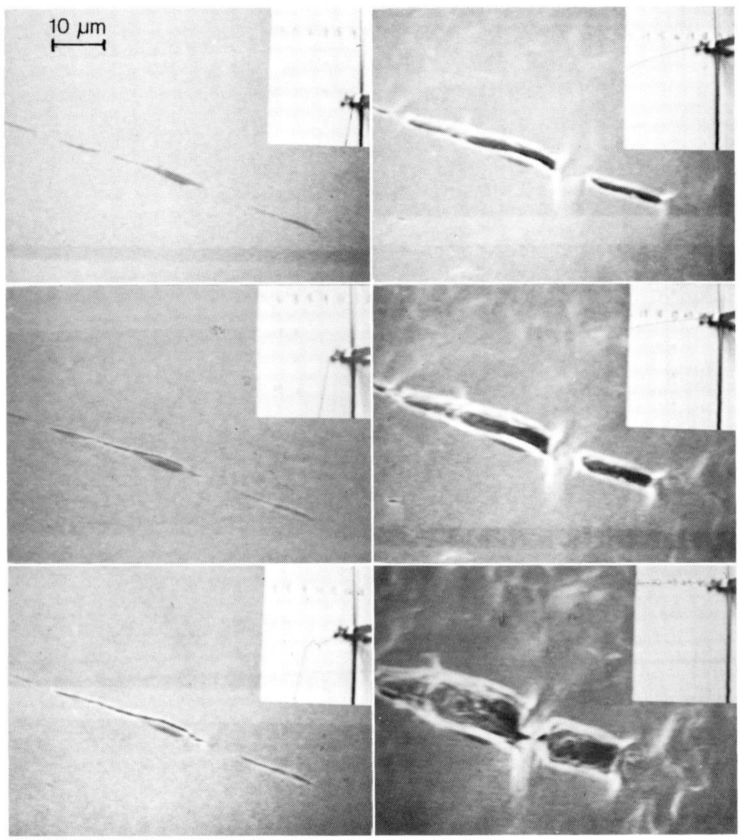

260b

COMMENT Note the coalescence of voids in the sequences of the elongated inclusions and the simultaneously recorded load–elongation curve

260 Void development around spherical (*a*) and elongated (*b*) sulphide inclusions in micro-alloyed steel (from Ref. 289)

defined when the failure process is known and the contributions from other structural factors are known.

A particularly important practical example of the value of understanding the influence of inclusions on properties has been the development of steels with improved through-thickness, or short-transverse, ductility. Numerous studies have demonstrated the importance of control of inclusion type, size, shape, and distribution to achieve the ductility levels required to resist lamellar tearing.

A recent example, demonstrating the importance of quantitative metallographic study, is given by the work of Simpson *et al.*[350] Several inclusion parameters were measured, but it was found that a significant correlation with short-transverse ductility could be obtained in which only those inclusions individually or as a cluster greater than about 50 μm in length influenced the through-thickness ductility in a way indicating lowering of the resistance to lamellar tearing (this could therefore be defined as a critical inclusion size, *see* above). The authors also used statistical models for both the shape and size distributions to estimate the spatial distribution of inclusions in the matrix. They found that on an individual particle basis, duplex silicates were more detrimental to short-transverse properties than type II sulphides. However, type II sulphides had a larger length : width ratio than silicates and occured in close-packed groups spread evenly throughout the matrix. The collective effect of type II sulphides therefore was more deleterious for lamellar tearing than the presence of individual silicates.

Attempts have also been made to develop a theory for determining the critical inclusion size with respect to void formation during hot working operations by Klevebring *et al.*[351] The theory was experimentally tested on silica inclusions in pure iron. A critical size of the order of 2·5–3·5 μm for strains of about 0·7 was found for inclusions which may cause microcrack formation at the inclusion/matrix interface in steels. Based on the theory, the strength of silica/iron and (Fe,Mn)O/iron interfaces has also been discussed.[352] Attempts were made to include the deformation indexes for the inclusions (Fig.211). The critical inclusion size for (Fe,Mn)O inclusions was determined to 4–7·5 μm, depending on both the strain and the deformation index.

Precipitation phenomena and grain growth

In III:110 it was mentioned that grain growth and grain boundary movements in steels could be influenced by non-metallic phases. It is well-known that carbides, nitrides, and borides are effective but far less was known about the role of inclusions, although a few references were given.[124, 268, 269] It is now well recognized that at high austenitizing temperatures manganese sulphide can be taken into solution and reprecipitated at austenite grain boundaries on cooling.[353–355] An investigation into the effect of MnS distribution on the fracture properties of heat-treated low-alloy steel has recently been published, which deals with reprecipitation of MnS from austenite.[356] The reprecipitation of sulphides can lead to intergranular ductile failure and some loss of toughness, a phenomenon called overheating. Manganese and sulphur are critically important in determining the austenitizing temperature at which overheating occurs. In a low-sulphur low-manganese steel all sulphur is in solution above 1240°C and there is no restraint by sulphides on grain growth. By contrast, in a high-sulphur high-manganese steel even at 1400°C there is a considerable amount of sulphide undissolved and particles are present which are capable of restricting grain growth. Overheating is proposed to occur at a critical ratio of MnS inclusion spacing at the boundary to MnS inclusion spacing within the grains. Overheated steels can be 'reclaimed' and toughness restored by reaustenitizing for prolonged periods below the overheating temperature. The distribution and size of the MnS inclusions can be altered in such a way that overheating is prevented and toughness improved. Alternatively, modification of the sulphide to a more stable form, for example by REM treatment, will eliminate the problem.

Q 7 Influence of inclusions on corrosion

Already at the beginning of this century it was known that localized corrosion of carbon steel, particularly pitting, was initiated at sulphide inclusions.[357, 358] For stainless steels also, there are several early indications that pitting starts at sulphide inclusions.[359–361] The initial corrosion process was not considered, however, to influence the general corrosion of iron and steel, and therefore not much attention was paid to the role of inclusions at that time. Further, a 'sulphide inclusion' at that time always meant MnS and it was difficult to explain the experimental results. Especially it was observed that some sulphides were 'active' whereas others were 'inactive', but it was not possible to distinguish between these sulphides by optical microscopy.

Since then both electron-probe analysis as well as more advanced electrochemical methods have become available and progress has been made. Wranglén pointed out the importance of inclusions in 1968 in connection with severe pitting corrosion damage to a ship's hull. He summarized the role of sulphides in the initiation of corrosion in a paper to the ASM Port Chester symposium (Ref. 299, pp. 361–379), which also gives extensive references.

Eklund has studied the initiation of pitting corrosion in stainless[362] as well as carbon[363] steels and concluded that the sulphides and their composition play a key role. As his conclusions are the only ones which at present give convincing explanations for the corrosion initiation in steels and also for the difference between 'active' and 'inactive' sulphides, they will be discussed in some detail.

In stainless steel, the sulphide inclusions are seldom pure MnS but are usually solid solutions of the (Mn,*Me*)S type (II:128). Solid solutions of the $(Cr,Me)_{1-x}S$ type (*see* page 18) as well as different Cr and Ti sulphides are also found. The sulphide inclusions are electronic conductors and can be polarized to the potential of a passive steel surface; their conductivity varies with the amount of *Me* in solid solution in (Mn,*Me*)S (Fig.249). Eklund has shown that the sulphides at the potential of the passive steel surface are not thermodynamically stable and tend to dissolve. The highest potential at which pure MnS can exist is about −100 mV and it is still lower for (Mn,*Me*)S. Since the potential of a passive steel surface is considerably higher and the total metal surface is several orders of magnitude larger than the sulphide area, the embedded sulphide particles tend to dissolve. Because the sulphides are formed directly from the melt during the solidification, no protective oxide film exists at the boundary between the sulphide and the matrix and a virgin metal surface is exposed as the sulphide particle is dissolved.

Initially, the exposed metal surface passivates, but as dissolution proceeds an acid solution is formed in the micro-area due to hydrolysis of the metal ions from the sulphides. When the solution has reached a certain composition the exposed metal surface can no longer passivate and a pit is initiated.

A growing pit will develop if the rate of formation of hydrogen ions from metal matrix hydrolysis exceeds that of the chemical dissolution of the sulphide. The bared metal matrix will, on the other hand, passivate if the hydrogen ions are neutralized through the formation of H_2S.

It is evident that the electrochemical dissolution properties of the sulphide inclusions depend on their composition, which in turn depends upon when and where in the steel ingot the inclusions were formed. Therefore, even in the same steel sample, there will be 'active' and 'inactive' sulphides. For carbon steels a similar mechanism was proposed, operating when they are immersed in water.

Both investigations deal with the *initiation* phase of a corrosion attack. The pitting, which is a result of this attack, is in itself of importance for such properties as the fatigue and surface appearance of stainless steel. Also, it has been suggested that the important phenomenon of crevice corrosion develops by the same mechanism as pitting corrosion[364] and therefore may be influenced by the sulphide inclusions.

The *propagation* mechanisms for most forms of corrosion attack are still unknown. Therefore it is at present not possible to relate models of initial corrosion to the general corrosion behaviour of a steel.

The mechanisms proposed by Eklund are still under some debate, but there is much experimental evidence in favour of the hypotheses.

TABLE LI Breakdown potentials of the ferritic resulphurized 18Cr–2Mo steels of Fig.250. ***A–C*** **arrows indicate the sample series with different Ti contents. Series** ***D*** **shows a series with different amounts of Ti sulphides**

Series	Composition Mn	Ti	S	E_{br} mV v. SCE
A	0·10	< ·01	0·19	500±40
A	0·38	< ·01	0·16	190±10
A	0·69	< ·01	0·17	130±15
A	1·67	< ·001	0·18	0±40
B	0·09	0·32	0·20	450±40
B	0·34	0·29	0·17	360±15
B	0·55	0·28	0·17	270±20
B	1·44	0·28	0·19	300±20
C	0·10	0·71	0·19	560±40
C	0·38	0·68	0·18	500±20
C	0·61	0·78	0·16	510±20
C	1·37	0·77	0·17	490±70
D	0·12	0·59	0·10	570±60
D	0·10	0·71	0·19	560±40
D	0·12	1·39	0·34	540±40

A conclusion would be that to eliminate pitting corrosion at the sulphides, MnS and (Mn,*Me*)S inclusions should be avoided in favour of more 'noble' phases like Cr and Ti sulphides, which have a wider thermodynamic stability range. For instance, if the sulphide structure in the resulphurized ferritic steels (Fig.250) could be kept within the Cr–Ti–S or Ti–S area, only Cr and Ti sulphides are present and the sulphides are Mn-free. Experimental evidence has shown that the breakdown potential for such resulphurized ferritic steels with Mn-free sulphides is nearly as high as for sulphur-free steels with the same composition. If, however, the sulphides contain Mn (the Mn areas) the breakdown potential decreases more or less drastically, as shown in Table LI.[365] Furthermore, the breakdown potential for a given alloy is nearly independent of the amount of the inert Ti sulphide, which is shown by series *D* in the Table.

There exist several other hypotheses for pit initiation,[366] which postulate alternative mechanisms based on a difference in the character of the passive film formed over dislocations and crystalline defects. In reducing acids penetration of the passive film by aggressive ions occurs at surface irregularities, for instance inclusions, and the sulphide inclusions are dissolved. These theories do not explain, however, why pit initiation also occurs in neutral Cl solutions, which Eklund's mechanisms cover. Whichever theory is right, the sulphide inclusion properties are of key importance for pit initiation in stainless steels.

Vermilyea[367] has characterized the present situation as follows. 'The question of pit initiation is also controversial. Some authors have given evidence for initiation at film defects such as inclusions, fractures or differences in the character of the film formed over dislocations or crystalline defects; others postulate a penetration of the passive film by aggresive ions. It is likely that pits do form at inclusions, but it is hard to discount the possibility of film penetration'.

REFERENCES

The first seven references listed below were given as 'to be published' in parts I–III and have since been published.

37. J. Bruch, U. Grisar and E. Müller: *Arch. Eisenh.*, 1965, **36**, 799–807.
165. R. Kiessling, B. Hässler and C. Westman: *JISI*, 1967, **205**, 531–534.
173. S. Baeckström: *Scand. J. Met.*, 1972, **1**, 121–126.
192. K. Torssell and M. Olette: *Rev. Mét.*, 1969, **66**, 813–822.
207. E. L. Morgan, J. R. Blank, W. J. M. Salter and F. B. Pickering: *JISI*, 1968, **206**, 987–1001.
244. J. Bellot, M. Hugo and E. Herzog: *Rev. Mét.*, 1969, **66**, 389–403.
276. T. Y. Tien, R. J. Martin and L. H. Van Vlack: *Trans. AIME*, 1968, **242**, 2153–2154.

---

282. L. Kiessling: *Swed. Inst. Met. Research*, 1977, I. M. 1207, Figs. 4 and 5.
283. R. A. Rege, E. S. Szekérés and W. D. Forgenz: *Met. Trans.*, 1970, **1**, 2652–2653.
284. K. Okohira, N. Sato and H. Mori: *Trans. ISIJ*, 1974, **14**, 102–109.
285. A. Fischer and H. Bertram: *Arch. Eisenh.*, 1973, **44**, 93.
286. M. Sumita, I. Uchiyama and T. Araki: *Trans. Nat. Res. Inst. Metals*, 1973, **15:1**, 1–19.
287. G. Eklund: 'Clean steel,' Rept. Royal Swed. Acad. Engin. Sci., 1971, **169:1**, 152–158.
288. B. Lehtinen and K. E. Easterling: '8th Int. Congr. Electron Microscopy, Canberra', Vol. 1, 160–161; 1974.
289. W. Roberts, B. Lehtinen and K. E. Easterling: *Acta Met.*, 1976, **24**, 745–758.
290. 'Developments in electron microscopy and analysis', Proc. EMAG 75, 1976, London–New York–San Francisco, Academic Press.
291. H. P. Hougardy: *The Microscope*, 1974, **22**, 5–26 and 1976, **24**, 7–23.
292. G. Berger, A. Gault, G. Guntz and A. Sulmont: *Rev. Mét.*, 1971, **68**, 839–847.
293. W. Bartholome, M. Fröhlke and H. J. Köstler: *Radex Rundschau*, 1971, 514–525.
294. L. Luyckx, B. N. Ferry and A. McLean: *J. Metals*, 1974, **26**, June, 35–43.
295. A. G. Franklin and D. H. Evans: *JISI*, 1971, **209**, 369–379.
296. G. König and Th. Ernst: *Radex Rundschau*, 1970, (1) and (2), 45–63 and 67–98.
297. W. G. Wilson, D. A. R. Kay and A. Vahed: *J. Metals*, 1974, **26**, May, 14–23.
298. A. McLean and S. K. Lu: *Metals and Materials*, 1974, **8**, 452–457.
299. 'Sulfide inclusions in steel', ASM Materials/Metalworking Technology Series, 1975, No. 6, Metals Park, Ohio, USA.
300. L.-I. Staffansson: *Met. Trans. B.* 1976, **7B**, 131–134.
301. H. Fredriksson and M. Hillert: *Scand. J. Met.*, 1973, **2**, 125–145.
302. L. K. Bigelow and M. C. Flemings: *Met Trans. B*, 1975, **6B**, 275–283.
303. R. Kiessling and L. Rohlin: *Scand. J. Met.*, 1977, **6**, 56–58.
304. G. S. Mann and L. H. Van Vlack: *Met. Trans. B*, 1976, **7B**, 469–475.
305. E. Doroschkewitsch, R. Kiessling and C. Westman: *JISI*, 1970, **208**, 698.
306. R. Kiessling: *Jernkont. Ann.*, 1969, **153**, 295–302.
307. R. Kiessling: 'Influence of metallurgy on machinability of steel', Proc. ISIJ/ASM Symposium, Tokyo, 1977, 253–261.
308. R. Kiessling and C. Westman: *JISI*, 1970, **208**, 699–700.

309. W. J. M. Salter and F. B. Pickering: *JISI*, 1969, **207**, 992–1002.
310. R. Kiessling: to be published.
311. E. Plöckinger, W. Holzgruber and G. Kühnelt: *Radex Rundschau*, 1969, (2), 508–517.
312. H. Malissa, E. Peisteiner, M. Grasserbauer and J. Kaltenbrunner: *Radex Rundschau*, 1975, (2), 386–392.
313. T. Malmberg, G. Runnsjö and B. Aronsson: *Scand. J. Met.*, 1974, **3**, 169–172.
314. G. S. Mann and L. H. Van Vlack: *Met. Trans. B*, 1977, **8B**, 47–51.
315. G. S. Mann and L. H. Van Vlack: *ibid.*, 53–57.
316. L. Lorenz: *VDI–Z*, 1975, **117**, 977–984.
317. S. Arwidson: *Scand. J. Met.*, 1972, **1**, 167–170.
318. R. B. Aucott, D. W. Gray and C. G. Holland: *J. West Scotland ISI*, 1971–72, **79**, 118–121.
319. D. A. R. Kay and R. J. Pomfret: *JISI*, 1971, **209**, 962–965.
320. W. E. Anable, R. H. Nafziger and D. D. Robinson: *J. Metals*, 1973, **25**, Nov., 55–61.
321. R. Diederichs and H. Valentin: *DEW–Technische Ber.*, 1971, **11**, 219–221.
322. H. Löwenkamp, A. Choudhury, R. Jauch and F. Regnitter: *Stahl Eisen*, 1973, **93**, 625–635.
323. K. R. Olen, L. S. Gonano and E. L. Heck: *J. Metals*, 1970, **22**, July, 36–41.
324. H. Mori, N. Tanaka and M. Hirai: *Trans. ISIJ*, 1971, **11**, 424–438.
325. F. Listhuber, K. Ecker and T. Fastner: *Iron and Steel Eng.*, 1974, **51**, Apr., 92–98.
326. T. Matsumoto, K. Saito and S. Miyoshi: *Nippon Kokan Techn. Rep. Overseas*, 1972, Dec., 23–41.
327. T. Ohno *et al.*: *Trans ISIJ*, 1975, **15**, 407–416.
328. Y. Habu *et al.*: *ibid.*, 246–251.
329. T. Emi: *Scand. J. Met.*, 1975, **4**, 1–8.
330. M. Kepka: *Neue Hütte*, 1976, **21**, 645–652.
331. I. G. Davies, M. Randle and R. Widdowson: *Met. Technol.*, 1974, **1**, 241–248.
332. S. E. Lunner: *JISI*, 1972, **210**, 168–178.
333. A. E. Voronin, A. A. Zborovskii and E. I. Rabinovich: *Steel USSR*, 1971, **1**, 448–450.
334. L. Luyckx: ASM Mat. Sci. Symp., Cincinnati, Nov. 75.
335. W. G. Wilson: *Goldschmidt informiert*, 1974, **31**, (4), 26–32.
336. S. Malm: *Scand. J. Met.*, 1976, **5**, 248–257.
337. P. E. Waudby: 'Steel Times Annual Review', 147–152; 1972.
338. S. Ekerot: *Scand. J. Met.*, 1974, **3**, 21–27.
339. T. J. Baker, K. B. Gove and J. A. Charles: *Met. Technol.*, 1976, **3**, 183–193.
340. L. Luyckx, J. R. Bell, A. McLean and M. Korchynsky: *Met. Trans.*, 1970, **1**, 3341–3350.
341. L. Luyckx *et al.*: US Patent 3.951.645, 1976, April 20.
342. H. W. Bennett and L. P. Sandell Jr.: *J. Metals*, 1974, **26**, Feb., 21–24.
343. T.-Y. Shih and T. Araki: *Trans. ISIJ*, 1973, **13**, 11–19.
344. 'Effect of second-phase particles on the mechanical properties of steel'; 1971, London, The Iron and Steel Institute.
345. 'Inclusions and their effects on steel properties', Proc. BSC conference 1974, in press.
346. R. Kiessling and H. Nordberg: 'Production and application of clean steels', 179–185; 1972, London, The Iron and Steel Institute.

347. D. Brooksbank and K. W. Andrews: *JISI*, 1972, **210**, 246–255.
348. R. Kiessling and H. Nordberg: 'Clean steel', *Rep. Roy. Swed. Acad. Engin. Sci.*, 1971, 169, (1), 159–169.
349. F. B. Pickering: 'Steel Times Annual Review', 99–110; 1973.
350. J. D. Simpson, L. Dyer and J. K. MacDonald: *BHP Techn. Bull*, 1976, 20, (1), 30–36.
351. B.-I. Klevebring, E. Bogren and R. Mahrs: *Met. Trans. A*, 1975, **6A**, 319–327.
352. B.-I. Klevebring: *Scand. J. Met.*, 1976, **5**, 63–68.
353. G. D. Joy and J. Nutting: 'Effect of second-phase particles on the mechanical properties of steel', 95–100; 1971, London, The Iron and Steel Institute.
354. T. J. Baker and R. Johnson: *JISI*, 1973, **211**, 783–791.
355. R. O. Richie and J. Knott: *Met. Trans.*, 1974, **5**, 782–785.
356. R. N. O'Brien, D. H. Jack and J. Nutting: 'Heat treatment 76', 161–168; 1976, London, The Metals Society.
357. J. W. Cobb: *JISI*, 1911, **83**, 170–190.
358. J. E. Stead: *JISI*, 1916, **94**, 74.
359. M. A. Streicher: *J. Elecrochem. Soc.*, 1956, **103**, 375–390.
360. N. D. Greene and M. G. Fontana: *Corrosion*, 1959, **15**, 25–31.
361. W. Schwenk: *Korrosion*, 1960, **13**, 20–25.
362. G. Eklund: *J. Electrochem. Soc.*, 1974, **121**, 467–473.
363. G. Eklund: *Scand. J. Met.*, 1972, **1**, 331–336.
364. G. Eklund: *J. Electrochem. Soc.*, 1976, **123**, (2), 170–173.
365. S.-O. Bernhardsson and R. Kiessling: to be published.
366. Z. Szklarska-Smialovska: *Corrosion*, 1971, **27**, 223–233.
367. D. A. Vermilyea: *Physics Today*, 1976, Sept., 23–31.
368. Proceedings of international symposium on influence of metallurgy on machinability of steel, organised by ASM and ISIJ, Tokyo, 26–28 Sept. 1977.
369. G. Eklund: *Scand. J. Met.*, 1977, **6**, 196–201.

Appendix

Summary of Tables (*continued from II:161*)

Table	*Contents*	*Part*
XXXVIII	Possible inclusion sources	III:3
XXXIX	Solubility of oxygen in liquid iron	III:12
XL	Phase analyses of isolated oxide inclusions from Fe–Cr ingots	III:28
XLI	Inclusion sources compared with *in situ* inclusions	III:43
XLII	Lowest plasticity temperature for spinels	III:66
XLIII	Deformability index for Fe–Mn silicate inclusions v. steel deformation and rolling temperature	III:68
XLIV	Cr sulphide inclusions with metals in solid solution	IV:18
XLV	Solid solubility limits for synthetic $(Cr,Me)_{1-x}S$	IV:19
XLVI	Influence of inclusion size on inclusion numbers	IV:24
XLVII	Rare-earth alloy compositions	IV:29
XLVIII	Inclusion types with REM	IV:36
XLIX	Melting temperature for silicate inclusions	IV:38
L	Critical defect size for different steels	IV:45
LI	Breakdown potentials of 18Cr2Mo steels with different sulphide composition	IV:54

Errata to parts I–III

		Reads	*Should read*
I:74	Line 11 from bottom	table p. 42	table p. 41
II:4	Fig.81	—	Pyrope, $3MgO \cdot Al_2O_3 \cdot 3SiO_2$ should be included in the ternary diagram
II:29	Table heading	P_2O	P_2O_5
II:99	Ordinate Fig.148	ΔG°	ΔG
II:115	Melting point	1655±5°C	*See* IV:8
II:122	Fig.168 sub-caption	0·18Cr	0·18C
II:149	Line 14	β-MnSe	γ-MnSe
II:158	Ref. 115a	*JISI* 1953	*JISI* 1935
III:viii	Line 9 from bottom	Fe, and Cr	Fe, Mn, and Cr
III:17	Line 14	pinhole porosity	subcutaneous blowholes
III:30	Line 15 from bottom	(Mn,Ca) (SK 4)	(Mn,Ca)S (K4)
III:30	Last line	(Mn,Ca)Si	(Mn,Ca)S
III:37	Line 10	ferrochromium	ferrochromium and steel
III:45	Last line	steel moulds	chill moulds
III:53	Second line	neverro ccur	never occur
III:54	Line 19	of different	from different
III:80	Fig.223	1% sulphur	0·1% sulphur
III:101	Line 9 from bottom	Section Q 5	Section Q 4
III:116	Ref. 172	Dickson	Dickenson